Datsun Automotive Repair Manual

by J Haynes
Member of the Guild of Motoring Writers

and Trevor Hosie

Models covered
Datsun/Nissan 810/Maxima Sedan and Station wagon
Gasoline models

ISBN 1 85010 053 5

Printed in the USA

(9D5 - 28025)
(376)

Haynes Publishing Group
Sparkford Nr Yeovil
Somerset BA22 7JJ England

Haynes North America, Inc
861 Lawrence Drive
Newbury Park
California 91320 USA

Acknowledgements

Thanks are due to Nissan Motor Company Limited of Japan for the supply of technical information and certain illustrations.

Special thanks are due to all those people at Sparkford who helped in the production of this manual. Amongst them are Ian Robson who planned the layout of each page, and Pete Ward who edited the text.

About this manual

Its aim

The aim of this manual is to help you get the best value from your car. It can do so in several ways. It can help you decide what work must be done (even should you choose to get it done by a service station or dealer), provide information on routine maintenance and servicing, and give a logical course of action and diagnosis when random faults occur. However, it is hoped that you will use the manual by tackling the work yourself. On simpler jobs it may even be quicker than booking the car into a service station or dealer, and going there twice to leave and collect it. Perhaps most important, a lot of money can be saved by avoiding the costs the service station or dealer must charge to cover its labor and overheads.

The manual has drawings and descriptions to show the function of the various components so that their layout can be understood. Then the tasks are described and photographed in a step-by-step sequence so that even a novice can do the work.

Its arrangement

The manual is divided into thirteen Chapters, each covering a logical sub-division of the vehicle. The Chapters are each divided into Sections, numbered with single figures, eg 5; and the Sections into paragraphs (or sub-sections), with decimal numbers following on from the Section they are in, eg 5.1, 5.2, 5.3 etc.

It is freely illustrated, especially in those parts where there is a detailed sequence of operations to be carried out. There are two forms of illustration: figures and photographs. The figures are numbered in sequence with decimal numbers, according to their position in the Chapter: eg, Fig. 6.4 is the 4th drawing/illustration in Chapter 6. Photographs are numbered (either individually or in related groups) the same as the Section or sub-section of the text where the operation they show is described.

There is an alphabetical Index at the back of the manual as well as a Contents List at the front.

References to the 'left' or 'right' of the vehicle are in the sense of a person sitting in the driver's seat facing forwards.

Introduction to the Datsun 810

The Datsun 810, which was introduced to the United States in 1977, is available as a Sedan or a Station Wagon. Being very similar in design to the earlier 610 Sedan and Station Wagon, the 810 has been given a lot more power in the form of the well tried and proved L24 engine used in the 240Z Coupe. The engine has been equipped with an electronically controlled fuel injection system, which provides optimum fuel/air ratios under all operating conditions, and is also largely responsible for the very low fuel and exhaust gas emission. Other forms of emission control are used, and for those vehicles operating in California a catalytic converter system is used.

Power is transferred to the rear wheels via a four-speed, or five speed, all synchromesh gearbox or optional automatic transmission unit. The Sedan has independent rear suspension, whilst a semi-elliptic leaf spring and rigid ('live') axle set-up is used on the Station wagon.

Other optional equipment, includes full air conditioning and power-assisted steering which, at this time, is only available on the basis that 'both systems must be purchased'.

The 810 is equipped with many comfort and convenience features, which, when considered along with the first-rate mechanical specifications, make the vehicle a very worthwhile purchase.

For additional information concerning late-model 810 and Maxima models, see the Chapter 13 supplement towards the back of this manual.

Contents

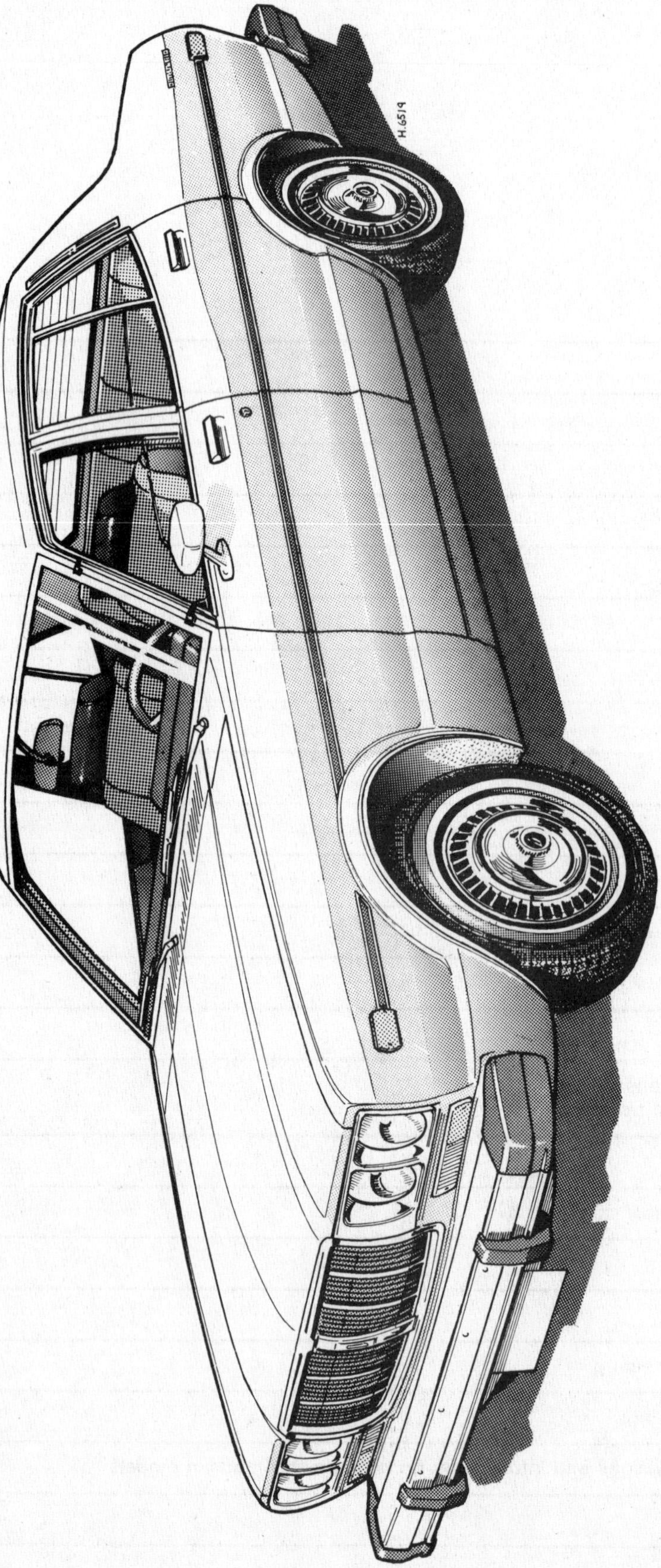

Datsun 810

Buying spare parts and vehicle identification numbers

Buying spare parts

Replacement parts are available from many sources, which generally fall into one of two categories – authorized dealer parts departments and independent retail auto parts stores. Our advice concerning these parts is as follows:

Retail auto parts stores: Good auto parts stores will stock frequently needed components which wear out relatively fast, such as clutch components, exhaust systems, brake parts, tune-up parts, etc. These stores often supply new or reconditioned parts on an exchange basis, which can save a considerable amount of money. Discount auto parts stores are often very good places to buy materials and parts needed for general vehicle maintenance such as oil, grease, filters, spark plugs, belts, touch-up paint, bulbs, etc. They also usually sell tools and general accessories, have convenient hours, charge lower prices and can often be found not far from home.

Authorized dealer parts department: This is the best source for parts which are unique to the vehicle and not generally available elsewhere (such as major engine parts, transmission parts, trim pieces, etc.).

Warranty information: If the vehicle is still covered under warranty, be sure that any replacement parts purchased – regardless of the source – do not invalidate the warranty!

To be sure of obtaining the correct parts, have engine and chassis numbers available and, if possible, take the old parts along for positive identification.

Vehicle identification numbers

Modifications are a continuing and unpublicized process in vehicle manufacture. Spare parts manuals and lists are compiled on a numerical basis, the individual vehicle numbers being essential to correctly identify the component required.

The *vehicle identification number* is located on the top surface of the instrument panel cowl and is visible through the windshield.

The *identification plate,* which is fixed to the center of the firewall, contains the car type, engine capacity, maximum horsepower, wheelbase, engine type and car serial numbers. The car serial number is also stamped on the firewall adjacent to the identification plate.

The *engine serial number* is stamped on the right side of the cylinder block.

The *color code number label* is stamped on a plate attached to the top surface of the radiator support crossmember.

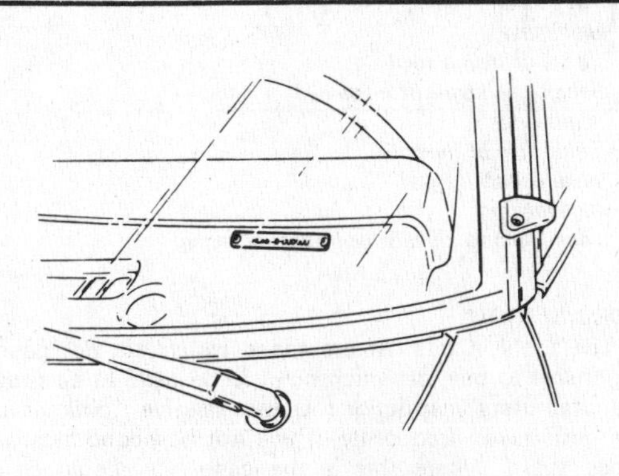

The vehicle identification number

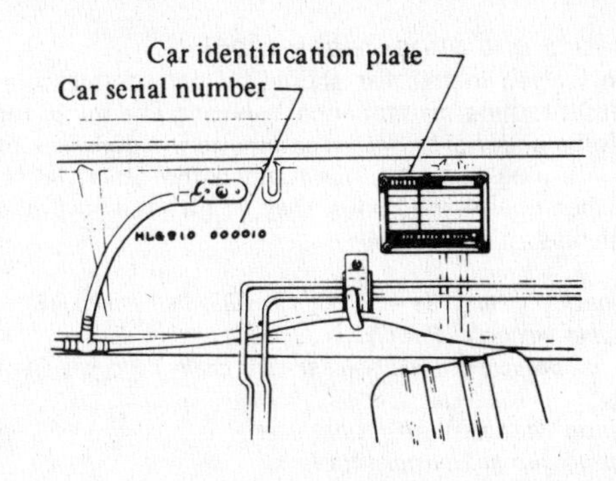

The identification plate

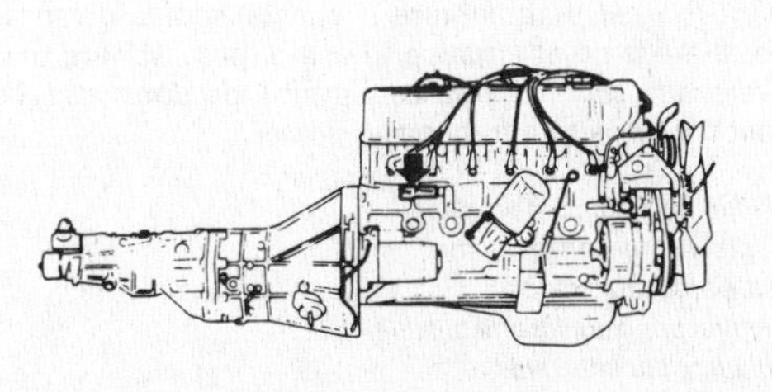

The engine serial number

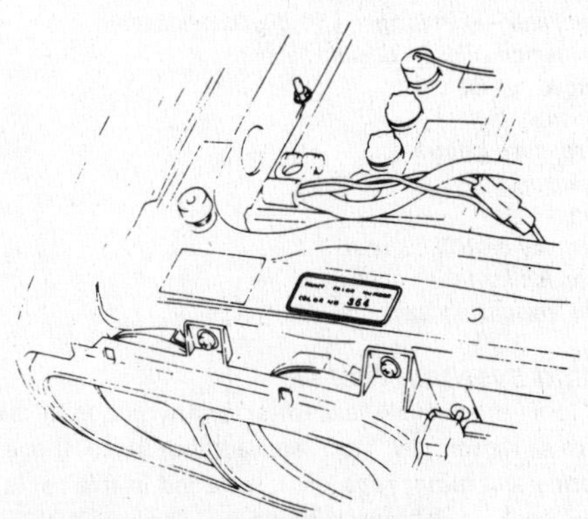

The color code number label

Tools and working facilities

Introduction

A selection of good tools is a fundamental requirement for anyone contemplating the maintenance and repair of a motor vehicle. For the owner who does not possess any, their purchase will prove a considerable expense, offsetting some of the savings made by doing-it-yourself. However, provided that the tools purchased meet the relevant national safety standards and are of good quality, they will last for many years and prove an extremely worthwhile investment.

To help the average owner to decide which tools are needed to carry out the various tasks detailed in this manual, we have compiled three lists of tools under the following headings: *Maintenance and minor repair, Repair and overhaul,* and *Special.* The newcomer to practical mechanics should start off with the *Maintenance and minor repair* tool kit and confine himself to the simpler jobs around the car. Then, as his confidence and experience grows, he can undertake more difficult tasks, buying extra tools as, and when, they are needed. In this way, a *Maintenance and minor repair* tool kit can be built-up into a *Repair and overhaul* tool kit over a considerable period of time without any major cash outlays. The experienced do-it-yourselfer will have a tool kit good enough for most repair and overhaul procedures and will add tools from the *Special* category when he feels the expense is justified by the amount of use these tools will be put to.

It is obviously not possible to cover the subject of tools fully here. For those who wish to learn more about tools and their use there is a book entitled *How to Choose and Use Car Tools* available from the publishers of this manual.

Maintenance and minor repair tool kit

The tools given in this list should be considered as a minimum requirement if routine maintenance, servicing and minor repair operations are to be undertaken. We recommend the purchase of combination spanners (ring one end, open-ended the other); although more expensive than open-ended ones, they do give the advantages of both types of spanner.

Combination wrenches – 9, 1C, 11, 13 and 17 mm AF
Adjustable wrench - 9 inch
Engine oil pan/transmission/rear axle drain plug key (where applicable)
Spark plug wrench (with rubber insert)
Spark plug gap adjustment tool
Set of feeler gauges
Brake adjuster wrench (where applicable)
Brake bleed nipple wrench
Screwdriver - 4 in long x $\frac{1}{4}$ in dia (flat blade)
Screwdriver - 4 in long x $\frac{1}{4}$ in dia (cross blade)
Combination pliers - 6 inch
Hacksaw, junior
Tire pump
Tire pressure gauge
Grease gun
Oil can
Fine emery cloth (1 sheet)
Wire brush (small)
Funnel (medium size)

Repair and overhaul tool kit

These tools are virtually essential for anyone undertaking any major repairs to a motor car, and are additional to those given in the *Maintenance and minor repair* list. Included in this list is a comprehensive set of sockets. Although these are expensive they will be found invaluable as they are so versatile - particularly if various drives are included in the set. We recommend the $\frac{1}{2}$ in square-drive type, as this can be used with most proprietary torque wrenches. If you cannot afford a socket set, even bought piecemeal, then inexpensive tubular box spanners are a useful alternative.

The tools in this list will occasionally need to be supplemented by tools from the *Special* list.

Sockets (or box wrenches) to cover the range in the previous list
Reversible ratchet drive (for use with sockets)
Extension piece, 10 inch (for use with sockets)
Universal joint (for use with sockets)
Torque wrench (for use with sockets)
'Self-grip' wrench - 8 inch
Ball pein hammer
Soft-faced hammer, plastic or rubber
Screwdriver - 6 in long x $\frac{5}{16}$ in dia (flat blade)
Screwdriver - 2 in long x $\frac{5}{16}$ in square (flat blade)
Screwdriver - $1\frac{1}{2}$ in long x $\frac{1}{4}$ in dia (cross blade)
Screwdriver - 3 in long x $\frac{1}{8}$ in dia (electricians)
Pliers - electricians side cutters
Pliers - needle nosed
Pliers - circlip (internal and external)
Cold chisel - $\frac{1}{2}$ inch
Scriber
Scraper
Center punch
Pin punch
Hacksaw
Valve grinding tool
Steel rule/straight edge
Allen keys
Selection of files
Wire brush (large)
Axle-stands
Jack (strong scissor or hydraulic type)

Special tools

The tools in this list are those which are not used regularly, are expensive to buy, or which need to be used in accordance with their manufacturers' instructions. Unless relatively difficult mechanical jobs are undertaken frequently, it will not be economical to buy many of these tools. Where this is the case, you could consider clubbing together with friends (or a motorists club) to make a joint purchase, or borrowing the tools against a deposit from a local garage or tool hire specialist.

The following list contains only those tools and instruments freely available to the public, and not those special tools produced by the car manufacturer specifically for its dealer network. You will find occasional references to these manufacturers' special tools in the text of this manual. Generally, an alternative method of doing the job without the car manufacturers' special tool is given. However, sometimes, there is no alternative to using them. Where this is the case and the relevant tool cannot be bought or borrowed you will have to entrust the work to a franchised dealer.

Valve spring compressor
Piston ring compressor
Balljoint separator
Universal hub/bearing puller
Impact screwdriver
Micrometer and/or vernier gauge
Dial gauge
Stroboscopic timing light
Dwell angle meter/tachometer

Universal electrical multi-meter
Cylinder compression gauge
Lifting tackle
Trolley jack
Light with extension lead

Buying tools

For practically all tools, a tool factor is the best source since he will have a very comprehensive range compared with the average garage or accessory shop. Having said that, accessory shops often offer excellent quality tools at discount prices, so it pays to shop around.

There are plenty of good tools around at reasonable prices, but always aim to purchase items which meet the relevant national safety standards. If in doubt, ask the proprietor or manager of the shop for advice before making a purchase.

Working facilities

Not to be forgotten when discussing tools, is the workshop itself. If anything more than routine maintenance is to be carried out, some form of suitable working area becomes essential.

It is appreciated that many an owner mechanic is forced by circumstances to remove an engine or similar item, without the benefit of a garage or workshop. Having done this, any repairs should always be done under the cover of a roof.

Wherever possible, any dismantling should be done on a clean flat workbench or table at a suitable working height.

Any workbench needs a vise; one with a jaw opening of 4 in (100 mm) is suitable for most jobs. As mentioned previously, some clean dry storage space is also required for tools, as well as the lubricants, cleaning fluids, touch-up paints and so on which become necessary.

Another item which may be required, and which has a much more general usage, is an electric drill with a chuck capacity of at least $\frac{5}{16}$ in (8 mm). This, together with a good range of twist drills, is virtually essential for installing accessories such as wing mirrors and additional lights.

Last, but not least, always keep a supply of old newspapers and clean, lint-free rags available, and try to keep any working area as clean as possible.

Care and maintenance of tools

Having purchased a reasonable tool kit, it is necessary to keep the tools in a clean serviceable condition. After use, always wipe off any dirt, grease and metal particles using a clean, dry cloth, before putting the tools away. Never leave them lying around after they have been used. A simple tool rack on the garage or workshop wall, for items such as screwdrivers and pliers is a good idea. Store all normal wrenches and sockets in a metal box. Any measuring instruments, gauges, meters, etc., must be carefully stored where they cannot be damaged or become rusty.

Take a little care when tools are used. Hammer heads inevitably become marked and screwdrivers lose the keen edge on their blades from time-to-time. A little timely attention with emery cloth or a file will soon restore items like this to a good serviceable finish.

Wrench jaw gap comparison table

Jaw gap (in)	Wrench size
0·250	$\frac{1}{4}$ in AF
0·275	7 mm AF
0·312	$\frac{5}{16}$ in AF
0·315	8 mm AF
0·340	$\frac{11}{32}$ in AF; $\frac{1}{8}$ in Whitworth
0·354	9 mm AF
0·375	$\frac{3}{8}$ in AF
0·393	10 mm AF
0·433	11 mm AF
0·437	$\frac{7}{16}$ in AF
0·445	$\frac{3}{16}$ in Whitworth; $\frac{1}{4}$ in BSF
0·472	12 mm AF
0·500	$\frac{1}{2}$ in AF
0·512	13 mm AF
0·525	$\frac{1}{4}$ in Whitworth; $\frac{5}{16}$ in BSF
0·551	14 mm AF
0·562	$\frac{9}{16}$ in AF
0·590	15 mm AF
0·600	$\frac{5}{16}$ in Whitworth; $\frac{3}{8}$ in BSF
0·625	$\frac{5}{8}$ in AF
0·629	16 mm AF
0·669	17 mm AF
0·687	$\frac{11}{16}$ in AF
0·708	18 mm AF
0·710	$\frac{3}{8}$ in Whitworth; $\frac{7}{16}$ in BSF
0·748	19 mm AF
0·750	$\frac{3}{4}$ in AF
0·812	$\frac{13}{16}$ in AF
0·820	$\frac{7}{16}$ in Whitworth; $\frac{1}{2}$ in BSF
0·866	22 mm AF
0.875	$\frac{7}{8}$ in AF
0·920	$\frac{1}{2}$ in Whitworth; $\frac{9}{16}$ in BSF
0·937	$\frac{15}{16}$ in AF
0·944	24 mm AF
1·000	1 in AF
1·010	$\frac{9}{16}$ in Whitworth; $\frac{5}{8}$ in BSF
1·023	26 mm AF
1·062	$1\frac{1}{16}$ in AF; 27 mm AF
1·100	$\frac{5}{8}$ in Whitworth; $\frac{11}{16}$ in BSF
1·125	$1\frac{1}{8}$ in AF
1·181	30 mm AF
1·200	$\frac{11}{16}$ in Whitworth; $\frac{3}{4}$ in BSF
1·250	$1\frac{1}{4}$ in AF
1·259	32 mm AF
1·300	$\frac{3}{4}$ in Whitworth; $\frac{7}{8}$ in BSF
1·312	$1\frac{5}{16}$ in AF
1·390	$\frac{13}{16}$ in Whitworth; $\frac{15}{16}$ in BSF
1·417	36 mm AF
1·437	$1\frac{7}{16}$ in AF
1·480	$\frac{7}{8}$ in Whitworth; 1 in BSF
1·500	$1\frac{1}{2}$ in AF
1·574	40 mm AF; $\frac{15}{16}$ in Whitworth
1·614	41 mm AF
1·625	$1\frac{5}{8}$ in AF
1·670	1 in Whitworth; $1\frac{1}{8}$ in BSF
1·687	$1\frac{11}{16}$ in AF
1·811	46 mm AF
1·812	$1\frac{13}{16}$ in AF
1·860	$1\frac{1}{8}$ in Whitworth; $1\frac{1}{4}$ in BSF
1·875	$1\frac{7}{8}$ in AF
1·968	50 mm AF
2·000	2 in AF
2·050	$1\frac{1}{4}$ in Whitworth; $1\frac{3}{8}$ in BSF
2·165	55 mm AF
2·362	60 mm AF

Jacking and towing

The pantograph-type jack supplied with the car is strong enough to enable the vehicle to be raised to change a wheel. *Never* go beneath the vehicle when only the pantograph jack is being used, but use some suitable means of secondary support in case the car rocker panel collapses. When working beneath the vehicle, ensure that it is safely supported and that, where necessary, the roadwheels are firmly chocked.

With regard to towing a vehicle equipped with automatic transmission, provided that the transmission is known to be serviceable, the vehicle may be towed with the selector lever in the 'N' position at no more than 20 mph (30 km/h) for distances up to 6 miles (10 km). If the transmission is defective in any way, or the distance to be towed is more than that previously mentioned, the propeller shaft must be removed.

The towing points shown in the relevant illustrations can also be used, if necessary, for tie-down points if the vehicle is to be transported by sea, air or car transporter.

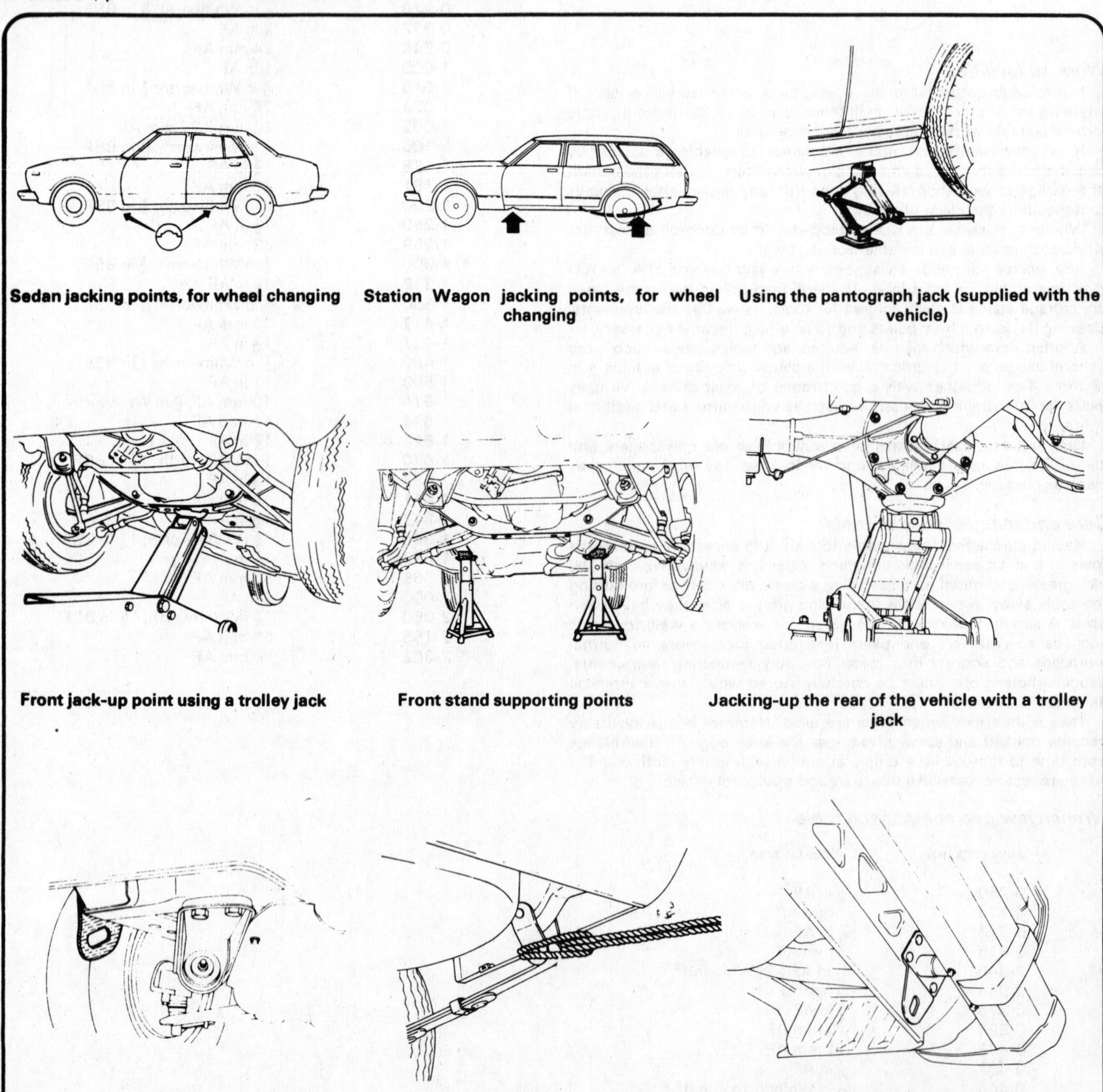

Sedan jacking points, for wheel changing

Station Wagon jacking points, for wheel changing

Using the pantograph jack (supplied with the vehicle)

Front jack-up point using a trolley jack

Front stand supporting points

Jacking-up the rear of the vehicle with a trolley jack

Front towing point

Rear towing point (Station Wagon)

Rear towing point (Sedan)

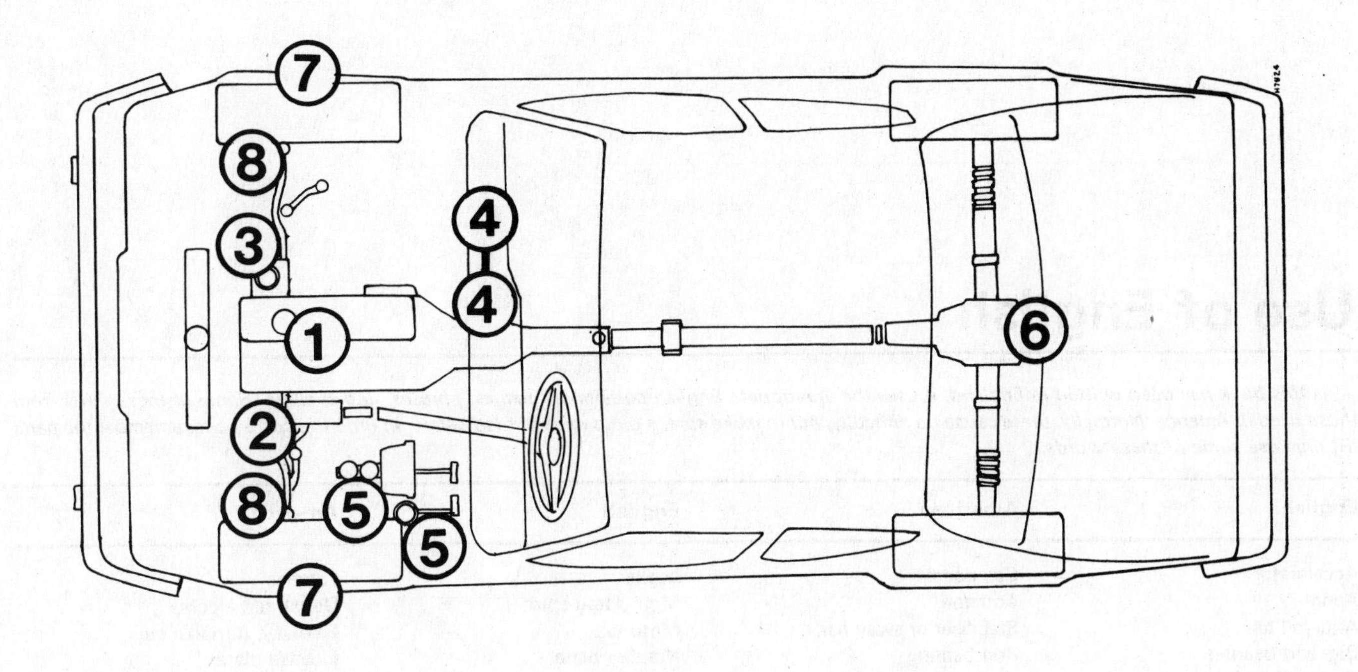

Recommended lubricants and fluids

Component or system	Lubricant or fluid specification
Engine (1)	**Multigrade engine oil, SAE classification SE**
Steering box-manual (2)	**SAE 90EP gear oil, API GL-4**
Power steering system (3)	**Dexron automatic transmission fluid**
Manual transmission (4)	**SAE 90EP gear oil, API GL-4**
Automatic transmission (4)	**Dexron automatic transmission fluid**
Brake and clutch reservoirs (5)	**DOT 3, DOT 4 or SAE J1703F**
Rear axle (6)	**SAE 90EP gear oil, API GL-5**
Front wheel bearings (7)	**Multipurpose lithium-based grease, NLGI 2**
Rear wheel bearings – Sedan only	**Multipurpose lithium-based grease, NLGI 2**
Steering balljoints (8)	**Multipurpose lithium-based grease, NLGI 2**

Note: *Listed here are manufacturer recommendations at the time this manual was written. Manufacturers occasionally upgrade their fluid and lubricant specifications, so check with your local auto parts store for current recommendations.*

Use of English

As this book has been written in England, it uses the appropriate English component names, phrases, and spelling. Some of these differ from those used in America. Normally, these cause no difficulty, but to make sure, a glossary is printed below. In ordering spare parts remember the parts list may use some of these words:

English	American	English	American
Accelerator	Gas pedal	Locks	Latches
Aerial	Antenna	Methylated spirit	Denatured alcohol
Anti-roll bar	Stabiliser or sway bar	Motorway	Freeway, turnpike etc
Big-end bearing	Rod bearing	Number plate	License plate
Bonnet (engine cover)	Hood	Paraffin	Kerosene
Boot (luggage compartment)	Trunk	Petrol	Gasoline (gas)
Bulkhead	Firewall	Petrol tank	Gas tank
Bush	Bushing	'Pinking'	'Pinging'
Cam follower or tappet	Valve lifter or tappet	Prise (force apart)	Pry
Carburettor	Carburetor	Propeller shaft	Driveshaft
Catch	Latch	Quarterlight	Quarter window
Choke/venturi	Barrel	Retread	Recap
Circlip	Snap-ring	Reverse	Back-up
Clearance	Lash	Rocker cover	Valve cover
Crownwheel	Ring gear (of differential)	Saloon	Sedan
Damper	Shock absorber, shock	Seized	Frozen
Disc (brake)	Rotor/disk	Sidelight	Parking light
Distance piece	Spacer	Silencer	Muffler
Drop arm	Pitman arm	Sill panel (beneath doors)	Rocker panel
Drop head coupe	Convertible	Small end, little end	Piston pin or wrist pin
Dynamo	Generator (DC)	Spanner	Wrench
Earth (electrical)	Ground	Split cotter (for valve spring cap)	Lock (for valve spring retainer)
Engineer's blue	Prussian blue	Split pin	Cotter pin
Estate car	Station wagon	Steering arm	Spindle arm
Exhaust manifold	Header	Sump	Oil pan
Fault finding/diagnosis	Troubleshooting	Swarf	Metal chips or debris
Float chamber	Float bowl	Tab washer	Tang or lock
Free-play	Lash	Tappet	Valve lifter
Freewheel	Coast	Thrust bearing	Throw-out bearing
Gearbox	Transmission	Top gear	High
Gearchange	Shift	Torch	Flashlight
Grub screw	Setscrew, Allen screw	Trackrod (of steering)	Tie-rod (or connecting rod)
Gudgeon pin	Piston pin or wrist pin	Trailing shoe (of brake)	Secondary shoe
Halfshaft	Axleshaft	Transmission	Whole drive line
Handbrake	Parking brake	Tyre	Tire
Hood	Soft top	Van	Panel wagon/van
Hot spot	Heat riser	Vice	Vise
Indicator	Turn signal	Wheel nut	Lug nut
Interior light	Dome lamp	Windscreen	Windshield
Layshaft (of gearbox)	Countershaft	Wing/mudguard	Fender
Leading shoe (of brake)	Primary shoe		

Routine maintenance

Maintenance is essential for ensuring safety and desirable for the purpose of getting the best in terms of performance and economy from the car. Over the years the need for periodic lubrication – oiling, greasing and so on – has been drastically reduced if not totally eliminated. This has unfortunately tended to lead some owners to think that because no such action is required, the items either no longer exist, or will last forever. This is a serious delusion. It follows therefore that the largest initial element of maintenance is visual examination. This may lead to repairs or renewals.

Every 250 miles (400 km) travelled or weekly – whichever comes first

Steering
- Check tire pressures
- Examine tires for wear or damage
- Is steering smooth and accurate?

Brakes
- Check reservoir fluid level
- Try an emergency stop. Is there any fall-off in braking efficiency?

Lights, wipers and horns
- Do all bulbs work at the front and rear?
- Are headlamp beams aligned properly?
- Check windshield washer fluid level
- Do wipers and horns work?

Engine
- Check oil level and top-up if required
- Check radiator coolant level and top-up if required
- Check battery electrolyte level and top-up to top of plates with distilled water, if required

Every 3000 miles (4800 km) travelled or 3 months – whichever comes first

Steering
- Examine all steering linkage rods, joints and bushes for signs of wear or damage
- Check front wheel hub bearings and adjust if necessary
- Check tightness of steering gear mounting bolts
- Check rear wheel hub bearings and adjust if necessary (Sedan)

Brakes
- Examine disc pads and drum shoes to determine amount of friction material left. Renew if necessary
- Examine all hydraulic pipes, cylinders and unions for signs of chafing, corrosion, dents or any other form of deterioration or leaks

Suspension
- Examine all nuts, bolts and mountings securing suspension units, front and rear. Tighten if necessary
- Examine the rubber bushes for signs of wear and play

Transmission (manual and automatic)
- Check oil level and top-up if necessary

Clutch
- Check fluid reservoir level and top-up if necessary

Body
- Lubricate all locks and hinges
- Check that water drain holes at bottom of doors are clear

Every 6000 miles (9600 km) travelled or 6 months – whichever comes first

Engine
- Check tension of engine drive-belts
- Check valve clearances and adjust if necessary
- Renew engine oil and filter
- Renew fuel line filter (if necessary after visual inspection)

Steering
- Rotate road-wheels and rebalance if necessary

Brakes
- Check pedal free movement, and for fluid leakage

Clutch
- Check pedal free movement, and for fluid leakage

Every 12 000 miles (19 000 km) travelled or 12 months – whichever comes first

Engine
- Check crankcase fume emission control system (Chapter 1)
- Check fuel storage evaporative emission control system (Chapter 3)
- Check exhaust emission control system (Chapter 3)
- Install new spark plugs
- Check ignition timing and adjust if necessary
- Check HT ignition leads for deterioration

Steering
- Check wheel alignment

Suspension
- Check shock absorber operation

Transmission
- Check security of propeller shaft bolts
- Check oil level in rear axle and top-up if necessary

Every 24 000 miles (38 000 km) travelled or 2 years – whichever comes first

Engine
- Flush cooling system and refill with antifreeze mixture
- Renew air cleaner elements
- Inspect vacuum servo unit hoses, connections and check valve
- Lubrication rear axle driveshaft joints
- Inspect air conditioning unit hoses and connections, and for leakage from the refrigerant unit

Every 30 000 miles (48 000 km) travelled or 2½ years – whichever comes first

Transmission
- Drain manual transmission and refill with fresh oil
- Drain rear axle and refill with fresh oil
- Check propeller shaft universal joints for wear

Steering

Lubricate balljoints (after removing grease plugs)
Check steering box oil level and top-up if necessary

Headlights

Check beams and adjust if required

Wheel hubs (except rear hubs on Station Wagon)

Dismantle, clean out old grease and repack with new

Every 48 000 miles (77 000 km) travelled or 4 years – whichever comes first

Brakes

Drain hydraulic system, renew all cylinder seals, and refill with fresh fluid. Bleed system. Overhaul vacuum servo unit

Clutch

Drain hydraulic system, renew master and operating cylinder seals, refill with fresh fluid. Bleed system

Transmission (Sedan)

Dismantle driveshafts and grease sliding joint (Chapter 8)

Additionally the following items should be attended to as time can be spared

Underbody

This should be cleaned by using a high pressure hose or steam-cleaner in order to detect rust and corrosion.

Exhaust system

An exhaust system must be leakproof, and the noise level below a certain maximum. Excessive leaks may cause carbon monoxide fumes to enter the passenger compartment. Excessive noise constitutes a public nuisance. Both these faults may cause the vehicle to be kept off the road. Repair or renew defective sections when symptoms are apparent.

Chapter 1 Engine

Refer to Chapter 13 for specifications and information applicable to 1980 through 1984 models.

Contents

Specifications

General

Engine code number	L24
Engine type	Six-cylinder, in-line, OHC
Displacement	146·0 in (2393 cc)
Bore	3·27 in (83 mm)
Stroke	2·90 in (73·7 mm)
Compression ratio	8·6 : 1
Oil pressure (hot at 2000 rev/min)	50 to 60 lbf/in^2 (3·5 to 4·2 kfg/cm^2)

Firing order 1-5-3-6-2-4

The blackened terminal shown on the distributor cap indicates the Number One spark plug wire position

Cylinder location and distributor rotation

Crankshaft

Journal diameter	2·1631 to 2·1636 in (54·942 to 54·955 mm)
Maximum taper or ovality of journal	Less than 0·0004 in (0·01 mm)
Maximum endfloat	0·002 to 0·007 in (0·05 to 0·18 mm)
Crankpin diameter	1·9670 to 1·9675 in (49·961 to 49·974 mm)
Maximum taper or ovality of crankpin	Less than 0·0004 in (0·01 mm)
Main bearing running clearance (normal)	0·0008 to 0·0028 in (0·020 to 0·072 mm)
Maximum main bearing running clearance	0·0047 in (0·12 mm)
Undersize main bearing availability	0·0098 in (0·25 mm), 0·0197 in (0·50 mm), 0·0295 in (0·75 mm), 0·0394 in (1·00 mm)

Connecting rod

Bearing thickness	0·0588 to 0·0593 in (1·493 to 1·506 mm)
Big-end side-play	0·0079 to 0·0118 in (0·20 to 0·30 mm)
Connecting rod bearing running clearance	0·0010 to 0·0022 in (0·025 to 0·055 mm)
Undersize connecting rod bearing availability	0·0024 in (0·06 mm), 0·0047 in (0·12 mm), 0·0098 in (0·25 mm), 0·0197 in (0·50 mm), 0·0295 in (0·75 mm), 0·0394 in (1·00 mm)

Cylinder block

Bore diameter	3·2677 to 3·2697 in (83·000 to 83·050 mm)
Ovality of bore	0·0006 in (0·015 mm)

Taper	0·0006 in (0·015 mm)
Difference between cylinders	0·0020 in (0·05 mm)
Surface flatness of cylinder block (to cylinder head)	Less than 0·0020 in (0·05 mm)

Piston

Piston diameter	3·2671 to 3·2691 in (82·985 to 83·035 mm)
Oversize piston availability	0·0197 in (0·50 mm), 0·0394 in (1·00 mm)
Ring groove width:	
Top	0·0799 to 0·0807 in (2·030 to 2·050 mm)
Second	0·0795 to 0·0803 in (2·020 to 2·040 mm)
Oil control	0·1581 to 0·1591 in (4·015 to 4·040 mm)
Piston-to-bore clearance	0·0010 to 0·0018 in (0·025 to 0·045 mm)
Piston pin hole diameter	0·8268 to 0·8271 in (21·001 to 21·008 mm)

Piston pin

Pin diameter	0·8265 to 0·8267 in (20·993 to 20·998 mm)
Pin length	2·8346 to 2·8445 in (72·00 to 72·25 mm)
Piston pin-to-piston clearance	0·0002 to 0·0005 in (0·006 to 0·013 mm)
Interference fit of piston pin-to-connecting rod bush	0·0006 to 0·0013 in (0·015 to 0·033 mm)

Piston ring

Ring width:	
Top	0·0778 to 0·0783 in (1·977 to 1·990 mm)
Second	0·0778 to 0·0783 in (1·977 to 1·990 mm)
Side clearance:	
Top	0·0016 to 0·0029 in (0·040 to 0·073 mm)
Second	0·0012 to 0·0028 in (0·030 to 0·070 mm)
Ring gap:	
Top	0·0098 to 0·0157 in (0·25 to 0·40 mm)
Second	0·0059 to 0·0118 in (0·15 to 0·30 mm)
Oil control	0·0118 to 0·0354 in (0·30 to 0·90 mm)

Camshaft

Camshaft endfloat	0·0031 to 0·0150 in (0·08 to 0·38 mm)
Camshaft lobe lift:	
Intake	0·2618 in (6·65 mm)
Exhaust	0·2756 in (7·00 mm)
Journal diameter	1·8878 to 1·8883 in (47·949 to 47·962 mm)
Bearing inside diameter	1·8898 to 1·8904 in (48·000 to 48·016 mm)
Running clearance	0·0015 to 0·0026 in (0·038 to 0·067 mm)

Valves

Clearance (Hot):	
Intake	0·010 in (0·25 mm)
Exhaust	0·012 in (0·30 mm)
Clearance (Cold):	
Intake	0·008 in (0·20 mm)
Exhaust	0·010 in (0·25 mm)
Head diameter:	
Intake	1·65 in (42 mm)
Exhaust	1·38 in (35 mm)
Seat width:	
Intake	0·055 to 0·063 in (1·4 to 1·6 mm)
Exhaust	0·071 to 0·087 in (1·8 to 2·2 mm)
Seat angle (intake and exhaust)	45°

Valve guide

Length (intake and exhaust)	2·323 in (59 mm)
Outer diameter (intake and exhaust)	0·4733 to 0·4738 in (12·023 to 12·034 mm)
Inner diameter (intake and exhaust)	0·3150 to 0·3157 in (8·000 to 8·018 mm)
Guide interference fit (intake and exhaust)	0·0011 to 0·0019 in (0·027 to 0·049 mm)
Height from head surface (intake and exhaust)	0·409 to 0·417 in (10·4 to 10·6 mm)
Guide-to-stem clearance:	
Intake	0·0008 to 0·0021 in (0·020 to 0·053 mm)
Exhaust	0·0016 to 0·0029 in (0·040 to 0·073 mm)

Valve spring

Free length:	
Outer	1·968 in (49·98 mm)
Inner	1·766 in (44·85 mm)

Oil pump

Type	Rotor
Oil pump drive	Helical gear on the crankshaft
Rotor side clearance (rotor-to-bottom cover)	0·0016 to 0·0031 in (0·04 to 0·08 mm)
Wear limit	0·0079 in (0·20 mm)

Rotor tip clearance	Less than 0·0047 in (0·12 mm)
Wear limit	0·0079 in (0·20 mm)
Outer rotor-to-body clearance	0·0059 to 0·0083 in (0·15 to 0·21 mm)
Wear limit	0·0197 in (0·5 mm)

Oil pressure regulator valve

Regulator valve spring free length	2·067 in (52·5 mm)

Oil filter

Type	Full flow, replaceable cartridge

Engine lubrication

Engine oil capacity (including filter)	6 US quarts (5·7 liters)
Lubricant type	Multigrade engine oil (see lubrication chart)

Torque wrench settings

	lbf ft	kgf m
Cylinder head bolts:		
Stage 1	29	4·0
Stage 2	43	6·0
Stage 3	51 to 61	7·0 to 8·5
Connecting rod big end nuts	33 to 40	4·5 to 5·5
Flywheel fixing bolts	94 to 108	13 to 15
Main bearing cap bolts	33 to 40	4·5 to 5·5
Camshaft sprocket bolt	94 to 108	13 to 15
Oil pan bolts	4·3 to 7·2	0·6 to 1·0
Oil pump bolts	8·0 to 10·8	1·1 to 1·5
Oil pan drain plug	14 to 22	2·0 to 3·0
Rocker pivot locknuts	36 to 43	5·0 to 6·0
Camshaft locating plate bolts	3·6 to 5·8	0·5 to 0·8
Manifold nuts and bolts:		
M8	10 to 13	1·4 to 1·8
M10	25 to 36	3·5 to 5·0
Throttle chamber securing bolts	11 to 14	1·5 to 2·0
Crankshaft pulley bolts	87 to 116	12 to 16
Front cover bolts:		
M6	2·9 to 5·8	0·4 to 0·8
M8	7·2 to 11·6	1·0 to 1·6
Oil strainer	5·8 to 8·0	0·8 to 1·1
Oil pump cover bolts	5·1 to 7·2	0·7 to 1·0
Regulator valve cap nut	29 to 36	4 to 5
Spark plug	11 to 14	1·5 to 2·0
Air compressor bracket mounting bolts	33 to 40	4·5 to 5·5
Torque converter housing-to-engine bolts	29 to 36	4·0 to 5·0
Driveplate-to-torque converter bolts	29 to 36	4·0 to 5·0
Driveplate-to-crankshaft	101 to 116	14·0 to 16·0
Clutch bellhousing-to-engine bolts	32 to 43	4·4 to 5·9

1 General description

The L24 engine, which is installed on both the Sedan and Station Wagon, is an in-line six-cylinder type with the valves operating from an overhead-mounted camshaft.

Beginning at the top, the cylinder head is made of light, strong aluminum alloy with good cooling efficiency. Brass cast seals are used for the intake valves, while heat-resistant steel is employed for the exhaust valve seats. The valves are operated by rockers in contact with a double-row roller chain-driven camshaft, mounted centrally on the cylinder head.

The distributor, which is mounted on the left-hand side of the cylinder block, is driven by a helical gear mounted on the front of the crankshaft; this gear also drives the oil pump.

The electronic fuel injection equipment is installed on to the intake manifold; the devices which comprise the emission control systems have been added to this, the ignition and the exhaust systems on the engine.

The cylinder block is a cast structure, and provides support for the crankshaft at seven main bearings.

The crankshaft, which is made of special forged steel, has internal oil passages to provide lubrication to the main and big-end bearings.

The oil pump, which is mounted low down on the right-hand side of the cylinder block, is on a common center-line to the distributor and is driven by the same helical gear. Oil is delivered, via the filter and pressure relief valve, to the main oil gallery from which it passes to the main bearing journals and then to the connecting-rod bearing journals through drillings in the crankshaft. Oil spillage from the connecting rod big-ends, as well as a jet hole drilled through the connecting rod into the big-ends, provides splash lubrication for the pistons and connecting rod small ends. At the top of the engine, galleries drilled in the camshaft supports provide oil for the five bearings, while a pipe that runs along the length of the camshaft, delivers oil to each cam pad surface, to provide lubrication for the rocker arm and pivot.

2 Major operations possible with the engine in the car

1 The following major tasks can be performed with the engine installed. However, the degree of difficulty varies and for anyone who has lifting tackle available it is recommended that items such as the pistons, connecting rods and crankshaft bearings are attended to after removal of the engine from the vehicle:

(a) Removal and installation of the cylinder head
(b) Removal and installation of the camshaft and bearings
(c) Removal and installation of the engine supports (mounts)
(d) Removal and installation of the oil pan
(e) Removal and installation of the oil pump
(f) Removal and installation of the main bearings
(g) Removal and installation of the pistons and connecting rods
(h) Removal and installation of the connecting rod big-end bearings

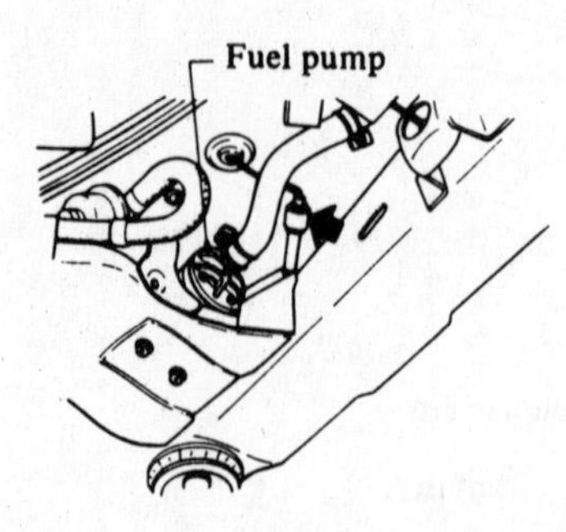

Fig. 1.1 The fuel pump ground lead (arrowed) (Sec 4)

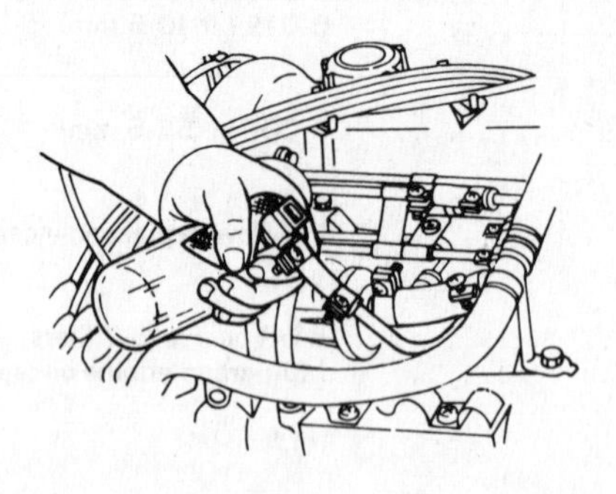

Fig. 1.2 Discharging fuel from the cold start valve into a container (Sec 4)

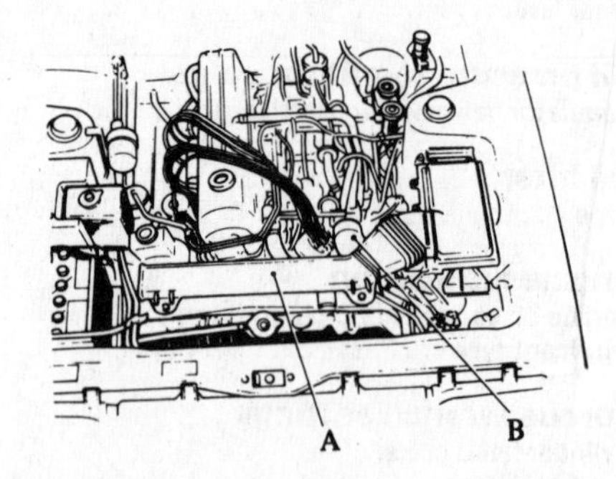

Fig. 1.3 The air cleaner ducts A and B (Sec 4)

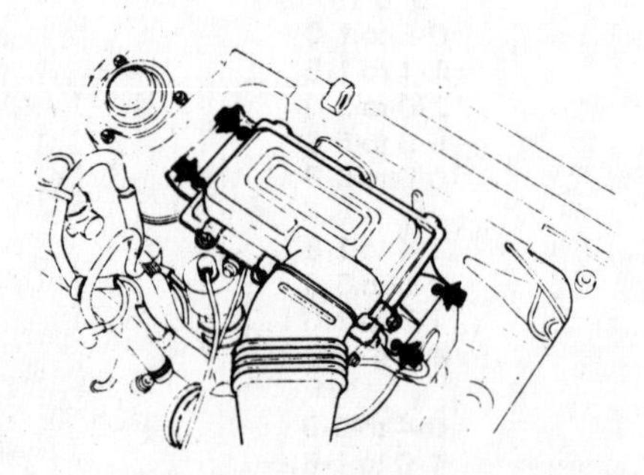

Fig. 1.4 The airflow meter mounting (Sec 4)

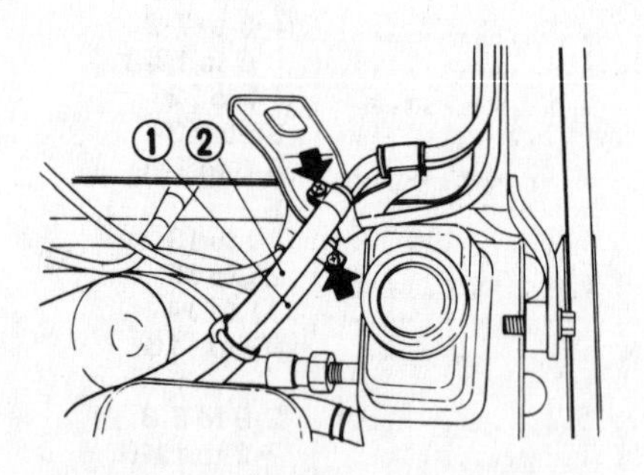

Fig. 1.5 Fuel hose retaining clips (arrowed) (Sec 4)

1 Fuel return hose 2 Fuel charge hose

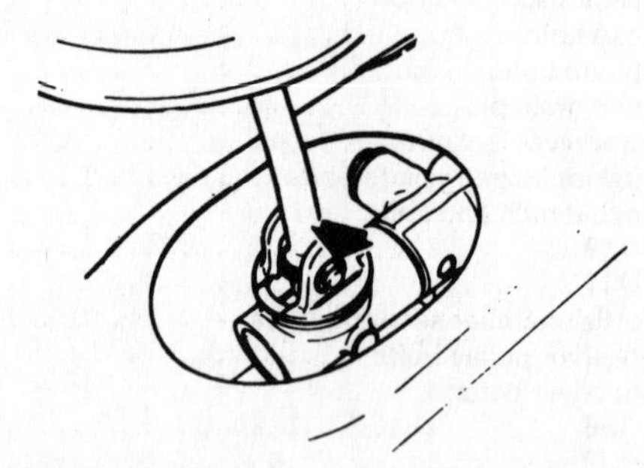

Fig. 1.6 The gear shift control lever E-ring (arrowed) (Sec 4)

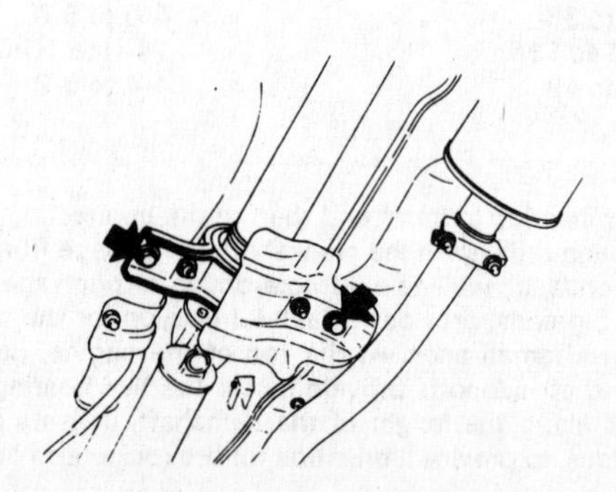

Fig. 1.7 The propeller shaft center bearing support bracket (arrowed) (Sec 4)

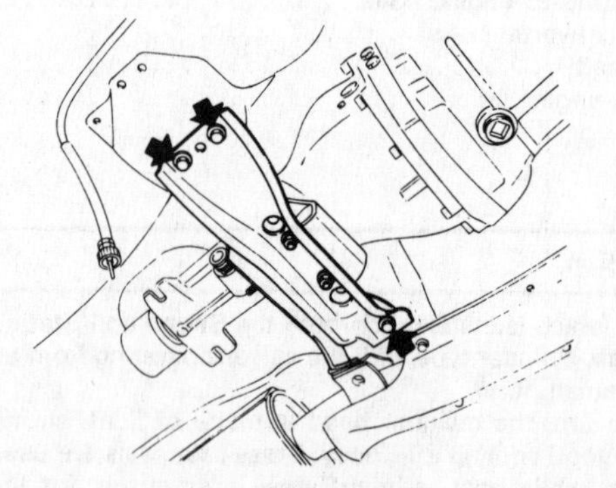

Fig. 1.8 The transmission rear mount (arrowed) (Sec 4)

3 Method of engine removal

1 The engine can either be removed complete with the transmission, or the two units can be removed separately. If the transmission is to be removed, it is preferable to remove it while still coupled to the engine.

2 Essential equipment includes a suitable jack or support(s) so that the vehicle can be raised whilst working underneath, and a hoist or lifting tackle capable of taking the necessary weight. If an inspection pit is available some problems associated with jacking will be alleviated, but at some time during the engine removal procedure a jack will be required beneath the transmission (a trolley jack is very useful for this application).

4 Engine/transmission – removal

1 Before commencing work it will be necessary to reduce the fuel line pressure to zero. To do this, proceed as described in paragraphs 2 through 7.

2 Disconnect the battery ground cable.

3 At the rear of the vehicle, locate the fuel pump under the car and disconnect the ground lead.

4 At the starter motor, disconnect the ground lead S terminal.

5 Undo and remove the two screws securing the cold start valve to the intake manifold. Place the valve into a suitable container of approximately 1.22 cu in (20cc) capacity.

6 Re-connect the battery ground cable. Now, with the assistance of a second person, turn the ignition switch to the Start position. Keep the ignition switch in this position until there are no signs of fuel coming from the cold start valve.
7 For safety, assemble the cold start valve to the intake manifold. Switch off the ignition. Provided that the ignition is not turned on again, the fuel pump may be reconnected.
8 Initially mark the position of the hood hinges, using a pencil or ball-point pen, to aid installation. Remove the hood, referring to Chapter 12 if necessary.
9 Detach the battery leads, and stow them to one side where they are out of the way.
10 Drain the cooling system (refer to Chapter 2 if necessary), retaining the coolant if it contains antifreeze.
11 Place a suitable container under the oil pan and drain the engine oil into it.
12 Undo and remove the bolts that attach the air cleaner ducts to the mounting brackets and the airflow meter. Remove the ducts from the engine compartment. (Refer to Chapter 3, if necessary).
13 Disconnect the top and bottom hoses from the radiator.
14 Remove the air cleaner and airflow meter, as an assembly, from the engine compartment. (Refer to Chapter 3, if necessary).
15 Pull off the rubber hoses to the carbon canister and remove the canister from the vehicle.
16 Remove the four screws attaching the radiator shroud to the front of the engine compartment. To remove the shroud, a series of manipulations back and forth whilst pressing inwards either side of the shroud should soon have it out.
17 Remove the four radiator retaining bolts and lift the radiator upwards and away from the engine compartment. Take care not to damage the radiator matrix.
18 Remove the cotter pin that connects the torsion shaft to the accelerator linkage.
19 *On vehicles equipped with air conditioning:* Loosen the idler pulley locknut, to relieve the drivebelt tension, then remove the compressor mounting bracket complete with the compressor. **Very important.** *The pipes to and from the compressor must not be disconnected except by a refrigeration specialist.* Using suitable wire, suspend the compressor from a convenient anchorage point. Never allow it to hang suspended by the pipes.
20 *On vehicles equipped with power steering:* Remove the three bolts and washers that attach the oil pump to the mounting bracket. Lift away the oil pump and suspend it as described for the air conditioning compressor.
21 Remove all the electrical connections to the engine and transmission, identifying each lead with a tag as it is detached. Note the routing of cables, and provisions for clipping and support.
22 Disconnect the fuel return and fuel charge hoses.
23 Disconnect the heater inlet and outlet hoses.
24 At the intake manifold, disconnect the vacuum hose that feeds the brake Master-Vac.
25 Working beneath the vehicle, unbolt the clutch operating cylinder and tie it up out of the way. There is no need to disconnect the hydraulic line.
26 Disconnect the speedometer cable from the rear extension of the transmission.
27 Inside the vehicle, remove the center console and sealing grommet from the gearshift lever.
28 Place the transmission control lever in the neutral position. Remove the E-ring and control lever pin from the transmission striking rod guide. Lift out the gearshift control lever.
29 Unscrew and remove the nuts which secure the flange of the exhaust downpipe to the manifold.
30 Tie the exhaust pipe to a convenient anchorage point and remove the pipe support bracket from the rear mounting insulator.
31 Remove the two bolts that secure the propeller shaft center bearing bracket to the bodyframe.
32 Mark the propeller shaft-to-rear axle pinion mating flanges (for exact replacement), unbolt the rear flange and withdraw the sliding sleeve from the transmission. Place a polythene bag over the open end of the rear extension, held in position with a rubber band, to prevent oil spillage.
33 Support the transmission with a jack, then unscrew and remove the bolts which secure the transmission rear mounting member to the bodyframe; then unbolt the mount from the transmission unit.
34 Using a hoist and suitable chains or slings, take the weight of the engine so that the engine mounting insulators can be disconnected.
35 With the engine safely held by the hoist, proceed to remove the nuts securing the engine mounts to the chassis.
36 Now, whilst alternately lowering the jack under the transmission, and lifting the engine by the hoist, proceed to lift the engine and transmission from the vehicle.

5 Engine/automatic transmission – removal

1 The procedure is similar to that described in the preceding Section, except that a clutch operating cylinder is not used.
2 In addition, the following operations are required:

(a) The splash board must be removed from under the power unit
(b) The leads must be disconnected from the starter inhibitor switch and the downshift solenoid
(c) Disconnect the speed selector linkage from the manual shaft
(d) Disconnect the vacuum pipe from the vacuum diaphragm nozzle on the transmission unit
(e) Disconnect the oil filler tube
(f) Disconnect the oil cooler inlet and outlet tubes at the transmission unit

6 Engine/transmission – separation

1 Unscrew and remove the starter motor securing bolts and withdraw the starter from the clutch bellhousing.
2 Unscrew and remove the bolts which secure the clutch bellhousing to the engine crankcase.
3 Withdraw the transmission in a straight line, so that it's weight does not hang up on the input shaft while it is still engaged in the driven plate of the clutch mechanism.

7 Engine/automatic transmission – separation

1 Unscrew and remove the dust cover from the lower half of the torque converter housing.
2 Unscrew and remove each of the bolts which secure the torque converter to the driveplate. Access to these bolts can be gained (one at a time) by turning the engine until each bolt comes into view through the lower half of the converter housing.
3 Withdraw the automatic transmission unit leaving the driveplate bolted to the crankshaft rear flange.

8 Engine dismantling – general

1 It is best to mount the engine on a dismantling stand but if one is not available, then stand the engine on a strong bench so as to be at comfortable working height. Failing this, the engine can be stripped down on the floor.
2 During the dismantling process the greatest care should be taken to keep the exposed parts free from dirt. As an aid to achieving this, it is a sound scheme to thoroughly clean down the outside of the engine, removing all traces of oil and congealed dirt.
3 Use kerosene or a good water soluble grease solvent. The latter compound will make the job easier, as, after the solvent has been applied and allowed to stand for a time, a vigorous jet of water will wash off the solvent and all the grease and filth. If the dirt is thick and deeply embedded, work the solvent into it with a wire brush.
4 Finally wipe down the exterior of the engine with a rag and only then, when it is quite clean should the dismantling process begin. As the engine is stripped, clean each part in a bath of kerosene or gasoline.
5 Never immerse parts with oilways in kerosene, i.e. the crankshaft, but to clean, wipe down carefully with a gasoline dampened rag. Oilways can be cleaned out with wire. If an air line is present all parts can be blown dry and the oilways blown through as an added precaution.
6 Re-use of old engine gaskets is false economy and can give rise to oil and water leaks, if nothing worse. To avoid the possibility of trouble after the engine has been reassembled ALWAYS use new gaskets throughout.

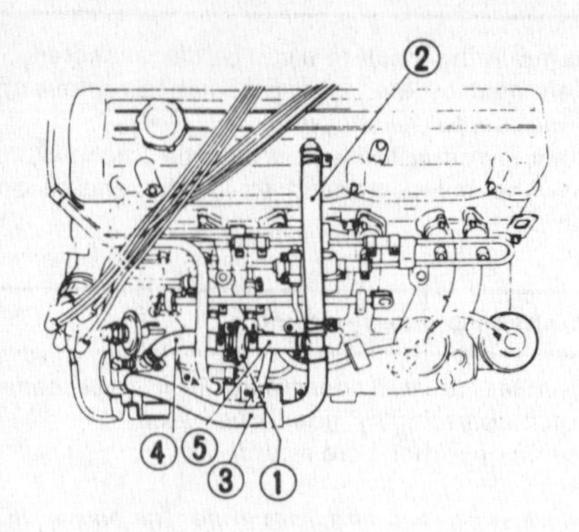

Fig. 1.9 Removing the air regulator (Sec 9)

1 *Air regulator*
2 *3-way connector*
3 *Throttle chamber pipe*
4 *Air regulator hose*
5 *3-way connector*

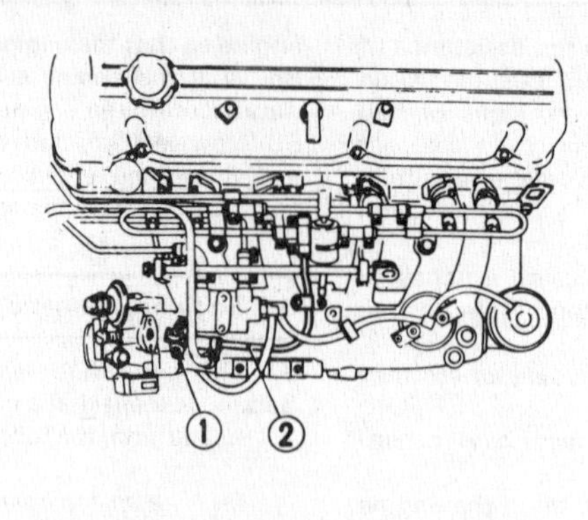

Fig. 1.10 Removing the cold start valve (Sec 9)

1 *Cold start valve* 2 *Fuel pipe*

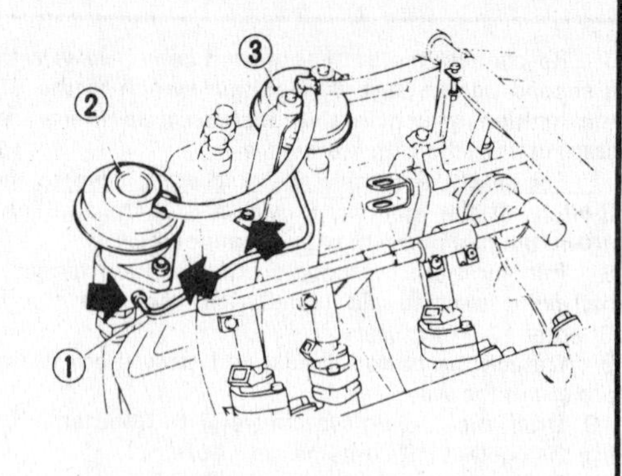

Fig. 1.11 Removing the EGR and BPT valves (Sec 9)

1 *BPT valve control tube* 2 *EGR valve*
3 *BPT valve*

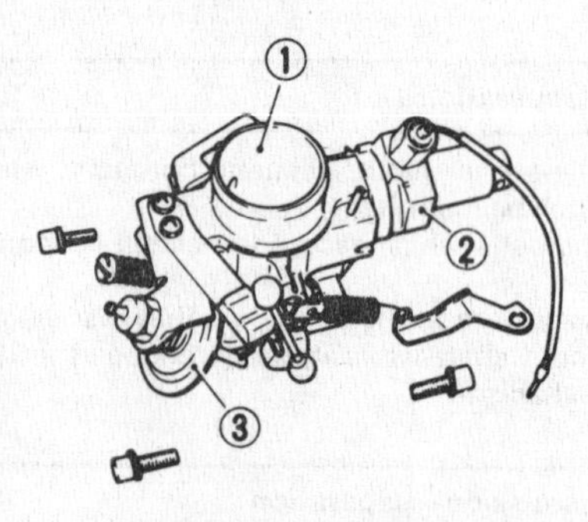

Fig. 1.12 The throttle chamber assembly (Sec 9)

1 *Throttle chamber* 2 *BCDD*
3 *Dashpot*

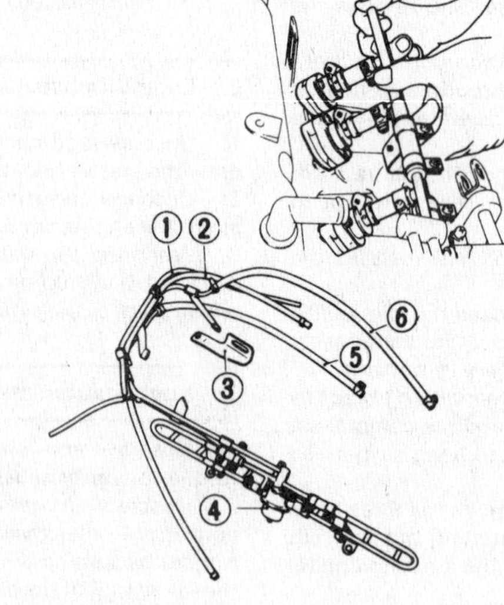

Fig. 1.13 The fuel line assembly (Sec 9)

1 *Canister control vacuum tube*
2 *Canister purge hose*
3 *Front engine slinger*
4 *Pressure regulator*
5 *Fuel feed hose*
6 *Fuel return hose*

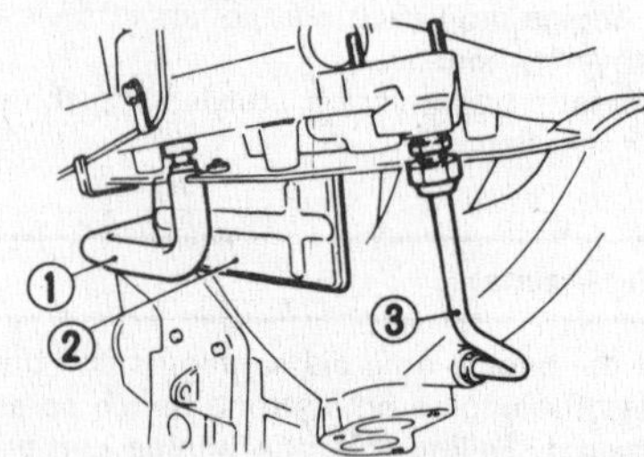

Fig. 1.14 Removing the hose and sub-heat shield plate (Sec 9)

1 *PCV to connector hose* 2 *Sub-heat shield plate*
3 *EGR tube*

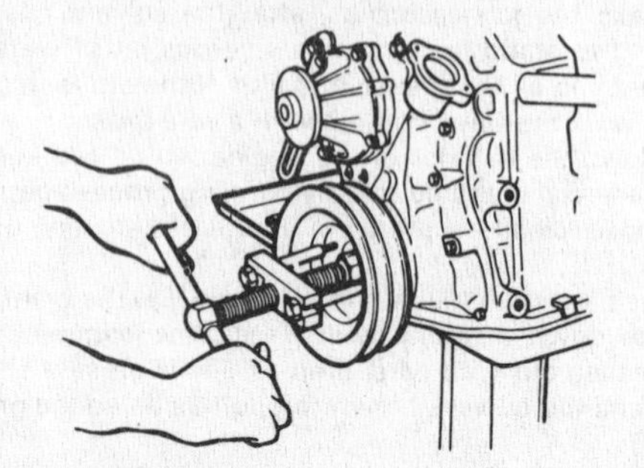

Fig. 1.15 Extracting the crankshaft pulley (Sec 9)

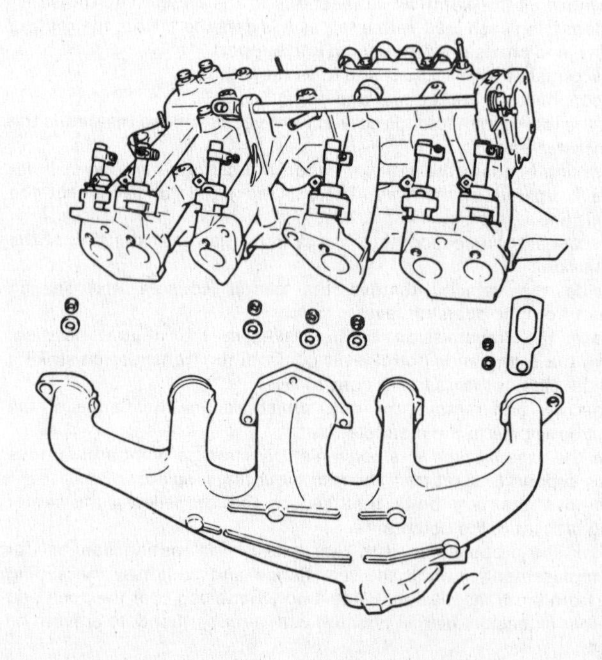

Fig. 1.16 The intake and exhaust manifolds

7 Do not throw the old gaskets away as it sometimes happens that an immediate replacement cannot be found and the old gasket is then useful as a template. Hang up the old gaskets as they are removed on a suitable hook or nail.
8 To strip the engine it is best to work from the top down. The oil pan provides a firm base on which the engine can be supported in an upright position. When the stage, where the oil pan must be removed, is reached, the engine can be turned on its side and all other work carried out with it in this position.
9 Wherever possible, re-install nuts, bolts and washers, fingertight from wherever they were removed. This helps avoid later loss and muddle. If they cannot be re-installed then lay them out in such a fashion that it is clear from where they came.

9 Engine – removal of ancillary components

1 With the engine removed from the vehicle and separated from the transmission, the ancillary components should be removed before dismantling of the engine begins.
2 On the right-hand side of the engine, slacken the alternator mounting and adjustment strap bolts and push the alternator in towards the block. The drivebelt can now be slipped off the alternator and the fan pulleys.
3 Remove the alternator.
4 *Vehicles fitted with air conditioning and power steering:* Unbolt and remove the compressor mounting bracket bolts. Lift away the bracket and idler pulley as an assembly. Remove the bolts that attach the power steering pump mounting bracket. Lift away the bracket together with the idler pulley.
5 Remove the fan and pulley assembly from the water pump hub.
6 Unscrew and remove the six bolts that attach the water pump assembly to the front of the block. Carefully remove the water pump.
7 Undo and remove the engine mountings.
8 Unscrew and remove the oil filter. This will probably require the use of a chain or strap-type wrench.
9 Unscrew and remove the oil pressure switch.
10 First identify, then remove, the spark plug leads. Remove the spark plugs from the engine.
11 Mark the position of the distributor retaining plate screw in it's elongated hole, then unscrew and remove the screw and withdraw the distributor.
12 Remove the air regulator, three-way connector-to-rocker cover hose, throttle chamber-to-three-way connector, air regulator-to-connector hose and the three-way connector-to-air regulator hose as a complete assembly.
13 Working in the same area, remove the cold start valve and fuel pipe as an assembly.
14 Now unscrew the back pressure transducer (BPT) control tube from the intake manifold. Undo and remove the nuts and bolts that attach the exhaust gas recirculation (EGR) valve and the BPT valve to the manifold. Lift away the BPT valve, EGR valve and the control tube as an assembly.
15 Remove the four bolts that attach the throttle chamber; remove the throttle chamber together with the dashpot and the boost controlled deceleration device (BCDD).
16 Unfasten each clip that retains the rubber fuel hoses to the injectors. Now remove, as an assembly, the fuel return hose, fuel feed hose, vacuum signal hose, canister purge hose, pressure regulator and the front engine slinger. Take great care not to twist or bend the rigid pipes in this assembly.
17 At the front left-hand side of the cylinder head, remove the thermostat housing, thermo-time switch, thermal transmitter and water temperature sensor, as an assembly.
18 Disconnect the positive crankcase valve (PCV) tube and the EGR tube. Remove the sub-heat shield plate, which is located under the PCV tube.
19 Undo and remove the bolts that attach the intake manifold, making a note of the different length bolts used. Lift away the intake manifold and the heat shield plate, as an assembly.
20 With the intake manifold removed, remove the exhaust manifold.
21 Undo and remove the large bolt that retains the crankshaft pulley to the crankshaft.
22 Using a suitable extractor, draw the pulley off the front end of the crankshaft.
23 Unbolt and remove the clutch pressure plate and driven plate from the flywheel (manual transmission vehicles). On vehicles fitted with automatic transmission, unbolt the driveplate from the crankshaft rear flange.
24 Undo and remove the screws that attach the rocker cover to the cylinder head. Lift away the cover and gasket.

10 Cylinder head – removal

1 Before removing the cylinder head, make sure that the engine is absolutely cool, otherwise the head may distort after it is removed.
2 Where the cylinder head only is to be removed, without further dismantling, then it is vital that the timing sprocket-to-chain position is not altered. If a major overhaul is being carried out then the retention of the timing position is not essential as this will be reset as described in Section 41.
3 Turn the engine until the marks on the timing chain and the camshaft sprocket are in alignment. The timing chain marks are bright link plates, the centers of which are the timing points. Where the cylinder head is being removed with the engine in the vehicle, if the tension on the chain is released, it is possible for the timing chain tensioner to be ejected from its housing and to replace it would necessitate the removal of the timing cover. To avoid this possibility, it is recommended that before the camshaft sprocket is removed, a wedge of hardwood 10 inches long $\frac{3}{4}$ inch thick and $1\frac{1}{2}$ inches wide at the top (254 mm long 19 mm thick and 38 mm wide) is inserted into the timing case to keep the two runs of the chain apart and to keep the chain tensioner plunger depressed into its housing.
4 Unscrew and remove the bolt which secures the camshaft sprocket to the end flange of the camshaft.
5 Withdraw the camshaft sprocket from the loop of the timing chain and if the timing is not to be upset then tension must be maintained on the chain so that it does not move its position on the crankshaft sprocket. A long piece of wire hooked into the top link of the chain is useful for this purpose. **Note:** *It is important that after this stage of dismantling, the camshaft or crankshaft are not turned, otherwise the valves may contact the piston crowns.*
6 Remove the oil supply pipe from the valve gear.
7 Loosen the cylinder head bolts a half turn at a time, working from the center bolts outwards. These bolts have socket heads and will require the use of an Allen-type wrench.
8 Lift the cylinder head complete with camshaft and valve gear from the cylinder block. If the timing is not to be upset then tension must be maintained on the timing chain by use of the wire hook as the cylinder head is withdrawn.
9 Remove the cylinder head gasket.

11 Camshaft and valve gear – removal

1 Loosen the rocker arm pivot locknuts.
2 Remove the rocker arms by depressing the valve springs, taking care to retain the valve rocker guides.
3 Withdraw the camshaft ensuring that the lobes do not scratch or damage the bearings as they pass through. *On no account disturb the camshaft bearing cap bolts or caps as they are in-line bored after assembly.*

12 Valves – removal

1 Remove each valve from the cylinder head by compressing each spring in turn with a valve spring compressor until the valve rocker guide (if not previously removed) and the two halves of the split collet can be extracted.
2 Release the compressor and remove the spring seats, springs and valve stem oil seals.
3 If, when the valve spring compressor is screwed down, the valve spring retaining cap refuses to free to expose the split collet, do not continue to screw down on the compressor as there is a likelihood of damaging it.
4 Gently tap the top of the tool directly over the cap with a light hammer. This will free the cap. To avoid the compressor jumping off the valve spring retaining cap when it is tapped, hold the compressor firmly in position with one hand.
5 Keep the valves in the correct numerical sequence (1 to 12) num-

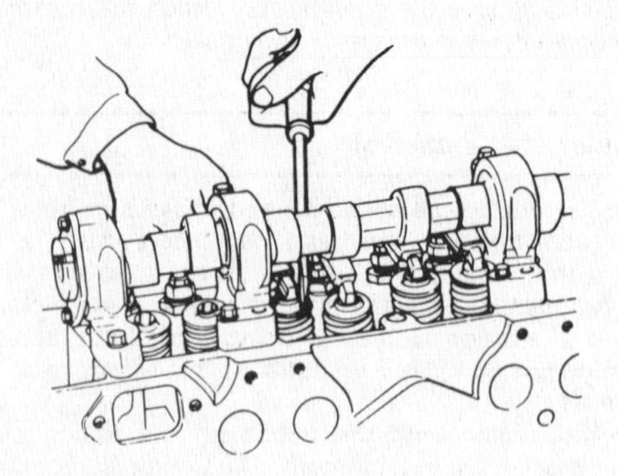

Fig. 1.17 Removing a rocker arm (Sec 11)

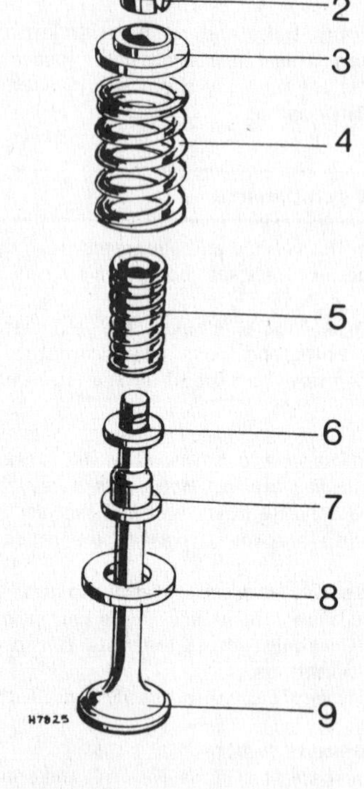

Fig. 1.18 Valve components (Sec 12)

1 *Valve rocker guide*
2 *Split collets*
3 *Spring retainer seat*
4 *Outer spring*
5 *Inner spring*
6 *Seal clip*
7 *Oil seal*
8 *Spring retainer seat*
9 *Valve*

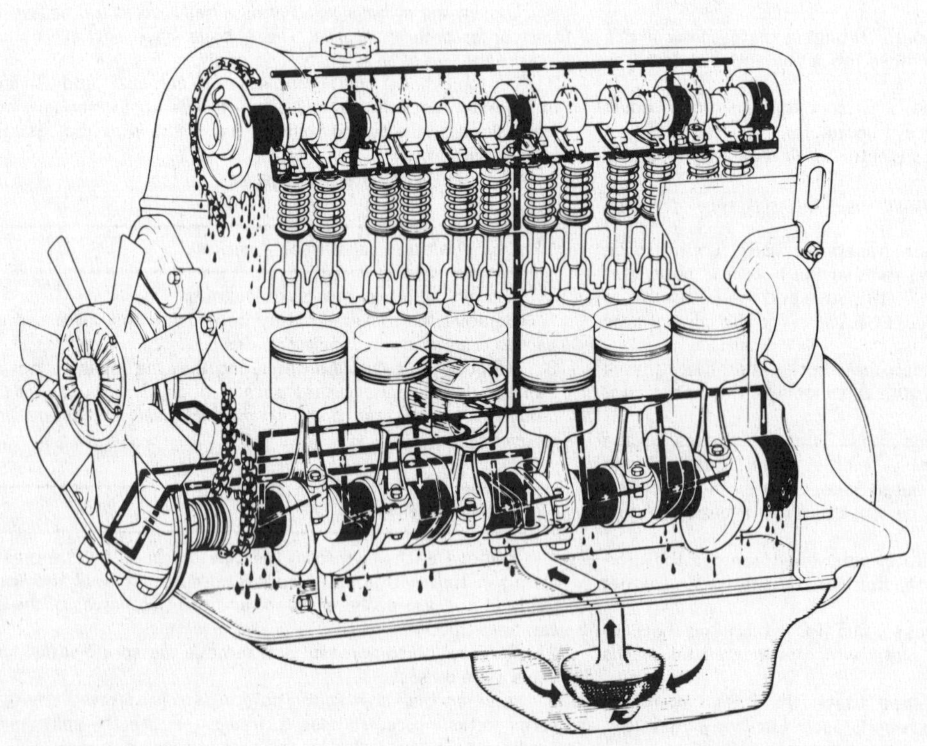

Fig. 1.19 The engine lubrication system (Sec 14)

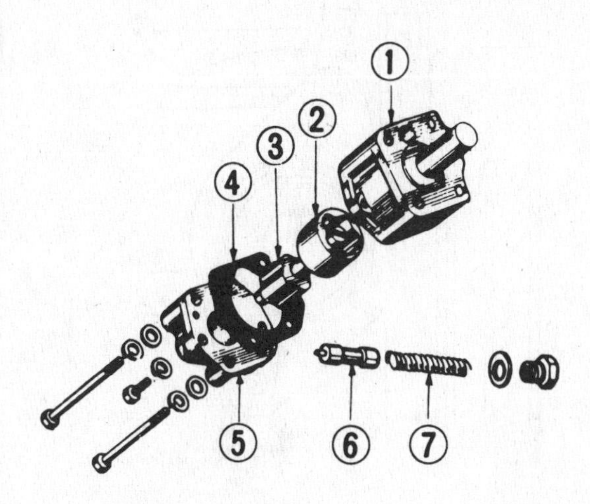

Fig. 1.20 Exploded view of the oil pump (Sec 15)

1 Body
2 Outer rotor
3 Inner rotor
4 Gasket
5 Cover
6 Pressure regulator valve
7 Spring

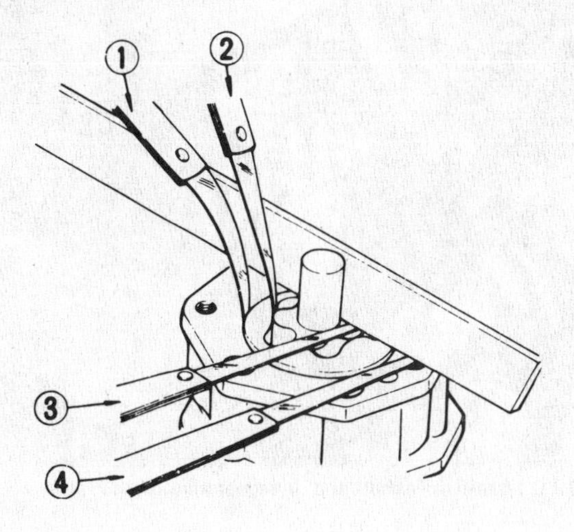

Fig. 1.21 Checking the oil pump rotor clearances (Sec 15)

1 Outer rotor-to-body clearance
2 Tip clearance
3 Gap between rotor and straight edge
4 Gap between body and straight edge

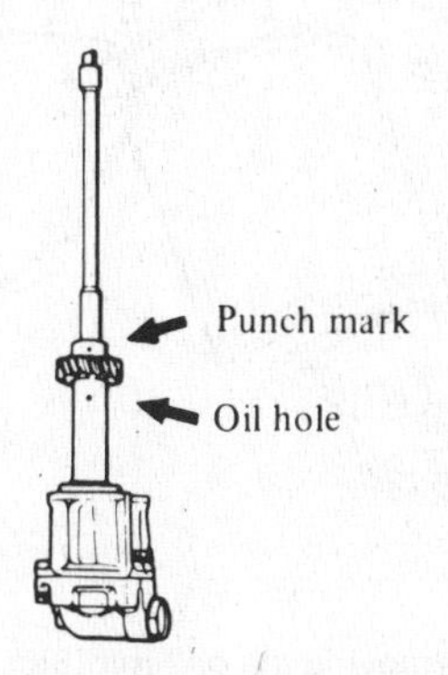

Fig. 1.22 Oil pump installation alignment marks (Sec 15)

bering from the front of the engine as originally installed.

13 Oil pan – removal

1 Unscrew and remove the retaining bolts and lift off the oil pan. Should it be stuck do not prise it off by inserting a lever in the joint but cut around the gasket by using a thin sharp knife. Pull the oil pan straight down as far as possible to clear the oil pump intake filter and tube.

14 Engine lubrication system

The pressurised lubrication system of the engine is dependent upon a trochoid type oil pump. The pump is externally-mounted and is driven by a helical gear on the crankshaft. The pump draws oil through a strainer from the oil pan and then supplies it, under pressure, through a full-flow type oil filter to the main oil gallery of the engine. The oil is then fed through a system of drillings to the bearing surfaces of the engine compartment.

A pressure regulator valve is incorporated in the oil pump.

15 Oil pump and distributor – removal, servicing of pump, refitting and installation

1 Turn the crankshaft until no 1 piston is at TDC (top-dead-center). To establish this position, if the engine is in the car, remove the rocker cover and observe the position of the inlet and exhaust valves of no 1 cylinder - the camshaft lobes should be clear of both valves. Alternatively, remove no 1 spark plug and feel the compression being generated with a finger. Remove the distributor cap and check when the rotor head points to no 1 segment in the cap.

2 The engine has static ignition timing marks on the engine front cover and crankshaft pulley. Align the notch on the crankshaft pulley with the TDC mark on the engine front cover but employ one of the methods previously described to ensure that the engine is on a compression stroke.

3 Mark the position of the securing screw in relation to the elongated holes in the distributor retaining plate, then remove the screw and withdraw the distributor.

4 Drain the engine oil from the oil pan.

5 Unscrew and remove the bolts which secure the oil pump body to the engine front cover and withdraw the oil pump/driveshaft assembly.

6 Unbolt the cover from the oil pump and extract the inner and outer rotors.

7 Unscrew and remove the pressure regulator valve plug, and extract the spring and valve.

8 Wash all components in clean gasoline and allow to dry.

9 Check all components for wear or scoring, and test the rotor clearances in the following manner:

(a) Using a straight-edge, check the clearance between the rotor faces and the straight-edge with a feeler gauge. Any differences between the two rotor face to straight-edge clearances must not exceed 0.0079 in (0.20 mm).
(b) Check the clearance between the tops (high points) of the inner and outer rotors. This should not exceed 0.0079 in (0.20 mm).
(c) Check the clearance between the outer rotor and the body. This should not exceed 0.0197 in (0.5 mm).
(d) Again using the straight-edge placed across the body, check the clearance between body and straight-edge; this should not exceed 0.0079 in (0.20 mm). The inner and outer rotors should always be assembled in the pump body so that the dotted marks on their end faces are not visible from the cover end.

10 Where the specified clearances are exceeded, renew the com-

1

15.13 Oil pump driveshaft and drivegear alignment marks

15.14 Installing the oil pump

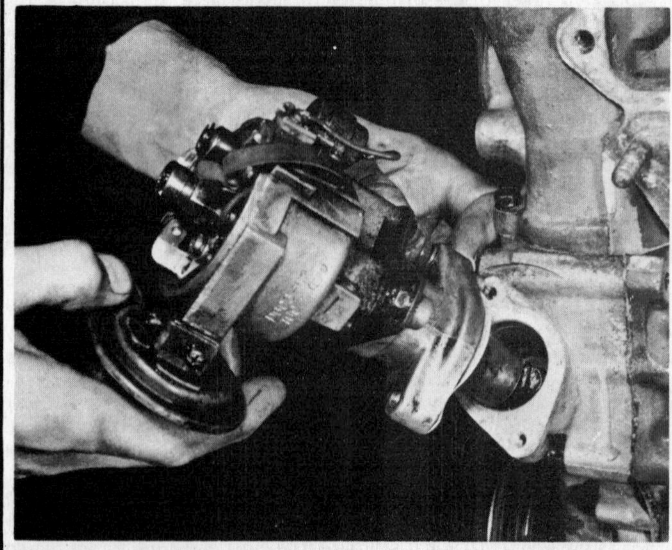

15.16 Installing the distributor

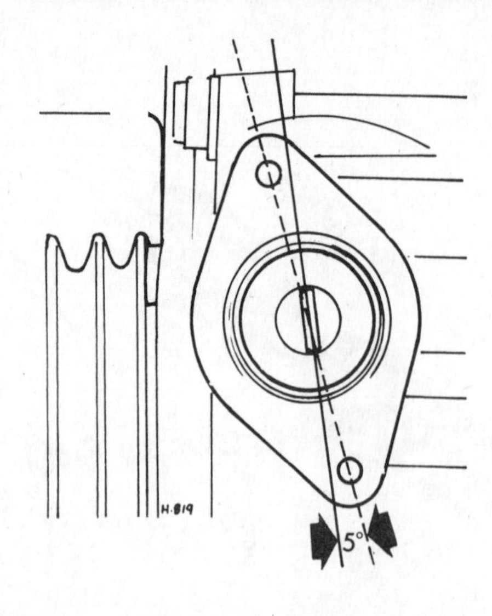

Fig. 1.23 Correct distributor drive alignment (Sec 15)

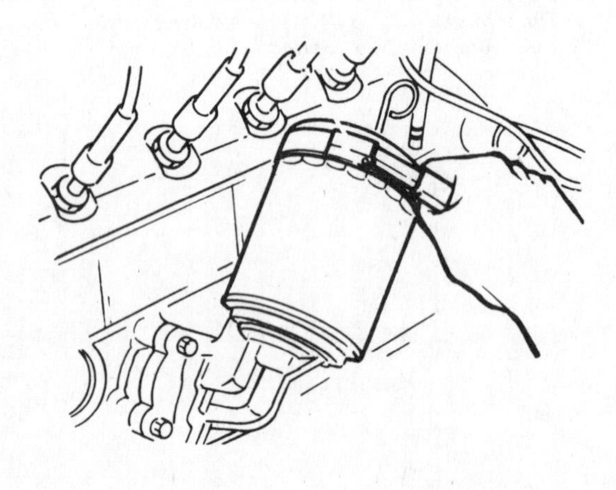

Fig. 1.24 Removing the oil filter (Sec 16)

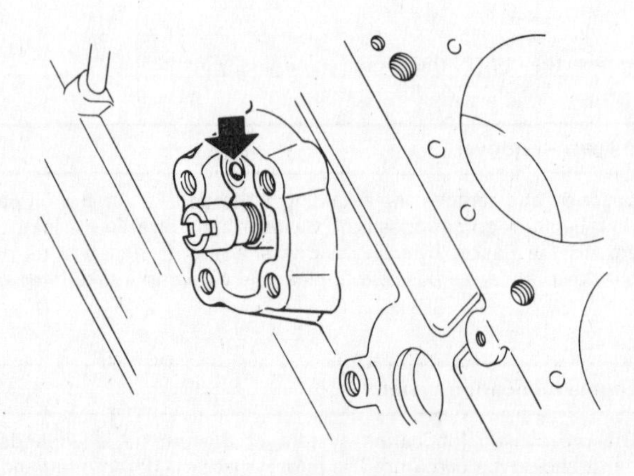

Fig. 1.25 Location of the oil filter pressure relief valve (arrowed) (Sec 16)

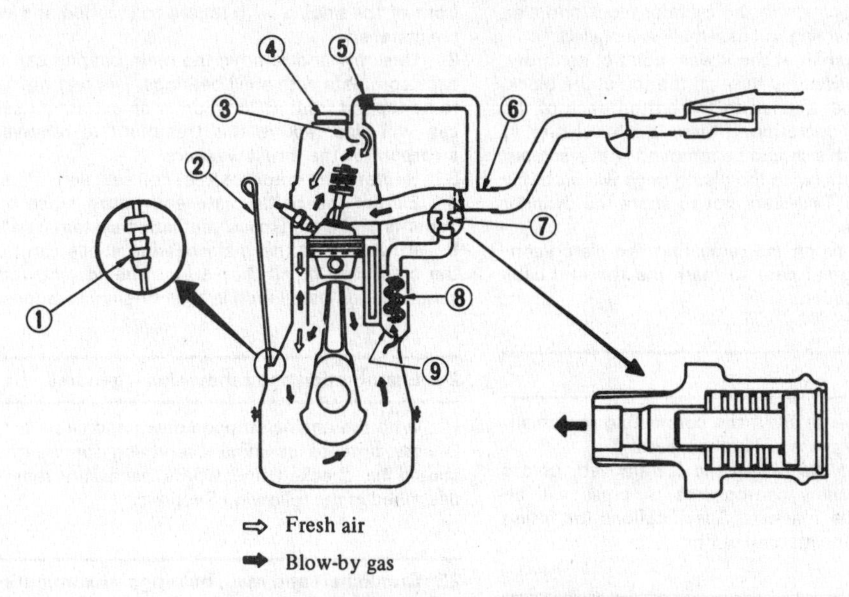

Fig. 1.26 The crankcase ventilation system (Sec 17)

1 Rubber seal (dipstick)
2 Dipstick
3 Baffle plate
4 Oil filler cap
5 Flame arrester
6 Throttle chamber
7 PCV valve
8 Steel net
9 Baffle plate

ponents. The inner and outer rotors are only supplied as matched sets. If the oil pump body is worn or damaged, renew the complete pump assembly.

11 Check the condition of the pressure regulator valve and spring, and renew if necessary.

12 Before installing the oil pump, check that the engine has not been moved from its original setting (No. 1 piston at TDC).

13 Fill the pump body with engine oil to prime it and align the punch mark on the driveshaft with the oil hole below the drive gear (photo).

14 Use a new flange gasket and insert the pump into its recess in the engine front cover so that as its driveshaft meshes with the gear on the crankshaft, the distributor drive tongue will be in the position shown (Fig. 1.23) (5° to a line drawn through the distributor mounting bolt hole centers) when looked down upon and have the smaller segment towards the front of the engine (photo).

15 Tighten the oil pump securing bolts.

16 Insert the distributor into its recess so that the large and small segments of the driveshafts engage, and tighten the distributor retaining plate screw at the original position in the elongated hole (photo).

17 It is recommended that the ignition timing is finally checked as described in later Chapters of this manual.

16 Oil filter and pressure relief valve

1 The oil filter is of the canister, throw-away type and will usually require the use of a chain or strap wrench to remove it.

2 A new filter should have its rubber sealing ring greased before installing and *should be screwed on with hand pressure only.*

3 Whenever the filter is removed, check the operation of the pressure relief valve which is located within the crankcase filter mounting bracket. If it is stuck or cracked, pry it out with a screwdriver and carefully tap in the new one using a piece of tubing as a drift.

17 Crankcase ventilation system

1 The system is of closed type, and returns blow-by gas from the engine crankcase and rocker cover back to the intake manifold to be re-burnt.

2 A positive crankcase ventilation valve (PCV valve) is incorporated, to regulate the flow of gas to the intake manifold under partial and full throttle conditions.

3 Regularly check the condition and security of the hoses and renew

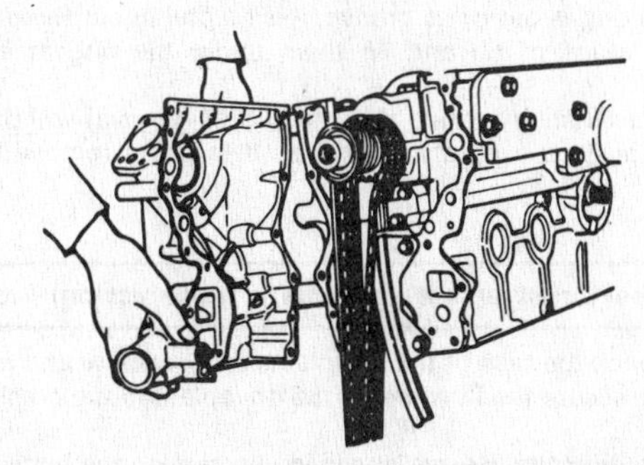

Fig. 1.27 Removing the engine front cover (Sec 18)

as appropriate.

4 Inserted into the hose that runs between the throttle chamber and rocker cover, is a flame arrester which should be checked from time to time, for security.

5 To check the PCV valve, run the engine at idle speed, remove the ventilator hose from the PCV valve, then check for a hissing noise as air passes through the valve. Alternatively, place a finger over the valve aperture to check for a strong vacuum. If neither is evident, renew the PCV valve.

18 Engine front cover and timing gear – removal

1 Remove the crankshaft pulley bolt and withdraw the pulley. Unbolt the engine front cover (oil pump and distributor already removed) and withdraw it from the cylinder block.

2 Remove the chain tensioner and the chain guides.

3 Lift the timing chain from the crankshaft sprocket.

4 From the end of the crankshaft pull off the oil thrower, the oil pump helical gear and the chain sprocket. The latter will almost certainly require the use of an extractor. Note that three Woodruff keys are used to secure these components.

19 Pistons and connecting rods – removal

1 Unscrew each of the two big-end bolts on No 1 connecting rod. Check that the big-end bearing cap is marked with its number on both cap and connecting rod. If they are unmarked, dot punch them so that

there can be no doubt of their location in the cylinder block and also which way round the piston/connecting rod assembly is installed.

2 With the journal of the crankshaft at the lowest point of its throw, push the piston/connecting rod assembly through the top of the block. The use of the stock of a hammer applied to the bottom face of the connecting rod will facilitate this operation. If there is a 'wear ring' at the top of the cylinder bore then this should be removed with a scraper before pushing out the pistons, otherwise the piston rings will probably break as they are forced over it. Take care not to score the cylinder bores during the scraping process.

3 Repeat the removal operations on the remaining five piston/connecting rod assemblies, taking great care to mark the rod end caps with regard to location and orientation.

20 Piston pin – removal

1 The piston pin is an interference fit in the connecting rod small-end bush and a push-fit (by finger pressure) in the piston.

2 It is not recommended that the piston pin is removed, unless essential due to the fitting of new components. A press will be required and reference should be made to Specifications for fitting tolerances and to Section 40 for the method of fitting.

21 Piston rings – removal

1 Each ring should be sprung open only just sufficiently to permit it to ride over the lands of the piston body.

2 Once a ring is out of its groove, it is helpful to cut three $\frac{1}{4}$ in (6.35 mm) wide strips of tin and lip them under the ring at equidistant points.

3 Using a twisting motion this method of removal will prevent the ring dropping into a empty groove as it is being removed from the piston.

22 Flywheel (or driveplate – automatic transmission) – removal

1 Bend back the tabs of the lockplates and unscrew and remove the bolts which secure the flywheel or the driveplate to the crankshaft rear flange.

2 If some difficulty is experienced in unscrewing the bolts due to the rotation of the crankshaft, wedge a block of wood between the crankshaft web and the inside of the crankcase. Alternatively, wedge the starter ring gear by inserting a cold chisel at the starter motor aperture.

3 Lift the flywheel or driveplate from the crankshaft flange.

4 Unbolt and remove the engine endplate. Now is a good time to check the cylinder block rear core plug for security and leakage.

23 Crankshaft and main bearings – removal

1 Mark each of the main bearing caps (1 to 7) commencing from the front of the engine with regard to location and which way round they are installed.

2 Unscrew and remove the main bearing cap bolts and remove the caps complete with shell bearings. The rear main bearing cap will have to be tapped from its location or an extractor used. The center bearing cap will also require this treatment to remove it, as the cap shell incorporates the thrust washers.

3 Remove the crankshaft rear oil seal and lift out the crankshaft.

4 Extract the baffle plate and gauze mesh block which are components of the crankcase ventilation system (see Section 17).

5 If necessary, the main bearing shells can now be removed from the crankcase but if they are not being renewed, keep them in strict sequence for installation in their original locations.

24 Examination and renovation – general

With the engine stripped down and all parts thoroughly cleaned, it is now time to examine everything for wear. The following items should be checked and where necessary renewed or renovated as described in the following Sections.

25 Crankshaft and main bearings – examination and renovation

1 Examine the crankpin and main journal surfaces for signs of scoring or scratches. Check the ovality of the crankpins at different positions with a micrometer. If more than 0.0004 in (0.01 mm) out of round, the crankpin will have to be reground. It will also have to be reground if there are any scores or scratches present. Also check the journals in the same fashion.

2 If it is necessary to regrind the crankshaft and fit new bearings your local Datsun dealer or automobile engineering works will be able to decide how much metal to grind off and the size of new bearing shells.

3 Full details of crankshaft regrinding tolerances and bearing undersizes are given in Specifications.

4 The main bearing clearances may be established by using a strip of Plastigage between the crankshaft journals and the main bearing/shell caps. Tighten the bearing cap bolts to a torque of between 33 and 40 lbf ft (4.5 to 5.5 kgf m). Remove the cap and compare the flattened Plastigage strip with the index provided. The clearance should be compared with the tolerances in Specifications.

5 Temporarily install the crankshaft to the crankcase having installed the upper halves of the shell main bearings in their locations. Install the center main bearing cap only, complete with shell bearing and tighten the securing bolts to between 33 and 40 lbf ft (4.5 to 5.5 kgf m). Using a feeler gauge, check the endfloat by pushing and pulling the crankshaft. Where the endfloat is outside the specified tolerance 0.0118 in (0.3 mm) the center shell bearing will have to be renewed (photo).

6 Check the condition of the spigot bush in the center of the crankshaft rear flange. If it is worn or damaged extract it by tapping a thread into it and screwing in a bolt or by filling it with grease and driving in a

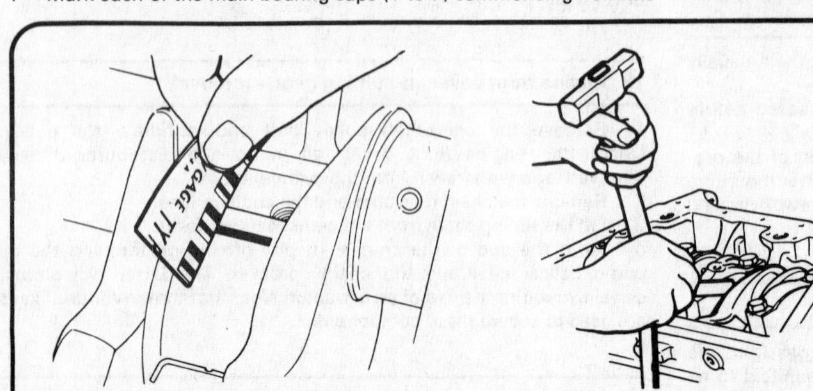

Fig. 1.28 Checking the main bearing running clearance (Sec 25)

Fig. 1.29 Checking crankshaft endfloat (Sec 25)

6.5 to 7.0 mm
(0.256 to 0.276)
Pilot bushing

Fig. 1.30 Spigot bush setting dimension (Sec 25)

close fitting rod which will create sufficient hydraulic pressure to eject it.

7 Tap in the new bush so that it stands proud of the crankshaft flange by between 0.256 and 0.276 in (6.5 to 7.0 mm). Do not lubricate the bush.

25.5 The center main bearing shell with integral thrust segments

26 Cylinder bores – examination and renovation

1 The cylinder bores must be examined for taper, ovality, scoring and scratches. Start by carefully examining the top of the cylinder bores. If they are at all worn a very slight ridge will be found on the thrust side. This marks the top of the piston ring travel. The owner will have a good indication of the bore wear prior to dismantling the engine, or removing the cylinder head. Excessive oil consumption accompanied by blue smoke from the exhaust is a sure sign of worn cylinder bores and piston rings.

2 Measure the bore diameter just under the ridge with a micrometer and compare it with the diameter at the bottom of the bore, which is not subject to wear. If the difference between the two measurements is more than 0.008 in (0.2032 mm) then it will be necessary to install special pistons and rings or to have the cylinders rebored and install oversize pistons. If no micrometer is available, remove the rings from a piston and place the piston in each bore about $\frac{3}{4}$ in (19 mm) below the top of the bore. If an 0.0012 in (0.3048 mm) feeler gauge slid between the piston and cylinder wall requires more than a pull of between 0.4 and 3.3 lbf (0.2 and 1.5 kgf) to withdraw it, using a spring balance, then remedial action must be taken.

3 Oversize pistons are available as listed in Specifications. These are accurately machined to just below the indicated measurements so as to provide correct running clearances in bores bored out to the exact oversize dimensions.

4 If the bores are slightly worn but not so badly worn as to justify reboring them, then special oil control rings and pistons can be installed which will restore compression and stop the engine burning oil. Several different types are available and the manufacturer's instructions concerning their installation must be followed closely.

5 If new pistons are being installed and the bores have not been reground, it is essential to slightly roughen the hard glaze on the sides of the bores with fine glass paper so the new piston rings will have a chance to bed in properly.

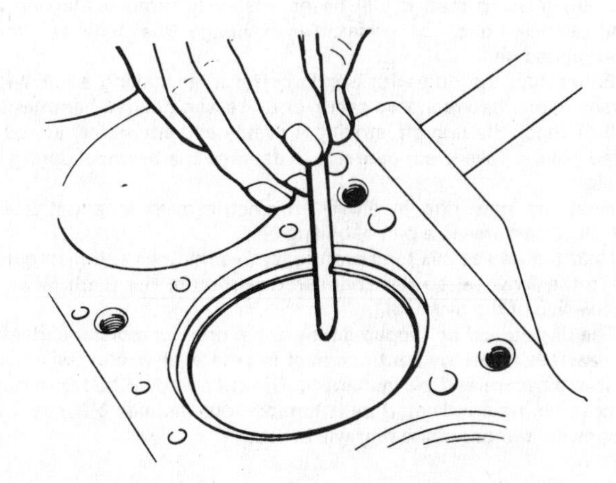

Fig. 1.31 Measuring piston ring end-gap (Sec 27)

27 Pistons and rings – examination and renovation

1 If new pistons are to be installed, they will be selected from grades available (see Specifications) after measuring the cylinder bores as described in the preceding Section, or will be provided in the appropriate oversize by the repairer who has rebored the cylinder block.

2 If the original pistons are to be re-used, carefully remove the piston rings as described in Section 21.

3 Clean the grooves and rings free from carbon, taking care not to scratch the aluminum surfaces of the pistons.

4 If new rings are to be installed, the top compression ring must be stepped to prevent it impinging on the 'wear ring' which will almost certainly have been formed at the top of the cylinder bore.

5 Before installing the rings to the pistons, push each ring in turn down to the part of its respective cylinder bore (use an inverted piston to do this and to keep the ring square in the bore) and measure the ring end-gap.

6 Using a feeler blade, the end-gap of the top compression ring should be 0.0098 to 0.0157 in (0.02 to 0.40 mm) and the second compression ring 0.0059 to 0.0118 in (0.15 to 0.30 mm). The oil control ring end-gap should be 0.0118 to 0.0354 in (0.3 to 0.9 mm).

7 The rings should now be tested in their respective grooves for side clearance which should be as follows:

Top ring 0.0016 to 0.0029 in (0.040 to 0.073 mm)
Second ring 0.0012 to 0.0028 in (0.030 to 0.070 mm)

8 Where necessary a piston ring which is slightly tight in its groove may be rubbed down holding it perfectly squarely on an oilstone or a sheet of fine emery cloth laid on a piece of plate glass. Excessive tightness can only be rectified by having the grooves machined out.

9 The piston pin should be a push-fit into the piston at room

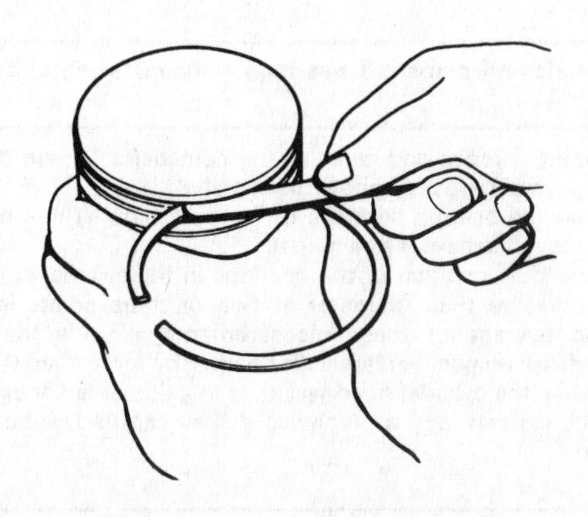

Fig. 1.32 Measuring piston ring groove clearance (Sec 27)

1

temperature. If it appears slack, then both the piston and piston pin should be renewed.

28 Connecting rods – examination and renovation

1 Big-end bearing failure is indicated by a knocking from with in the crankcase and a slight drop in oil pressure.
2 Examine the big-end bearing surfaces for pitting or scoring. Renew the shells in accordance with the sizes specified in Specifications. Where the crankshaft has been reground, the correct undersize big-end shell bearings will be supplied by the repairer (photo).
3 Should there be any suspicion that a connecting rod is bent or twisted or the small end bush no longer provides an interference-fit for the piston pin then the complete connecting rod assembly should be exchanged for a reconditioned one but ensure that the comparative weight of the two rods is within 0.247 oz (7.0g).
4 Measurement of the big-end bearing clearances may be carried out in a similar manner to that described for the main bearings in Section 25 but tighten the securing nuts on the cap bolts to 30 lbf ft (4.1 kgf m).
5 When each connecting rod is installed on its crankpin and the bolts tightened, check for side-play using a feeler blade. This should not exceed 0.0118 in (0.30 mm) otherwise the connecting rod will have to be renewed.

29 Flywheel and starter ring gear – examination and renovation

1 If the teeth on the flywheel starter ring gear are badly worn, or if some are missing then it will be necessary to remove the ring and install a new one, or preferably exchange the flywheel for a reconditioned unit.
2 Either split the ring with a cold chisel after making a cut with a hacksaw blade between two teeth, or use a soft headed hammer (not steel) to knock the ring off, striking it evenly and alternately at equally spaced points. Take great care not to damage the flywheel during this process.
3 Heat the new ring in either an electric oven to about 200°C (392°F) or immerse in a pan of boiling oil.
4 Hold the ring at this temperature for five minutes and then quickly fit it to the flywheel so the chamfered portion of the teeth faces the gearbox side of the flywheel.
5 The ring should be tapped gently down onto its register and left to cool naturally when the contraction of the metal on cooling will ensure that it is a secure and permanent fit. Great care must be taken not to overheat the ring, indicated by it turning light metallic blue, as if this happens the temper of the ring will be lost.

30 Driveplate and starter ring gear – examination and renovation

1 Where the starter ring teeth on the driveplate (automatic transmission) are worn or chipped then the complete driveplate should be renewed.

31 Camshaft and camshaft bearings – examination and renovation

1 Examine the lobes and cams of the camshaft. Where these are obviously worn or grooved, renew the camshaft.
2 Examine the bearing surfaces of the camshaft. Where these are pitted or grooved, renew the camshaft.
3 Examine the condition of the bearings in the cylinder head housings and measure their diameter at two or three points internally. Where the average of these measurements exceeds the outside diameter of the respective camshaft bearing by more than 0.0039 in (0.1 mm) then the cylinder head will have to be renewed or exchanged as the bearings can only be renewed if they can be line-bored after installation.

32 Timing sprockets and chain – examination and renovation

1 Examine the teeth on both the crankshaft gear wheel and the camshaft gear wheel for wear. Each tooth forms an inverted V with the gearwheel periphery, and if worn the side of each tooth under tension will be slightly concave in shape when compared with the other side of the tooth. i.e. one side of the inverted V will be concave when compared with the other. If any sign of wear is present the gearwheels must be renewed.
2 Examine the links of the chain for side slackness and renew the chain if any slackness is noticeable when compared with a new chain. It is a sensible precaution to renew the chain at about 30 000 miles (48 000 km) and at a lesser mileage if the engine is stripped down for a major overhaul. The actual rollers on a very badly worn chain may be slightly grooved.
3 When the camshaft sprocket is bolted in position, the end-float should not exceed 0.015 in (0.38 mm) - otherwise renew it.
4 Adjustment to compensate for a slightly stretched chain is described in Section 41.
5 Examine the components of the chain tensioner and guides. Renew any item which is worn or badly grooved.

33 Valves and valve seats – examination and renovation

1 Examine the heads of the valves for pitting and burning, especially the heads of the exhaust valves. The valve seatings should be examined at the same time. If the pitting on valve and seat is very slight the marks can be removed by grinding the seats and valves together with coarse, and then fine, valve grinding paste.
2 Where bad pitting has occurred to the valve seats it will be necessary to recut them and install new valves. If the valve seats are so worn that they cannot be recut, then it will be necessary to install new valve seat inserts. These latter two jobs should be entrusted to the local Datsun dealer or automobile engineering works. In practice it is very seldom that the seats are so badly worn that they require renewal. Normally, it is the valve that is too badly worn, and the owner can easily purchase a new set of valves and match them to the seats by valve grinding.
3 Valve grinding is carried out by smearing a trace of coarse carborundum paste on the seat face and applying a suction grinder tool to the valve head. With a semi-rotary motion, grind the valve head to its seat, lifting the valve occasionally to redistribute the grinding paste. When a dull matt even surface finish is produced on both the valve seat and the valve, wipe off the paste, lifting and turning the valve to redistribute the paste as before. A light spring placed under the valve head will greatly ease this operation. When a smooth unbroken ring of light gray matt finish is produced, on both valve and valve seat faces, the grinding operation is completed.
4 Scrape away all carbon from the valve head and the valve stem. Carefully clean away every trace of grinding compound, taking great care to leave none in the ports or in the valve guides. Clean the valves and valve seats with a kerosene soaked rag then with a clean rag, and finally, if an air line is available, blow the valves, valve guides and valve ports clean.

34 Valve guides and springs – examination and renovation

1 Severe wear in a valve guide may be due to a bent valve stem. To estimate the wear, fully insert the valve and then deflect its tip in opposite directions. If the deflection is more than about 0.008 in (0.2 mm) then the guide must be renewed.
2 Immerse the cylinder head in hot water and drive the old guide from the combustion chamber side towards the rocker cover.
3 New valve guides are available in an 0.008 in (0.2 mm) oversize outside diameter to provide an interference fit in the cylinder head when pressed in. The cylinder head should be heated evenly in an oven to a temperature of between 150° and 200° C (302 to 392° F) for this operation.
4 The guide should be pressed in to allow the necessary projection as shown in Fig. 1.34.
5 When installed, the guide should be reamed to between 0.3150 to 0.3157 in (8.000 to 8.018 mm).
6 The valve springs should be measured against the free lengths quoted in Specifications and renewed if they prove shorter as a result of continuous compression. In any event it is worthwhile to renew the valve springs if they have been in use for 20 000 miles or more (32 000 km).

28.2 Correct installation of big-end cap shell bearing

35 Rocker arms and pivots – examination and renovation

1 Examine the contact surfaces of the rocker pivot arm for wear, pitting or scoring and renew as necessary.
2 Where either component is defective, renew the rocker arm and pivot as a pair.

36 Cylinder head – decarbonising and examination

1 With the cylinder head removed, use a blunt scraper to remove all traces of carbon and deposits from the combustion spaces and ports. Remember that the cylinder head is aluminum alloy and can be damaged easily during the decarbonising operations. Scrape the cylinder head free from scale or old pieces of gasket or jointing compound. Clean the cylinder head by washing it in kerosene and take particular care to pull a piece of rag through the ports and cylinder head bolt holes. Any grit remaining in these recesses may well drop onto the gasket or cylinder block mating surfaces as the cylinder head is lowered into position and could lead to a gasket leak after reassembly is complete.
2 With the cylinder head clean, test for distortion if a history of coolant leakage has been apparent. Carry out this using a straight-edge and feeler gauges or a piece of plate glass. If the surface shows any warping in excess of 0.0039 in (0.1 mm) then the cylinder head will have to be resurfaced which is a job for a specialist automobile engineering works.
3 Clean the pistons and top of the cylinder bores. If the pistons are still in the block then it is essential that great care is taken to ensure that no carbon gets into the cylinder bores as this could scratch the cylinder walls or cause damage to the piston and rings. To ensure that this does not happen, first turn the crankshaft so that two of the pistons are at the top of their bores. Stuff rag into the other two bores or seal them off with paper and masking tape. The waterways should also be covered with small pieces of masking tape to prevent particles of carbon entering the cooling system and damaging the water pump. Press a little grease into the gap between the cylinder walls and the two pistons which are to be worked on. With a blunt scraper carefully scrape away the carbon from the piston crown, taking care not to scratch the aluminum. Also scrape away the carbon from the surrounding lip of the cylinder wall. When all carbon has been removed, scrape away the grease which will now be contaminated with carbon particles, taking care not to press any into the bores. To assist prevention of carbon build-up the piston crown can be polished with a metal polish. Remove the rags or masking tape from the other two cylinders and turn the crankshaft so that the two pistons which were at the bottom are now at the top. Place rag or masking tape in the cylinders which have been decarbonised and proceed as just described.

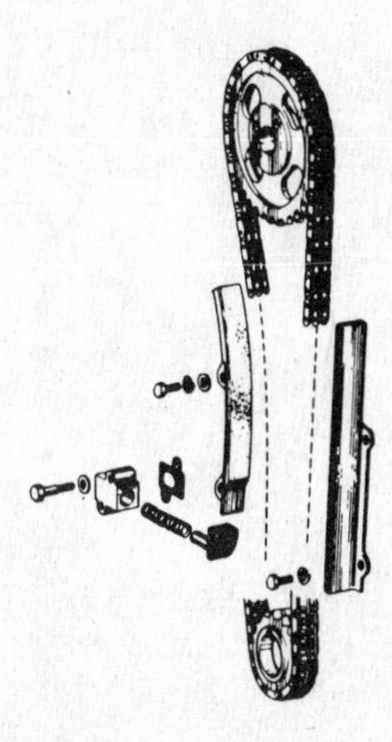

Fig. 1.33 Exploded view of the timing chain tensioner and guide (Sec 32)

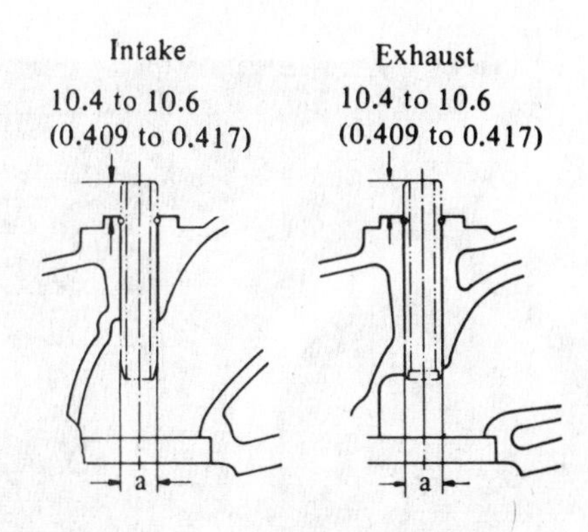

Fig. 1.34 Valve guide protrusion dimensions (Sec 34)

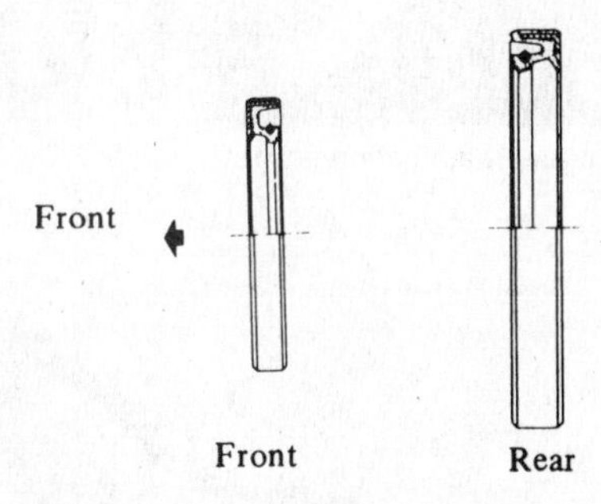

Fig. 1.35 Crankshaft oil seal fitting diagram (Sec 37)

37 Oil seals – renewal

1 The oil seal located in the engine front cover and the crankshaft rear oil seal should be renewed as a matter of routine whenever the engine is stripped for major overhaul.
2 Removal of the front oil seal is best carried out using a piece of tubing as a drift.
3 When installing the new seal, this must be done with the cover removed. Lubricate the oil seal periphery with clean engine oil prior to assembly. It should be noted that the seal is a threaded type; therefore do not lubricate the seal lip.
4 The rear oil seal is installed after installation of the crankshaft (see Section 39).

39.3a Lubricating the crankcase main bearing shells

39.3b Lowering the crankshaft into the crankcase

39.4 Fitting a center main bearing cap

39.5 Fitting the rear main bearing cap

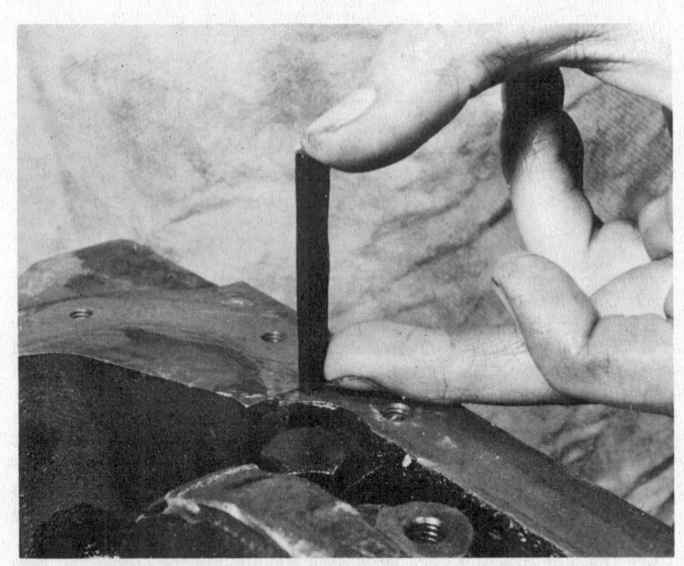

39.8 Fitting a rear main bearing cap side seal

39.9 Correct installation of the crankshaft rear oil seal

39.10 The engine rear endplate located on its dowels

39.11 Tightening a flywheel bolt

38 Engine reassembly – general

1 To ensure maximum life with minimum trouble from a rebuilt engine, not only must everything be correctly assembled, but everything must be spotlessly clean, all the oilways must be clear, locking washers and spring washers must always be installed where indicated, and all bearing and other working surfaces must be thoroughly lubricated during assembly.
2 Observe the clearances and torque wrench settings given in Specifications; guessing will not do where modern engines are concerned.

39 Crankshaft, main bearings and flywheel – installation

1 Install the main bearing shells into their recesses in the crankcase, with the oil holes in correct alignment.
2 Install the baffle plate and mesh block which are components of the crankcase ventilation system (emission control)
3 Oil the main bearing shells liberally with clean engine oil and lower the crankshaft into position in the crankcase (photos).
4 Install each of the main bearing caps, complete with shell bearings, ensuring that they are replaced in their original locations with the arrows pointing towards the front of the engine. The center shell bearing has integral thrust segments with oil grooves (photo).
5 The corners of the rear main bearing cap should have an application of jointing compound at the points indicated (photo) (Fig. 1.36).
6 Tighten the main bearing cap bolts progressively and in the order shown to the specified torque.
7 When installation is complete check that the crankshaft turns smoothly and theendfloat is as previously established (Section 25).
8 Apply jointing compound to the rear main bearing cap side seals and tap them into position (photo).
9 Grease the crankshaft rear oil seal and tap it into position using a piece of tubing as a drift (see Section 37) (photo).
10 Bolt on the engine rear endplate (photo).
11 Locate the flywheel (or driveplate – automatic transmission) on the crankshaft rear flange and tighten the bolts to the specified torque (photo).

40 Pistons and connecting rods – installation

1 The piston should be assembled to its connecting rod by using a press to install the piston pin to the small-end bush.
2 When correctly fitted, the side of the piston marked F (alternatively there may be a notch on the piston crown) will be towards the front of the engine and the oil hole in the connecting rod will be towards the right-hand side of the cylinder block ((photos).

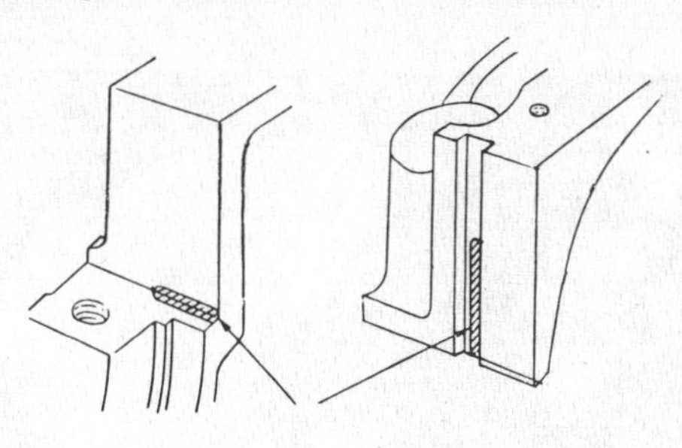

Fig. 1.36 Jointing compound application points on crankshaft rear main bearing cap and cylinder block (arrowed) (Sec 39)

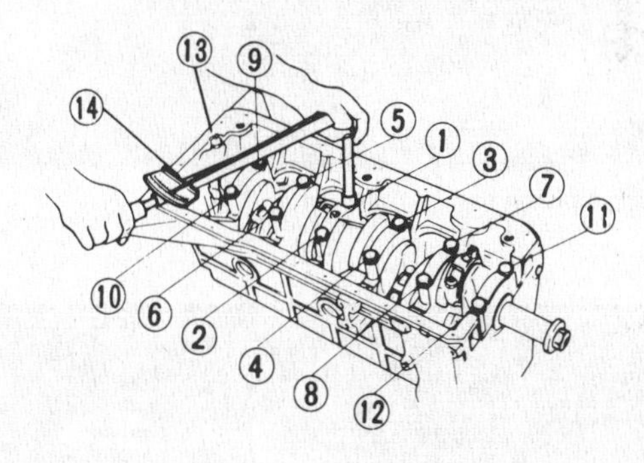

Fig. 1.37 Main bearing cap bolt tightening sequence diagram (Sec 39)

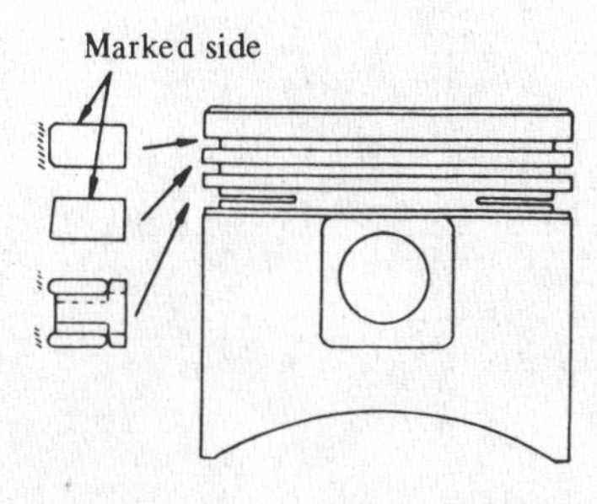

Fig. 1.38 Piston ring installation diagram (Sec 40)

40.2a Location of pistons F (front-facing) mark

40.2b Location of connecting rod oil hole

40.4 Using a piston ring compressor

40.5 Piston/connecting rod assembly ready for installation

40.7a Fitting a big-end cap

40.7b Connecting rod and cap matching numbers

40.8 Tightening a big-end cap bolt

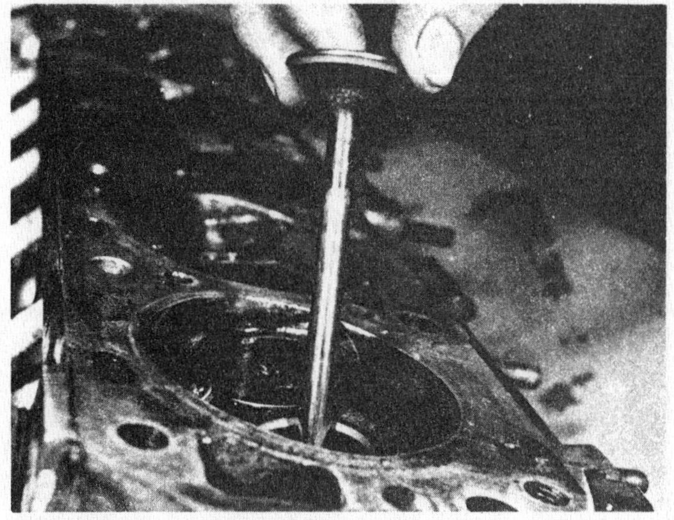
41.1 Inserting a valve into its guide

3 Install the piston rings to their grooves by reversing the removal procedure given in Section 21. The top two compression rings are marked TOP. Stagger the ring end-gaps at 120° to each other so that they are not in alignment to cause blow-by of gas. Make sure that the gaps do not align with the piston pin axis or the direction of piston thrust.
4 Liberally oil the piston rings and apply a ring compressor (photo).
5 Insert the connecting rod into its respective cylinder bore and locate the bottom of the piston skirt on the face of the cylinder block. Check that the FRONT mark on the piston faces the front of the engine (photo).
6 With the wooden handle of a hammer positioned in the center of the piston crown, strike the head of the hammer smartly with the hand and the piston/connecting rod assembly will pass into the cylinder bore.
7 Arrange the crankshaft so that the crankpin is at the bottom of its throw, and install the big-end caps complete with shell bearing. Ensure that the matching numbers or marks made at dismantling are adjacent on the same sides of the connecting rod and cap (photos).
8 Tighten the big-end cap bolts to the specified torque. The side-play will have been checked as described in Section 28 (photo).

41.2a Installing the valve springs

41 Cylinder head and timing gear – reassembly and installation

1 Insert each valve in turn into its respective guide, applying a little engine oil to the stem (photo).
2 Fit a new oil seal to the valve stem and with the aid of a spring compressor, assemble the springs (outer spring has close coils nearest cylinder head), retainers and split collets. The latter can be retained in the valve stem cut-out with a dab of thick grease (photos).
3 Oil the camshaft bearings and insert the camshaft carefully into position.
4 Screw the valve rocker pivots complete with locknuts into the pivot bushes.
5 Install the rocker arms by depressing the valve springs with a screwdriver.
6 Engage the rocker springs (photo).
7 Install the camshaft locating plate to the camshaft so that the horizontally engraved line is visible from the front and is positioned at the top of the plate.
8 Rotate the camshaft until the valves of number 1 cylinder are fully closed (equivalent to number 1 piston at TDC), then turn the crankshaft (by means of the flywheel or driveplate) until number 1 piston is at TDC.
9 Bolt the two timing chain guides into position (photo).
10 Clean the mating surfaces of the cylinder block and head, and locate a new gasket on the face of the block; do not use gasket cement (photo).
11 Lower the cylinder head into position and insert the two center bolts finger-tight only at this stage (photo).

41.2b Inserting the valve collets with the aid of a spring compressor

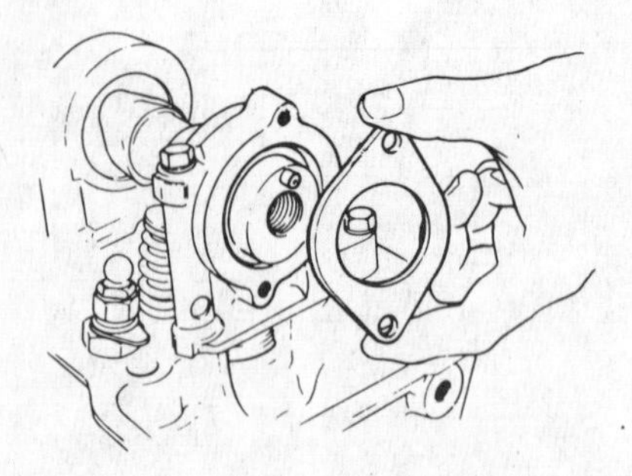

Fig. 1.39 Fitting camshaft locating plate (Sec 41)

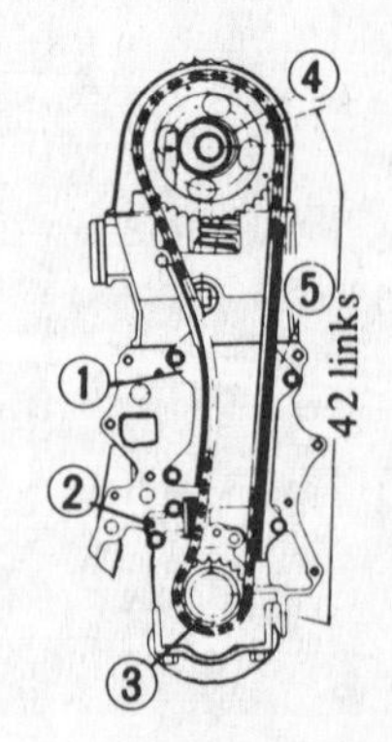

Fig. 1.40 Timing chain and sprocket alignment diagram (Sec 41)

1 Chain guide
2 Chain tensioner
3 Crank sprocket
4 Cam sprocket
5 Chain guide

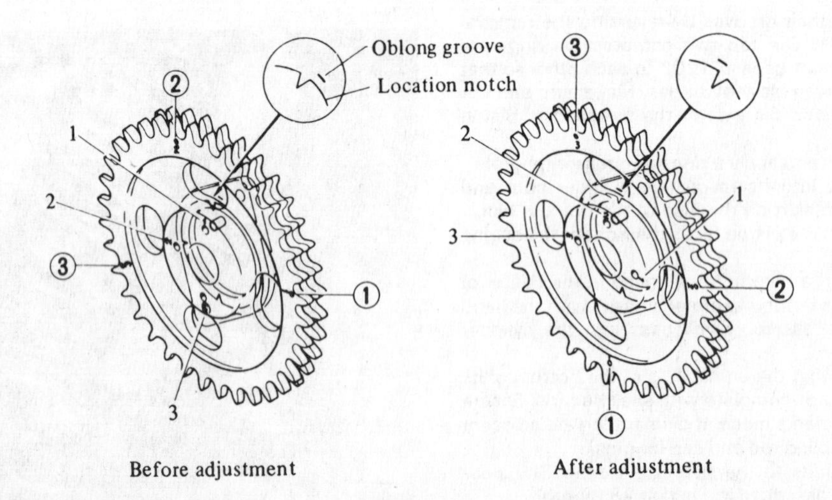

Fig. 1.41 Camshaft sprocket adjustment positions to compensate for timing chain stretch (Sec 41). Numbers in circles indicate timing marks; other numbers indicate dowel positioning holes

41.6 Correct engagement of rocker springs

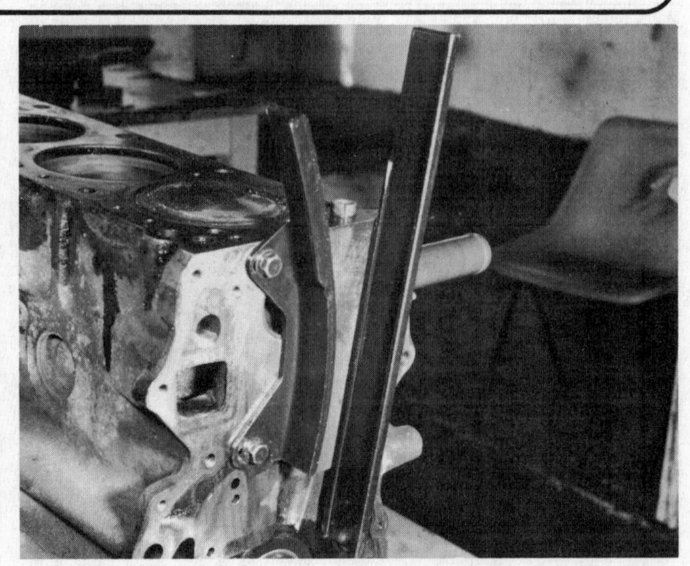

41.9 Installation of timing chain guides

41.10 Correct location of cylinder head gasket

41.11 Installing the two center cylinder head bolts

41.12 Sprocket, drivegear and oil thrower fitted to front end of crankshaft

41.14a Engaging camshaft sprocket with timing chain

41.14b Bolting sprocket to camshaft

41.14c Camshaft sprocket No 1 mark opposite bright chain link

41.14d Crankshaft sprocket punch mark opposite bright chain link

41.14e Alignment of camshaft sprocket notch with engraved line on locating plate

41.16 Installation of the chain tensioner

41.18 Installation of engine front cover gasket

41.20 Positioning the engine front cover

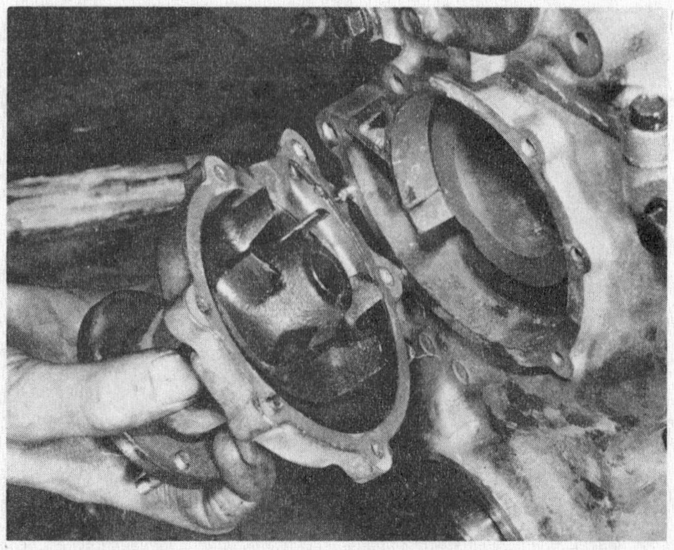

41.24 Installing the water pump

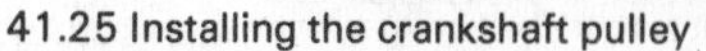
41.25 Installing the crankshaft pulley

41.27 Installation of the oil distribution tube

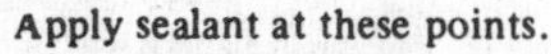

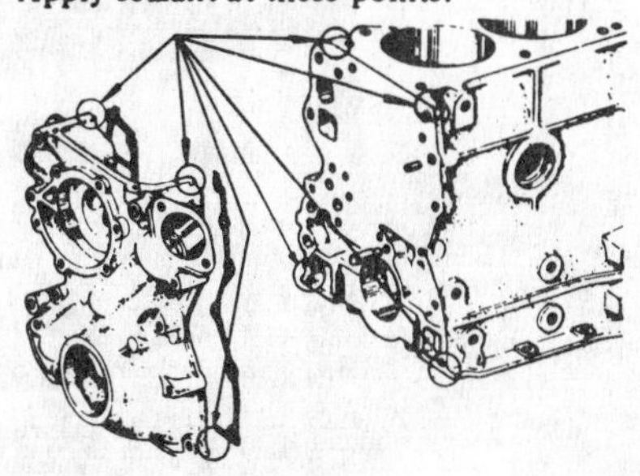

Fig. 1.42 Jointing compound application points prior to installation of front cover (Sec 41)

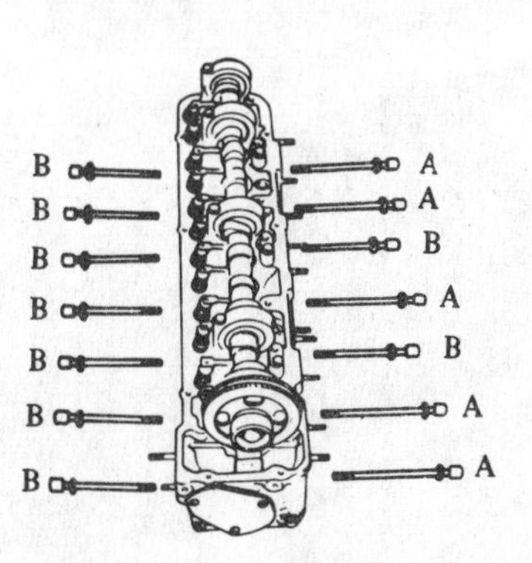

Fig. 1.43 Location of different length cylinder head bolts (Sec 41)

A Longer type *B Shorter type*

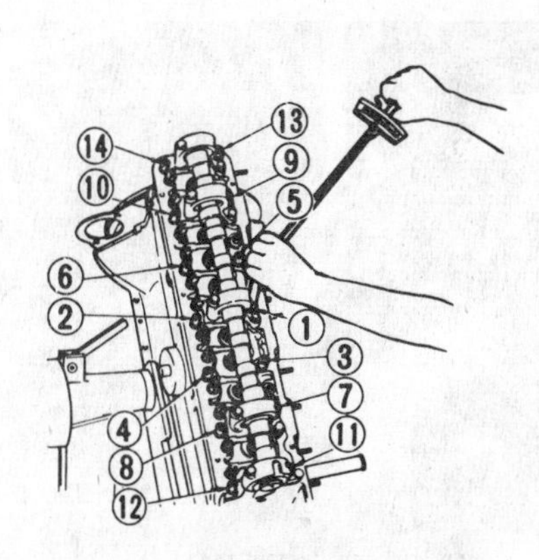

Fig. 1.44 Cylinder head bolt tightening sequence (Sec 41)

12 To the front of the crankshaft, fit the sprocket, oil pump, distributor drive gear and the oil thrower. Make sure that the timing marks on the sprocket are visible from the front (photo).

13 On no account turn the crankshaft or camshaft until the timing chain is installed, otherwise the valves will contact the piston crowns.

14 Engage the camshaft sprocket within the upper loop of the timing chain; then engage the chain with the teeth of the crankshaft sprocket and bolt the camshaft sprocket to the camshaft ensuring that the following conditions are met (photos):

- *(a) The keyway of the crankshaft sprocket should point vertically*
- *(b) The timing marks (bright link plates) on the chain should align with those on the two sprockets and be positioned on the right-hand side when viewed from the front, with 20 black plates between the two bright plates (photos)*
- *(c) Where a timing chain has stretched, this can upset the valve timing and provision is made for this by alternative dowel holes drilled in the camshaft sprocket*

15 With number 1 piston at TDC (compression stroke) check whether the notch in the camshaft sprocket (with chain correctly engaged) appears to the left of the engraved line on the locating plate (photo). If this is the case disengage the camshaft sprocket from the chain and move the sprocket round so that when it is re-engaged with the chain it will locate with the camshaft flange dowel in its number 2 hole. Where this adjustment does not correct the chain slack, repeat the operation using number 3 hole of the camshaft sprocket to engage with the flange dowel. Where number 2 or 3 sprocket holes are used then the number 2 or 3 timing marks must be used to position the chain. Where this adjustment procedure still will not correct or compensate for the slackness in the timing chain then the chain must be renewed.

16 When the timing is satisfactory, tighten the camshaft sprocket bolt to the specified torque (photo).

17 Install the chain tensioner so that there is the minimum clearance between the spindle/slipper assembly and the tensioner assembly (photo).

18 Thoroughly clean the mating faces of the front cover and cylinder block (photo).

19 Locate a new gasket on the front face of the engine, applying gasket cement to both sides of it (photo).

20 Apply gasket cement to the front cover and cylinder block as indicated (photo) (Fig. 1.42).

21 Offer-up the front cover to the engine and insert the securing bolts finger-tight. Take care not to damage the head gasket which is already in position (photo).

22 The top face of the front cover should be flush with the top surface of the cylinder block or certainly not more than 0.006 in (1.5 mm) difference in level.

23 Tighten the front cover bolts to the specified torque.

24 Install the water pump (photo).

25 Oil the lips of the front cover oil seal and push the pulley onto the crankshaft. Tighten its securing nut to the specified torque (photo).

26 Insert the remaining twelve cylinder head bolts and then tighten

42.3 Oil pump strainer and intake tube

42.4 Installation of oil pan gasket

42.5 Installing the oil pan

42.6a Checking a valve lash (clearance)

42.6b Adjusting a valve lash (clearance)

all fourteen bolts in the sequence shown (Fig. 1.44). The tightenings should be carried out progressively in three stages. Refer to the Specifications for the correct torque.

27 Install the oil distribution tube assembly to the rocker gear (photo).

42 Engine – final assembly and adjustments

1 Install the oil pump as described in Section 15.

2 Assemble the thermostat housing, thermotime switch, thermal transmitter and water temperature sensor.

3 Install the oil strainer to the oil pump intake tube (photo).

4 Locate a new gasket on the crankcase flange after applying gasket cement to both sides of it (photo).

5 Bolt on the oil pan, tightening the bolts evenly and in a diagonal sequence to the specified torque (photo).

6 Check the valve lash (clearance). To do this, turn the crankshaft until number 1 piston is at TDC on it's compression stroke. In this position the high points of the cam lobes will be furthest from the rocker arms. Check the clearances between the ends of the valve stems and the rocker arm tips by inserting the appropriate feeler blade; the blade should be a stiff sliding fit (photos).

7 To adjust the clearance, release the locknut and turn the pivot screw. The valve clearances *cold* are given in the Specifications.

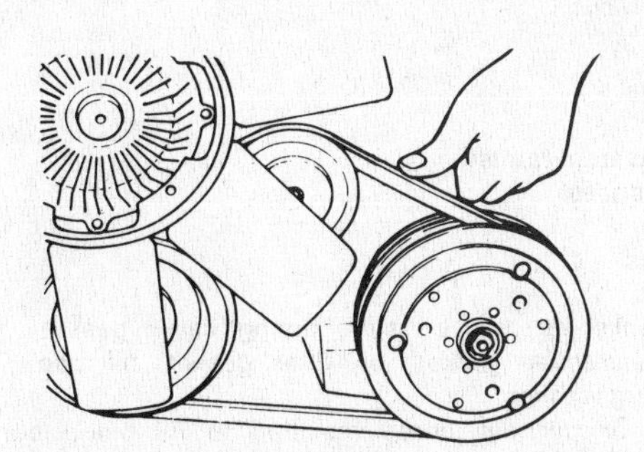

Fig. 1.45 Checking the air conditioner compressor drive-belt tension (Sec 43)

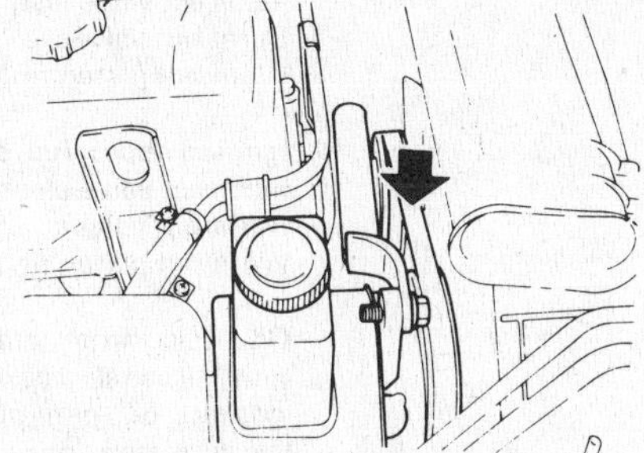

Fig. 1.46 The power steering oil pump drive-belt (Sec 43)

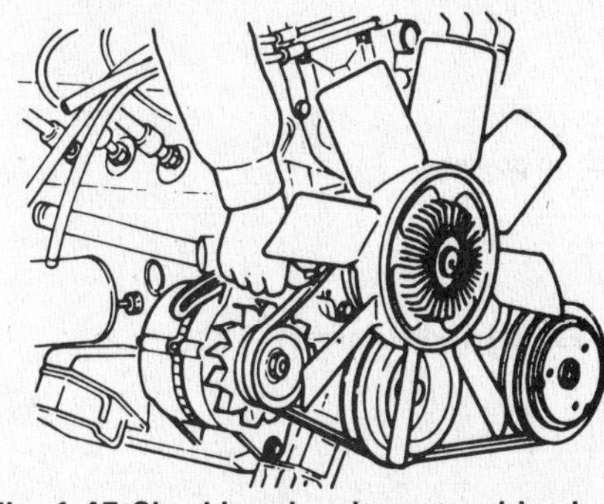

Fig. 1.47 Checking the alternator drive-belt tension (Sec 43)

8 As the firing order is 1-5-3-6-2-4, it will reduce the amount of crankshaft rotation required if the valve clearances are adjusted in accordance with the firing order. To obtain a better appreciation of the valve clearance it is recommended that the rocker arm springs are detached. Numbering from the front, the inlet valves are 2,3,5,8,10 and 11 and the exhaust valves 1,4,6,7,9 and 12.
9 When carrying out valve clearance adjustment with the engine in the car, the crankshaft can most easily be turned by engaging top gear, and jacking-up and turning a rear roadwheel (manual transmission). On cars fitted with automatic transmission a wrench will have to be applied to the crankshaft pulley bolt which makes the adjustment procedure somewhat more protracted. With either method, the work will be facilitated if the spark plugs are first removed.
10 Refit all the ancillary components by reversing the operations described in Section 9. Reference should be made to Chapter 5 for the method of centralizing the clutch driven plate, and to Section 15 for installation of the distributor.

43 Engine/transmission – installation

1 Reconnect the engine and transmission by reversing the separation procedure in Section 6 or 7 according to type. Tighten the securing bolts to the specified torque.
2 Using the hoist and slings, install the engine/transmission in the vehicle by reversing the removal procedure given in Sections 4 and 5.
3 When installation is complete, check and adjust the tension of all the drivebelts to the following recommended deflections:

(a) Air conditioner drivebelt: total deflection 0.315 to 0.472 in (8 to 12 mm) checked at midway position between idler pulley and compressor pulley. Adjust tension by turning idler pulley bolt in or out.

(b) Power steering oil pump drivebelt: total deflection 0.315 to 0.472 in (8 to 12 mm) checked midway between idler pulley and oil pump pulley. Adjust tension by turning idler pulley bolt in or out.

(c) Alternator drivebelt: total deflection, with approximately 22lbf (10 kgf) force applied downwards between alternator pulley and fan pulley, 0.315 to 0.472 in (8 to 12 mm). Adjust tension by moving alternator on its mountings.

4 Fill the engine with the correct grade and quantity of oil.
5 Refill the cooling system; refer to Chapter 2 for further information.
6 Check the oil level in the manual or automatic transmission, and top-up if required.

44 Engine – adjustment after major overhaul

1 With the engine installed in the vehicle, make a final visual check to see that everything had been reconnected and that no loose rags or tools have been left inside the engine compartment.
2 Turn the idling speed adjusting screw in about $\frac{1}{2}$ turn to ensure that the engine will have a faster-than-usual idling speed during initial start-up and operation.
3 Check that the fuel pump ground lead has been reconnected.
4 As soon as the engine starts, allow it to run at a fast idle, adjusting the idle speed screw as necessary. Examine all hoses and pipe connections for leaks.
5 Operate the vehicle on the road until normal engine temperature is reached, then remove the rocker cover and adjust the valve lash (clearance) *hot* to that specified, as described in Section 42.
6 When the engine has cooled completely, check the cylinder head bolt torque settings.
7 Where the majority of engine internal bearings or components (pistons, rings, etc) have been renewed, then the operating speed should be restricted for the first 500 miles (800 km) and the engine oil changed at the end of this period.
8 Check and, if necessary, adjust the ignition timing (Chapter 4).

1

45 Fault diagnosis – engine

Symptom	Reason/s
Engine will not turn over when starter switch is operated	Flat battery Bad battery connections Bad connections at solenoid switch and/or starter motor Defective solenoid Starter motor defective
Engine turns over normally but fails to start	No spark at plugs No fuel reaching engine Too much fuel reaching the engine (flooding)

Engine starts but runs unevenly and misfires	Ignition and/or fuel system faults Incorrect valve lash Burnt out valves Worn out piston rings
Lack of power	Ignition and/or fuel system faults Incorrect valve clearances Burnt out valves Worn out piston rings
Excessive oil consumption	Oil leaks from crankshaft rear oil seal, timing cover gasket and oil seal, rocker cover gasket, oil filter gasket, oil pan gasket, oil pan plug washer Worn piston rings or cylinder bores resulting in oil being burnt by engine Worn valve guides and/or defective valve stem seals
Excessive mechanical noise from engine	Wrong valve to rocker clearances Worn crankshaft bearings Worn cylinders (piston slap) Slack or worn timing chain and sprockets

NOTE: *When investigating starting and uneven running faults do not be tempted into snap diagnosis. Start from the beginning of the check procedure and follow it through. It will take less time in the long run. Poor performance from an engine in terms of power and economy is not normally diagnosed quickly. In any event the ignition, fuel and emission control systems must be checked first before assuming any further investigation needs to be made.*

Chapter 2 Cooling system

Contents

Specifications

System type

Pressurized and sealed, with centrifugal pump, thermostat, fan and radiator

Radiator

Type	Corrugated fin and tube
Pressure cap setting	13 lbf/in^2 (0·91 kgf/cm^2)

Thermostat

Type	Wax pellet
Opening temperature:	
Standard	82°C (180°F)
Cold areas	88°C (190°F)
Tropical areas	76·5°C (170°F)

Water pump

Type	Centrifugal impeller

Fan

Type	Temperature-controlled fan coupling
Number of blades	8

Coolant capacity (including heater and reservoir tank)

11 US quarts/9$\frac{1}{8}$ Imp quarts/10·4 liters

Fan/alternator belt tension

Check midway between fan and alternator pulleys, with a downward force of 22 lbf (10 kgf)	0·315 to 0·472 in (8 to 12 mm)

1 General description

The cooling system comprises the radiator, top and bottom water hoses, reservoir tank, water pump, cylinder block water jackets, radiator cap with pressure relief valve – and flow and return heater hoses. The thermostat is located in a recess at the front of the cylinder head. The principle of the system is that cold water in the bottom of the radiator circulates upwards through the lower radiator hose to the water pump, where the pump impeller pushes the water round the cylinder block and head through the various cast-in passages to cool the cylinder bores, combustion surfaces and valve seats. When sufficient heat has been absorbed by the cooling water, and the engine has reached an efficient working temperature, the water moves from the cylinder head past the now open thermostat into the top radiator hose and into the radiator header tank.

The water then travels down the radiator tubes when it is rapidly cooled by the in-rush of air, when the vehicle is in forward motion. A multi-bladed fan, mounted on the water pump pulley, assists this cooling action.The water, now cooled, reaches the bottom of the radiator and the cycle is repeated.

When the engine is cold the thermostat remains closed until the coolant reaches a pre-determined temperature (see Specifications). This assists rapid warming-up.

The reservoir tank, which is located on the left-hand side of the radiator, is in effect a reserve coolant tank to maintain the radiator coolant level. In operation, the increasing coolant temperature in the cooling system will create a pressure build-up in the radiator. The pressure relief valve in the radiator cap will open to allow excess coolant to pass into the reservoir tank. When the coolant temperature lowers and pressure decreases, the vacuum valve in the radiator cap opens to allow the coolant in the reservoir tank to syphon back into the radiator, thus maintaining the correct radiator coolant level at all times.

An electrosensitive capsule located in the cylinder head measures the water temperature.

The cooling system also provides the heat for the car interior heater and heats the inlet manifold.

On vehicles equipped with automatic transmission, the transmission fluid is cooled by a cooler attached to the base of the radiator.

On cars equipped with air conditioning systems, a condenser is placed ahead of the radiator and is bolted in conjunction with it.

The radiator cooling fan incorporates a temperature controlled

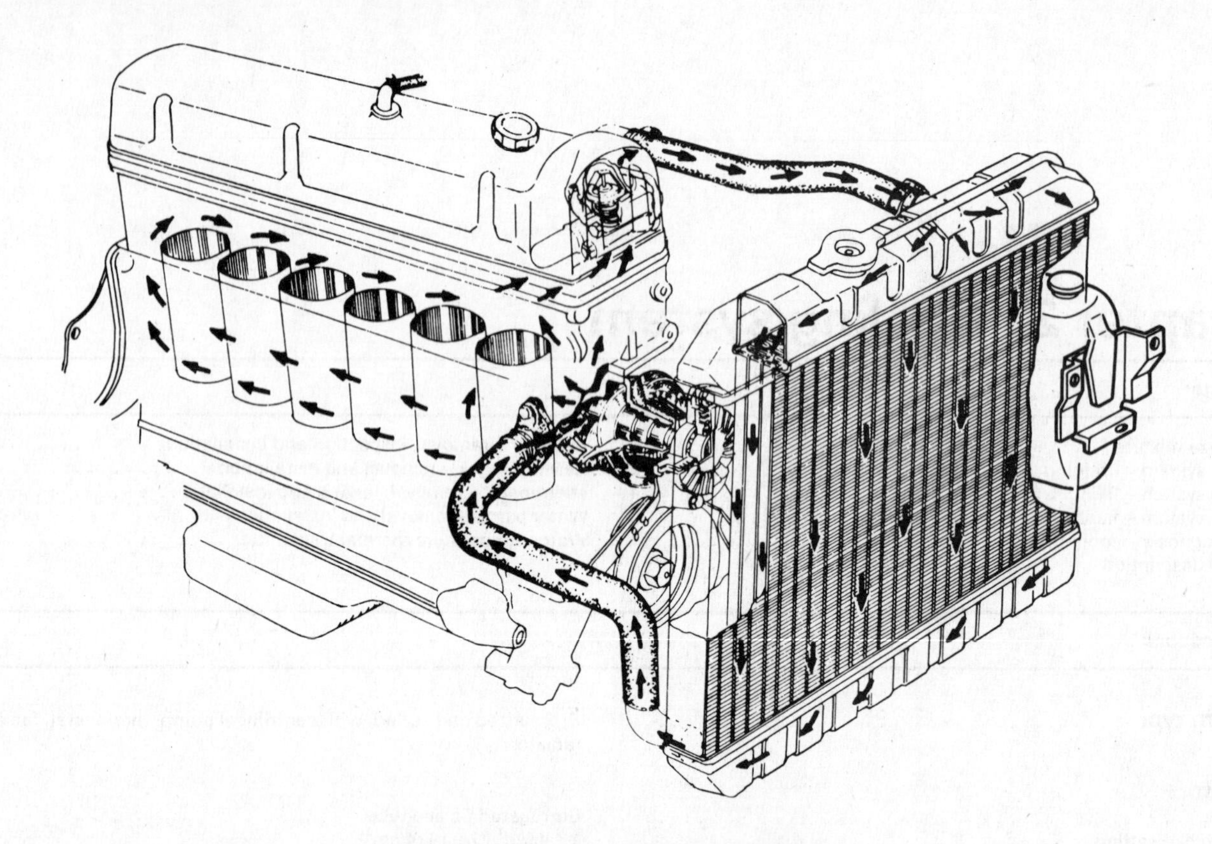

Fig. 2.1 The cooling system water-flow diagram

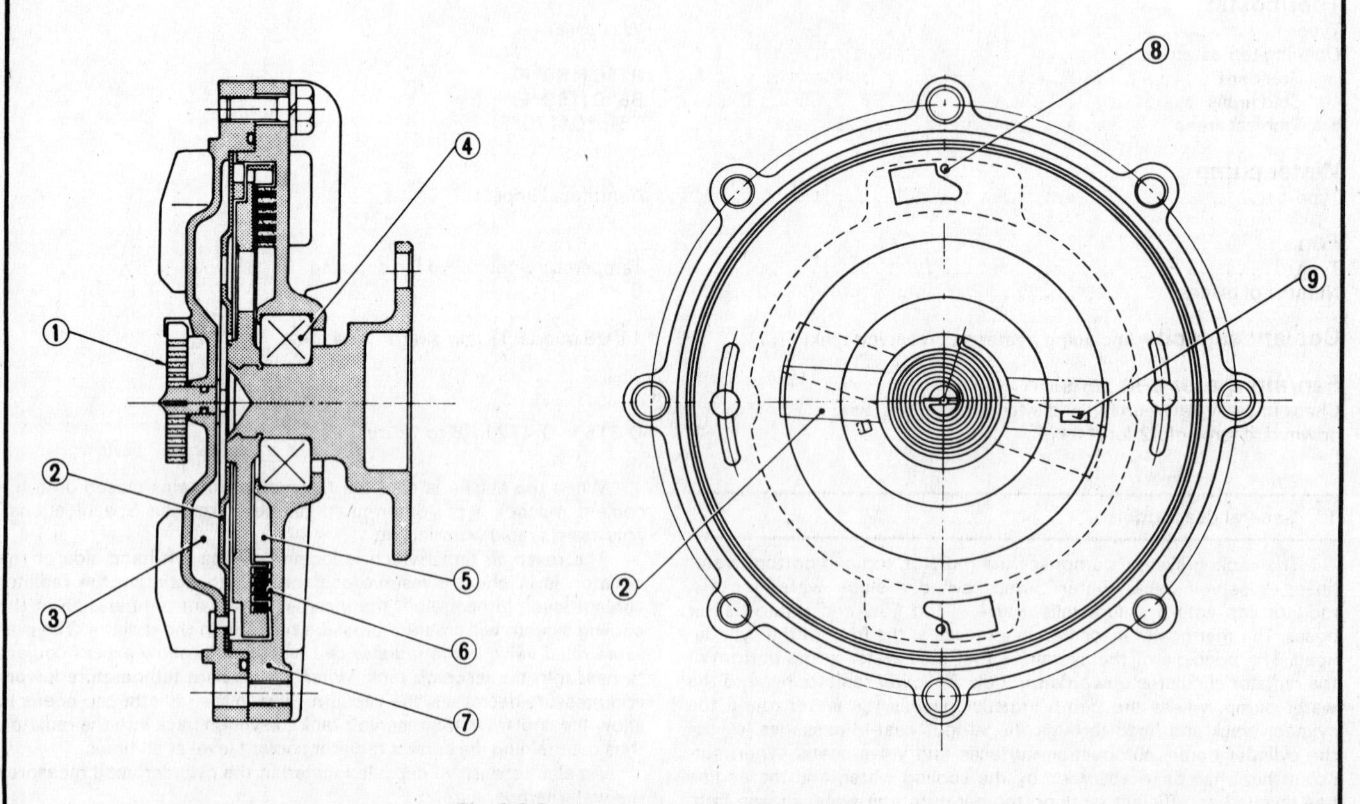

Fig. 2.2 Sectional view of the fan coupling

1 Bi-metallic thermostat
2 Slide valve
3 Oil reservoir (coupling unlocked)
4 Bearing
5 Hydraulic coupling locking device
6 Hydraulic coupling locking device
7 Driven component of fan assembly
8 Oil outlet
9 Oil inlet

coupling. This device comprises an oil operated clutch and a coiled bi-metallic thermostat which functions jointly to permit the fan to slip when the engine is below normal operating temperature level and does not require the supplementary air flow provided by the fan at normal running speed. At higher engine operating temperature, the fan is locked and rotates at the speed of the water pump pulley. The fan coupling is a sealed unit and requires no periodic maintenance.

2 Cooling system – draining

1 Should the system have to be left empty for any reason, both the cylinder block and radiator must be drained, otherwise with a partly drained system corrosion of the water pump impeller seal face may occur with subsequent early failure of the pump seal and bearing.
2 Place the car on a level surface and have ready a container of a suitable capacity which will slide beneath the radiator and oil pan.
3 Move the heater control on the facia to *Hot* and unscrew and remove the radiator cap. If hot, unscrew the cap very slowly, first covering it with a cloth to remove the danger of scalding when the pressure in the system is released.
4 Unscrew the radiator drain tap at the base of the radiator and then when coolant ceases to flow into the receptacle, repeat the operation by unscrewing the cylinder block plug located on the left-hand side of the engine. Retain the coolant for further use, if it contains antifreeze.

3 Cooling system – flushing

1 The radiator and waterways in the engine after some time may become restricted or even blocked with scale or sediment which reduces the efficiency of the cooling system. When this condition occurs, or the coolant appears rusty or dark in color, the system should be flushed. In severe cases reverse flushing may be required as described later.
2 Place the heater controls to the *Hot* position and unscrew fully the radiator and cylinder block drain taps.
3 Remove the radiator filler cap and place a hose in the filler neck. Allow water to run thru the system until it emerges from the drain taps.
4 In severe cases of contamination of the coolant or in the system, reverse-flush by first removing the radiator cap and disconnecting the lower radiator hose at the radiator outlet pipe.
5 Remove the top hose at the radiator connection end and remove the radiator as described in Section 6.
6 Invert the radiator and, with the hose in the bottom outlet pipe, continue flushing until clear water comes from the radiator top tank.
7 To flush the engine water jackets, remove the thermostat as described later in this Chapter and place a hose in the thermostat location until clear water runs from the water pump inlet. Cleaning by the use of chemical compounds is not recommended.

4 Cooling system – filling

1 Place the heater control to the *Hot* position.
2 Screw in the radiator drain tap and close the cylinder block drain tap.
3 Pour coolant slowly into the radiator so that air can be expelled through the thermostat pin hole without being trapped in a waterway.
4 Fill to the correct level which is 1½ in (38 mm) below the radiator filler neck and replace the filler cap.
5 Run the engine, check for leaks , and recheck the coolant level.
6 Check that the reservoir tank level is between the *Min* and *Max* level marks.

5 Antifreeze mixture

1 The cooling system should be filled with antifreeze solution in early autumn. The heater matrix and radiator bottom tank are particularly prone to freezing if antifreeze is not used in air temperatures below freezing. Modern antifreeze solutions of good quality will also prevent corrosion and rusting and they may be left in the system to advantage all year round, draining and refilling with fresh solution each year.
2 Before adding antifreeze to the system, check all hose connections and check the tightness of the cylinder head bolts as such solutions are searching. The cooling system should be drained and refilled with clean water as previously explained, before adding antifreeze.
3 The quantity of antifreeze which should be used for various levels of protection is given in the table below, expressed as a percentage of the system capacity.

Antifreeze volume	Protection to	Safe pump circulation
25%	–26°C (–15°F)	–12°C (10°F)
30%	–33°C (–28°F)	–16°C (3°F)
35%	–39°C (–38°F)	–20°C (–4°F)

4 Where the cooling system contains an antifreeze solution, any topping-up should be done with a solution made up in similar proportions to the original in order to avoid dilution.

6 Radiator – removal, inspection and installation

1 Unscrew the radiator drain plug, and drain the coolant into a suitable container. Retain the coolant if it contains antifreeze mixture. There is no need to drain the cylinder block when removing the radiator.
2 Disconnect the radiator top and bottom hoses.
3 *On vehicles fitted with automatic transmission,* disconnect the inlet and outlet pipes from the fluid cooler at the bottom of the radiator.
4 Unscrew and remove the radiator shroud attaching screws.
5 To remove the shroud, press both lower sides inwards (towards each other) then lever the lower part of the shroud lip away from the inner lip until it can be manipulated away from the fan.
6 Undo and remove the bolts that attach the radiator to the vehicle front frame. Note that the reservoir tank is secured by the upper left-hand bolt and a mounting bracket, which should also be removed.
7 Lift the radiator upwards until it is clear of the engine compartment. Take great care with this operation so that the radiator matrix is not damaged.
8 Inspect the radiator seams for leaks; if evident, it is recommended that the repair is left to a specialist or the radiator is exchanged for a reconditioned one.
9 Whenever the radiator is removed, take the opportunity of brushing all bugs and accumulated dirt from the radiator fins, or by applying air from a tire air compressor in the reverse direction to normal airflow.
10 The radiator pressure cap should be tested by a service station and, if it leaks or it's spring has weakened, it must be renewed with one of the specified pressure rating.
11 Installing the radiator is a reversal of removal. Refill the cooling system as described in Section 4.

7 Thermostat – removal, testing and installation

1 Partially drain the cooling system (about ½ gal/2.2 liters drawn off through the radiator drain plug will be sufficient).

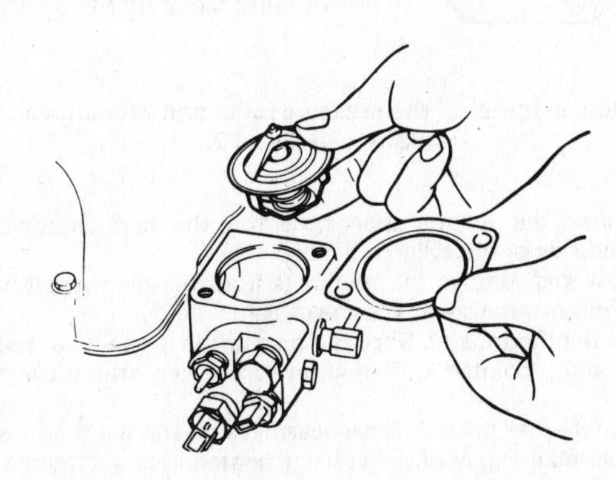

Fig. 2.3 Removing the thermostat and its gasket (Sec 7)

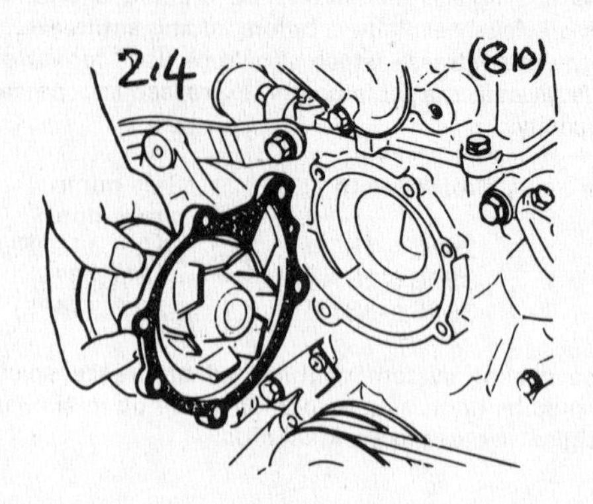

Fig. 2.4 Removing the water pump from the engine front cover (Sec 8)

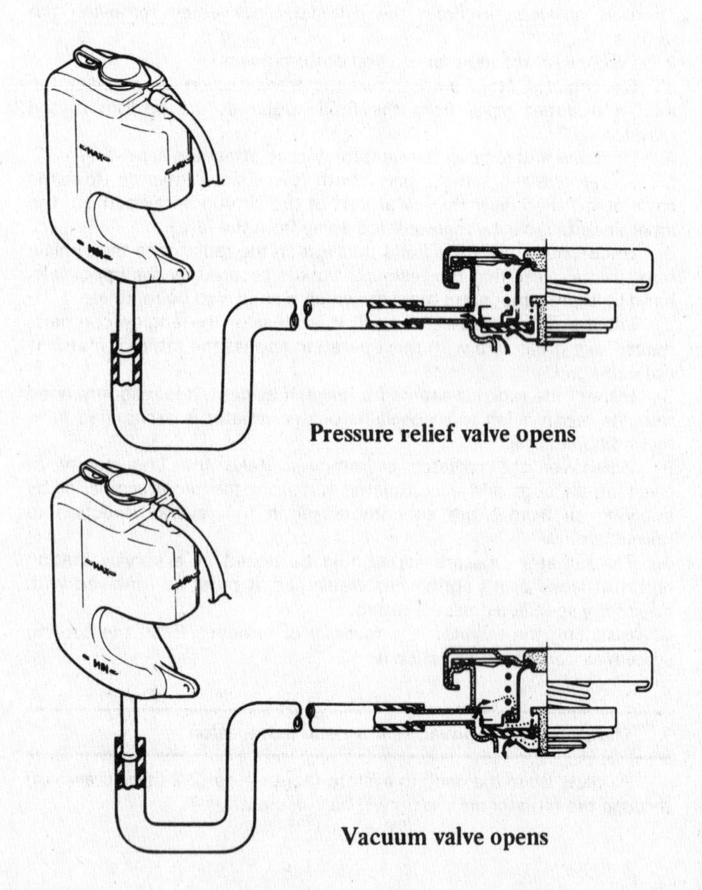

Fig. 2.6 Flow diagram of the pressure relief and vacuum valve in the radiator cap (Sec 10)

2 Disconnect the radiator upper hose from the thermostat elbow on the left-hand side of the cylinder head.
3 Unscrew and remove the two bolts from the thermostat housing cover, and remove the cover and the gasket.
4 Extract the thermostat. Should it be stuck in it's seat cut round it's rim with a sharp pointed knife and on no account attempt to lever it out.
5 To test whether the unit is serviceable, suspend the thermostat on a piece of string in a pan of water being heated. Using a thermometer, with reference to the opening temperature in Specifications, its operation may be checked. The thermostat should be renewed if it is stuck open or closed, or it fails to operate at the specified temperature. The

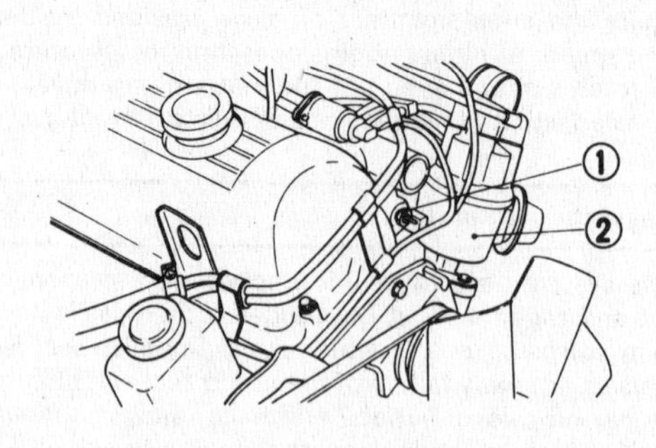

Fig. 2.5 The water temperature thermal transmitter (1) is located at the front of the cylinder head, near the distributor (2) (Sec 9)

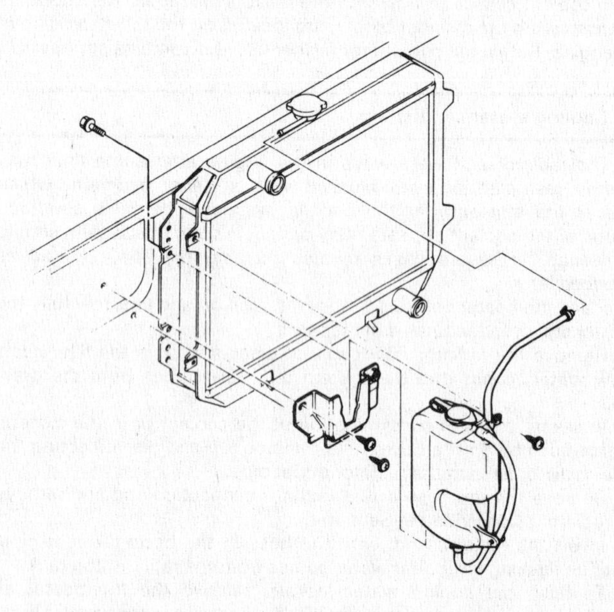

Fig. 2.7 The reservoir tank and its mounting points (Sec 10)

operation of the thermostat is not instantaneous and sufficient time must be allowed for movement during testing. Never replace a faulty unit – leave it out if no replacement is available immediately.
6 Installation of the thermostat is a reversal of the removal procedure. Ensure the mating faces of the housing are clean. Use a new gasket with jointing compound. The word *Top* which appears on the thermostat face must be visible from above.

8 Water pump – removal and installation

1 Drain the cooling system, retaining the coolant for further use.
2 Remove the shroud from the radiator.
3 Loosen the alternator mounting and adjustment bolts, and push the alternator in towards the engine so that the driving belt can be slipped off the alternator and fan pulleys.
4 Unbolt and remove the fan blade/pulley assembly from the water pump hub.
5 Unscrew evenly and then remove the bolts which secure the water pump to the engine front cover. If the water pump is stuck tight, do not lever it off but tap it gently with a hammer and hardwood block to break the gasket seal.
6 Where there is evidence of a leaking seal, or where severe corrosion of the impeller blades has occurred, do not attempt to repair the

water pump but renew it for a reconditioned exchange unit.

7 Installation is a reversal of removal but always use a new sealing gasket and then adjust the fan belt tension as described in Chapter 1 Section 43. Refill the cooling system.

9 Water temperature thermal transmitter

1 The water temperature thermal transmitter is screwed into the front of the cylinder head, on the right-hand side near the distributor. It is connected to the gauge located on the instrument panel; the circuit incorporates a voltage stabilizer.

2 In the event of the fuel level gauge and water temperature gauge both becoming inoperative at the same time, then the instrument voltage stabilizer should be suspected.

3 If unsatisfactory water temperature gauge readings are being obtained, the thermal transmitter may be tested by detaching the lead from the transmitter and connecting the end of the lead to a good ground. Switch on the ignition and observe the gauge needle. If the needle moves to the *Hot* position, the transmitter is faulty.

4 If the needle does not move at all, then the connecting cable or the gauge itself will be at fault.

5 Renewal of the transmitter unit can be carried out after partially draining the cooling system.

10 Reservoir tank – removal and installation

1 If only the reservoir tank is to be removed, this is quite simply carried out by disconnecting the hose at the radiator filler neck and sliding the tank upwards and out of the mounting bracket.

2 To remove the mounting bracket. Remove the reservoir tank as previously described; then undo and remove the bracket attaching bolts.

3 Installation, in both cases, is a straightforward reversal of the removal procedure, but ensure that the reservoir tank coolant level is maintained between the *Min* and *Max* marks.

11 Fault diagnosis – cooling system

Symptom	Reason/s
Heat generated in cylinders not being successfully disposed of by radiator	Insufficient water in cooling system Fan belt slipping (accompanied by a shrieking noise on rapid engine acceleration) Radiator core block or radiator grille restricted Bottom water hose collapsed, impeding flow Thermostat not opening properly Ignition advance and retard incorrectly set (accompanied by loss of power and perhaps misfiring) Fuel incorrectly adjusted Exhaust system partially blocked Oil level in oil pan too low Blown cylinder head gasket (water/steam being forced into the reservoir tank under pressure) Engine not yet run-in Brakes binding
Too much heat being dispersed by radiator	Thermostat jammed open Incorrect grade of thermostat fitted allowing premature opening of valve Thermostat missing
Leaks in system	Loose clips on water hoses Top or bottom water hoses perished and leaking Radiator core leaking Thermostat gasket leaking Pressure cap spring worn or seal ineffective Blown cylinder head gasket (pressure in system forcing water/steam into reservoir tank) Cylinder wall or head cracked

Chapter 3 Fuel, exhaust and emission control systems

Refer to Chapter 13 for specifications and information applicable to 1980 through 1984 models.

Contents

Specifications

System type ..... Bosch L-jectronic electronic fuel injection

Air cleaner element ..... Viscous paper type

Fuel pump

Type	Electrical (Bosch)
Operating pressure	36·3 lbf/in² (2·55 kgf/cm²)

Engine idle speed

Manual transmission	700 rpm
Automatic transmission (selector in 'D')	650 rpm

CO emission at idle speed (normal operating temperature)

California models	0·5% or lower
Non-California models	1·0% or lower

Recommended fuel

Non-California models	91 RON (research octane rating) unleaded or low lead
California models	Unleaded gasoline only

Fuel tank capacity

Sedan	15⅞ US gallons (60 liters)
Station Wagon	14½ US gallons (55 liters)

Torque wrench settings	**lbf ft**	**kgf m**
Manifold nuts or bolts:		
Size 8M	10 to 13	1·4 to 1·8
Size 10M	25 to 36	3·5 to 5·0
Throttle chamber securing bolts	11 to 14	1·5 to 2·0

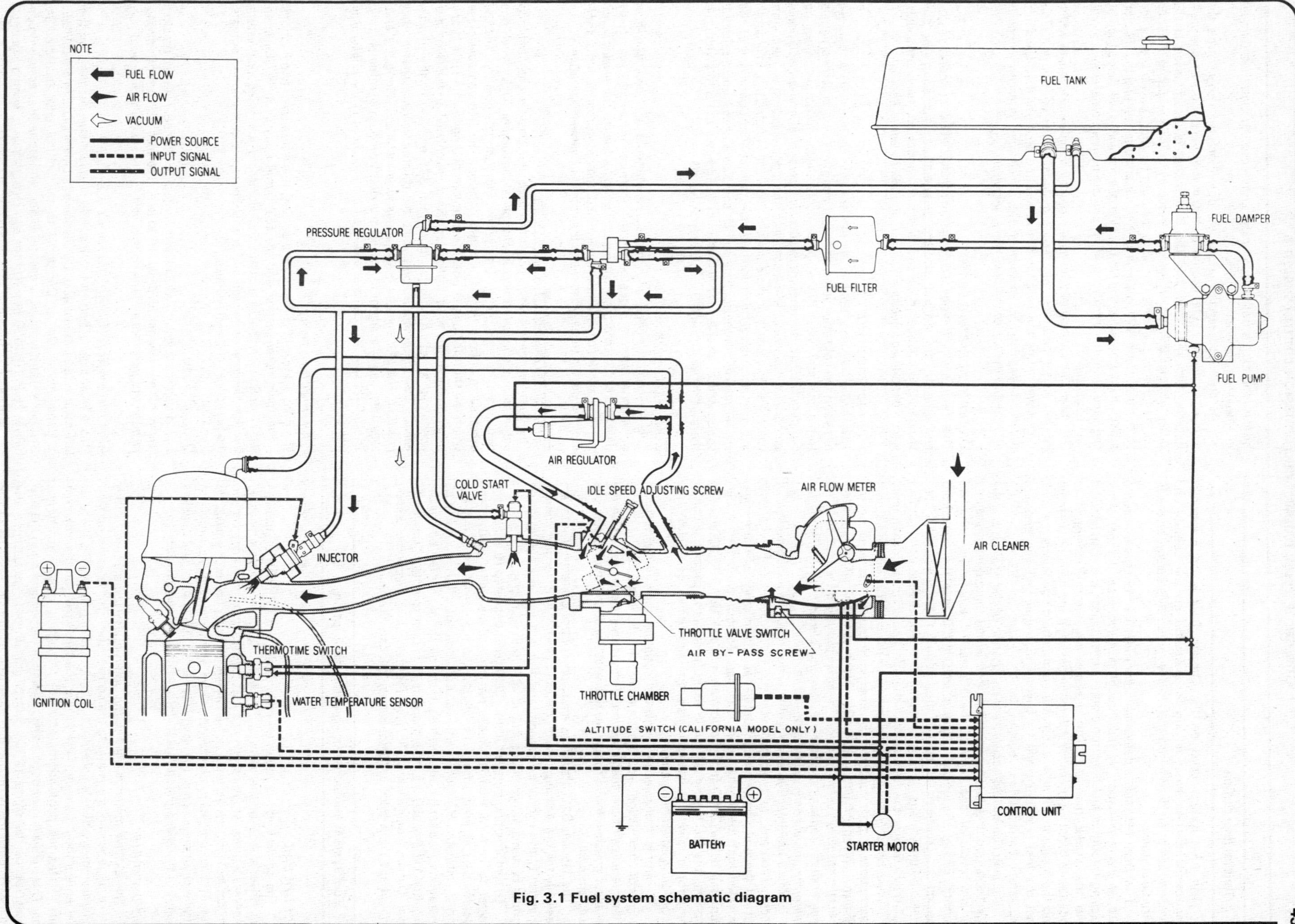

Fig. 3.1 Fuel system schematic diagram

1 General description

The fuel system used on the Datsun 810 is an electronic-controlled injection type, relying for metering on the volume and temperature of air drawn into the airflow meter. It is known as the L-jetronic system.

Fuel from the tank is delivered under pressure from the fuel pump and, to avoid pulsation, is fed through a mechanical damper. After passing through a fuel filter and a pressure regulator, it is then injected into the intake manifold. The pressure regulator is designed to maintain a constant pressure difference between the fuel line pressure and the intake manifold vacuum. Where manifold conditions are such that the fuel pressure could be beyond that specified, the pressure regulator returns surplus fuel to the tank.

An injection of fuel occurs at each rotation of the engine, and the injected amount of fuel for each injection is half the quantity required for one cycle of operation of the engine. To supply the complete quantity of fuel, the ignition coil is utilized. In this case, the signal from the ignition coil does not specify the timing for injection. This is controlled by the various signals computed by the control unit; the ignition coil only controls the frequency of the injections. The following units are all responsible for supplying signals to the control unit, and a brief description of their function, together with other units used on the system, is given in Section 2: *Air flow meter, ignition coil (negative terminal), throttle valve switch, water temperature sensor, air temperature sensor, thermotime switch, starting switch and, for California models only, an altitude switch.*

With a fuel injection system, fuel is not required during periods of deceleration and, to avoid fuel wastage, the throttle valve switch will cut off the fuel supply over a certain revolution range of the engine, provided that the throttle valve is closed. As soon as engine revolutions fall below 2800 (non-Californian models), or 1300 (Californian models), the fuel supply will cut in again, irrespective of the position of the throttle valve. This switching system is extremely beneficial when negotiating down-hill gradients.

Electronic fuel injection provides optimum mixture ratios at all stages of combustion, and this, together with the immediate response characteristics of the fuel injection, permits the engine to run on the weakest possible fuel/air mixture; this vastly reduces the exhaust gas toxic emission.

The Datsun 810 also uses emission control systems. These are basically the crankcase emission control system, which is described in Chapter 1, the exhaust emission, and the fuel evaporative emission control systems, which are described later in this Chapter.

The information detailed in this Chapter is designed to enable the owner to carry out basic checks on the various units used in the system. The majority of checks will involve the use of an ohmmeter and a voltmeter. Where a particular unit is suspect but proves to be operational, the fault could well be in the electronic control box or the circuitry. When such a problem occurs, it is best to entrust the job to your Datsun dealer who will have the necessary electronic testing equipment, knowledge and experience, to locate the fault.

2 Fuel system components – general description

Control unit

The control unit is mounted on a dash panel support bracket on the driver's side of the vehicle. The essential role of this unit is to generate a pulse. Upon receiving an electrical signal from each sensor, the control unit generates a pulse whose duration (injector open time period) is controlled to provide the exact amount of fuel, according to engine characteristics at that particular time. The control unit consists mainly of three integrated circuits formed on a printed circuit board.

Airflow meter

The airflow meter measures the volume and temperature of the intake air, and sends a signal to the control unit. This is achieved by a potentiometer which is linked to the air intake flap shaft. The more air that enters the flow meter, the further the flap valve rotates, which in turn rotates the pontentiometer wiper through a variable resistance coil. This increasing or decreasing resistance (dependent upon the flap angle) sends the signal to the control unit. In order to dampen any excessive movement of the flap, due to vacuum depressions in the intake manifold, a helical spring and compensating plate in a damper chamber are provided.

As an added safety feature, the potentiometer also controls the current supply to the fuel pump, so that in the event of the engine stalling, the flap will close, and the switch that is incorporated in the potentiometer will cut off the power supply to the fuel pump. As soon as the ignition switch is turned to the *Start* position, power is supplied to the fuel pump. When the engine has started to run and the ignition switch has returned to the *On* position, the flap valve will be opened by the vacuum in the intake manifold, and the pontentiometer switch will be in operation.

Also built into the airflow meter is an air temperature sensor which senses air temperatures and sends a signal to the control unit. This signal will define the duration of the injection time. Air that flows into the meter is first passed through the air cleaner assembly which is attached to the top of the flow meter assembly.

Water temperature sensor

This device, which is housed in the thermostat housing, monitors any changes in the engine coolant temperature. As soon as any temperature change is sensed, a signal is sent to the control unit, where a modified injector pulse duration will be computed.

Thermotime switch

This switch is also housed in the thermostat housing. A harness between the cold start valve and the thermotime switch carries signals to the cold start valve. If the engine coolant is below normal operating temperature, the thermotime switch is operational and an additional amount of fuel will be injected into the engine from the cold start valve. This is especially useful for starting a cold engine but, in order to avoid an excessively rich mixture due to repeated operations of the starter motor, the thermotime switch contains a heater and a bi-metal spring which, after a specified amount of time, warm up to switch the thermotime switch and the cold start valve off.

Cold start valve

This valve is screwed into the intake manifold and only becomes operational after a signal from the thermotime switch, previously described. The cold start valve supplies extra fuel to the engine, until the coolant reaches a pre-determined temperature; then the thermotime switch turns off to make the cold start valve inoperative.

Throttle valve switch

The throttle valve switch is attached to the throttle chamber and actuates in response to accelerator pedal movement. This switch has two sets of contact points. One set monitors the idle position, and the other set monitors the full throttle position. The idle contacts close when the throttle valve is positioned at idle, and open when it is at any other position. The full throttle contacts close only when the throttle is positioned at full throttle (or more than 34° opening of the throttle valve). The contacts are open while the throttle valve is at any other position. The idle switch compensates for enrichment during idle and, after idle, sends a signal to the control unit to modify the fuel supply. The full throttle switch compensates for enrichment at the full throttle position.

Fuel injection relay

The relay is located in front of the battery. It is made up of two sections – the main relay section and the fuel pump relay section. The main relay section serves to actuate the electronic fuel injection system through the ignition switch and the fuel pump relay section to actuate the fuel pump and air regulator.

Dropping resistor

The dropping resistor, which is mounted to the air cleaner bracket below the air cleaner, is provided to keep the electric current flowing through the injectors and the control unit to the absolute minimum.

Altitude switch (California models only)

This switch is attached to the instrument panel support stay, on the left-hand side of the vehicle. Consisting of a bellows and micro-switch, it transmits an *On* or *Off* signal to the control unit according to changes in atmospheric pressure. When atmospheric pressure drops below 26 in Hg (660 mm Hg) the switch activates to send a signal to the control unit, to reduce the fuel injected by 6%.

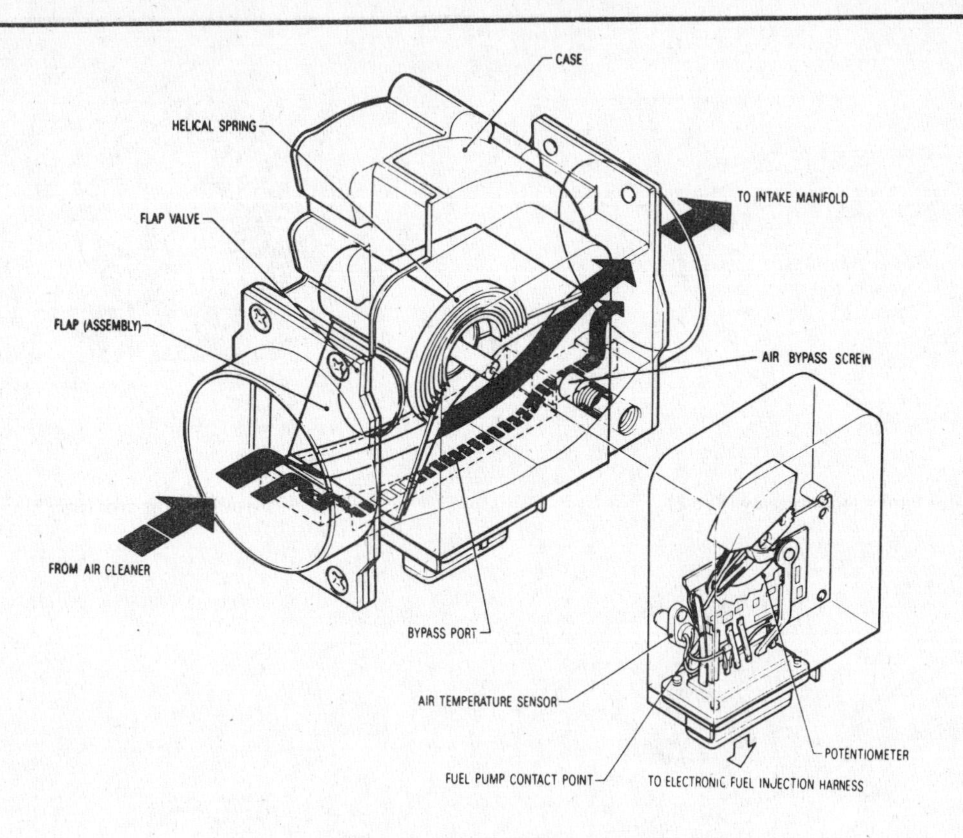

Fig. 3.2 The airflow meter (Sec 2)

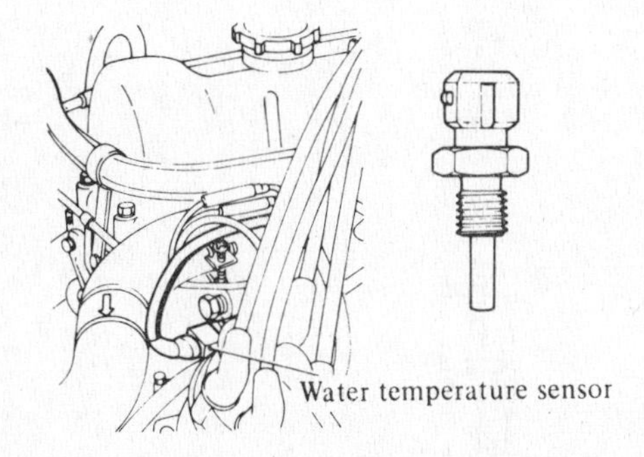

Fig. 3.3 Water temperature sensor (Sec 2)

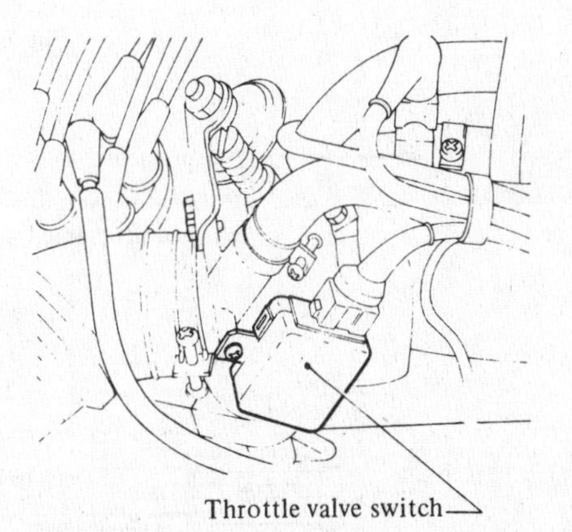

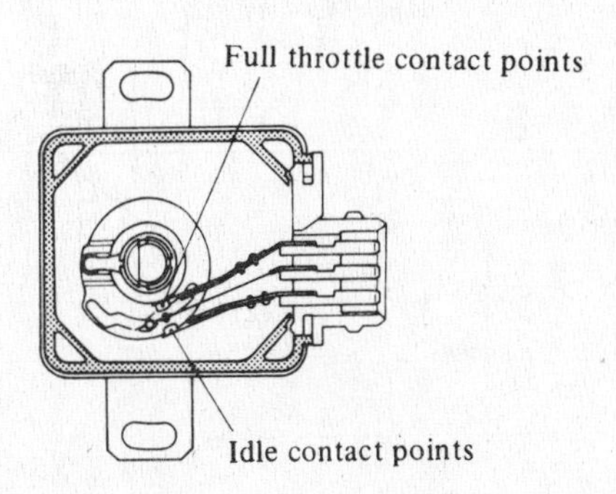

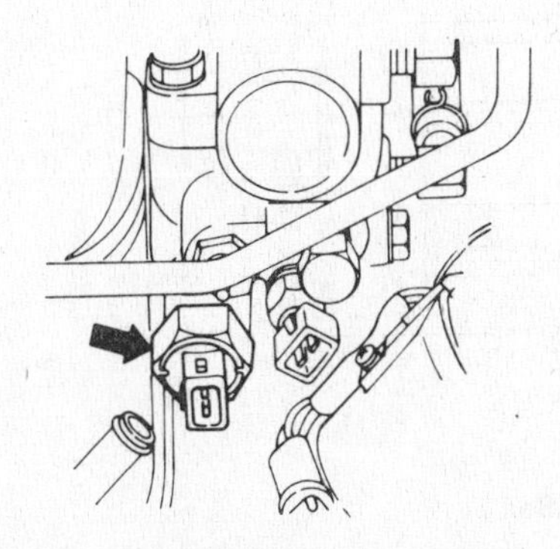

Fig. 3.4 Thermotime switch (Sec 2)

Fig. 3.5 Throttle valve switch (Sec 2)

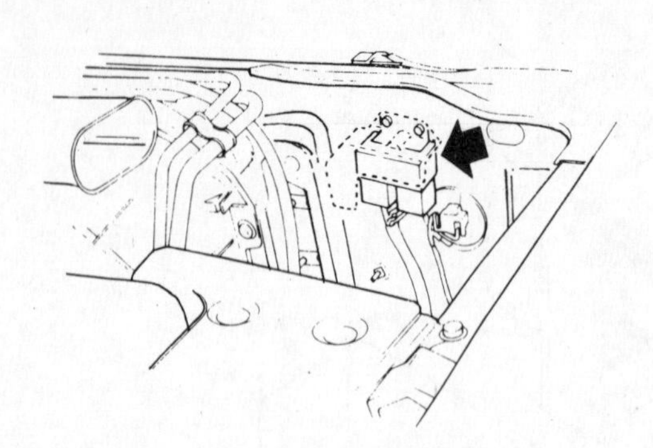

Fig. 3.6 Fuel injection relay (Sec 2)

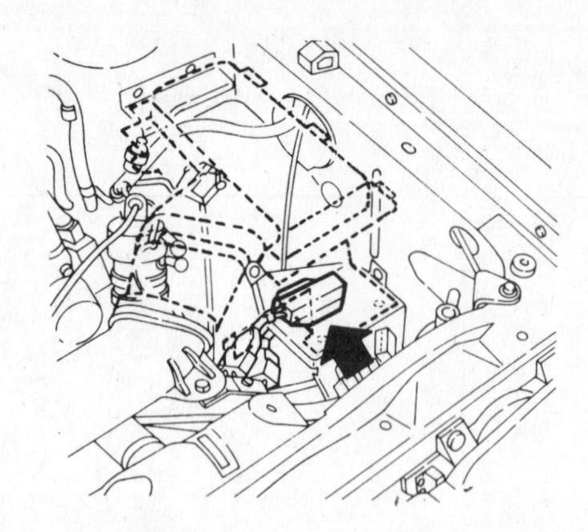

Fig. 3.7 Dropping resistor (Sec 2)

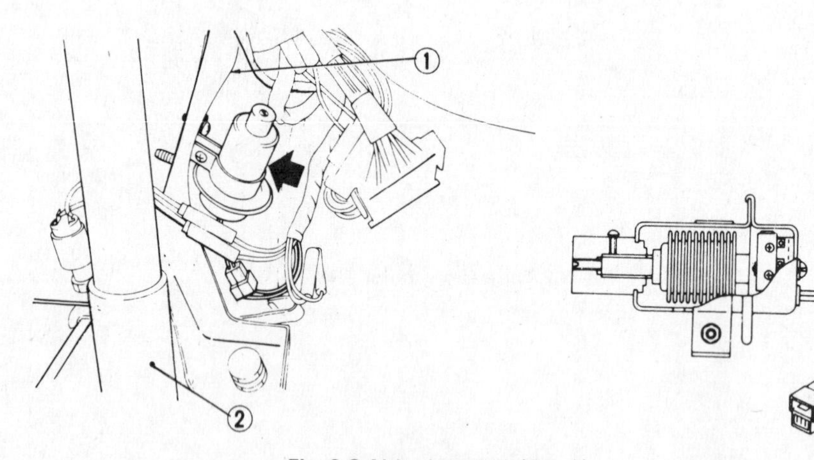

Fig. 3.8 Altitude switch (Sec 2)

1 Pedal bracket
2 Steering column

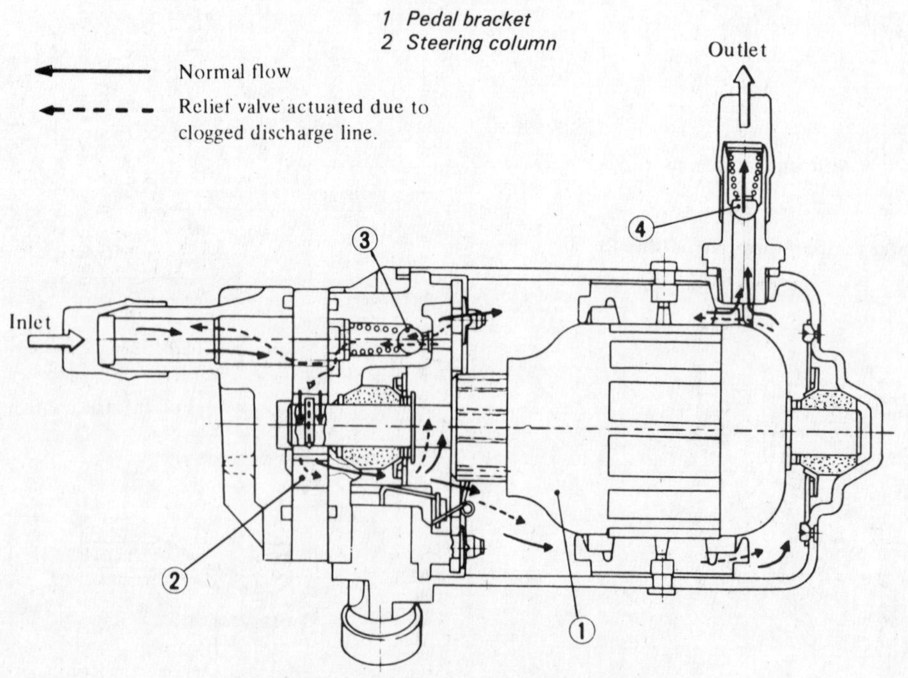

Fig. 3.9 Sectional view of the fuel pump (Sec 2)

1 Motor
2 Pump
3 Relief valve
4 Check valve

Fuel pump

The fuel pump is mounted near the fuel tank at the right-hand rear end of the vehicle. Built into the outlet pipe of the pump is a check valve, and at the inlet pipe, a relief valve. The relief valve is designed to open when pressure in the fuel line rises to 43 to 64 lbf/in^2 (3 to 4·5 kgf/cm^2) due to any blocking in the line. The check valve prevents any abrupt drop in pressure when the engine is stopped. When the ignition switch is turned to the *Start* position, the fuel pump is actuated irrespective of the position of the airflow meter potentiometer switch. After the engine starts, the ignition switch will be at the *On* position, which in effect breaks the current supply to the fuel pump; this is immediately overcome by the engine running, because the potentiometer switch in the airflow meter now takes over the control of power supply to the fuel pump.

Fuel damper

The fuel damper, which is mounted to the fuel pump at the rear of the vehicle, is designed to suppress any pulsation in fuel flow from the fuel pump.

Pressure regulator

The pressure regulator, which is mounted adjacent to the intake manifold in the fuel supply pipe, maintains a constant fuel pressure at all stages of acceleration and deceleration. Under extreme manifold vacuum conditions, the full pressure delivered by the fuel pump, combined with a high vacuum, could cause excessive pressure in the fuel line. Where such a condition occurs, the pressure regulator opens to return excess fuel to the fuel tank.

Fuel filter

The fuel filter, which is mounted on the right-hand side of the engine compartment, ensures that only clean fuel reaches the injectors.

Air regulator

The air regulator bypasses the throttle valve, to control the quantity of air required for increasing the engine idling speed when starting the engine at an underhood temperature of below 80°C (176°F). A bi-metal spring and a heater are built into the regulator. When the ignition switch is turned to the *Start* position, or if the engine is running, electric current flows through the heater and the bi-metal spring and, as the heater warms up, the bi-metal spring will close the air passage. The air passage will remain closed until the engine is stopped, or the underhood temperature drops below 80°C (176°F).

Injectors

An injector is mounted on each branch portion of the intake manifold. Each injector is actuated by a small solenoid valve, built into the injector body. Actuating the solenoid valve pulls the needle valve into the open position to allow the fuel to inject. The duration of the pulse sent from the control unit, defines the period of time that the solenoid valve is actuated.

Throttle chamber

The throttle chamber, located between the intake manifold and the airflow meter is equipped with a valve. This valve controls the intake airflow in response to accelerator pedal movement. The shaft of this valve is connected to the throttle valve switch. The valve remains closed during engine idling, and the air required for idling passes through a bypass port, into the intake manifold. Idle adjustment is made by the idle speed adjusting screw, located in the bypass port. There is also another bypass line in the throttle chamber, to pass sufficient air through the air regulator into the intake manifold, when a cold engine is started.

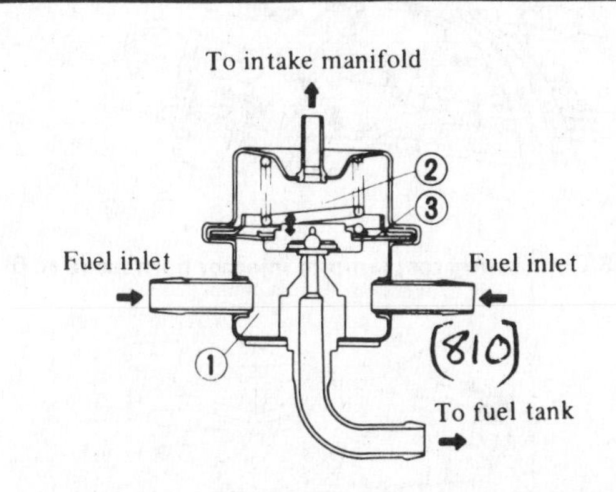

Fig. 3.10 Sectional view of the pressure regulator (Sec 2)

1 Fuel chamber
2 Spring chamber
3 Diaphragm

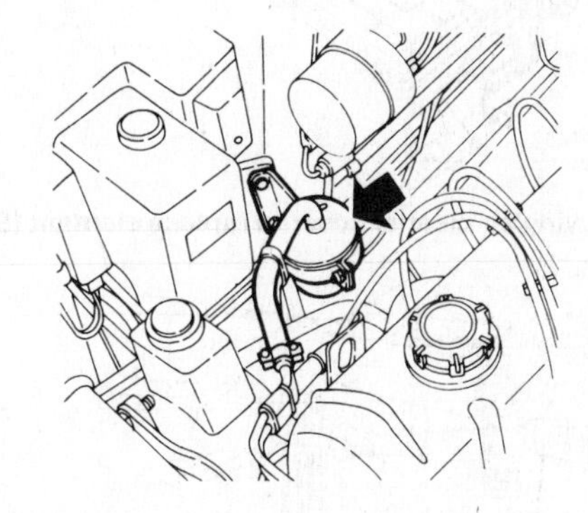

Fig. 3.11 Fuel filter (Sec 2)

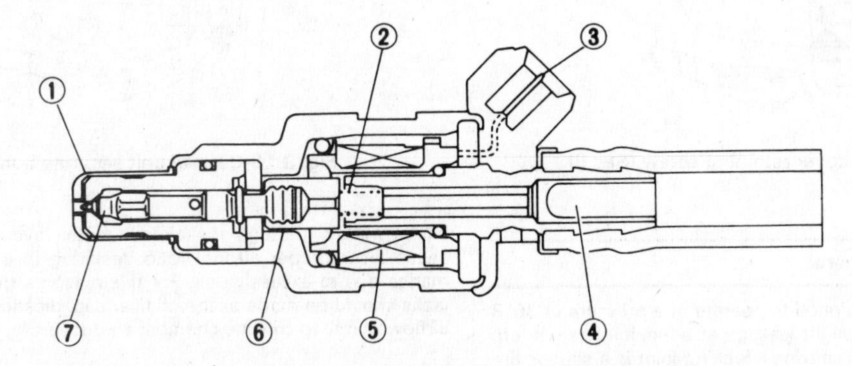

Fig. 3.12 Sectional view of an injector (Sec 2)

1 Nozzle
2 Return spring
3 Electric terminal
4 Fuel filter
5 Magnetic coil
6 Core
7 Needle valve

3

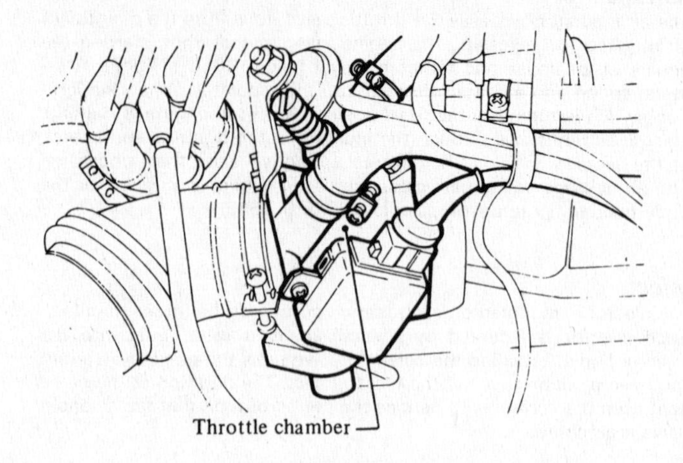

Fig. 3.13 Location of throttle chamber (Sec 2)

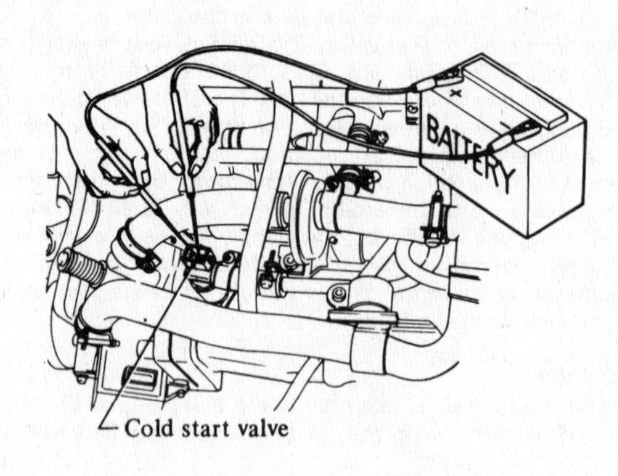

Fig. 3.14 Releasing fuel line pressure (Sec 4)

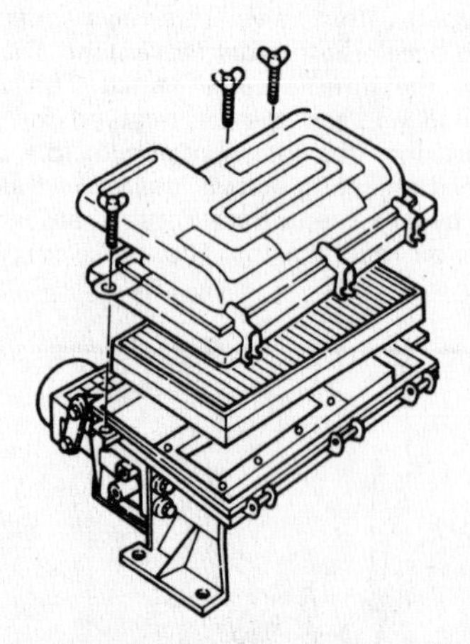

Fig. 3.15 Removing air cleaner cover to replace element (Sec 5)

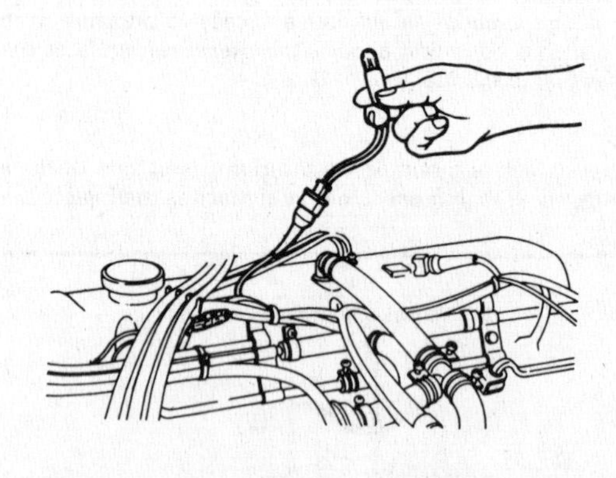

Fig. 3.16 Connecting test lamp to injector harness (Sec 6)

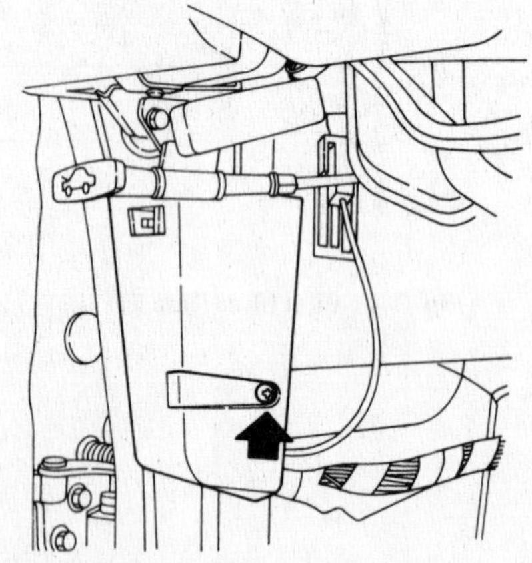

Fig. 3.17 Control unit cover retaining screw (Sec 6)

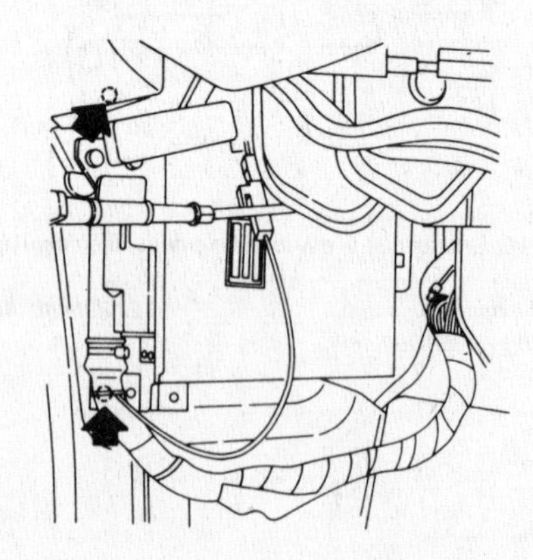

Fig. 3.18 Control unit securing bolts (Sec 6)

3 Fuel lines and hoses – general

The fuel injection system is designed to operate at a pressure of 36·3 lbf/in^2 (2·55 kgf/cm^2) and any slight leakage at a fuel joint can, if left, soon become a major problem. Leaving a leaking joint is a serious fire risk and the pressure loss is certain to effect the system's efficiency. From time-to-time, check the security of all the fuel joints; check also that the rigid pipes are not kinked or bent in any way. If a rubber hose shows signs of deterioration, renew it.

Since the electronic fuel injection system accurately meters the intake airflow through the airflow meter, even a slight air leak will cause an improper air/fuel ratio, resulting in a faulty engine performance due to excessive air. For this reason a thorough inspection for leaks should be made at the oil filler cap, dipstick seal, blow-by hoses, airflow meter-to-throttle chamber air duct etc.

4 Fuel system – releasing pressure

Note: *Because many of the units used in the fuel system operate under pressure, it is necessary, before removing them, to relieve the*

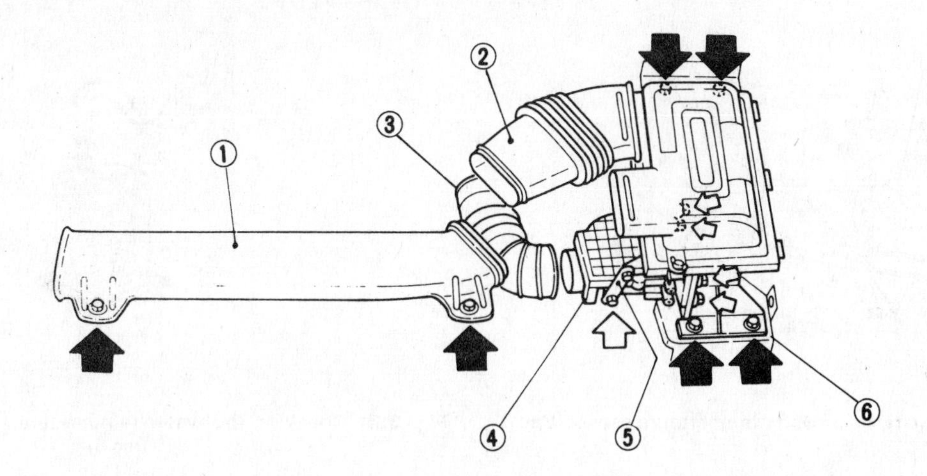

Fig. 3.19 Airflow meter assembly detachment points arrowed (Sec 7)

1 Air duct
2 Air duct
3 Air duct
4 Air flow meter
5 Stay
6 Air cleaner bracket

fuel line pressure. To do this, carry out the operations described in the following paragraphs.

1 Disconnect the ground cable from the battery.
2 Pull off the harness connector from the cold start valve.
3 Connect two jumper leads to the battery, one to the positive terminal and the other to the negative terminal.
4 Now connect the other ends of the jumper leads to the terminals of the cold start valve. Allow three or four seconds to pass; this will release the pressure in the system.
5 When connecting the jumper leads to the cold start valve, take care to keep them away from each other, to avoid short circuiting.

5 Air cleaner element – renewal

1 The air cleaner element is of the viscous paper type and requires no cleaning. Renewal of the element must be carried out at the time recommended in the Routine Maintenance Section.
2 First, disconnect the air duct at the air cleaner horn.
3 Unscrew and remove the three wing nuts that retain the cover to the air cleaner body. Lift off the cover.
4 Lift out the air cleaner element.
5 To install the new element, reverse the removal procedure.

6 Control unit – checking, removal and installation

Checking

1 This check employs a miniature test lamp, to check whether the open-injector pulse for cranking the engine is actually applied to the injectors, should the engine fail to start. To carry out this check, the engine must be cranked at a speed of more than 80 rpm; also the battery must be known to be in a good state of charge. The test lamp should be a 3-volt miniature type, with a suitable connection to its end.
2 Turn the ignition switch to the *Off* position.
3 Disconnect the harness connector of number 1 cylinder injector.
4 Disconnect the cold start valve harness connector.
5 Now connect the test lamp to the injector harness of the number 1 cylinder.
6 Turn the ignition switch to the *Start* position to crank the engine, and observe whether or not the lamp flashes. If the lamp flashes, all is well; if it doesn't, the control unit is faulty and must be renewed.
7 As a further check, if all is well at paragraph 6, disconnect the harness from the water temperature sensor and crank the engine as before. The test lamp should now be flashing a lot brighter than before. If the lamp doesn't flash brighter, the water temperature sensor is probably faulty, and should be checked as described in Section 8.
8 Because the power unit uses two separate power transistors (one is for numbers 1, 2 and 3 cylinders and the other for 4, 5 and 6 cylinders), the same check must be carried out on number 4 cylinder, to prove that the other transistor is functioning.
9 It is emphasized that this check only proves whether or not a signal pulse is reaching the injectors. If the control unit is still under suspicion, there are numerous other circuits inside the unit that could be at fault, and it best to remove it to have it checked professionally.

Remove and installation

10 To remove the control unit, make sure that the ignition switch is at the *Off* position.
11 The control unit is located just below the hood release operating lever. Remove the single screw that retains the control unit cover, and remove the cover.
12 Disconnect the 35-pin coupler from the control unit.
13 Unscrew and remove the two bolts that secure the control unit to the side panel, and lift away the control unit.
14 Installation of the control unit is a straightforward reversal of the removal procedure, but ensure that the 35-pin connector is secure before switching on the ignition.

3

7 Airflow meter – checking, removal and installation

Note: *The following checks can be carried out without removing the airflow meter assembly from the vehicle.*

1 Remove the air cleaner cover and element, as described in Section 5.

Checking the potentiometer

2 Disconnect the battery ground cable.
3 Disconnect the harness connector from the airflow meter.
4 Now, using an ohmmeter, check the resistance between terminals 6 and 8. The resistance should be approximately 180 ohms.
5 Measure the resistance between terminals 8 and 9, which should be approximately 100 ohms.
6 Connect a 12 volt dc supply across terminals 9 (positive) and 6 (negative).
7 Connect the positive lead of a voltmeter to terminal 8, and the negative lead to terminal 7.
8 Gradually open the air intake flap by hand to ensure that the voltmeter reading decreases proportionately. If the reading is erratic, the potentiometer may require renewal. The airflow meter should be renewed as an assembly, if any of the previously described checks are not as specified.

Checking insulation resistance of the airflow meter

9 To check the insulation resistance between the airflow meter body and the potentiometer terminals, simply hold one of the ohmmeter probes in contact with the airflow meter casing and, with the other probe, touch each of the terminals (terminals 6, 7, 8 and 9). If any con-

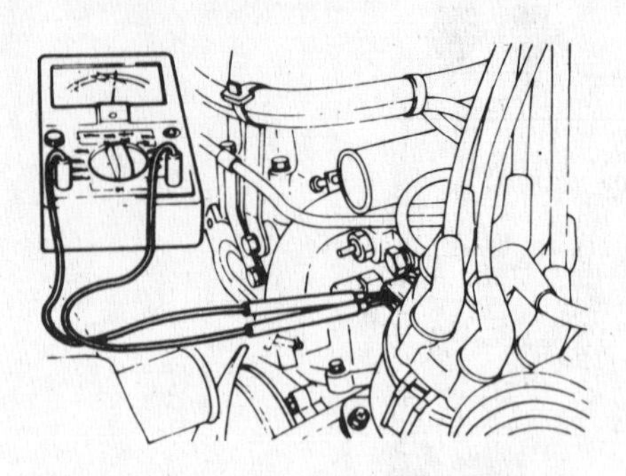

Fig. 3.20 Checking resistance of water temperature sensor (Sec 8)

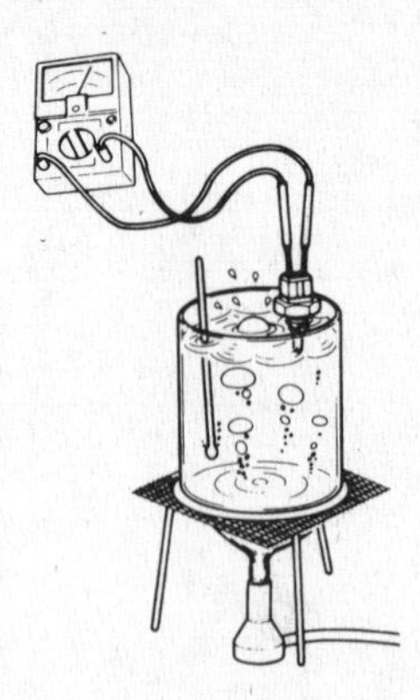

Fig. 3.21 Checking the water temperature sensor off the engine (Sec 8)

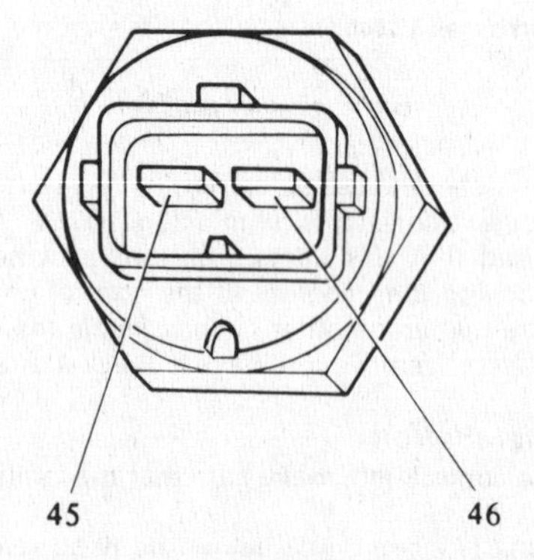

Fig. 3.22 Terminals 45 and 46 on thermotime switch (Sec 9)

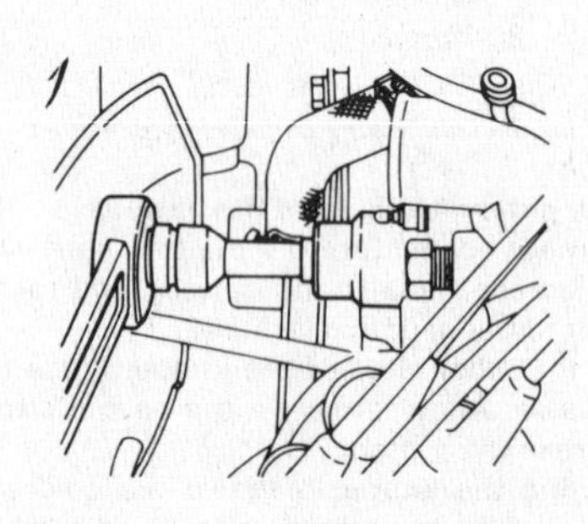

Fig. 3.23 Removing thermotime switch (Sec 9)

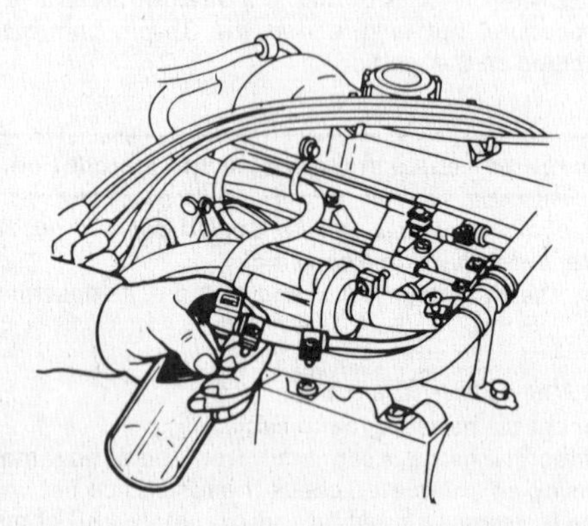

Fig. 3.24 Checking cold start valve (Sec 10)

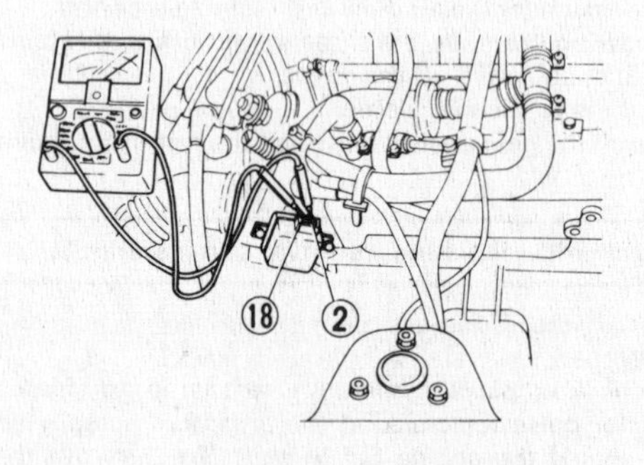

Fig. 3.25 Checking idle switch (Sec 11)

tinuity exists, the airflow meter is faulty, and should be renewed as an assembly.

Checking fuel pump contact points

10 Connect the ohmmeter probes to terminals 36 and 39; with the air intake flap in the closed position, there should be no continuity. Now gradually open the flap by hand until a position of approximately 8° is achieved, when the contact points should become operational to produce a continuity reading on the ohmmeter. If the intake flap does not operate smoothly, or continuity exists when the flap is closed, the airflow meter must be renewed as an assembly. Also, if the points do not become operational when the flap is in the 8° position, the potentiometer is obviously faulty, and again, the airflow meter must be renewed.

Checking the air temperature sensor

11 Connect the ohmmeter probes to terminals 6 and 27.

12 Using a normal household thermometer, measure the air temperature as near to the sensor as possible.

13 The following table gives the resistance valve over a given temperature range. The air temperature sensor is part of the airflow meter assembly and, where the resistance readings are outside the limits listed in the chart, the airflow meter must be renewed as an assembly.

Air temperature	Resistance
–30°C (–22°F)	20.3 to 33.0k ohms
–10°C (–14°F)	7.6 to 10.8k ohms
10°C (50°F)	3·25 to 4·15k ohms

20°C (68°F)	2·25 to 2·75k ohms
50°C (122°F)	0·74 to 0·94k ohms
80°C (176°F)	0·29 to 0·36k ohms

14 Finally, check the insulation resistance between terminal 27 and the airflow meter body. If continuity exists, the air temperature sensor is faulty; renew the airflow meter as an assembly.
15 To remove the airflow meter assembly, first ensure that the battery ground cable has been disconnected.
16 Disconnect the harness connector from the airflow meter.
17 Unfasten the clamp that secures the air duct running between the airflow meter and the throttle chamber. Disengage the duct at the airflow meter.
18 Remove the air cleaner cover and element as described in Section 5.
19 Remove the bolts securing the air flow meter and stay, and lift away the airflow meter.
20 Installation is a direct reversal of the removal procedure, but ensure that the harness connector has been correctly inserted into the airflow meter before reconnecting the battery ground cable.

8 Water temperature sensor – checking, removal and installation

Checking

Note: The water temperature sensor is designed to give a specific resistance reading over a given temperature range. It can be checked after removal from the engine compartment using containers of hot and cold water, but the procedure described is for one which is installed.
1 Disconnect the battery ground cable.
2 Disconnect the water temperature sensor harness connector.
3 By placing a thermometer into the radiator filler neck, establish the temperature of the coolant, with the engine cold.
4 Connect the probes of the ohmmeter to the terminals of the water temperature sensor. Record the resistance reading obtained, and compare it with the following chart of resistance valves:

Engine coolant	Resistance
–30°C (–22°F)	20·3 to 33·0k ohms
–10°C (–14°F)	7·6 to 10·8k ohms
10°C (50°F)	3·25 to 4·15k ohms
20°C (68°F)	2·25 to 2·75k ohms
50°C (122°F)	0·74 to 0·94k ohms
80°C (176°F)	0·29 to 0·36k ohms

5 Now reconnect the water temperature sensor harness connector and the battery ground cable. Run the engine until normal operating temperature is reached, then repeat the operations described in paragraphs 1, 2, 3 and 4. By first carrying out a cold check, then a hot check, will give a good indication of the condition of the sensor over its operating range. If the resistance readings do not come within the range specified, renew the water temperature sensor.

Removal and installation

6 To remove the water temperature sensor will involve draining approximately 1⅝ US quarts (1·5 liters) of coolant from the radiator. If necessary, refer to Chapter 2 for further information.
7 For ease of access, remove the radiator top hose.
8 Disconnect the harness connector from the water temperature sensor.
9 Unscrew and remove the water temperature sensor.
10 Installation is a direct reversal of the removal procedure, but ensure that a new copper washer is installed between the sensor and the thermostat housing. When topping-up the radiator coolant, always ensure that the correct antifreeze quantity is used.

9 Thermotime switch – checking, removal and installation

Static check

1 Disconnect the battery ground cable.
2 Disconnect the electrical connector from the thermotime switch.
3 Using an ohmmeter, measure the resistance between terminal 46, and the switch body. The readings should be as follows:

Coolant temperature less than 14°C (57°F) . . . Zero
Coolant temperature between 14 to 22°C (57 to 72°F) . . . . Zero or infinity
Coolant temperature more than 22°C (72°F) . . . Infinity

4 Now, measure the resistance between terminal 45, and the switch, which must be between 70 and 86 ohms. If the thermotime switch does not conform to any of the previously mentioned readings, it must be renewed.
5 The following dynamic check will require removal of the thermotime switch. To do this, proceed as described in the following paragraphs.
6 Disconnect the battery ground cable.
7 Drain approximately 1⅝ US quarts (1·5 liters) of coolant from the radiator. Refer to Chapter 2, if necessary.
8 For ease of access, remove the radiator top hose.
9 Disconnect the wires from the thermal transmitter, and remove the thermal transmitter.
10 Disconnect the electrical connector from the thermotime switch.
11 Unscrew and remove the thermotime switch from the thermostat housing.
12 The dynamic check will involve the use of a container of water, which can be heated up very quickly (1°C (1·8°F) per second).
13 Dip the heat-sensing portion of the thermotime switch into the container of water, which should be maintained at 10°C (50°F). Allow a couple of minutes to pass, in order to ensure that the thermotime switch stabilizes.
14 Measure the resistance between terminals 45 and 46. The reading should be approximately 78 ohms.
15 Now heat the water at a rate of 1°C (1·8°F) per second until the temperature is more than 25°C (77°F). Now check the continuity between terminals 45 and 46. If the reading increases from 78 ohms to infinity, the thermotime switch is operating correctly. If the previously mentioned readings are not achieved, renew the thermotime switch.
16 To install the thermotime switch, reverse the order of removal described in paragraph 6 through 11. When topping-up the radiator coolant, always make sure that the correct antifreeze quantity is used.

10 Cold start valve – checking, removal and installation

3

Checking

1 At the starter motor, disconnect the S terminal lead.
2 To ensure that the fuel pump is functioning, turn the ignition switch to the *Start* position, where an operating sound should be heard from the rear end of the vehicle.
3 Disconnect the battery ground cable.
4 Unscrew and remove the two screws securing the cold start valve to the intake manifold. Remove the valve from the manifold.
5 Disconnect the electrical connecter from the cold start valve.
6 Put the cold start valve into a glass container of 1·22 cu in (20 cc) capacity. Plug the container opening around the fuel pipe with a clean rag. This will help to keep the fuel in the container and minimize fire risk.
7 Reconnect the battery ground cable.
8 Turn the ignition switch to the *Start* position. The cold start valve should not inject fuel.
9 Turn the ignition switch to the *Off* position. Using jumper leads connected to the battery negative and positive terminals, touch the cold start valve terminals. The valve should now inject fuel into the glass container. If there is no evidence of any fuel being injected, renew the cold start valve as described in the following paragraphs.

Removal and installation

10 Disconnect the battery ground cable.
11 Release the pressure in the fuel system as described in Section 4.
12 Undo and remove the two screws securing the cold start valve to the intake manifold. Remove the valve from the manifold.
13 Unfasten the hose clip and disengage the cold start valve from the fuel hose.
14 When installing the cold start valve, reverse the procedures

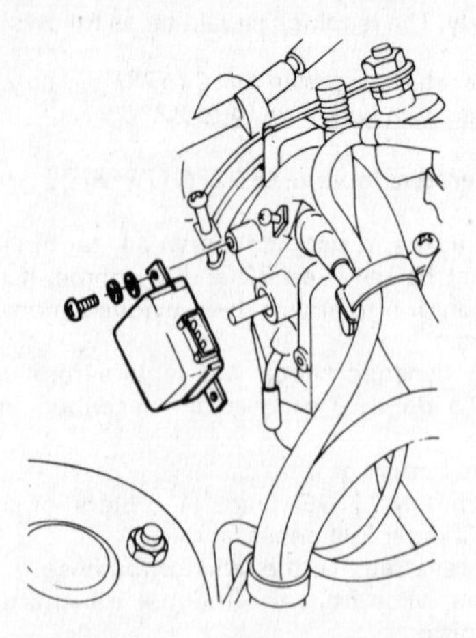

Fig. 3.26 Throttle valve switch ready for installation (Sec 11)

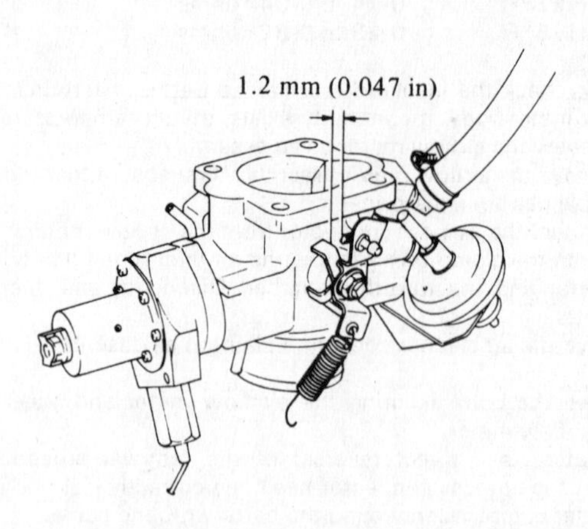

Fig. 3.27 Throttle valve stop screw-to-throttle valve shaft lever clearance

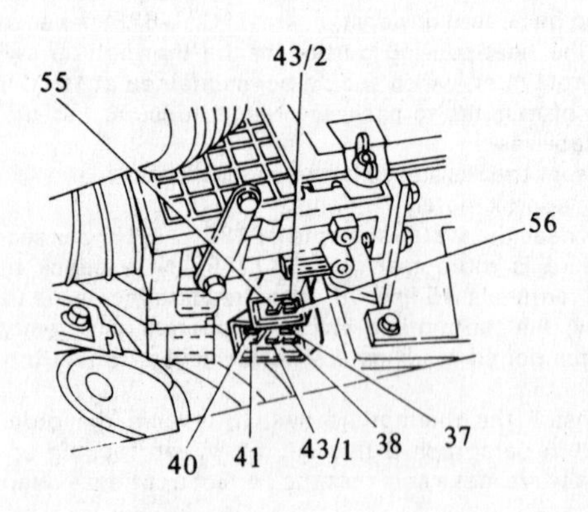

Fig. 3.28 Dropping resistor terminal numbers (Sec 12)

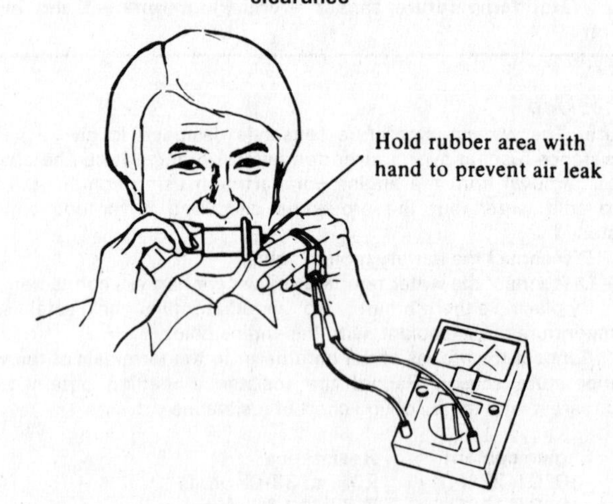

Fig. 3.29 Checking altitude switch (Sec 13)

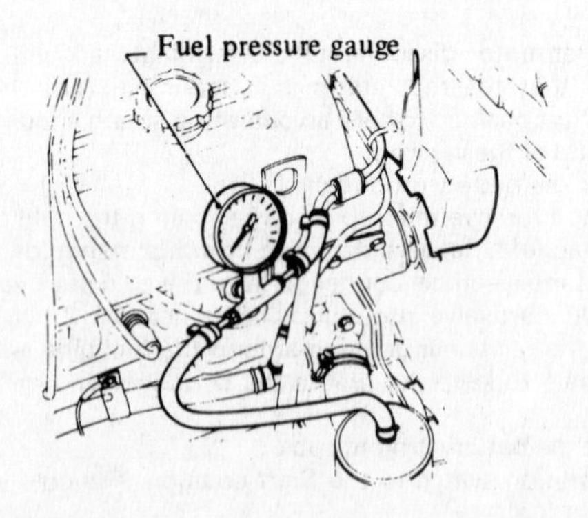

Fig. 3.30 Pressure gauge connected into fuel line (Sec 14)

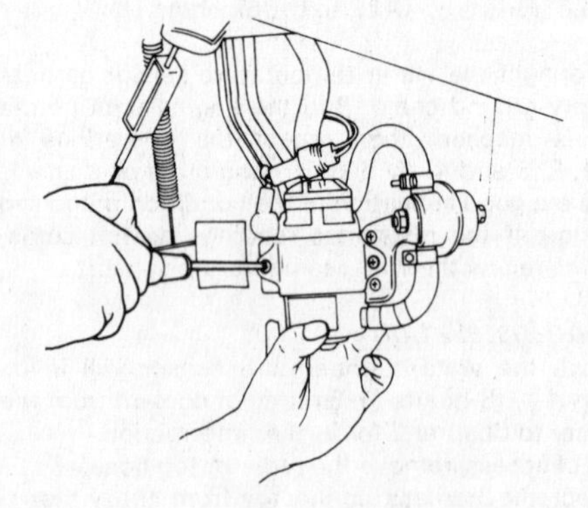

Fig. 3.31 Removing fuel pump (Sec 14)

described in paragraph 10 through 13.

11 Throttle valve switch – checking, removal and installation

1 Disconnect the battery ground cable.
2 Remove the throttle valve switch connector.

Idle switch check

3 Using an ohmmeter with probes connected to terminals 2 and 18, ensure that continuity exists when the throttle valve is in the idle position. Now open the throttle valve approximately 4°, where there should be no continuity. If this is not as specified, the throttle valve switch must be renewed as an assembly.

Full throttle switch check

4 Connect the ohmmeter probes to terminals 3 and 18
5 Gradually open the throttle valve from the fully closed position. Observe the ohmmeter reading when the valve is opened approximately 34°. If the ohmmeter reading at all other positions is greater than that obtained at the 34° position, the switch is functioning properly. If the throttle valve switch does not conform to the pre-

viously mentioned readings, it will have to be renewed.

Throttle valve switch insulation check

6 Connect one of the ohmmeter probes to a convenient point on the vehicle frame (ground).

7 With the other ohmmeter probe, touch terminals 2, 3 and 18 in turn, and observe that the ohmmeter reading is infinite at each terminal. If it is not, there's obviously a short- circuit somewhere, that can only be rectified by renewing the throttle valve switch as described in the following paragraphs.

Removal and installation

8 Disconnect the battery ground cable

9 Disconnect the throttle valve switch harness connector.

10 Undo and remove the two screws securing the switch to the throttle chamber.

11 Carefully pull the throttle valve switch off the throttle valve shaft.

12 To install the throttle valve switch, reverse the removal procedure.

13 After installation, the position of the throttle valve switch will need to be adjusted so that the idle switch turns from *On* to *Off* at approximately 1400 rpm under a *no-load* condition (ie engine running but vehicle stationary). This is best carried out by repeating the checks described in paragraph 1 through 3 and turning the throttle valve switch in the required direction of rotation until, at the 4° throttle valve position, no continuity exists. When this setting has been achieved, ensure that the throttle valve switch securing screws are firmly tightened.

14 As a further guide, when the throttle valve switch is correctly adjusted, the clearance between the throttle valve stop screw and the throttle valve shaft lever should be 0·047 in (1·2 mm).

12 Dropping resistor – checking, removal and installation

Checking

1 Disconnect the battery ground cable.

2 Disconnect the 4-pin and 6-pin connectors from the dropping resistors.

3 At the 6-pin connector, carry out the following resistance checks:

(a) Between terminal 43/1 and terminal 41 (number 4 cylinder resistor).

(b) Between terminal 43/1 and terminal 40 (number 3 cylinder resistor).

(c) Between terminal 43/1 and terminal 38 (number 2 cylinder resistor).

(d) Between terminal 43/1 and terminal 37 (number 1 cylinder resistor).

4 At the 4-pin connector, carry out the following resistance checks:

(a) Between the 43/2 terminal and terminal 56 (number 6 cylinder resistor).

(b) Between the 43/2 terminal and terminal 55 (number 5 cylinder resistor).

5 The resistance reading for all of the above mentioned checks should be approximately 6 ohms. If one particular terminal appears to differ from the others, the whole resistor should be renewed. To do this, proceed as described in the following paragraphs.

Removal and installation

6 Disconnect the battery ground cable.

7 Disconnect the harness connector at the airflow meter.

8 Disconnect the rubber hose from the airflow meter.

9 Disengage the air ducts from the airflow meter.

10 Unscrew and remove the bolts securing the air cleaner to the air cleaner bracket. Lift away the air cleaner and airflow meter as an assembly.

11 Disconnect the two electric connectors from the dropping resistor.

12 Undo and remove the two screws attaching the dropping resistor to the air cleaner bracket. Lift away the dropping resistor.

13 To install the dropping resistor is a direct reversal of the removal procedure, but before reconnecting the battery ground cable, ensure that all electrical connections have been correctly assembled.

13 Altitude switch (California models only) – removal, checking and installation

1 Disconnect the battery ground cable.

2 Working under the instrument panel on the driver's side of the vehicle, disconnect the electric connector from the altitude switch.

3 Undo and remove the bolts securing the altitude switch bracket to the instrument panel support bracket. Lift out the altitude switch.

4 Connect the probes of an ohmmeter to the altitude switch terminals.

5 At the discharge port, either blow or suck (by mouth), until a *click* is heard and the ohmmeter registers continuity.

6 The altitude switch is set during manufacture and there are no provisions for adjustment, so if the switch is faulty it must be renewed as an assembly.

7 Installation of the altitude switch is a reversal of the removal procedure.

14 Fuel pump – checking, removal and installation

The only tests that can be carried out on the fuel pump, are a basic-function test and a discharge pressure check. The discharge pressure check will involve the use of a pressure gauge capable of measuring up to 40 lbf/in²(2·81 kgf/cm²) accurately, and a suitable T-piece.

Basic-function check

1 To carry out the basic-function test, first disconnect the starter motor S terminal lead.

2 Turn the ignition switch to the *Start* position, where an operating sound from the rear of the vehicle should be clearly audible. If no operating sound is audible, check that the electrical wiring to the fuel pump is in order. This could involve the use of special circuit testing equipment, and should be entrusted to your Datsun dealer.

3 If the fuel pump appears satisfactory in paragraph 2 proceed to check the fuel discharge pressure, as described in the following paragraphs.

Discharge pressure check

4 Disconnect the battery ground cable.

5 Disconnect the harness connector at the cold start valve.

6 Release the pressure in the fuel line, as described in Section 4.

7 Connect the pressure gauge into the fuel hose immediately after the fuel filter.

8 Disconnect the S terminal lead from the starter motor.

9 Disconnect the electrical connector at the cold start valve.

10 Now reconnect the battery ground cable.

11 Seek the assistance of a second person, to turn the ignition switch to the *Start* position.

12 When the ignition switch is in the *Start* position, observe the pressure reading on the pressure gauge.

13 If the reading is below 36·3 lbf/in² (2·55 kgf/cm²), check for clogged or deformed fuel lines, and if necessary, renew the fuel pump as an assembly.

14 If all is well in the previously described checks, it does not necessarily follow that the fuel pump is serviceable. The fuel delivery pressure at full engine revolutions could possibly be below the specified figure, but to check this requires the use of a rolling road test, or alternatively, jacking-up the rear of the vehicle in order to run the vehicle in a stationary position. This operation requires the use of specially designed axle-stands, and a great deal of care must be taken whilst carrying out the operation. In view of this, the job is really best entrusted to your Datsun dealer.

Removal and installation

15 To remove the fuel pump, first ensure that the battery ground cable is disconnected.

16 Release the pressure in the fuel line, as described in Section 4.

17 Raise the rear of the vehicle and safely support it with axle-stands or suitable blocks of wood. As an added precaution, firmly chock both front wheels.

18 Working inboard of the right-hand rear wheel, locate the fuel pump, and remove the cover.

19 With a suitable clamp, (a self-grip wrench with the jaws protected

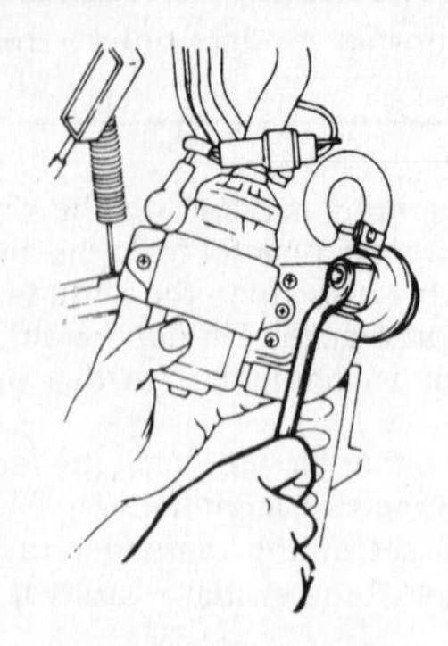

Fig. 3.32 Removing fuel damper (Sec 15)

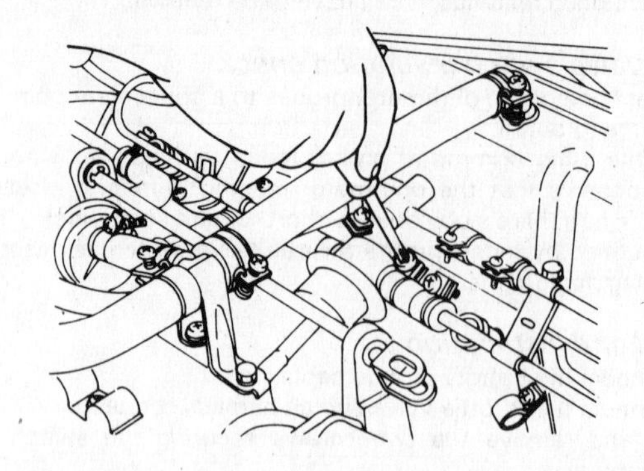

Fig. 3.33 Removing pressure regulator (Sec 16)

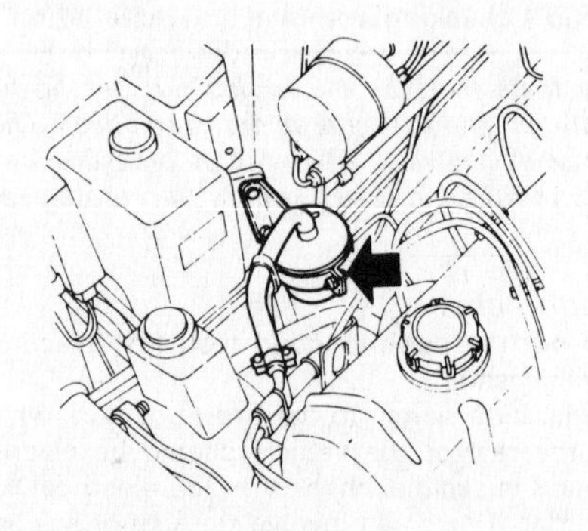

Fig. 3.34 The fuel filter securing bolt (Sec 17)

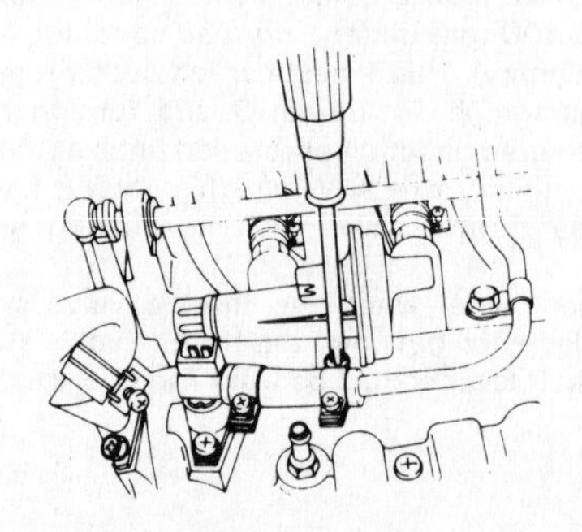

Fig. 3.35 Removing the air regulator (Sec 18)

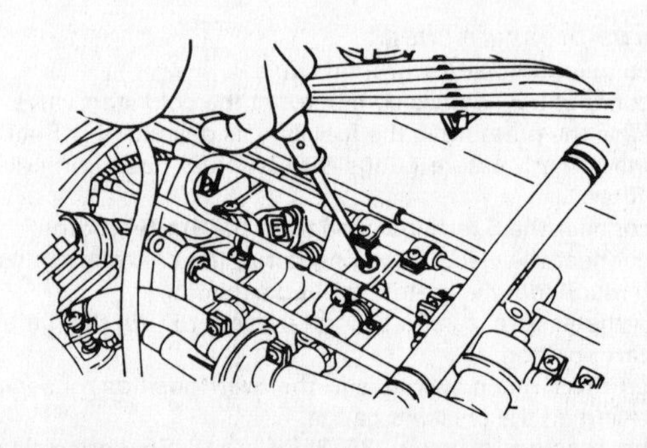

Fig. 3.36 Removing a clamp retaining screw which secures the front rigid fuel pipe (Sec 19)

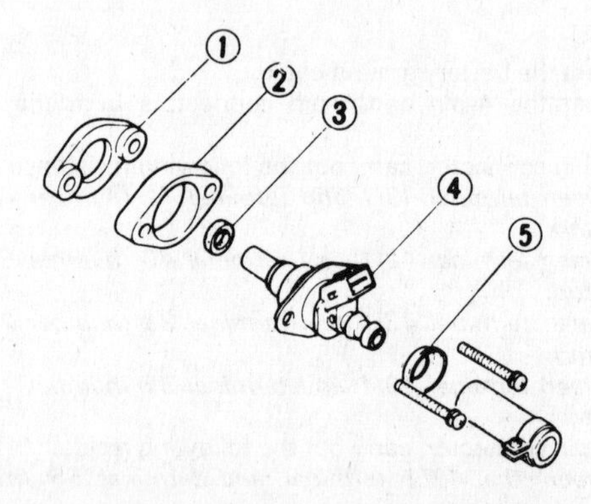

Fig. 3.37 Component parts of injector mounting (Sec 19)

1 Heat insulator
2 Injector holder
3 O-ring
4 Injector
5 Socket

will do), close the rubber fuel hose somewhere between the fuel tank and the fuel pump.

20 Unfasten the hose clamps at the suction and outlet sides of the fuel pump. Place a suitable clean container beneath the pump, to catch the inevitable fuel spillage. Pull off the two hoses. It is wise to seal the open ends of these pipes to prevent the ingress of dirt.

21 Undo and remove the two screws securing the fuel pump bracket; lift away the bracket, and remove the fuel pump.

22 When installing the fuel pump, reverse the removal procedure.

15 Fuel damper – checking, removal and installation

The only check that can be carried out on the fuel damper is a fuel pressure fluctuation check. This check is basically the same as the fuel discharge pressure check, described in Section 14.

Checking

1 Refer to Section 14, and carry out the operations described in paragraph 4 through 12.

2 If the reading at the pressure gauge fluctuates excessively, the fuel damper will have to be renewed as described in the following paragraphs.

Removal and installation

3 Disconnect the battery ground cable.

4 Disconnect the harness connector from the cold start valve.

5 Release the pressure in the fuel line, as described in Section 4.

6 Raise the rear end of the vehicle and safely support it with axle-stands or suitable wooden blocks. As an added safety precaution, firmly chock both front wheels.
7 Working inboard of the right-hand rear wheel, remove the fuel pump cover.
8 Using a suitable clamp, close the fuel suction hose between the fuel tank and the fuel pump.
9 Unfasten the fuel hose clamps at the inlet and outlet of the fuel damper. Pull off the fuel hoses.
10 Unscrew and remove the nuts securing the fuel damper to the mounting bracket and lift the damper away.
11 Installation of the fuel damper is a direct reversal of the removal procedure.

16 Pressure regulator – checking, removal and installation

Checking

1 Initially proceed as described in Section 14, paragraph 4 through 12 to check the fuel line pressure. If a pressure of 36·3 lbf/in^2 (2·55 kgf/cm^2) is not obtained, renew the pressure regulator as described in the following paragraphs.

Removal and installation

2 Disconnect the battery ground cable.
3 Disconnect the harness connector at the cold start valve.
4 Release the pressure in the fuel line, as described in Section 4.
5 Disengage the vacuum tube connecting the pressure regulator to the intake manifold.
6 Place a rag beneath the pressure regulator to catch any fuel spillage. Unfasten the hose clamp at each of the three connection pipes; disengage the three fuel hoses and remove the pressure regulator from the engine compartment.
7 When installing the pressure regulator, reverse the removal sequence.

17 Fuel filter – removal and installation

1 The fuel filter should be periodically renewed at the intervals recommended in the Routine Maintenance Section.
2 To do this, first disconnect the battery ground cable.
3 Release the fuel line pressure, as described in Section 4.
4 Unfasten the hose clamps at the outlet and inlet sides of the filter and pull off the fuel hoses. It is wise to place a rag beneath the filter body before removing the hoses, in order to prevent spilling fuel in the engine compartment.
5 Unscrew and remove the bolt securing the mounting bracket to the fuel filter. Remove the filter from the mounting bracket.
6 Installation of the fuel filter is a direct reversal of the removal procedure.

18 Air regulator – checking, removal and installation

Checking

1 As an initial check, with the engine just started and running at a very cold operating temperature, grasp the hose between the air regulator and the throttle chamber, and by squeezing it with your fingers, cut off the airflow through the hose. The engine idle should become very erratic and be noticeably reduced in idle speed.
2 As a further check, allow the engine to run until the normal operating temperature has been reached, and the underhood temperature is above 80°C (176°F). Repeat the check described in paragraph 1, and, because the by-pass air passage should now be closed, the engine idle speed should remain constant.
3 If the air regulator still proves to be suspect after carrying out the checks in paragraphs 1 and 2, it is possible to visually check to observe whether the air regulator valve is working.
4 With the engine stationary and at a cold temperature, disconnect the hoses from each end of the air regulator.
5 By looking through the air regulator it is possible to see the by-pass port open. If all is well, reconnect the hoses and run the engine until normal operating temperature is reached.
6 Now disconnect the hoses again as described in paragraph 4. This time the by-pass port should be completely closed.
7 If the checks previously mentioned prove unsatisfactory, disconnect the electrical connector at the air regulator and, using an ohmmeter, check that there is continuity in the air regulator. If there is no continuity, or the air regulator is proved to be inoperative in the checks described in paragraph 1 through 6, a replacement unit will have to be installed as described in the following paragraphs.

Removal and installation

8 Disconnect the battery ground cable.
9 Disconnect the electrical connector at the air regulator.
10 Unfasten the hose clamps and remove the air hoses from the air regulator.
11 Undo and remove the two screws securing the air regulator and remove it.
12 To install the air regulator is a direct reversal of the removal procedure.

19 Injectors – checking, removal and installation

Checking

1 If more than one injector is suspect, the control unit should be checked, as described in Section 6, to establish whether both power transistors are functioning. One of the transistors could have failed, rendering three of the injectors inoperative.
2 To trace a single faulty injector with the engine running, use a screwdriver as a stethoscope. Attach the blade of the screwdriver to the injector and hold the handle against the ear. It should be possible to hear an operational click every time the injector operates. Compare the noise of the click on the suspect injector with the other five injectors, if the noise of the operational click is noticeably less than at the other injectors, that particular injector is faulty and should be renewed.
3 If, for any reason, the engine will not run, disconnect the electrical connector from the cold start valve and, with the assistance of a second person operating the ignition switch to crank the engine, use the screwdriver method described in paragraph 2, to trace the faulty injector or injectors.
4 Once you have traced a possibly faulty injector, this can be proved by carrying out a continuity check. To do this, first disconnect the battery ground cable.
5 Disconnect the electrical connector from the suspect injector.
6 With the probes of an ohmmeter connected to the terminals on the injector, ensure that continuity exists. If there is no continuity, the solenoid coil windings could be open-circuit, or the terminal leads supplying the coil with current may be broken inside the injector. In such cases the only remedy is to renew the faulty injector. To do this, proceed as described in the following paragraphs.

Removal and installation

7 Disconnect the battery ground cable and release the pressure in the fuel line, as described in Section 4.
8 The rubber hoses that are attached to the injectors are sealed, so that no hose clamp is required; however, where the injector hoses attach to the rigid fuel supply pipes, hose clamps are used. Because of this, removal of one injector requires removal of the rigid supply pipe which feeds three injectors. The front rigid pipe, which supplies the front three injectors, is attached to the intake manifold by simple clamps, held in position by screws. The rear rigid pipe, which supplies the three rear injectors, is attached to the intake manifold by similar clamps, but these are retained by bolts. Once the necessary hoses and pipes have been removed, the faulty injector is quite simply removed by removing the two screws that secure it to the intake manifold. With the two screws removed, lift away the injector, the injector holder and the heat insulator.
9 Replacement injectors are available (with rubber hoses attached), and installation is a straightforward reversal of the removal procedure. Before turning the engine over, it is wise to check all the fuel joints to ensure that they are leakproof.

20 Throttle chamber – removal, checking and installation

1 Disconnect the battery ground cable.

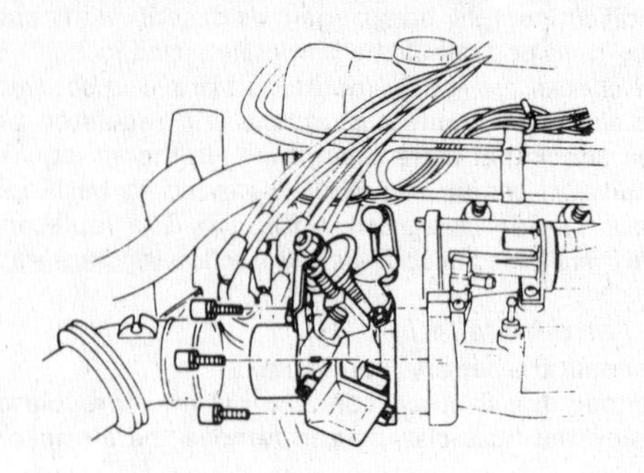

Fig. 3.38 Removing throttle chamber (Sec 20)

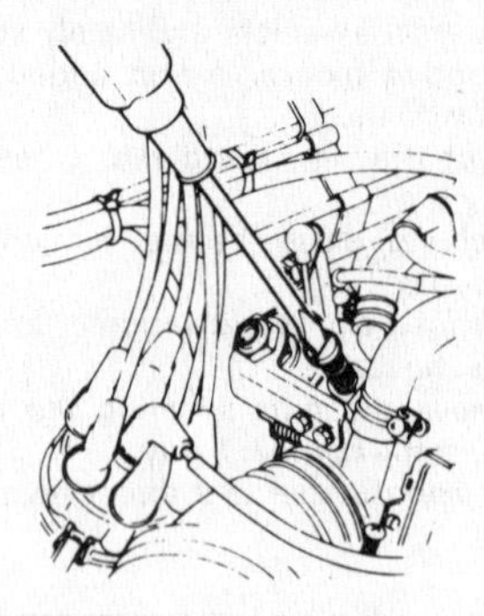

Fig. 3.39 Idle speed adjusting screw (Sec 21)

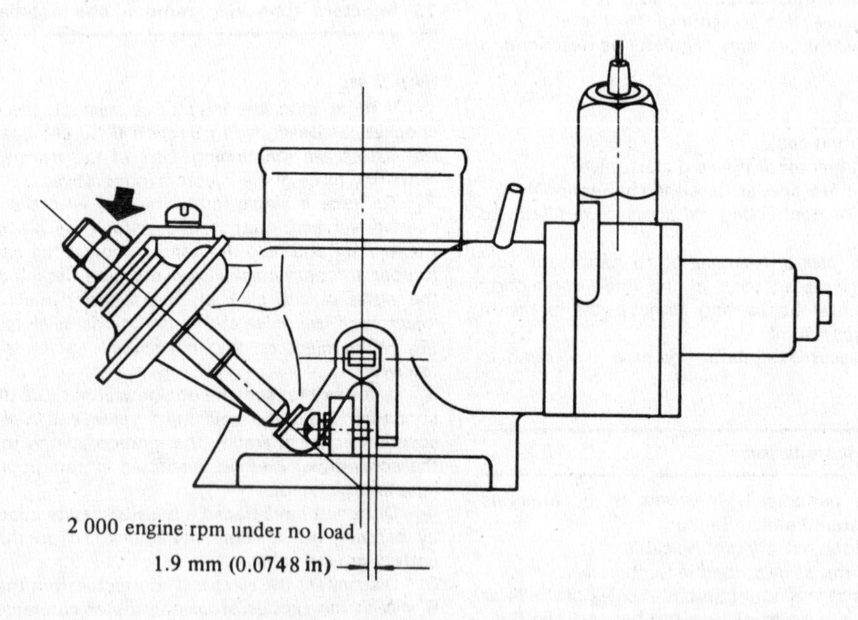

Fig. 3.40 Dashpot adjusting locknut arrowed (Sec 21)

2 To provide easier access, remove the distributor cap.
3 Remove the rubber hoses from the throttle chamber.
4 Refer to Section 11, and remove the throttle valve switch.
5 Disconnect the boost control deceleration device (BCDD) harness connector, which is underneath the throttle chamber.
6 Disconnect the rod connector at the auxiliary throttle shaft.
7 Undo and remove the four screws securing the throttle chamber to the intake manifold. The throttle chamber can now be removed, together with the BCDD and the dashpot, (manual transmission vehicles only are fitted with a dashpot).
8 With the throttle chamber removed from the engine, proceed to carry out the following checks.
9 Ensure that the throttle valve moves smoothly through its operational rotation. Make sure that the valve is undamaged around its sealing edge (slight nicks in the valve plate sealing edge will allow excess air to enter the intake manifold). Check for excessive side-play of the throttle shaft, which will also allow extra air into the intake manifold. If the throttle chamber assembly is obviously generally worn, it must be replaced.
10 Before installing the throttle chamber, wash it thoroughly in clean gasoline. Make sure that no gasoline is allowed to enter the solenoid valve on the BCDD unit. (The BCDD unit is part of the emission control system, and is discussed later in this Chapter).
11 Commence installation of the throttle chamber, by assemblying it to the intake manifold and securing it with the four screws.
12 Reconnect the rod connector at the auxiliary throttle shaft.
13 Connect the electrical harness connector to the BCDD unit.
14 Install the throttle valve switch; refer to Section 11, and carry out the throttle valve switch setting procedure.
15 Install the distributor cap.
16 Reconnect the battery ground cable.
17 Finally, start the engine, and if necessary adjust the idle rpm, as described in the next Section.

21 Idle mixture setting and dashpot adjustment

1 To ensure that the CO percentage contained in the exhaust gases is within the figures given in the Specifications, any adjustment of the fuel/air mixture will need to be carried out on a CO analyzer. If, after servicing a particular part of the fuel system, or if the fuel/air mixture is suspect, the vehicle should be taken to a Datsun dealer, who will have the necessary equipment to check the mixture settings and the CO content of the exhaust gas.
2 The fuel injection system is so designed to reduce the CO content of the exhaust gases and, if operating correctly, the percentage of CO will be so small that the analyzer will be unable to measure it. So, when using an analyzer, the throttle valve switch has to be short-circuited, in order to obtain a fuel/air mixture that is rich enough to give a reading on the analyzer. This alone is a complicated operation, and due to the stringent legislation regarding exhaust gas emissions, provided that the engine will run in a satisfactory manner, the job should be entrusted to your Datsun dealer.
3 If the engine will not run in a satisfactory manner, all that can be

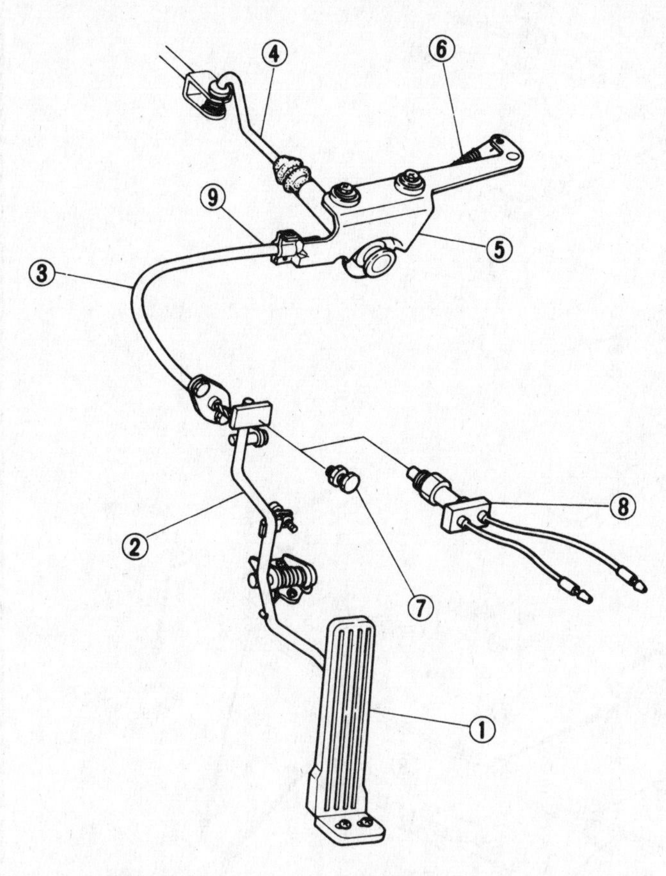

Fig. 3.41 Components of the accelerator linkage (Sec 22)

1 Accelerator pedal
2 Pedal arm
3 Accelerator cable
4 Torsion shaft
5 Torsion shaft support
6 Return spring
7 Stopper bolt B
8 Kickdown switch
9 Cable clip

done to correct it, is to turn the idle speed adjusting screw in or out, until a suitable engine idle speed has been obtained. Once a satisfactory engine speed has been obtained, the vehicle should be taken along to your Datsun dealer at the earliest possible convenient time, for a CO analyzer check.

Dashpot adjustment (manual transmission only)

4 Periodically check the clearance between the throttle valve stop screw and the throttle valve shaft lever in the following manner.
5 After warming up the engine to normal operating temperature, seek the assistance of a second person, to depress the gas pedal sufficiently to obtain a steady 2000 rpm.
6 Using feeler gauges, check the clearance between the throttle valve stop screw and the throttle valve shaft lever, which should be 0·0748 in (1·9 mm) at 2000 rpm.
7 If the clearance is not as specified, loosen the locknut at the dashpot mounting bracket and screw the dashpot in the required direction, until the correct clearance is obtained. The dashpot rod end must only be lightly touching the throttle valve shaft lever. Finally, when the correct adjustment has been obtained, tighten the locknut.

22 Accelerator linkage – adjustment, removal and installation

Adjustment

1 To adjust the accelerator linkage, remove the cotter pin at the torsion shaft connection within the engine compartment, and remove the return spring.

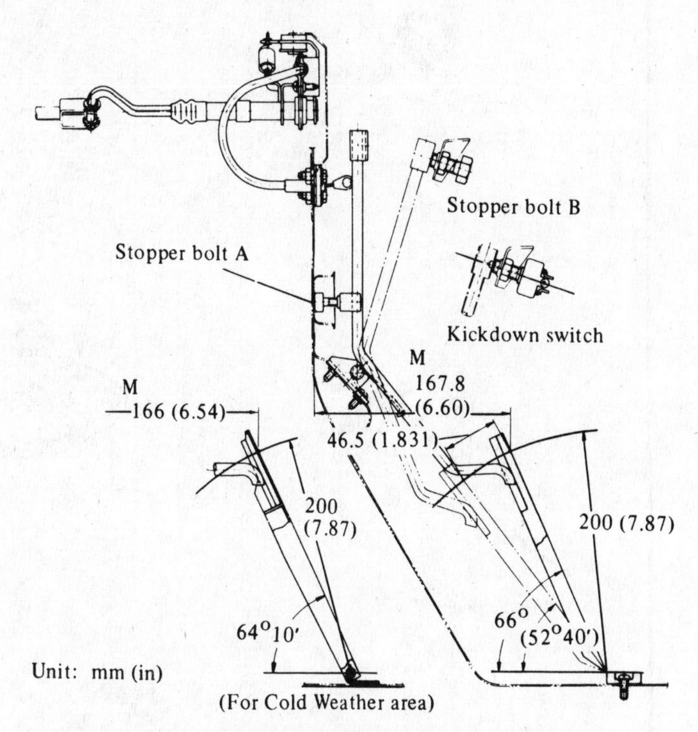

Fig. 3.42 Pedal height setting dimensions (Sec 22)

2 By adjusting the stop bolt A (Fig. 3.42), set the pedal height M.
3 Reconnect the torsion shaft linkage, leaving the return spring off for the time being, and loosen the cable clamp at the torsion shaft support.
4 With the torsion shaft held in the idling position, pull the outer case of the accelerator cable away from the torsion shaft support, until the torsion shaft just starts to move; at this point, return the outer cable by 0·040 in (1 mm) and tighten the cable clamp. Reconnect the return spring.
5 Press the accelerator pedal to the floor and, if the stop bolt A has been adjusted, adjust the stop bolt B to ensure that the full throttle position is being achieved.
6 Automatic transmission vehicles also employ a kickdown switch, which could need adjustment, to actuate when the accelerator is pressed to the floor.

Removal and installation

7 To remove the accelerator cable, remove the cotter pin at the torsion shaft connection and remove the torsion shaft from its support.
8 Undo and remove the two screws retaining the accelerator pedal and remove the pedal.
9 At the upper end of the pedal arm, remove the E-ring and pivot pin, and disengage the cable from the pedal arm.
10 Remove the three screws that secure the pedal arm pivot point and remove the pedal arm.
11 Detach the cable from the torsion shaft and remove it by pulling it into the passenger area.
12 Installation of the accelerator cable is a reversal of the removal procedure, but check and, if necessary adjust, the linkages as previously described.

23 Fuel tank – removal and installation

1 The fuel tank is either located in the rear trunk (Sedan), or under the rear of the vehicle (Station Wagon).
2 Disconnect the battery ground cable.

Sedan

3 Remove the trunk front finisher; then remove the spare tire. Place

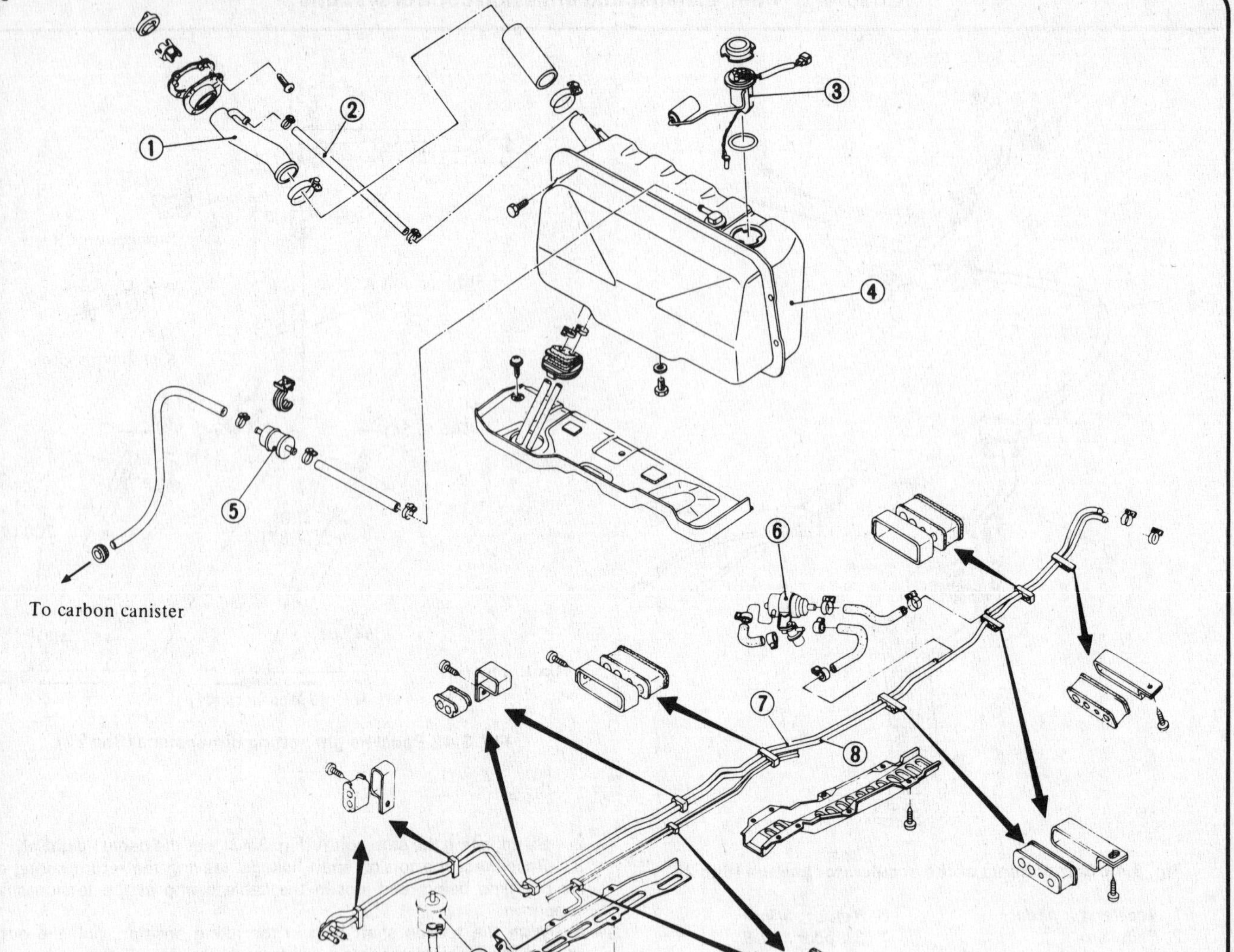

To carbon canister

Fuel piping

Sedan

Evaporation tube

Station Wagon

Fig. 3.43 Fuel tank (Sedan) and associated fuel lines (Sec 23)

1 *Filler hose*
2 *Ventilation hose*
3 *Fuel gauge unit*
4 *Fuel tank*
5 *Check valve*
6 *Fuel pump*
7 *Fuel outlet pipe*
8 *Fuel return pipe*
9 *Fuel filter*
10 *Carbon canister*

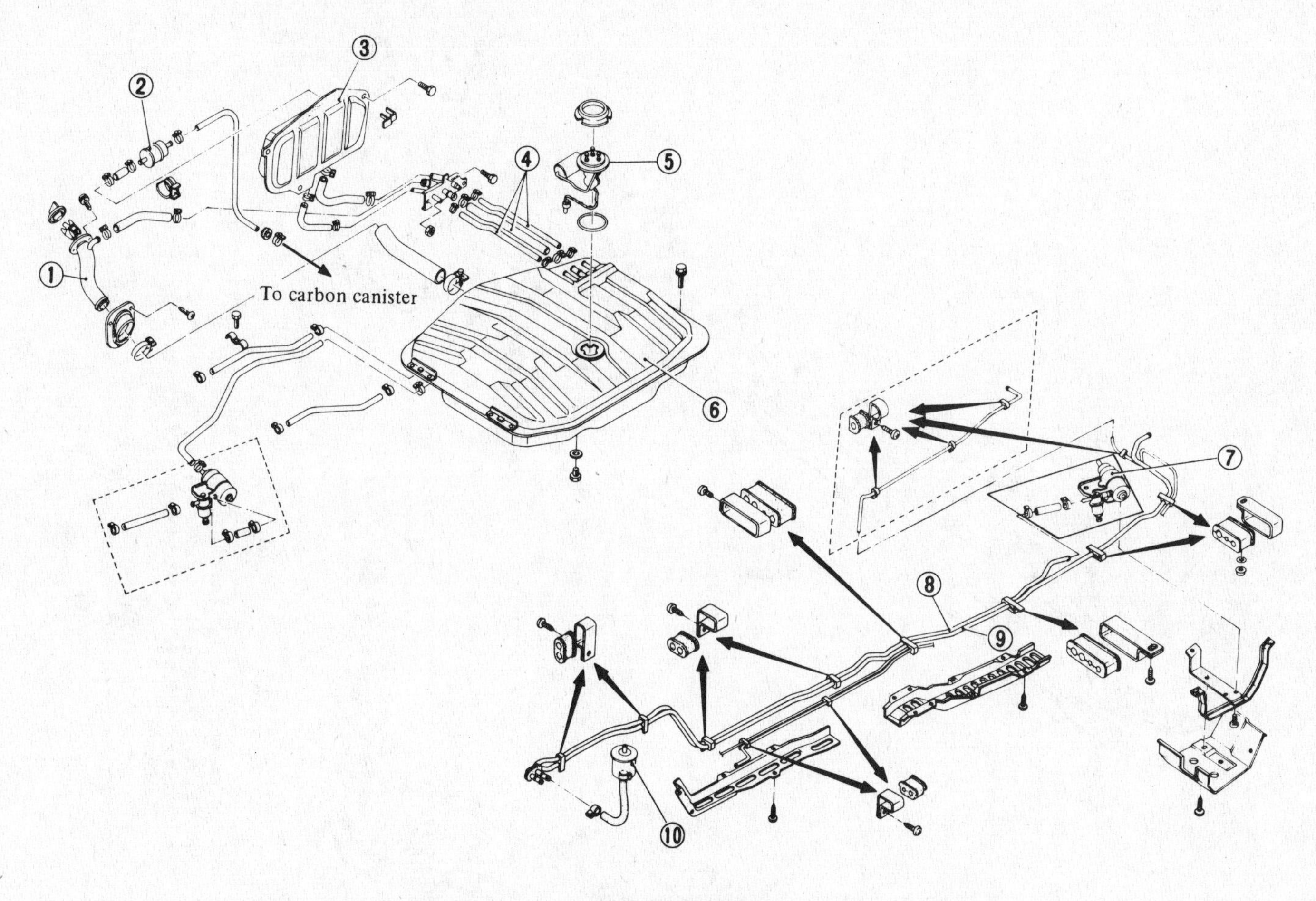

Fig. 3.44 Fuel tank and associated fuel lines (Station Wagon)

1 Filler hose
2 Check valve
3 Vapor/liquid separator
4 Ventilation hose
5 Fuel gauge unit
6 Fuel tank
7 Fuel pump
8 Fuel outlet
9 Fuel return pipe
10 Fuel filter

a suitable container beneath the drain plug in the bottom of the tank, and drain the fuel into it.
4 Disconnect the filler hose, ventilation hose and the outlet hose.
5 Disconnect the wires at the fuel tank gauge unit.
6 Undo and remove the four bolts securing the tank in place, and carefully remove the tank.

Station Wagon

7 Remove the spare tire; and whilst working beneath the rear of the vehicle, loosen and remove the drain plug in the bottom of the tank, and allow the fuel to drain into a suitable container.
8 Disconnect the filler hose, ventilation hose, evaporation hose and outlet hose.
9 Remove the tire stopper.
10 Disconnect the harness connector from the fuel tank gauge unit.
11 Undo and remove the four bolts attaching the fuel tank and remove it from the vehicle.

All vehicles

12 The tank gauge unit is of the bayonet type and it can be removed by turning the lockplate with a screwdriver in a counter-clockwise direction.
13 Do not be tempted to repair a fuel tank but leave it to professionals, or better still, purchase a new one.
14 Installation is a reversal of removal but check carefully the security of all the hose connections.

24 Exhaust system – removal and installation

1 There are considerable differences in the exhaust systems used on Datsun 810 models which are determined by the vehicle's operating territory.
2 Vehicles that operate in California are fitted with a catalytic converter and the entire system is encased with heat shields.
3 The main components used are a front exhaust tube, catalytic converter, center tube, main muffler assembly and a diffuser. The system is suspended by flexible hangers at the center and the rear. Both the Sedan and the Station Wagon use similar components, but, due to different pipe runs, there are differences.
4 Vehicles that operate in areas other than California, have a somewhat less complex system, and comprise a front exhaust tube with pre-muffler, and a main muffler assembly; the Sedan model also uses a center tube. The system is suspended at the center and rear by flexible hangers.
5 Examination of the exhaust tubes and mufflers at regular intervals is worthwhile, as small defects may be repairable; if left, they will almost certainly require renewal of one of the sections of the system. Also, any leaks, apart from the noise factor, may cause poisonous exhaust gases to get inside the car which can be unpleasant, to say the least, even in mild concentrations. Prolonged inhalation could cause sickness and giddiness.
6 As the sleeve connections and clamps are usually very difficult to

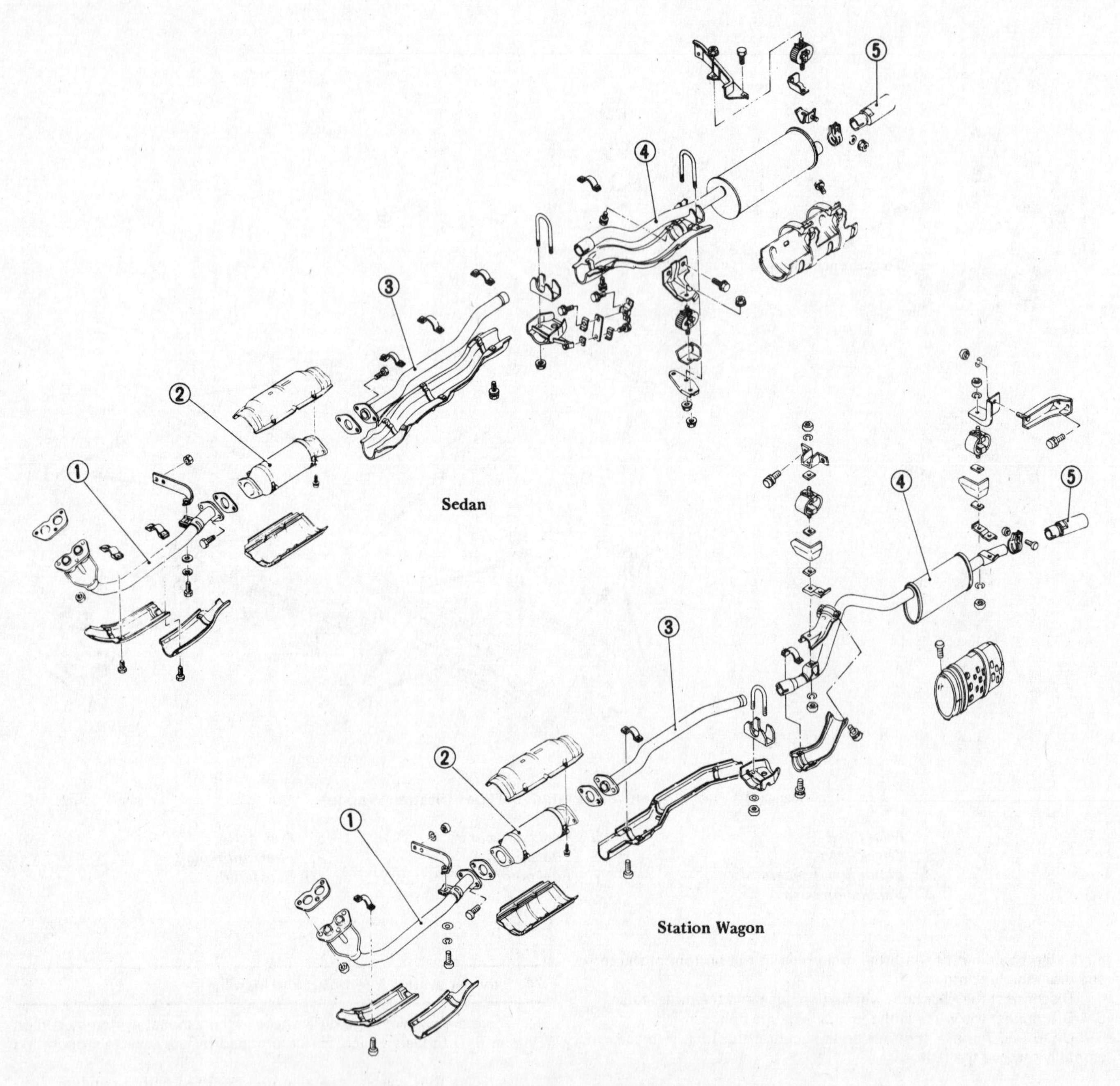

Fig. 3.45 Exhaust systems used on Californian vehicles (Sec 24)

1 Front exhaust tube
2 Catalytic converter
3 Center tube
4 Main muffler assembly
5 Diffuser

separate, it is quicker and easier in the long run to remove the complete system from the car when renewing a section. It can be expensive if another section is damaged when trying to separate another section from it. To remove the system, proceed as described in the following paragraphs.

Californian models

7 In order to make this job as easy as is possible, it is wise to drive the vehicle over an inspection pit, or raise the vehicle and support it firmly with suitable axle-stands. It is certain that assistance will be required to help with installation, to ensure correct suspension and alignment.

8 Undo and remove the nuts and bolts that retain the various heat shields in position.

9 At the main muffler clamp, remove the nuts and bolts that form the clamp and hanger assembly.

10 Working beneath the hood, disconnect the exhaust tube flange from the exhaust manifold, then remove the bolt from the rear engine mounting bracket. It should now be possible to lower the exhaust system at the front of the vehicle, but it will still be attached to the rear of the vehicle by the rear hanger. Once the rear hanger has been detached, the system can be withdrawn from beneath the vehicle.

11 The catalytic converter, which is located between the front exhaust tube and the center tube, can be removed by removing the flange bolts at the front and rear of the converter.

12 To separate the sleeve connections at either the center tube-to-

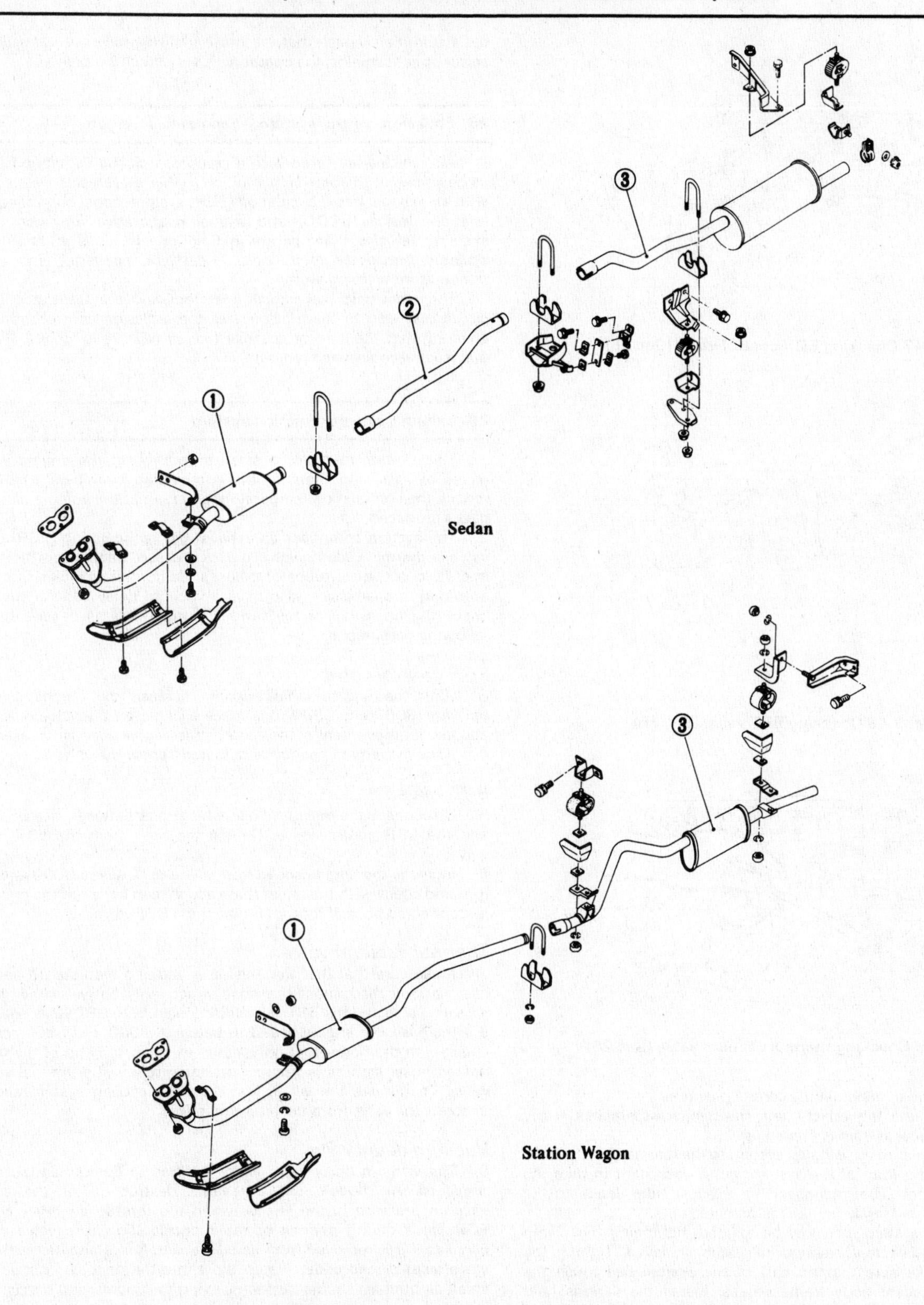

Fig. 3.46 Exhaust systems used on non-Californian vehicles (Sec 24)

1 Front exhaust tube with pre-muffler *2 Center tube* *3 Main muffler assembly*

main muffler assembly or the main muffler-to-diffuser, will involve light tapping with a hammer around the sleeve connection to break the sealant-type joint.

13 Carefully examine all components for signs of excessive rusting; renew any component that appears to have served its useful life. Renew any hangers that may have deteriorated; also check the clamps for excessive rusting, which may cause fractures and subsequent early failure.

14 To commence the installation procedure, first assemble the entire system before putting it beneath the vehicle. (All the sleeve connections should have been cleaned up and lightly greased to make assembly as easy as possible). Leave all the clamps loose so that each

Fig. 3.47 Checking EGR control valve (Sec 26)

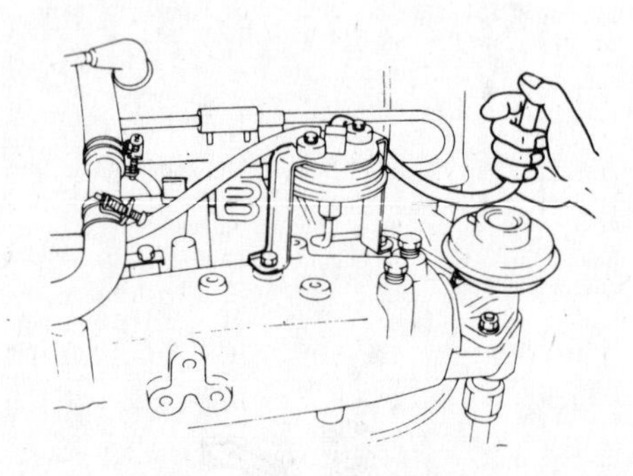
Fig. 3.48 Checking BPT valve (Sec 26)

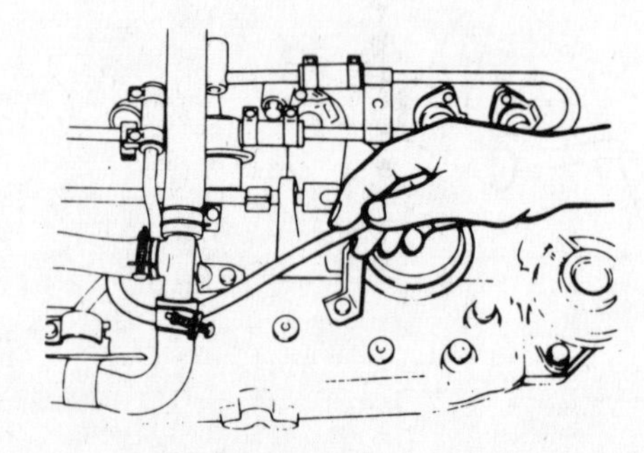
Fig. 3.49 Checking thermal vacuum valve (Sec 26)

joint can be rotated to maintain the correct pipe run.
15 Working beneath the vehicle, refit the hanger assemblies, ready for when the exhaust system is offered up.
16 Connect the rear of the exhaust system to the rear hanger.
17 Working at the front of the system, get an assistant to take the weight of the front tube, reconnect the exhaust tube flange to the exhaust manifold and firmly tighten the manifold nuts.
18 The exhaust system can now be aligned, tightening the clamp bolts when a satisfactory position has been achieved. Tighten the hangers and make sure that no part of the system can touch the driveshaft or adjacent body frame details. Install the various heat shields.
19 Finally, when the correct position has been obtained, exhaust sealant should be injected into each sleeve connection. This is carried out with a special injector tube, which forms part of an exhaust sealant kit; sealant should be injected into the sleeve connection until it begins to flow out of the slit of the tube.
20 To harden the sealant, start the engine and allow it to idle for ten minutes; the heat of the exhaust gases accelerates this process.

Non-Californian vehicles

21 The removal and installation procedure is basically the same as for the Californian vehicle but, as these vehicles have no catalytic converter or heat shields, the operation is very much simpler.

25 Emission control system – general description

All vehicles are fitted with a crankcase closed ventilation system as described in Chapter 1. In addition to this, all vehicles are also fitted with an exhaust gas recirculation (EGR) system, boost controlled deceleration device (BCDD) and a fuel evaporative emission control system. Vehicles that operate in California have, in addition to the systems previously mentioned, a catalytic converter and a floor temperature warning system.

The electronic fuel injection system and the transistor ignition circuit also add to the efficiency of the emission control system, by ensuring that the leanest possible fuel/air mixture is ignited at a very precise pre-determined moment.

26 Exhaust gas recirculation system

1 This system re-cycles a small proportion of the engine exhaust gases by returning them to the combustion chambers where they reduce the combustion temperature and restrict the volume of noxious gases produced.
2 The system comprises an exhaust gas recirculation (EGR) control valve, a thermal vacuum valve, a back pressure transducer (BPT) valve, and various control tubes and hoses. Californian vehicles fitted with automatic transmission also have a vacuum delay valve. Periodically check the operation of the various components as described in the following paragraphs.

EGR control valve

3 With the engine initially idling increase the engine speed to between 3000 and 3500 rpm. Place a finger on the valve diaphragm and feel for movement of the valve as the engine speed increases.
4 If the movement cannot be detected, renew the valve.

BPT valve

5 Disconnect the vacuum hose which runs between the BPT valve and the EGR control valve. Detach the hose from the EGR control valve.
6 Increase the engine speed from idling to between 3000 and 3500 rpm and check with the finger that vacuum can be felt at the end of the disconnected hose. If it cannot, renew the BPT valve.

Thermal vacuum valve

7 Make sure that that the engine is cold and then start it and let it idle. Quickly disconnect the hose which runs between the thermal vacuum valve and the BPT valve (disconnect from BPT valve end).
8 Increase the engine speed to between 3000 and 3500 rpm and check with the finger that no vacuum exists at the end of the disconnected hose. If there is vacuum pressure, renew the thermal vacuum valve. To remove the valve, first drain the cooling system and then unscrew the valve from the intake manifold.

Vacuum delay valve

9 The vacuum delay valve which is fitted to Californian (automatic transmission) models only, prevents destruction of the existing vacuum pressure in the line between the throttle chamber and the EGR valve during periods of rapid deceleration. The valve can be checked if it is removed from its connecting hoses and its nozzle (BPT valve side) placed under water. Blow into the opposing nozzle when small air bubbles should be seen leaving the submerged nozzle. If this is not the case, renew the valve. Take great care that the valve is installed so that its brown side is towards the thermal vacuum valve and the black side is towards the BPT valve.

27 Boost controlled deceleration device (BCDD)

1 This device is designed to reduce emission of hydrocarbons during periods when the car is coasting. The unit is installed under the throttle chamber and supplies extra air to the intake manifold in order to maintain the manifold vacuum at its correct operating pressure 19·29

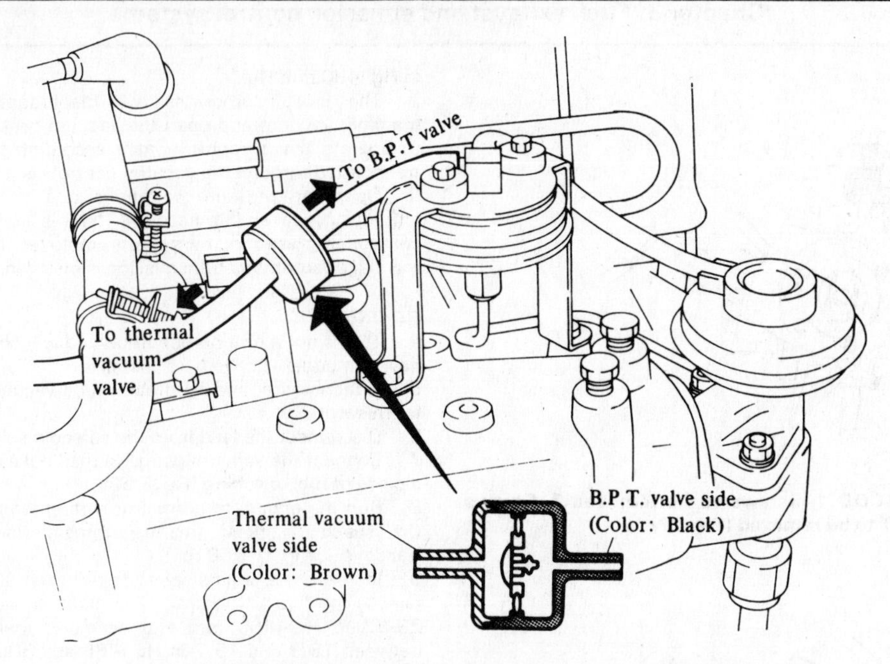

Fig. 3.50 Correct installation of vacuum delay valve (Sec 26)

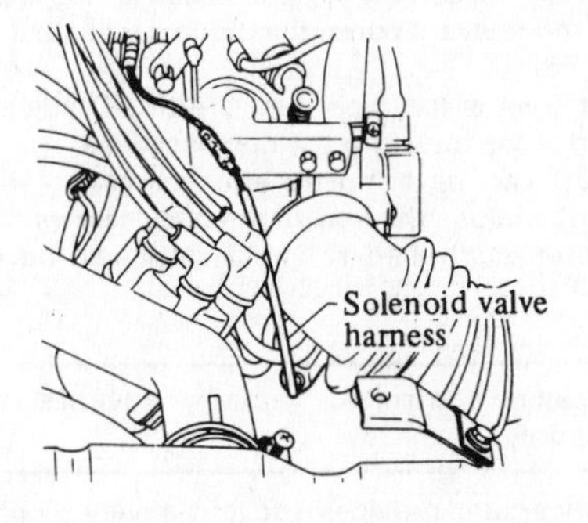

Fig. 3.51 Solenoid valve harness (BCDD)

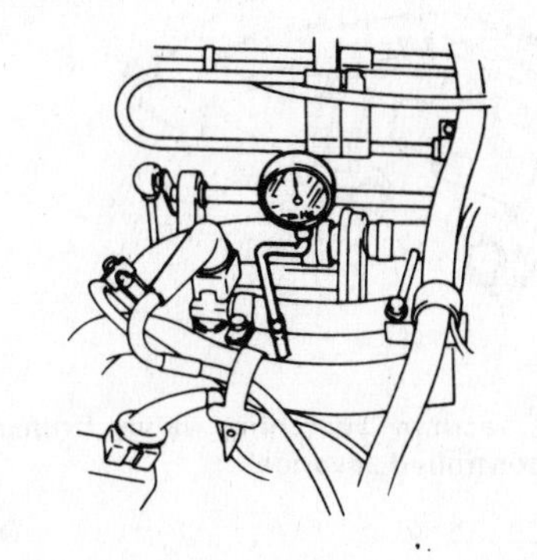

Fig. 3.52 Vacuum gauge connected for adjustment of BCDD

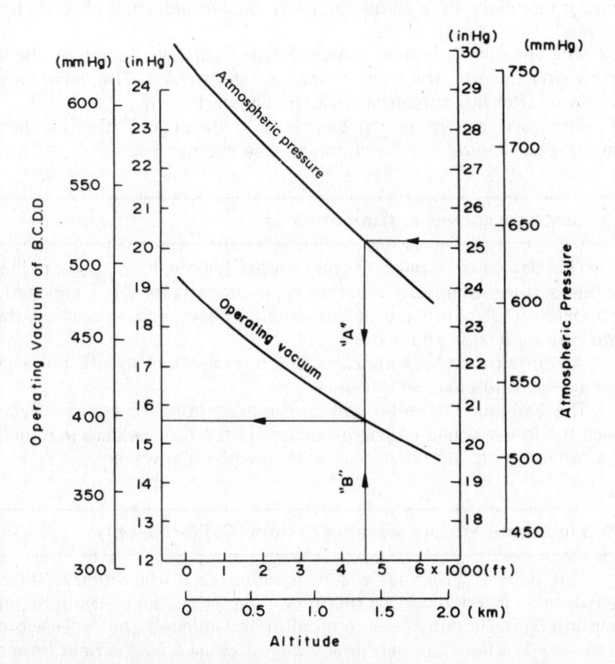

Fig. 3.53 BCDD setting altitude correction table. Example is for a vehicle at 4600 ft (1400 m) (Sec 27)

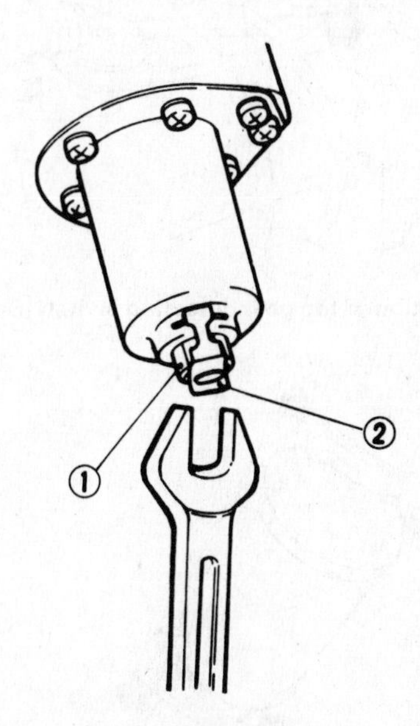

Fig. 3.54 Adjusting BCDD 1 Adjusting nut 2 Lockspring (Sec 27)

3

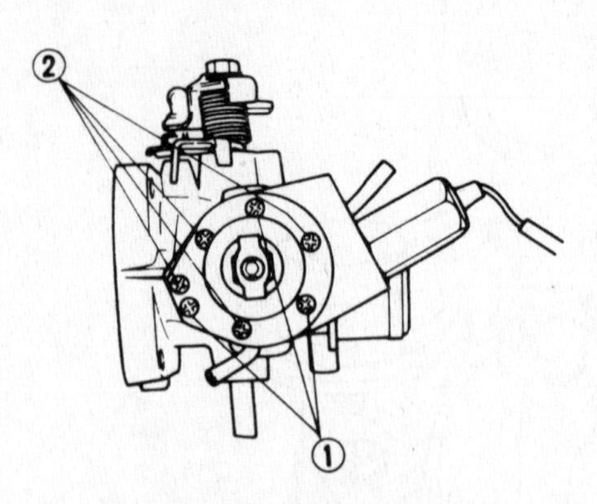

Fig. 3.55 Removing BCDD 1 Screws to be removed 2 Screws NOT to be removed (Sec 27)

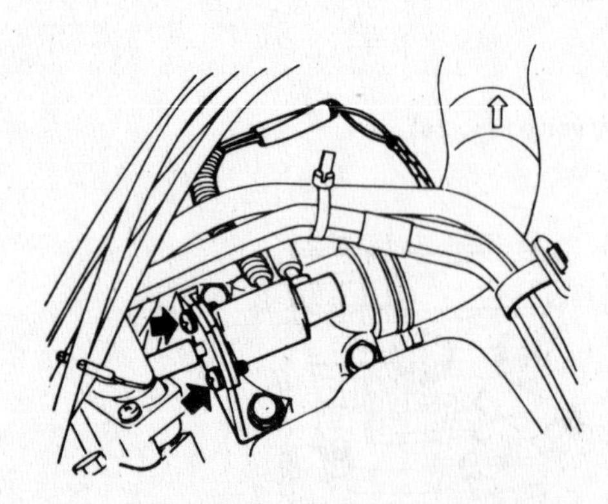
Fig. 3.56 Location of vacuum switching valve (transmission controlled advance)

Fig. 3.57 Location of top gear detecting switch (Sec 28)

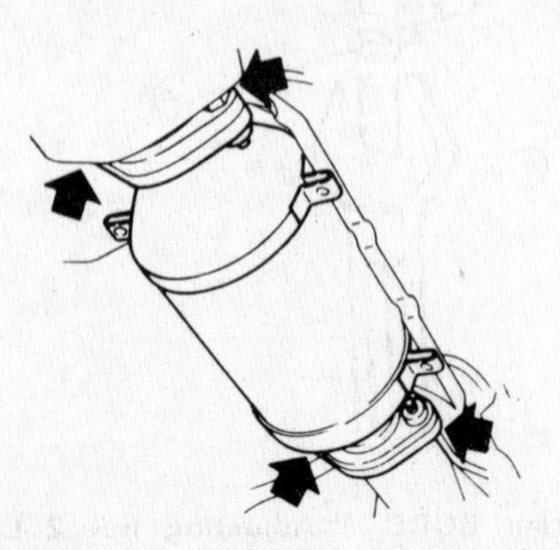
Fig. 3.58 Catalytic converter attachment points (Sec 29)

in Hg (490 mm Hg).

2 The unit incorporates two diaphragms, one to monitor the manifold vacuum and open the vacuum control valve, and the second to operate the air control valve according to the degree of vacuum transmitted through the vacuum control valve.

3 Operating pressure variations due to differences in operating altitudes are taken into account. There is a difference in the number of components used in the system employed, dependent upon whether manual or automatic transmission is installed.

Adjustment

4 This is not a routine operation and will only normally be required if new components have been installed.

5 A tachometer and a Bourdon tube vacuum gauge will be required for this work.

6 Disconnect the lead from the solenoid valve.

7 Connect the vacuum gauge to the intake manifold as shown, using a piece of rubber tubing (Fig. 3.52).

8 Run the engine to normal operating temperature and let it idle at the specified speed (manual transmission 700 rpm, automatic transmission in 'D' 650 rpm).

9 Increase the engine speed to between 3000 and 3500 rpm, then release the throttle abruptly. The manifold vacuum should increase to 23·62 in Hg (600 mm Hg) or more, then gradually decrease to between 18·9 and 19·7 in Hg (480 and 500 mm Hg). This is known as the specified set pressure (operating vacuum) at a sea level atmospheric pressure of 30 in Hg (760 mm Hg). The vacuum gauge reading will vary according to the local atmospheric pressure and altitude, and the relevant correction figures will need to be read off the graph (Fig. 3.53).

10 If the set level is too high, turn the BCDD adjusting nut counter-clockwise and, if too low, turn the nut clockwise.

11 The BCDD can be removed after extracting the three securing screws. Do not confuse these with the body screws.

12 The vacuum controlled solenoid valve can be removed using a spanner.

28 Transmission controlled vacuum advance system (manual transmission only)

1 This arrangement provides vacuum advance only when top gear has been selected. By means of a switch fitted in the transmission, a vacuum switching valve is opened to allow air into the distributor vacuum capsule, thus eliminating any vacuum advance which is being applied at that time.

2 The vacuum switching valve is located adjacent to and on the left-hand side towards the front of the camshaft cover. The valve can be removed after disconnecting leads and hoses from it.

3 The top gear detecting switch is screwed into the side of the gearbox just forward of the clutch release mechanism.

29 Catalytic converter (California only)

1 This device is installed in the exhaust system, its purpose being to accelerate the chemical reaction of hydrocarbons (HC) and carbon monoxide (CO) contained in the exhaust gases and to convert them into carbon dioxide and water.

2 A warning device is installed (see next Section) to indicate excessive temperature rises in this device.

3 The catalytic converter can be removed from the exhaust system once the lower shield has been removed from it. Take care in handling the converter and do not contaminate it with oil or water.

30 Floor temperature warning system (California only)

1 This device comprises a sensor mounted to the vehicles floor, a sensor relay fitted under the instrument panel and an instrument panel warning light. Its purpose is to monitor and indicate any excessive rise in the level of floor temperature, which is caused by the heat from the exhaust system and catalytic converter, due to an engine fault, or to unusually severe operating conditions.

2 To remove the floor-mounted sensor, open the trunk, remove the protective cover from the sensor, disconnect the harness connector

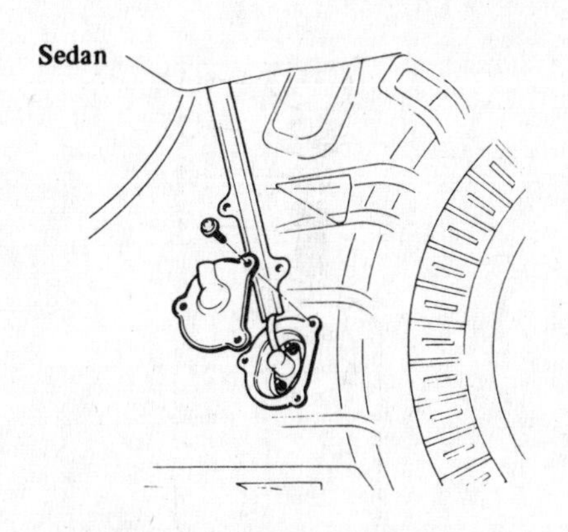

Fig. 3.59 Floor temperature warning sensor (Sedan illustrated) (Sec 30)

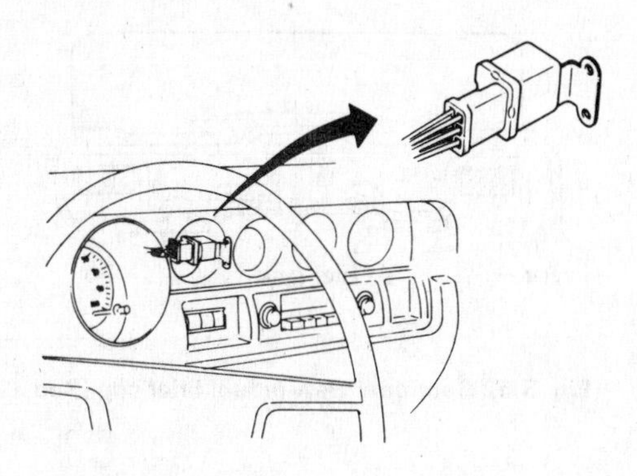

Fig. 3.60 Location of floor temperature warning system relay (Sec 30)

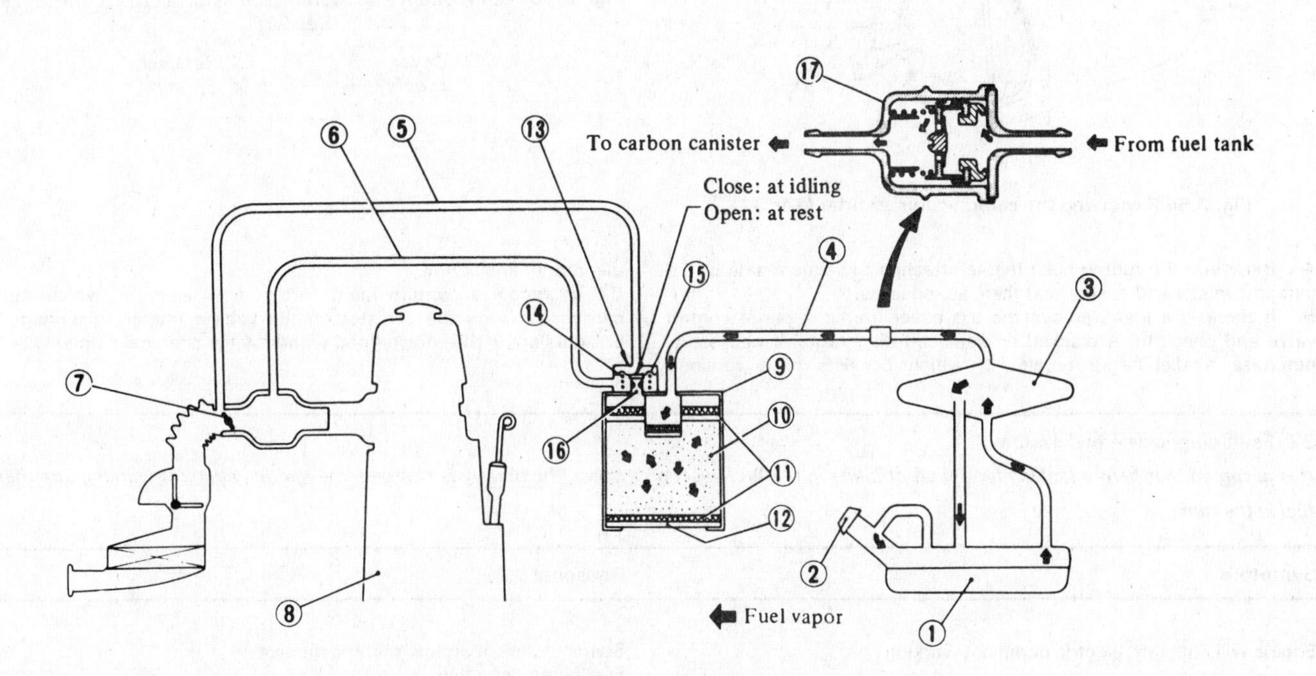

Fig. 3.61 Layout of fuel evaporative emission system (Sec 31)

1 Fuel tank
2 Filler cap and relief valve
3 Separator (Station Wagon only)
4 Vapor vent line
5 Vacuum signal line
6 Canister purge line
7 Throttle valve
8 Engine
9 Carbon canister
10 Carbon element
11 Screen
12 Filter
13 Purge control valve
14 Diaphragm spring
15 Diaphragm
16 Fixed orifice
17 Check valve

and remove the sensor from the floor. On Station Wagons the rear seat cushion will have to be removed to gain access to the sensor.

3 The relay and the warning light are both accessible after the cluster lid has been removed; for details of the cluster lid, refer to Chapter 10.

31 Fuel evaporative emission control system

1 This system comprises a fuel tank with a positive sealing filler cap, a vapor liquid separator (Station Wagon only), a check valve, a vapor vent line, carbon canister, vacuum signal line and a canister purge line.

2 In operation, fuel vapors from the sealed fuel tank are led into the carbon canister. On the Station Wagon, fuel vapors are channelled into the vapor liquid separator, to be returned to the fuel tank, before they are fed to the canister. The canister is filled with charcoal to absorb fuel vapors when the engine is idling or at rest. As the throttle valve opens, vacuum pressure in the vacuum signal line forces the purge control valve to open, to admit fuel to the intake manifold through the canister purge line. Testing the evaporative emission control system requires the use of a manometer. As this type of equipment is not usually found amongst the belongings of a DIY motorist, this Section is limited to supplying information on the carbon canister. If a leak is suspected in the evaporative fuel lines, take the vehicle to a Datsun dealer to check it out.

3 To check the carbon canister purge control valve, disconnect the rubber hose in the line between the T-connector and the carbon canister (at the T-connector).

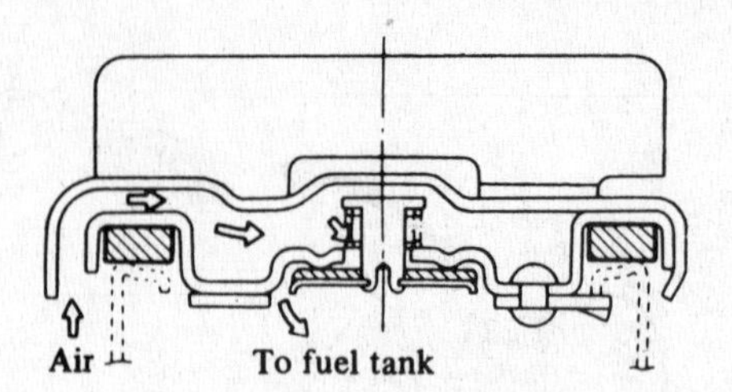

Fig. 3.62 Sectional view of fuel filler cap (Sec 31)

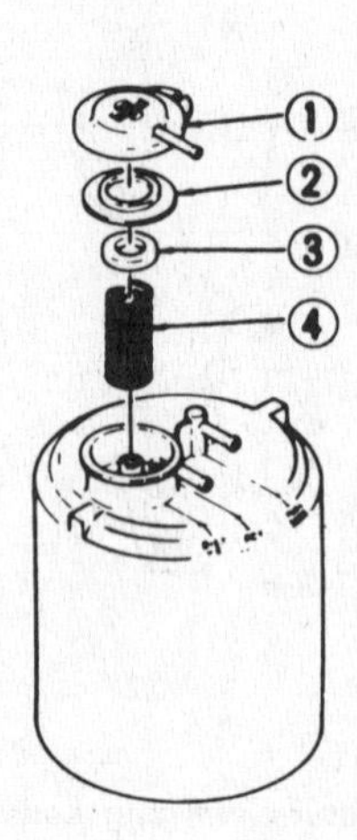

Fig. 3.63 Components of carbon canister purge control valve (Sec 31)

1 Cover
2 Diaphragm
3 Retainer
4 Spring

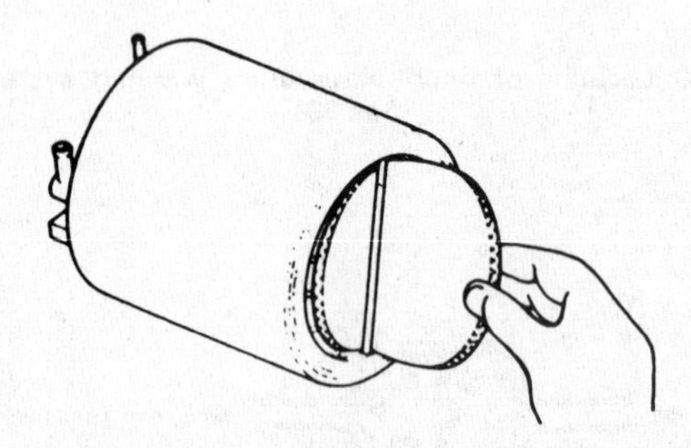

Fig. 3.64 Renewing the carbon canister filter (Sec 31)

4 Blow into the rubber hose that is attached to vacuum side of the carbon canister and ensure that there are no leaks.
5 If there is a leak, remove the top cover from the purge control valve and check for a cracked or displaced diaphragm. If necessary, purchase a diaphragm repair kit, which consists of a retainer, diaphragm and spring.
6 To renew a contaminated carbon filter element, which can be carried out with the canister on the vehicle, merely disengage the bottom flap on the canister and withdraw the charcoal element.

32 Fault diagnosis – fuel system

It is assumed that before looking for a fault or faults in the fuel injection system, the battery is charged, the ignition circuit is working and there is fuel in the tank.

Symptom	Reason/s
Engine will not start, electric pump not working	Switch points in airflow meter defective Fuel pump defective Circuit to pump broken
Engine will not start, electric pump operative	Engine flooded with fuel
Engine starts but then stalls	Cold start valve defective Water temperature sensor defective Thermotime switch defective Pressure regulator defective, causing low fuel line pressure Airflow meter sending incorrect signals to the control unit* Throttle chamber valve or throttle valve switch sticking
Engine misfires on one or more cylinders	Sticking injector Power transistor in control unit defective
Engine 'hunts' excessively at idling speed	Throttle valve switch points on full throttle position Faulty injector(s) Air regulator defective Water temperature sensor resistance defective Faulty altitude switch (California) Cold start valve defective Fuel line pressure low Airflow meter defective Control unit giving incorrect signals*

Lack of power	Faulty injector(s) Airflow meter flap sticking causing incorrect resistance Throttle valve switch defective Altitude switch (California) defective Cold start valve defective Low fuel pressure
Abnormal fuel consumption	Cold start valve defective Control unit defective* Airflow meter potentiometer resistance incorrect Air temperature sensor resistance incorrect Throttle valve switch faulty Fuel line leakage

When it is necessary to check the control unit for faults other than faulty power transistors (described in Section 6), special equipment will be needed, which only a Datsun dealer or a competent fuel injection specialist will have.

Chapter 4 Ignition system

Refer to Chapter 13 for specifications and information applicable to 1980 through 1984 models.

Contents

Specifications

System type	12 volt, battery, coil and transistor unit
Distributor	
Distributor type:	
Manual transmission (California models)	D6F5–02
Manual transmission (non-California models)	D6F6–01
Automatic transmission (all areas)	D6F4–03
Firing order	1 – 5 – 3 – 6 – 2 – 4
Rotational direction	Counter-clockwise
Air-gap	0·008 to 0·016 in (0·2 to 0·4 mm)
Cap carbon brush length	0·39 in (10 mm)
Ignition timing (BTDC) at idle speed	
Manual transmission	10° at 700 rpm (California models) or 8° at 700 rpm (non-California models)
Automatic transmission (selector in 'D' position)	10° at 650 rpm (California models) or 8° at 650 rpm (non-California models)
Centrifugal advance	
All models	0° at 600 rpm or 8·5 at 1250 rpm
Ignition coil	
Type	Hitachi CIT–17 or STC–10 (1977 models) CIT–30 or STC–30 (1978 models)
Primary resistance at 20°C (68°F)	0·45 at 0·55 ohms (1977 models) 0·84 to 1·02 ohms (1978 models)
Secondary resistance at 20°C (68°F)	8·5 to 12·7 k ohms
External resistor resistance at 20°C (68°F)	1·15 to 1·45 ohms (1977 models) 8·2 to 12·4 ohms (1978 models)
Spark plugs	
Type:	
Hot	B5ES–11 or L46W–11
Cold	B7ES–11 or L44W–11
Standard	B6ES–11 or L45W–11
Electrode gap	0·039 to 0·043 in (1·0 to 1·1 mm)

Torque wrench settings	**lbf ft**	**kgf m**
Spark plug	11 to 14	1·5 to 2·0

1 General description

In order that the engine can run correctly, it is necessary for an electrical spark to ignite the fuel/air mixture in the combustion chamber at exactly the right moment, in relation to engine speed and load. The ignition system is based on feeding low tension (LT) voltage from the battery to the ignition coil, where it is converted to high tension (HT) voltage. The high tension voltage is powerful enough to jump the spark plug gap in the cylinders many times a second under high compression pressures, provided that the system is in good condition and that all adjustments are correct.

The distributor employed on the Datsun 810 is of the contactless type, and is equipped with a reluctor and a pick-up coil. The pick-up coil electrically detects the ignition timing signal, which is sent from the transistor unit, obviating the need for contact breaker points.

The low tension (sometimes known as the primary) circuit consists of the ignition switch, the primary winding of the ignition coil, the transistor ignition unit and the inter-connecting low tension wires. The high tension (or secondary) circuit consists of the high tension or

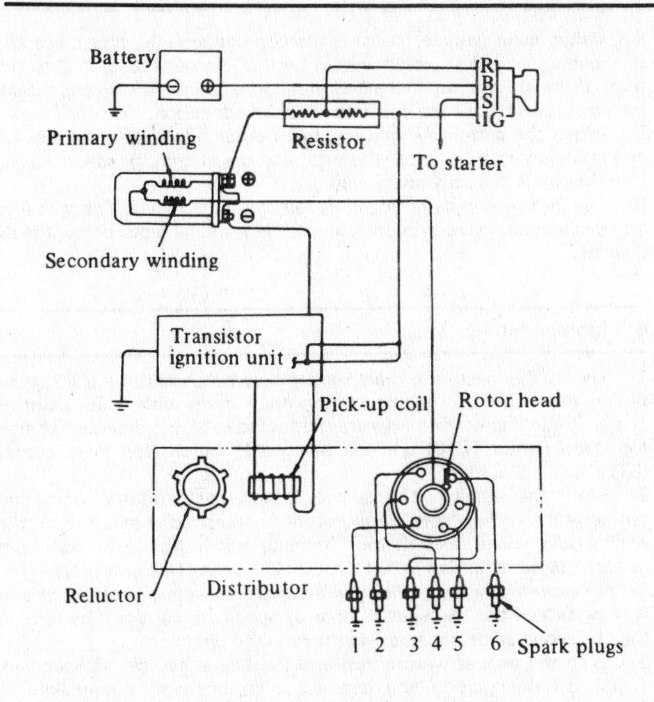

Fig. 4.1 The ignition circuit

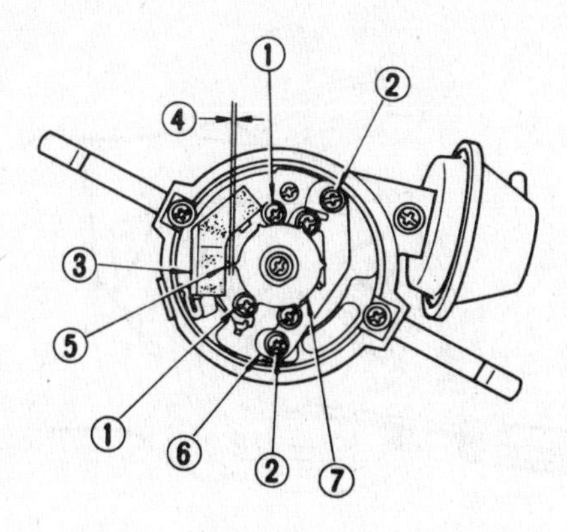

Fig. 4.2 Adjusting the air-gap (Sec 3)

1 Pick-up coil set screws
2 Adjuster plate set screws
3 Pick-up coil
4 Air-gap
5 Pole piece
6 Adjuster plate
7 Reluctor

secondary winding of the ignition coil, the heavy ignition lead from the center of the coil to the center of the distributor cap, the rotor, and the spark plug leads and spark plugs. The system functions in the following manner: When the ignition switch is turned on and the distributor reluctor rotates, current flows through the primary winding of the coil and through the transistor ignition unit to ground. When the primary circuit is opened by the transistor unit, the magnetic field built up in the primary winding collapses and induces a very high voltage in the secondary winding. The high voltage current then flows from the coil, along the heavy ignition lead, to the carbon brush in the distributor cap. From the carbon brush, current flows to the distributor rotor which distributes the current to one of the terminals in the distributor cap. The spark occurs while the high tension voltage jumps across the spark plug gap. This process is repeated for each power stroke of the engine. The ignition is advanced and retarded automatically, to ensure that the spark occurs at just the right instant for the particular load at the prevailing engine speed.

The ignition advance is controlled both mechanically and by a vacuum operated system. The mechanical governor mechanism comprises two weights, which move out from the distributor shaft, and so advance the spark. The weights are held in position by two light springs and it is the tension of the springs which is largely responsible for correct spark advancement. The vacuum control consists of a diaphragm, one side of which is connected via a small bore tube to the intake manifold, and the other side to the breaker plate assembly. Depression in the intake manifold, which varies with engine speed and throttle opening, causes the diaphragm to move, so moving the breaker plate assembly, and advancing or retarding the spark. A very fine degree of control is achieved by a spring in the vacuum assembly.

Note: *Never disconnect any of the ignition HT leads when the engine is running, or the transistor ignition unit will be permanently damaged.*

2 Routine maintenance

1 The ignition system is one of the most important and most neglected systems in any vehicle, and attention to routine maintenance cannot be over-emphasized. The maintenance intervals are given in the Routine Maintenance Section at the beginning of this Manual.

Spark plugs

2 Remove the plugs and thoroughly clean away all traces of carbon. Examine the porcelain insulation round the central electrode inside the plug, and if damaged discard the plug. Reset the gap between the electrodes. Do not use a set of plugs for more than 8000 miles; it is false economy. For further information on spark plugs, see Section 11.

Air-gap

3 Remove the distributor cap and rotor, and adjust the air-gap as described in Section 3.

Distributor cap, rotor and HT leads

4 Remove the distributor cap and HT leads from the ignition coil and spark plugs. Wipe the end of the coil, the HT leads and spark plug caps, and the internal and external surfaces of the distributor cap with a lint-free cloth moistened with gasoline or a cleaning solvent. Ensure that all traces of dirt and oil are removed, then carefully inspect for cracked insulation on the leads, spark plug caps, distributor cap and ignition coil end. At the same time, check that the carbon brush in the center of the distributor cap is intact and returns under the action of its spring when pressed in. Carefully scrape any deposits from the distributor cap electrodes and from the rotor; if any serious erosion has occurred, replacement parts should be installed.

Lubrication

5 Remove the distributor cap and pull off the rotor from the end of the rotor shaft. Pry out the rubber cap and, if necessary, add a little general purpose molybdenum disulphide grease.

Ignition timing

6 After completing the aforementioned, check and adjust the ignition timing as described in Section 4.

3 Air-gap – adjustment

To ensure that the ignition system functions correctly, the air-gap (distance between the pick-up coil and reluctor) must be maintained as specified. To do this, proceed as follows:

1 Disengage the two spring retaining clips and remove the cap from the distributor.

2 Remove the rotor from the end of the distributor shaft.

3 Position one of the raised segments of the reluctor directly opposite the pole piece protruding from the pick-up coil. This is best carried out by removing the spark plugs (to relieve compression) and rotating the engine by pulling on the alternator drivebelt.

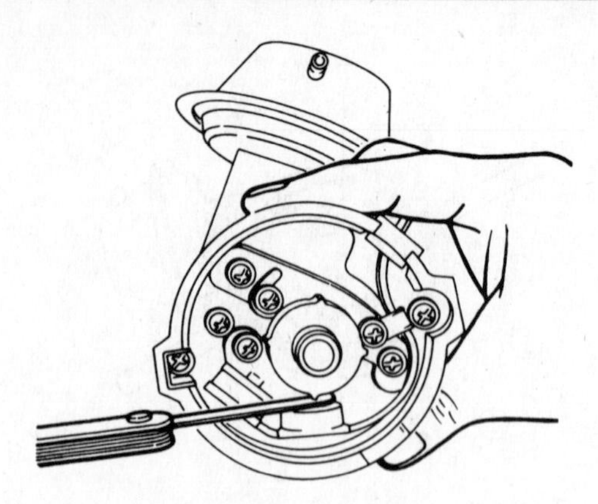

Fig. 4.3 Measuring the air-gap (Sec 3)

Fig. 4.4 The primary leads at the terminal block (Sec 3)

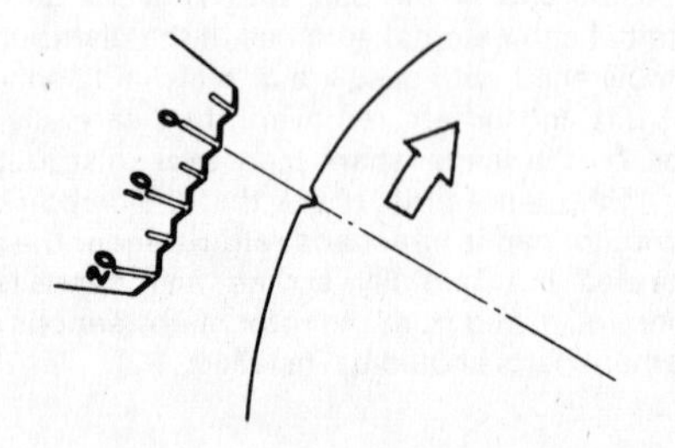

Fig. 4.5 The timing marks on the engine front cover and the crankshaft pulley notch (Sec 4)

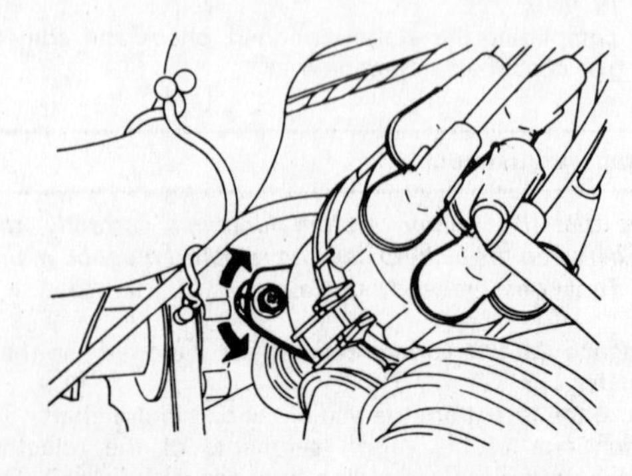

Fig. 4.6 The distributor clamp-plate screw (Sec 4)

4 Using feeler gauges, measure the gap between the pole piece and the reluctor segment which should be 0.008 to 0.016 in (0.2 to 0.4 mm). If the air-gap requires adjustment, loosen the pick-up coil retaining screws and move the coil in the required direction.
5 When the correct air-gap has been obtained, tighten the pick-up coil retaining screws; install the rotor and the distributor cap, ensuring that the cap is correctly positioned.
6 The pick-up can be removed by removing the two retaining screws and disconnecting the primary leads at the terminal block below the air cleaner.

4 Ignition timing

1 Thoroughly clean the crankshaft pulley notch and the front cover timing marks and, to ensure clarity, mark them with white paint or chalk. The timing cover marks are graduated in 5° increments, O being top dead center (TDC) and 20 being 20° before top dead center (BTDC).
2 Start the engine and allow it to reach normal operating temperature. When this temperature is reached, ensure that the engine idle speed is 700 rpm (manual transmission) or 650 rpm (automatic transmission in 'D'). **Note**: *When the selector lever is in the 'D' position, firmly apply the parking brake and chock both front and rear wheels.* If the idle speed is not as specified, correct it by turning the idle speed adjusting screw in the required direction.
3 With the engine warmed up and the idle speed set as specified, switch off the ignition and connect a stroboscopic timing light in accordance with the makers instructions.
4 Start the engine and allow it to idle. Point the timing light at the ignition timing marks; they will appear stationary and in alignment if the ignition timing is correct. If the marks are not in alignment, loosen the distributor clamp-plate screw and, very slowly, turn the distributor until the correct timing is achieved. Rotating the distributor body counter-clockwise will advance the ignition timing and rotating it clockwise will retard the timing.
5 Finally, when a satisfactory adjustment has been obtained, switch off the ignition, tighten the distributor clamp-plate screw and remove the timing light connections.

5 Distributor – removal and installation

1 To remove the distributor from the engine, begin by identifying the spark plug leads and pulling them from their respective spark plugs.
2 In order to aid reassembly, rotate the crankshaft until number 1 cylinder piston is on the firing stroke. This can be carried out by setting the engine to TDC using the timing marks described in Section 4. To check that number 1 piston is on the firing stroke, remove the distributor cap and observe the position of the rotor, which should be pointing towards the number 1 spark plug lead.
3 Remove the HT lead from the center of the ignition coil by undoing the lead retaining cap.
4 Pull off the small bore vacuum pipe to the advance and retard capsule.
5 Disconnect the primary lead wires at the terminal block (see Section 3).
6 Undo and remove the distributor clamp-plate screw and lift out the distributor.
7 Installation of the distributor is essentially the reverse of the removal procedure but if the engine has been rotated since the distributor was removed, refer to Chapter 1, Section 15, for further information. When installing the HT lead between the ignition coil and the distributor, ensure that the maximum possible clearance exists between the HT lead and the fuel injection harness. If too close, HT current can upset the signals passing through the harness. After the distributor has been installed check and adjust the ignition timing as described in Section 4.

6 Distributor – dismantling

1 Remove the distributor cap and pull off the rotor.
2 Remove the two screws which secure the vacuum unit, then tilt it slightly to disengage the operating rod from the baseplate pivot.
3 Unscrew and remove the screws which hold the pick-up coil and

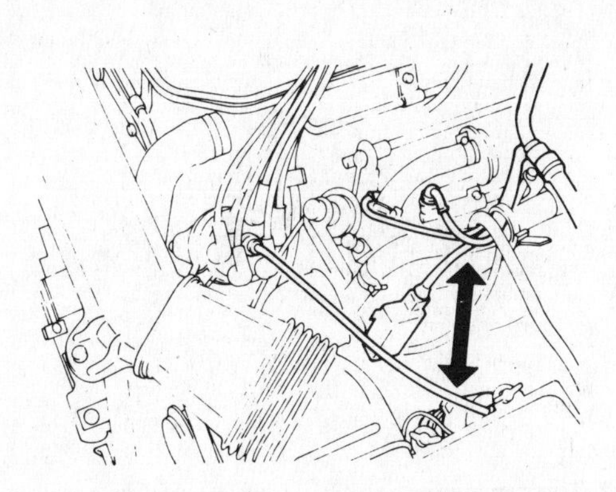

Fig. 4.7 Maximum clearance is required between the ignition coil HT lead and the fuel system wiring harness

remove the pick-up coil.

4 Using two suitable small pry bars, carefully pry the reluctor from the distributor driveshaft. Take particular care not to pry under the teeth of the reluctor, they are very easily broken.

5 Now that the reluctor has been removed, pull out the roll pin.

6 Unscrew and remove the screws which secure the baseplate and lift off the baseplate.

7 Drive out the pin from the lower end of the shaft; remove the collar and then withdraw the distributor driveshaft, together with the counterweight assembly, from the distributor body.

8 If the counterweights and springs are to be dismantled, proceed as described in the following paragraphs taking great care not to stretch the springs.

9 Mark the relative position of the distributor driveshaft and the rotor shaft; also mark the positions of the springs and their attaching brackets. Where the governor weights pivot between the rotor shaft and the driveshaft, take particular note of their pivot point positions.

10 Remove the packing from the top of the rotor shaft and unscrew and remove the rotor shaft-to-driveshaft retaining screw.

11 Very carefully unhook and remove the governor springs.

12 Remove the governor weights.

7 Distributor – inspection and repair

1 Check the distributor cap for signs of tracking, indicated by a thin black line between the segments. Renew the cap if any signs of tracking are found.

2 If the metal portion of the rotor is badly burnt or loose, renew the rotor. If slightly burnt clean the arm with fine abrasive paper.

3 Check that the carbon brush moves freely in the center of the distributor cap.

4 Examine the governor weights and pivot pins for wear, and renew the weights or rotor shaft/driveshaft if a degree of wear is found.

5 Place the driveshaft into the rotor shaft and check for excessive side movement. Renew parts as necessary.

6 If the driveshaft is a loose fit in the distributor bush and can be seen to be worn, it will be necessary to fit a new shaft and bush.

7 Examine the length of the governor weight springs, if possible compare them with new ones. If they have stretched they must be renewed.

8 Distributor – reassembly

1 If the counterweights and springs have been dismantled, apply a little grease to the weights and reassemble them to the pivot pins on the driveshaft. Carefully install the springs to their correct attachment points, whilst locating the rotor shaft to the driveshaft.

2 Secure the rotor shaft to the driveshaft using the washers and the retaining screw. Install the packing.

3 Slide the driveshaft/governor weight assembly into the distributor body. Re-install any shims that were used and the O-ring.

4 Locate the collar to the bottom of the driveshaft, aligning the roll

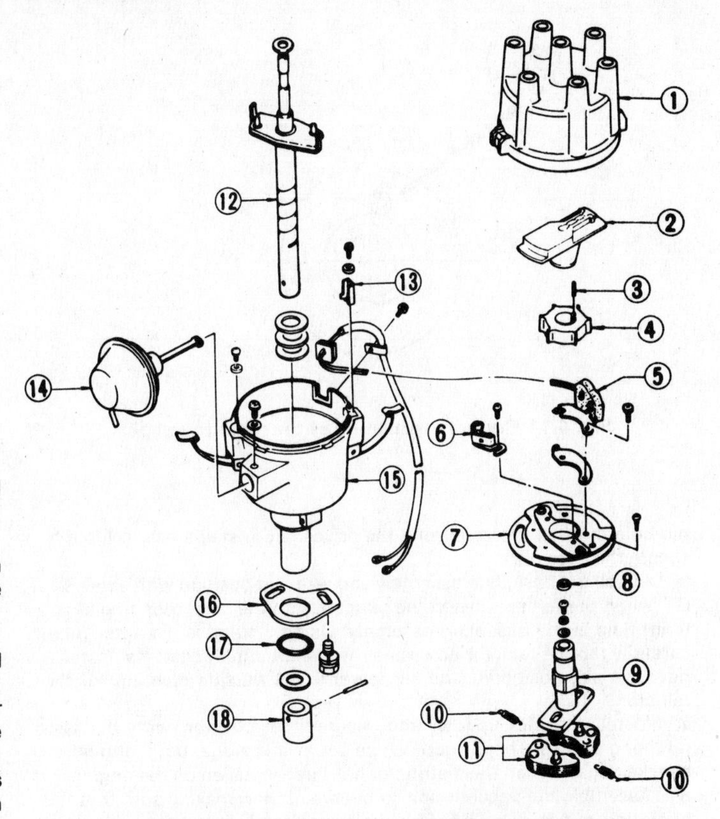

Fig. 4.8 The component parts of the distributor (Sec 6)

1	*Distributor cap*	10	*Governor spring*
2	*Rotor*	11	*Governor weight*
3	*Roll pin*	12	*Shaft assembly*
4	*Reluctor*	13	*Cap setter*
5	*Pick-up coil*	14	*Vacuum capsule*
6	*Contactor*	15	*Distributor body*
7	*Baseplate assembly*	16	*Fixing plate*
8	*Packing*	17	*O-ring*
9	*Rotor shaft*	18	*Collar*

4

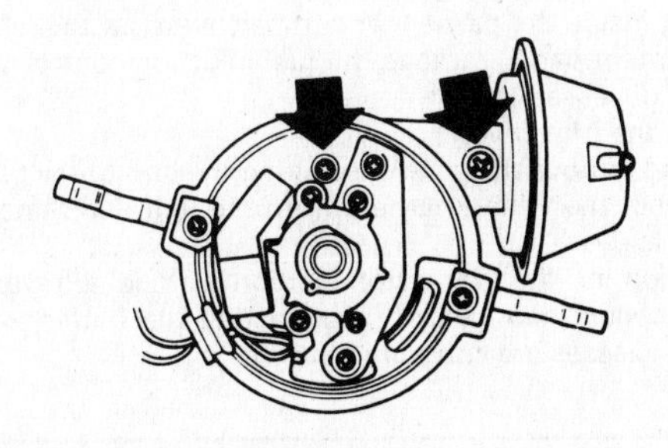

Fig. 4.9 The vacuum unit retaining screws arrowed (Sec 6)

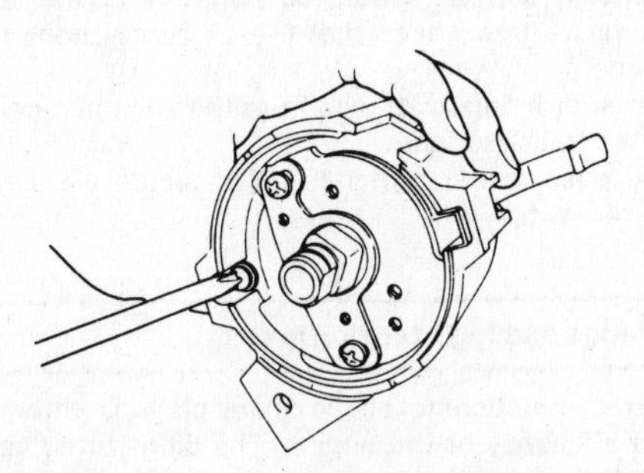

Fig. 4.10 Removing a baseplate retaining screw (Sec 6)

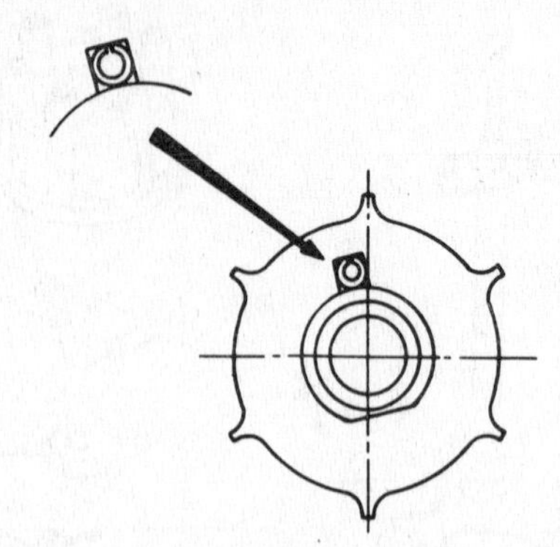

Fig. 4.11 Correct installation of the reluctor roll pin

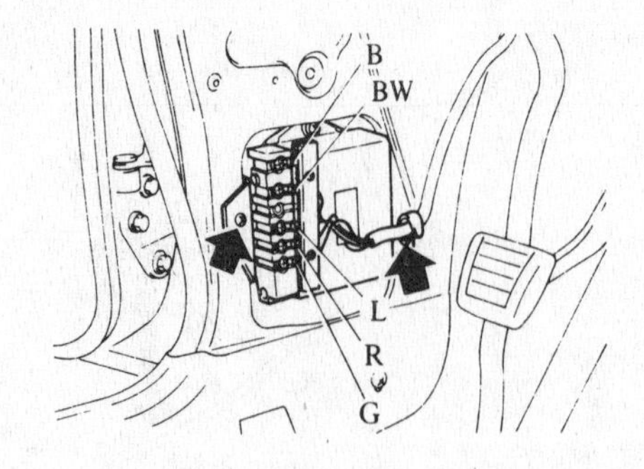

Fig. 4.12 The transistor ignition unit mounting screws arrowed (Sec 9)

pin hole through the collar and the driveshaft. Install a new roll pin to secure the assembly.
5 Install the baseplate assembly and retain in position with screws.
6 Align the roll pin hole in the reluctor with the corresponding alignment hole in the baseplate assembly. Using a suitable diameter tube, carefully tap the reluctor down the rotor shaft until it bottoms. Install a new roll pin, positioning its slit towards the outside diameter of the reluctor.
7 Install the pick-up coil and secure it in position with the two retaining screws. The air-gap can be set at this stage, but it should be checked again when the distributor has been installed on the engine.
8 Assemble the vacuum unit to the distributor, making sure that the operating rod is engaged with the baseplate before assembling the retaining screws.
9 Install the rotor and the distributor cap, and install the distributor on the engine as described in Section 5.

9 Transistor ignition unit – removal and installation

Due to the complexity of the transistor ignition unit, and the special equipment required to trace faults, any suspected malfunction should be checked by your Datsun dealer. The information contained in this Section is limited to removal and installation of the unit.
1 Disconnect the battery ground cable.
2 Working inside the passenger compartment, on the left-hand side of the instrument panel, remove the fuel injection control unit (refer to Chapter 3, if necessary)
3 Remove the trim cover.
4 Undo and remove the screws securing the unit to the panel.
5 Disconnect the wiring harness from the unit and remove the unit from the vehicle.
6 Installation of the transistor ignition unit is a reversal of the removal procedure, but before reconnecting the battery, ensure that the wiring harnesses are correctly installed.

10 Ignition coil – description and polarity

1 High tension current should be negative at the spark plug terminals. To ensure this, check that the LT connections to the coil are correctly made.
2 The primary (or negative) wire from the coil must connect to the B terminal at the transistor unit.
3 The coil positive (+) terminal is connected via a resistor to the ignition/starter switch.

11 Spark plugs and high tension leads

1 The correct functioning of the spark plugs is vital for the correct running and efficiency of the engine. The plugs fitted as standard are listed on the Specifications page.
2 At intervals of 3000 miles (5000 km) the plugs should be removed, examined, cleaned and, if worn excessively, renewed. The condition of the spark plug will also tell much about the overall condition of the engine.
3 If the insulator nose of the spark plug is clean and white, with no deposits, this is indicative of a weak mixture, or too hot a plug. (A hot plug transfers heat away from the electrode slowly – a cold plug transfers it away quickly).
4 If the top and insulator nose is covered with hard black looking deposits, then this is indicative that the mixture is too rich. Should the plug be black and oily, then it is likely that the engine is fairly worn, as well as the mixture being too rich.
5 If the insulator nose is covered with light tan to grayish-brown deposits, then the mixture is correct and it is likely that the engine is in good condition.
6 If there are any traces of long brown tapering stains on the outside of the white portion of the plug, then the plug will have to be renewed, as this shows that there is a faulty joint between the plug body and the insulator, and compression is being allowed to leak away.
7 Plugs should be cleaned by a sand blasting machine, which will free them from carbon more thoroughly than cleaning by hand. The machine will also test the condition of the plugs under compression. Any plug that fails to spark at the recommended pressure should be renewed.
8 The spark plug gap is of considerable importance, as, if it is too large or too small the size of the spark and its efficiency will be seriously impaired. The spark plug gap should be set to between 0.039 and 0.043 in (1.0 and 1.1 mm) for the best results.
9 To set it, measure tha gap with a feeler gauge, and then bend open, or closed, the outer plug electrode until the correct gap is achieved. The center electrode should never be bent as this may crack the insulation and cause plug failure, if nothing worse.
10 When renewing the plugs, remember to install the leads from the distributor in the correct firing order 1–5–3–6–2–4, number 1 cylinder being the one nearest the radiator.
11 The plug leads require no attention other than being kept clean and wiped over regularly.

12 Ignition system – testing and fault finding

Engine fails to start

1 If the engine fails to start and the car was running normally when it was last used, first check there is fuel in the fuel tank. If the engine turns over normally on the starter motor and the battery is evidently well charged, then the fault may be in either the high or low tension circuits. **Note:** *If the battery is known to be fully charged, the ignition light comes on, and the starter motor fails to turn the engine check the tightness of the leads on the battery terminals and also the secureness of the ground lead to its connection to the body. It is quite common for the leads to have worked loose, even if they look and feel secure. If one of the battery terminal posts gets very hot when trying to operate the starter motor this is a sure indication of a faulty connection to that ter-*

Common spark plug conditions

NORMAL

Symptoms: Brown to grayish-tan color and slight electrode wear. Correct heat range for engine and operating conditions.
Recommendation: When new spark plugs are installed, replace with plugs of the same heat range.

WORN

Symptoms: Rounded electrodes with a small amount of deposits on the firing end. Normal color. Causes hard starting in damp or cold weather and poor fuel economy.
Recommendation: Plugs have been left in the engine too long. Replace with new plugs of the same heat range. Follow the recommended maintenance schedule.

CARBON DEPOSITS

Symptoms: Dry sooty deposits indicate a rich mixture or weak ignition. Causes misfiring, hard starting and hesitation.
Recommendation: Make sure the plug has the correct heat range. Check for a clogged air filter or problem in the fuel system or engine management system. Also check for ignition system problems.

ASH DEPOSITS

Symptoms: Light brown deposits encrusted on the side or center electrodes or both. Derived from oil and/or fuel additives. Excessive amounts may mask the spark, causing misfiring and hesitation during acceleration.
Recommendation: If excessive deposits accumulate over a short time or low mileage, install new valve guide seals to prevent seepage of oil into the combustion chambers. Also try changing gasoline brands.

OIL DEPOSITS

Symptoms: Oily coating caused by poor oil control. Oil is leaking past worn valve guides or piston rings into the combustion chamber. Causes hard starting, misfiring and hesitation.
Recommendation: Correct the mechanical condition with necessary repairs and install new plugs.

GAP BRIDGING

Symptoms: Combustion deposits lodge between the electrodes. Heavy deposits accumulate and bridge the electrode gap. The plug ceases to fire, resulting in a dead cylinder.
Recommendation: Locate the faulty plug and remove the deposits from between the electrodes.

TOO HOT

Symptoms: Blistered, white insulator, eroded electrode and absence of deposits. Results in shortened plug life.
Recommendation: Check for the correct plug heat range, over-advanced ignition timing, lean fuel mixture, intake manifold vacuum leaks, sticking valves and insufficient engine cooling.

PREIGNITION

Symptoms: Melted electrodes. Insulators are white, but may be dirty due to misfiring or flying debris in the combustion chamber. Can lead to engine damage.
Recommendation: Check for the correct plug heat range, over-advanced ignition timing, lean fuel mixture, insufficient engine cooling and lack of lubrication.

HIGH SPEED GLAZING

Symptoms: Insulator has yellowish, glazed appearance. Indicates that combustion chamber temperatures have risen suddenly during hard acceleration. Normal deposits melt to form a conductive coating. Causes misfiring at high speeds.
Recommendation: Install new plugs. Consider using a colder plug if driving habits warrant.

DETONATION

Symptoms: Insulators may be cracked or chipped. Improper gap setting techniques can also result in a fractured insulator tip. Can lead to piston damage.
Recommendation: Make sure the fuel anti-knock values meet engine requirements. Use care when setting the gaps on new plugs. Avoid lugging the engine.

MECHANICAL DAMAGE

Symptoms: May be caused by a foreign object in the combustion chamber or the piston striking an incorrect reach (too long) plug. Causes a dead cylinder and could result in piston damage.
Recommendation: Repair the mechanical damage. Remove the foreign object from the engine and/or install the correct reach plug.

minal.

2 One of the commonest reasons for bad starting is wet or damp spark plug leads and distributor. Remove the distributor cap. If condensation is visible internally, dry the cap with a rag and also wipe over the leads. Re-install the cap.

3 If the fault cannot be readily diagnosed it is recommended that an investigation is carried out by a Datsun dealer using the correct test equipment. *Do not disconnect HT leads with the engine running, or the transistor ignition unit will be permanently damaged.*

Engine misfires

4 If the engine misfires regularly, this is probably a faulty spark plug or spark plug lead. Due to the fact that a spark plug lead must not be disconnected whilst the engine is running, the only solution is to renew both the spark plugs and leads. This will be costly and there is obviously no need to renew all of these components, so it is really far more beneficial to let your Datsun dealer find the fault.

5 Before doing this, you can carry out several checks to ensure that the cause of the problem is in fact in the area of the spark plugs or leads.

6 Remove the distributor cap and examine the inside for signs of electrical tracking. Check the carbon brush for smooth movement; also check the rotor for signs of burning or damage. Visually check the segments in the cap for excessive burning; ensure that each HT lead is firmly connected to each segment. Make sure that the distributor cap is not cracked.

7 Check the air-gap as described in Section 3.

8 Check the security of all electrical connections and examine the condition of the insulation. A wire may have chafed causing a short-circuit.

9 If the engine still misfires, the problem again could be in the LT circuit or the transistor unit, in which case take the car along to your Datsun dealer for a check on the transistor unit and the LT circuit.

Chapter 5 Clutch

Refer to Chapter 13 for specifications and information applicable to 1980 through 1984 models.

Contents

Specifications

Type	Single dry plate, diaphragm spring; hydraulic operation	
Driven plate (friction lining)		
Outside diameter	8·86 in (225·0 mm)	
Inside diameter	5·91 in (150·0 mm)	
Thickness	0·307 in (7·8 mm)	
Minimum depth of rivet head (below friction lining)	0·012 in (0·3 mm)	
Clutch pedal		
Height from toe-board	6·91 in (175·5 mm)	
Free-play	0·04 to 0·20 in (1 to 5 mm)	
Master cylinder		
Diameter	5/8 in (15·88 mm)	
Operating cylinder		
Diameter	3/4 in (19·05 mm)	
Torque wrench settings	**lbf ft**	**kgf m**
Clutch assembly-to-flywheel bolts	12 to 15	1·6 to 2·1
Operating cylinder-to-clutch housing bolts	22 to 30	3·1 to 4·1
Master cylinder mounting bolts	6 to 9	0·8 to 1·2
Master cylinder pipe connection	11 to 13	1·5 to 1·8
Operating cylinder hose connection	12 to 14	1·7 to 2·0
Pushrod adjusting nut	6 to 9	0·8 to 1·2

1 General description

The major components of the clutch comprise a pressure plate and cover assembly, diaphragm spring and a driven plate (friction disc) which incorporates torsion coil springs to cushion rotational shock when the drive is taken up.

The clutch release bearing is of sealed ball type and clutch actuation is hydraulic.

Depressing the clutch pedal moves the piston in the master cylinder forwards, so forcing hydraulic fluid through the clutch hydraulic pipe to the operating cylinder.

The piston in the operating cylinder moves forward on the entry of the fluid and actuates the clutch release arm by means of a short pushrod.

The release arm pushes the release bearing forward to bear against the release plate, so moving the center of the diaphragm spring inwards. The spring is sandwiched between two annular rings which act as fulcrum points. As the center of the spring is pushed in, the outside of the spring is pushed out, so moving the pressure plate backward and disengaging the pressure plate from the clutch disc.

When the clutch pedal is released the diaphragm spring forces the pressure plate into contact with the high friction linings on the clutch disc and at the same time pushes the clutch disc a fraction of an inch forward on its splines so engaging the clutch disc with the flywheel. The clutch disc is now firmly sandwiched between the pressure plate and the flywheel so the drive is taken up.

No adjustment is required to the clutch, as the design of the operating cylinder compensates for wear in the clutch friction linings.

2 Clutch pedal – removal, installation and adjustment

1 Working at the pedal pivot point, remove the cotter pin and withdraw the clevis pin; disconnect the master cylinder pushrod from the pedal assembly.
2 Remove the pedal return spring.
3 Unscrew and remove the fulcrum bolt and remove the pedal assembly.
4 Clean the following parts in a bowl of gasoline and examine them for wear or damage:

(a) Return spring
(b) Pedal sleeve

5

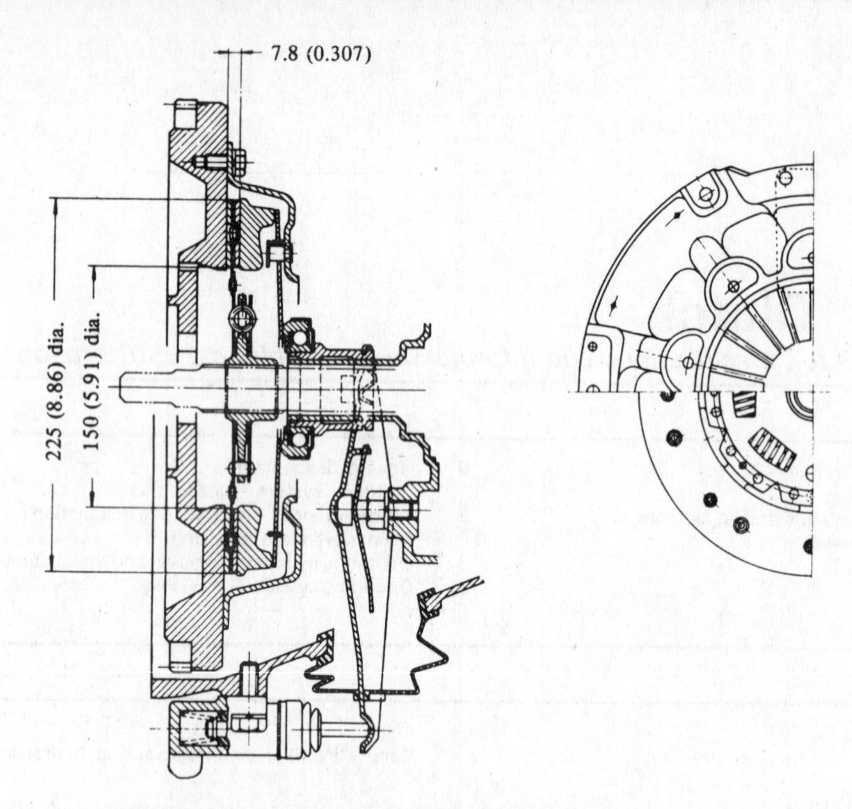

Fig. 5.1 Sectional drawing of the clutch assembly. Dimensions in mm (in)

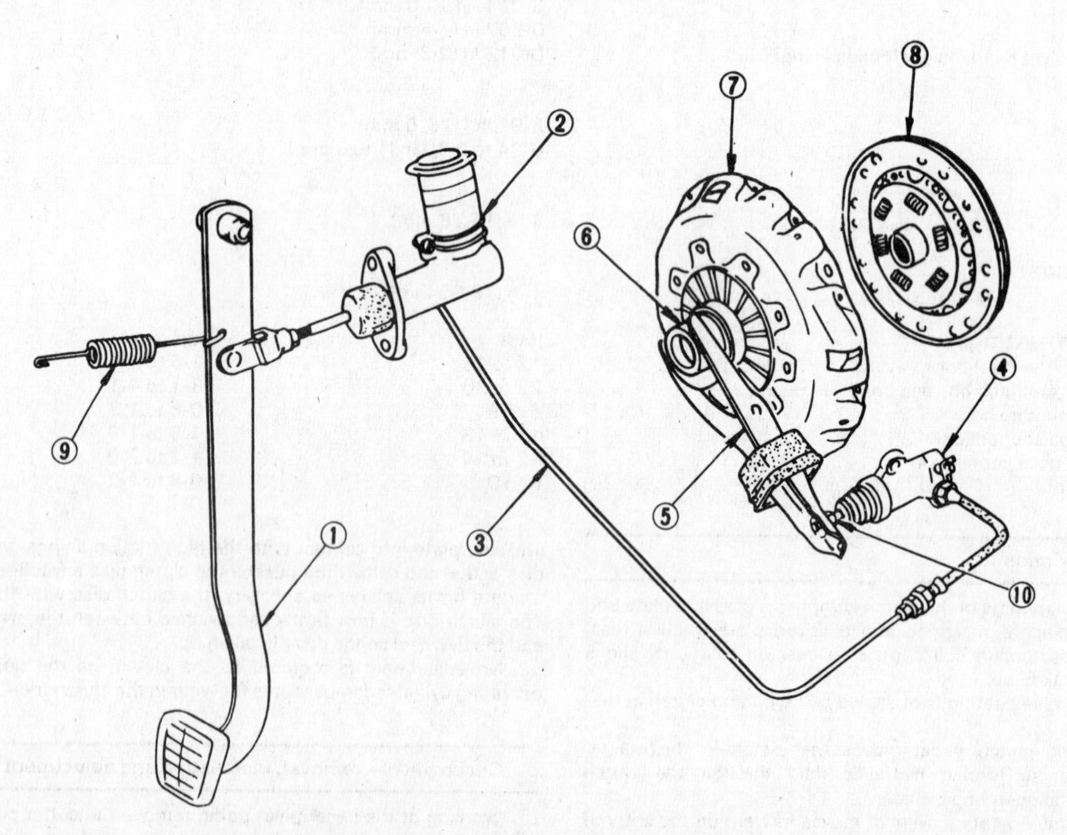

Fig. 5.2 The component parts of the clutch operating system (Sec 2)

1 Clutch pedal
2 Master cylinder
3 Hydraulic pipe
4 Operating cylinder
5 Withdrawal lever
6 Release bearing
7 Clutch cover assembly
8 Friction disc
9 Return spring
10 Operating cylinder pushrod

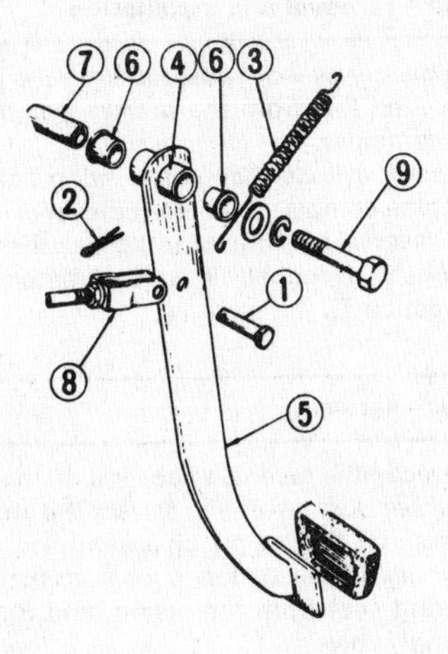

Fig. 5.3 The component parts of the clutch pedal assembly (Sec 2)

1 Clevis pin
2 Cotter pin
3 Return spring
4 Pedal boss
5 Pedal assembly
6 Bush
7 Sleeve
8 Pushrod
9 Fulcrum bolt

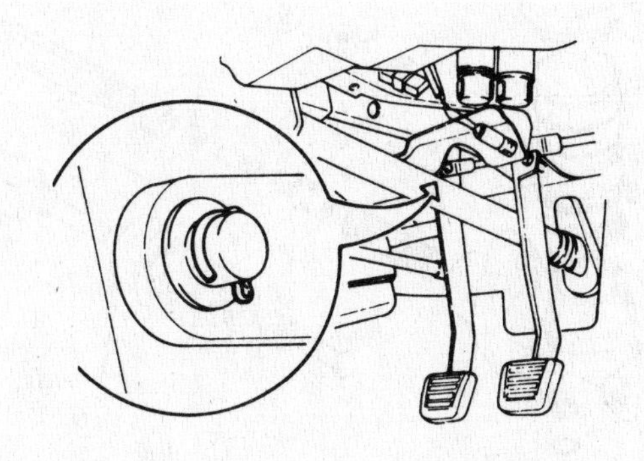

Fig. 5.4 Location of the cotter pin (Sec 2)

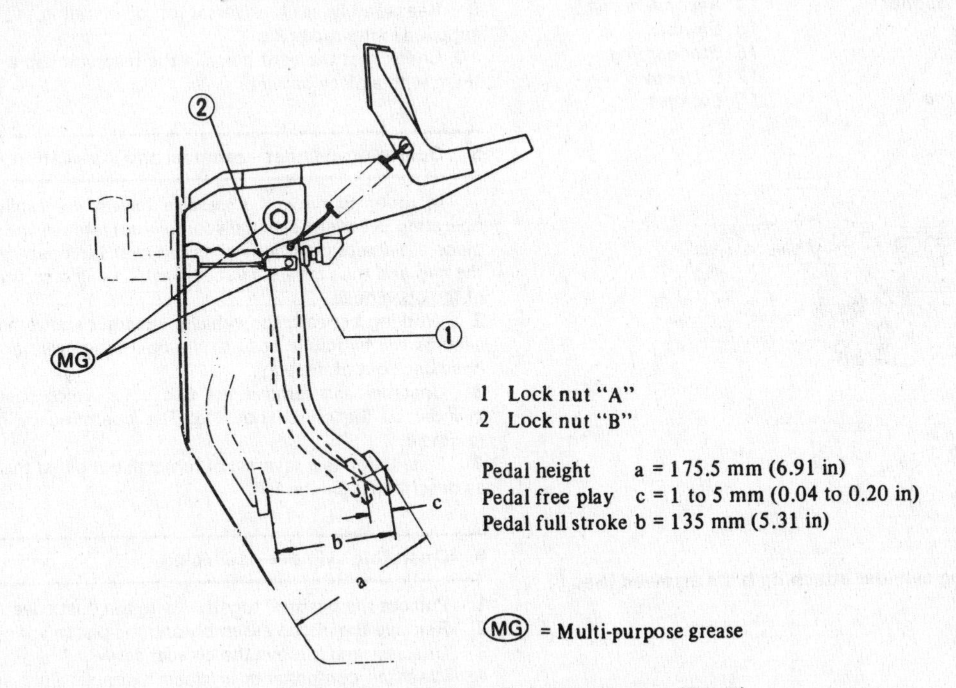

Fig. 5.5 Clutch pedal adjustment diagram (Sec 2)

1 Lock nut A *2 Lock nut B*

(c) Pedal boss
(d) Bushes
(e) Fulcrum bolt

Renew any parts which show signs of wear.

5 Installation is a reversal of removal but grease the bearing surfaces of the the pedal boss, sleeve, bushes and fulcrum bolt. When connecting the master cylinder pushrod to the pedal assembly, smear a little grease on the clevis pin; adjust the pedal height in the following manner.

6 Refer to Fig. 5.5 and loosen the pedal stop locknut A. By adjusting the pedal stop bolt, establish the pedal height from the toe-board which should be 6·91 in (175·5 mm). When this dimension has been achieved, tighten the locknut A to the specified torque.

7 Loosen the locknut B and, by screwing the master cylinder pushrod in the required direction, maintain a pedal free-play of 0·04 to 0·20 in (1 to 5 mm). Tighten the locknut B and re-check the pedal free-play. Make absolutely sure that the pedal free-play is as specified. If the play is too little the master cylinder pushrod will not travel the correct distance, which could result in a partially blocked hydraulic port.

8 When a satisfactory adjustment has been obtained check for satisfactory operation of the clutch.

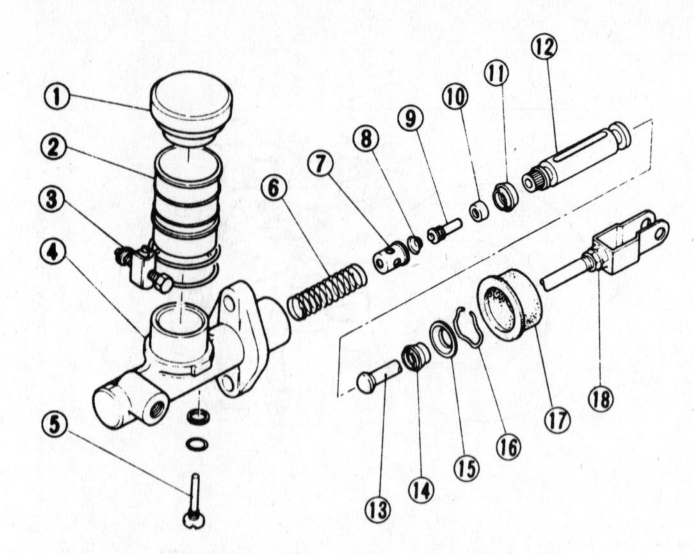

Fig. 5.6 An exploded view of the master cylinder (Sec 4)

1 Reservoir cap
2 Reservoir
3 Clip
4 Body
5 Supply valve stopper
6 Return spring
7 Spring seat
8 Valve spring
9 Supply valve rod
10 Supply valve
11 Primary cup
12 Piston
13 Pushrod
14 Secondary cup
15 Stopper
16 Stopper ring
17 Dust cover
18 Locknut

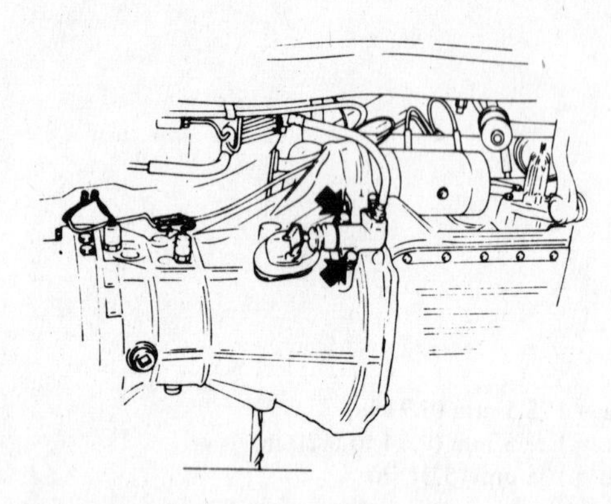
Fig. 5.7 The operating cylinder attaching bolts arrowed (Sec 5)

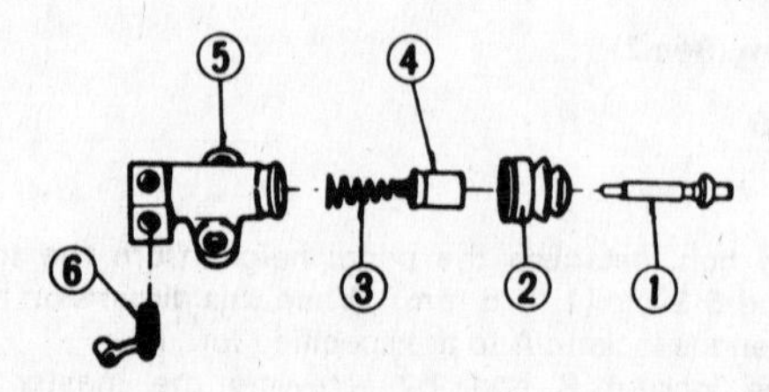

Fig. 5.8 The component parts of the operating cylinder (Sec 6)

1 Pushrod
2 Dust cover
3 Spring
4 Piston
5 Body
6 Bleed screw

3 Master cylinder – removal and installation

1 Disconnect the master cylinder pushrod from the pedal arm.
2 Disconnect the fluid line from the master cylinder and drain the fluid into a suitable container.
3 Remove the master cylinder flange mounting bolts and withdraw the unit from the engine compartment rear bulkhead.
4 Installation is a reversal of removal, but check the pedal height and free-play as previously described in this Chapter, and bleed the hydraulic system (Section 7).

4 Master cylinder – servicing

1 Unscrew and remove the reservoir cap, and drain any fluid.
2 Peel back the rubber dust cover and extract the stopper ring.
3 Withdraw the pushrod and the piston assembly.
4 Remove the primary and secondary cups from the piston.
5 Remove the spring seat from the piston, and remove the supply valve rod and the supply valve.
6 Wash all components in clean hydraulic fluid and examine the surface of the piston and the bore of the cylinder for scoring or 'bright' wear areas. Where these are evident, renew the master cylinder complete.
7 Where the components are in good condition, discard the rubber seals and obtain a repair kit.
8 Dip the new seals in clean hydraulic fluid and manipulate them into position using the fingers only. Check that their lips and chamfers face the correct way.
9 Reassembly is a reversal of dismantling, using all the parts supplied in the repair kit.
10 Check that the vent holes in the reservoir cap are clear by probing them with a piece of wire.

5 Operating cylinder – removal and installation

1 In order to prevent excessive loss of hydraulic fluid, when the operating cylinder hose is disconnected, remove the reservoir cap and place a piece of polythene sheeting over the open reservoir. Screw on the cap and thus create a vacuum which will stop the fluid running out of the open hose.
2 Working beneath the vehicle, unscrew and remove the nut which secures the hydraulic hose to the operating cylinder. Tie the hydraulic hose back out of the way.
3 Unscrew and remove the two bolts which secure the operating cylinder to the clutch housing. The operating cylinder can now be removed.
4 Installation is a reversal of removal but bleed the hydraulic system as described in Section 7.

6 Operating cylinder – servicing

1 Pull out the pushrod together with the dust cover.
2 Remove the piston assembly and the piston spring.
3 Unscrew and remove the bleeder screw.
4 Wash all components in clean hydraulic fluid and then examine the surfaces of the piston and cylinder bore. If these are scored or any 'bright' wear areas are evident, renew the operating cylinder complete.
5 If the components are in good order, discard the piston seal and obtain a repair kit.
6 Manipulate the new seal into position, ensuring that its lip faces the correct way.
7 Reassembly is a reversal of dismantling but take care not to rip the piston seal as it enters the cylinder bore. Always dip the piston assembly in clean hydraulic fluid before commencing to assemble it.

7 Hydraulic system – bleeding

The need for bleeding the cylinders and fluid lines arises when air gets into it. Air gets in whenever a joint or seal leaks or part has to be dismantled. Bleeding is simply the process of venting the air out again.
1 Make sure that the reservoir is filled and obtain a piece of $\frac{3}{16}$ inch (5

mm) bore diameter rubber or plastic tube about 3 feet (1 m) long and a clean glass jar. A small quantity of fresh, clean hydraulic fluid is also necessary.

2 Detach the cap (if installed) on the bleed nipple and surrounding area. Unscrew the nipple $\frac{1}{4}$ turn and fit the tube over it. Put about $\frac{1}{2}$ inch (15 mm) of fluid in the jar and put the other end of the pipe in it. The jar can be placed on ground under the car.

3 The clutch pedal should then be depressed quickly and released slowly until no more air bubbles come from the pipe. Quick pedal action carries the air along rather than leaving it behind. Keep the reservoir topped up.

4 When the air bubbles stop, tighten the nipple at the end of a down stroke.

5 Check that the operation of the clutch is satisfactory. Even though there may be no exterior leaks it is possible that the movement of the pushrod from the clutch cylinder is inadequate because fluid is leaking internally past the seals in the master cylinder. If this is the case, it is best to renew all seals in both cylinders.

6 Always use clean hydraulic fluid which has been stored in an air-tight container and has remained unshaken for the preceding 24 hours.

8 Clutch – removal

1 Remove the engine/transmission as a unit as fully described in Chapter 1, or alternatively remove the transmission as described in Chapter 6.

2 Separate the transmission from the engine by removing the clutch bellhousing to crankcase securing bolts.

3 The pressure plate need not be marked in relation to the flywheel as it can only be fitted one way due to the positioning dowels.

4 Unscrew the clutch assembly securing bolts a turn at a time in a diametrically opposite sequence until the tension of the diaphragm spring is released. Remove the bolts and lift the pressure plate assembly away.

5 At this stage, the driven plate (friction disc) will fall from its location between the pressure plate and the flywheel (photo).

9 Clutch – inspection and renovation

1 Due to the slow-wearing qualities of the clutch, it is not easy to decide when to go to the trouble of removing the transmission in order to check the wear on the friction lining. The only positive indication that something needs doing is when it starts to slip or when squealing noises on engagement indicate that the friction lining has worn down to the rivets. In such instances it can only be hoped that the friction surfaces on the flywheel and pressure plate have not been badly worn or scored.

2 Examine the surfaces of the pressure plate and flywheel for signs of scoring. If this is only light it may be left, but if very deep the pressure plate unit will have to be renewed. If the flywheel is deeply scored it should be taken off and advice sought from an automobile engineering firm. Providing it can be machined completely across the face, the overall balance of engine and flywheel should not be upset. If renewal of the flywheel is necessary the new one will have to be balanced to match the original.

3 The friction plate lining surfaces should be at least $\frac{1}{32}$ in(0·8 mm) above the rivets, otherwise the disc is not worth putting back. If the lining material shows signs of breaking up or black areas where oil contamination has occured it should also be renewed. If facilities are readily available for obtaining and fitting new friction pads to the existing disc this may be done but the saving is relatively small compared with obtaining a complete new disc assembly which ensures that the shock absorbing springs and the splined hub are renewed also. The same applies to the pressure plate assembly which cannot be readily dismantled and put back together without specialised riveting tools and balancing equipment. An allowance is usually given for exchange units.

10 Clutch release bearing – renewal

1 The sealed release bearing, although designed for long life, is worth renewing at the same time as the other clutch components are being renewed or serviced.

8.5 The clutch mechanism dismantled

10.3 The clutch release bearing

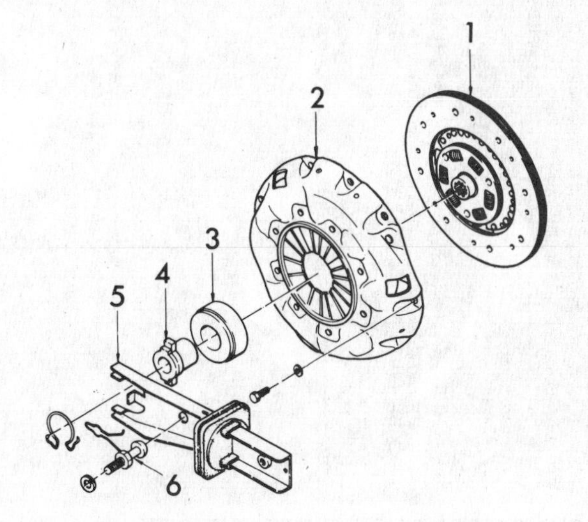

Fig. 5.9 An exploded view of the clutch release mechanism (Sec 10)

1 Clutch friction disc
2 Pressure plate assembly
3 Release bearing
4 Hub
5 Withdrawal lever
6 Ball pin

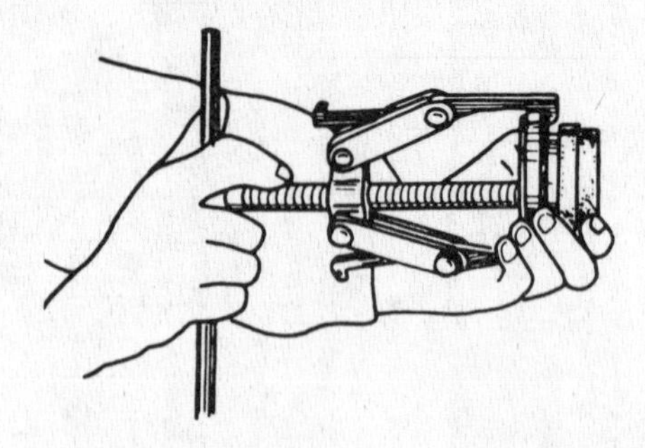

Fig. 5.10 Drawing the release bearing from its hub (Sec 10)

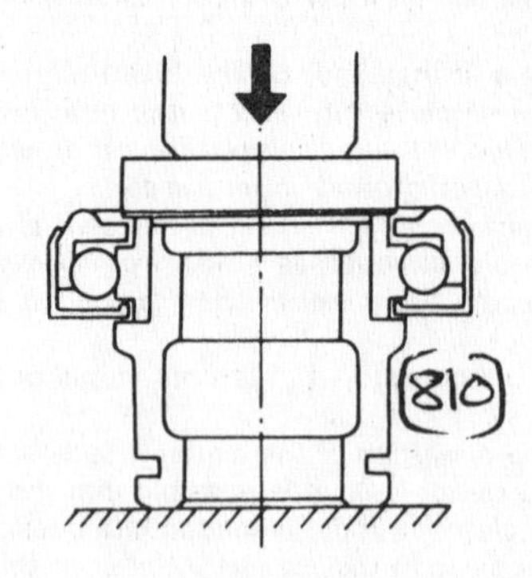

Fig. 5.11 Pressing the release bearing onto its hub (Sec 10)

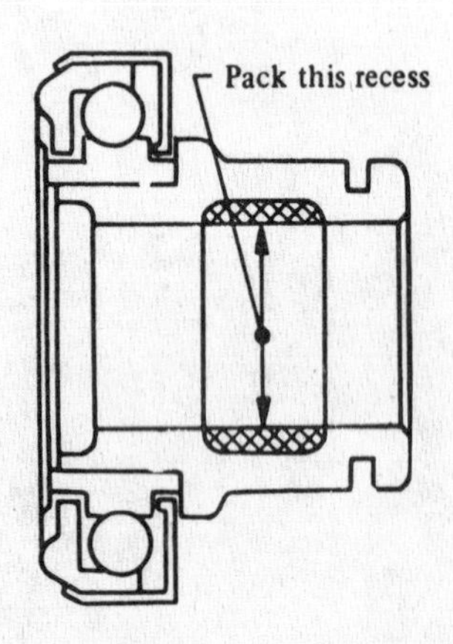

Fig. 5.12 Grease packing point on the release bearing hub (Sec 10)

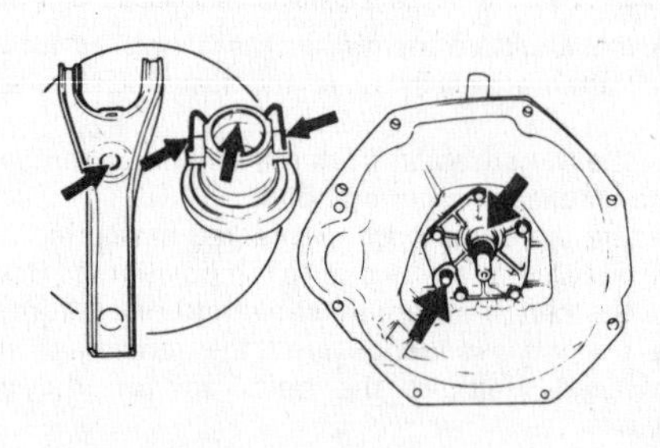

Fig. 5.13 Grease application points of withdrawal mechanism (Sec 10)

11.5 Using an old input shaft to centralize the clutch

2 Deterioration of the release bearing should be suspected when there are signs of grease leakage or the unit is noisy when spun with the fingers.

3 Remove the rubber dust excluder which surrounds the withdrawal lever at the bellhousing aperture (photo).

4 Using a screwdriver, unhook and detach the retainer spring from the ball-pin in the front transmission cover.

5 Remove the withdrawal lever together with the clutch release bearing and holding spring.

6 Unhook the holding spring from the withdrawal lever and the release bearing hub. The clutch release bearing and hub assembly can now be removed from the lever.

7 Remove the release bearing from its hub using a two or three legged puller and a bridge piece across the end-face of the hub.

8 Press on the new bearing but apply pressure only to the centre track.

9 Reassembly is a reversal of dismantling but apply high melting point grease to the internal recess of the release bearing hub.

10 Also apply similar grease to the pivot points of the clutch withdrawal lever.

11 Clutch – installation

1 Clean the face of the flywheel and the pressure plate.

2 Apply a smear of high melting point grease to the splines of the input shaft.

3 Locate the driven plate against the flywheel so that its larger projecting boss is to the rear.

4 Position the pressure plate assembly on the flywheel so that the positioning dowels engage and mate.

5 Screw in each of the pressure plate bolts finger-tight and then centralize the driven plate. This is accomplished by passing an old input shaft or stepped dowel rod through the splined hub of the driven

plate and engaging it in the spigot bush. By moving the shaft or rod in the appropriate directions, the position will be established where the centralizing tool can be withdrawn without any side pressure from the driven plate, proving that the driven plate is centralized (photo).

6 Without disturbing the setting of the driven plate, tighten the pressure plate bolts, a turn at a time, in diametrically opposite sequence to the specified torque.

7 Install the transmission to the engine (Chapter 6) when, if the driven plate has been correctly centralized, the input shaft of the transmission will pass easily through the splined hub of the driven plate to engage with the spigot bush in the centre of the flywheel. Do not allow the weight of the transmission to hang upon the input shaft while it is passing through the clutch mechanism or damage to the clutch components may result.

12 Fault diagnosis – clutch

Symptom	Reason/s
Judder when taking up drive	Loose engine or transmission mountings Badly worn friction surfaces or contaminated with oil Worn splines on transmission input shaft or driven plate hub Worn input shaft spigot bush in flywheel
Clutch spin (failure to disengage) so that gears cannot be meshed	Driven plate sticking on input shaft splines due to rust. May occur after vehicle standing idle for long period Damaged or misaligned pressure plate assembly
Clutch slip (increase in engine speed does not result in increase in vehicle road speed – particularly on gradients)	Friction surfaces worn out or oil contaminated
Noise evident on depressing clutch pedal	Dry, worn or damaged release bearing Insufficient pedal free travel Weak or broken pedal return spring Weak or broken clutch release lever return spring Excessive play between driven plate hub splines and input shaft splines
Noise evident as clutch pedal released	Distorted driven plate Broken or weak driven plate cushion coil springs Insufficient pedal free travel Weak or broken clutch pedal return spring Weak or broken release lever return spring Distorted or worn input shaft Release bearing loose on retainer hub

Chapter 6 Manual and automatic transmission

Refer to Chapter 13 for specifications and information applicable to 1980 through 1984 models.

Contents

Specifications

Manual transmission

Type	F4W71B, four forward speeds and reverse
Synchromesh	Warner type on all forward speeds
Gear ratios:	
1st	3·321 : 1
2nd	2·077 : 1
3rd	1·308 : 1
4th	1·000 : 1
Reverse	3·382 : 1
Oil capacity	3⅝ US pints (1·7 liters)

Automatic transmission

Type	JATCO 3N71B, three forward speeds and one reverse
Gear ratios:	
1st	2·458 : 1
2nd	1·458 : 1
3rd	1·000 : 1
Reverse	2·182 : 1
Idle speed	650 rpm in 'D' (800 rpm with air conditioning)
Stall speed	2000 to 2200 rpm
Fluid capacity	5⅞ US quarts (5·5 liters)

Torque wrench settings	**lbf ft**	**kgf m**
Manual transmission		
Adaptor plate bearing retainer screws	14 to 18	1·9 to 2·5
Detent ball plugs	14 to 18	1·9 to 2·5
Rear extension housing bolts	12 to 15	1·6 to 2·1
Back-up lamp switch	14 to 22	2·0 to 3·0
Mainshaft nut	101 to 123	14·0 to 17·0
Clutch bellhousing-to-engine bolts	32 to 43	4·4 to 5·9
Front cover bolts	12 to 15	1·6 to 2·1
Drain and filler plugs	18 to 25	2·5 to 3·5

Automatic transmission

Driveplate-to-crankshaft bolts	101 to 116	14·0 to 16·0
Driveplate-to-torque converter	29 to 36	4·0 to 5·0
Converter housing-to-engine	29 to 36	4·0 to 5·0
Converter housing-to-transmission case	33 to 40	4·5 to 5·5

PART 1: MANUAL TRANSMISSION

1 General description

The manual transmission installed on the Datsun 810 is of the four-forward speed and one reverse type, with synchromesh on all forward gears. The forward gears are of a helical gear formation, and the reverse gear a sliding mesh type using spur gears.

The main driveshaft gear is meshed with the counter drivegear. The forward speed gears on the countershaft are in constant mesh with the main gears. Each of the main gears rides on the mainshaft on needle roller bearings, rotating freely.

When the gearchange lever is operated, the relevant coupling sleeve is caused to slide on the synchronizer hub and engages its inner teeth with the outer teeth formed on the mainshaft gear. The synchronizer hub is splined to the mainshaft so enabling the parts to rotate in unison. Moving the gearchange lever to the reverse gear position moves the mainshaft reverse gear into engagement with the reverse idler gear.

The gear selector mechanism is controlled from a floor-mounted lever. Movement of the lever is transferred through a striking rod to dogs on the ends of the selector rods, and then through the medium of shift forks which are permanently engaged in the grooves of the synchro unit sleeves.

2 Transmission – removal and installation

1 Disconnect the battery ground cable.

2 Working inside the vehicle, remove the center console and sealing grommet from the gearshift lever.

3 Place the gearshift lever in the neutral position. Remove the E-ring and control lever pin from the transmission striking rod guide. Lift out the gearshift control lever.

4 Unscrew and remove the nuts which secure the flange of the exhaust downpipe to the manifold.

5 Tie the exhaust pipe to a convenient anchorage point and remove the pipe support bracket from the rear mounting insulator.

6 Jack-up the vehicle and safely support its weight on suitable blocks or stands.

7 Disconnect the wires from the reverse lamp switch and top detecting switch.

8 Disconnect the speedometer cable from the rear extension housing.

9 If desired, the oil can be drained into a suitable container, after removing the magnetic drain plug, whilst the disconnecting operations are carried out. Alternatively, the oil may be drained from the unit after removal.

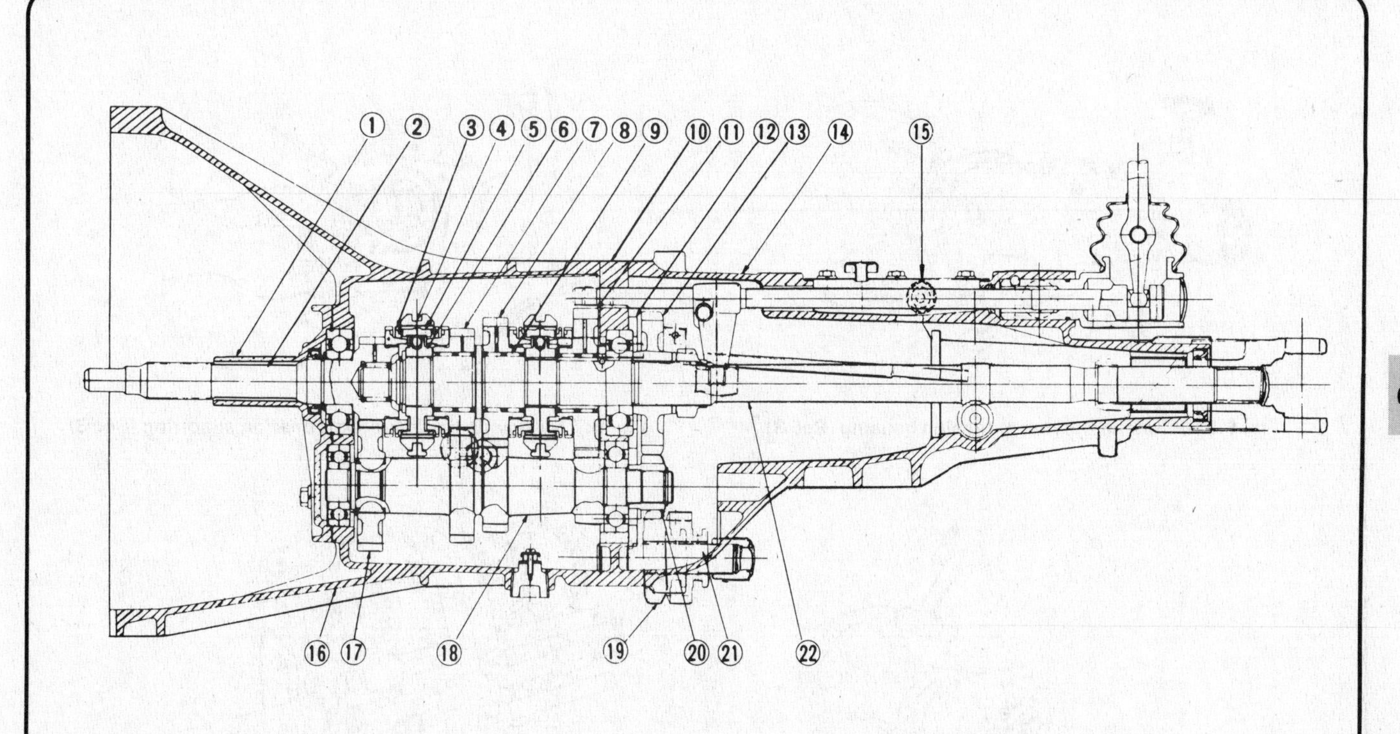

Fig. 6.1 Sectional view of the transmission

1 Front cover
2 Input shaft
3 Baulk ring
4 3rd/4th synchro sleeve
5 Synchro ring
6 Synchro hub
7 3rd gearwheel
8 2nd gearwheel
9 Needle bearing
10 Adaptor plate
11 1st gear
12 Bearing retainer
13 Reverse gear
14 Rear extension housing
15 Top detecting switch
16 Casing
17 Countershaft drive gear
18 Countershaft
19 Reverse idler gear
20 Reverse gear (countershaft)
21 Reverse idler shaft
22 Mainshaft

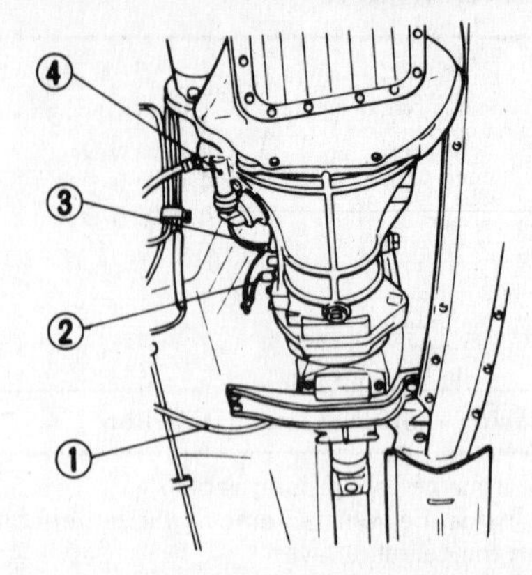

Fig. 6.2 Location of the speedometer cable (1), top detecting switch (2), reverse lamp switch (3), and the clutch operating cylinder (4)

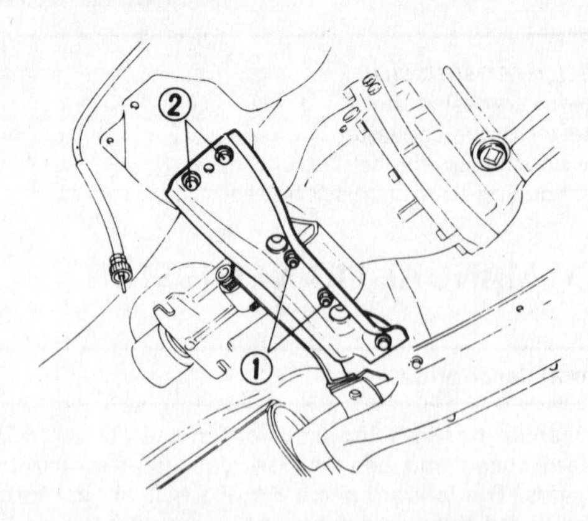

Fig. 6.3 The transmission mounting (Sec 2)

1 Mounting-to-insulator nuts 2 Mounting-to-bodyframe bolts

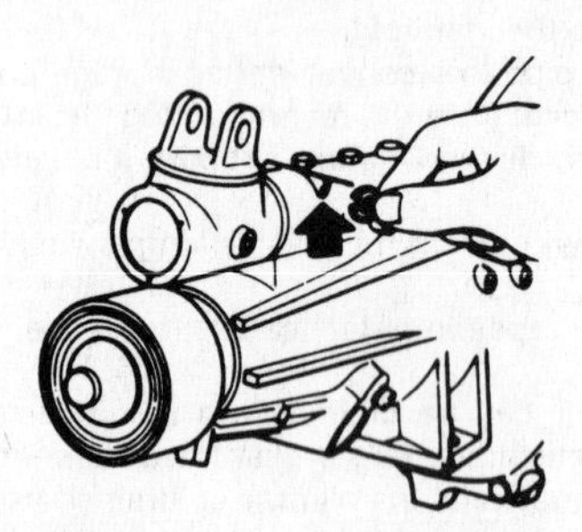

Fig. 6.4 Removing the striking rod snap-ring and stop pin (Sec 3)

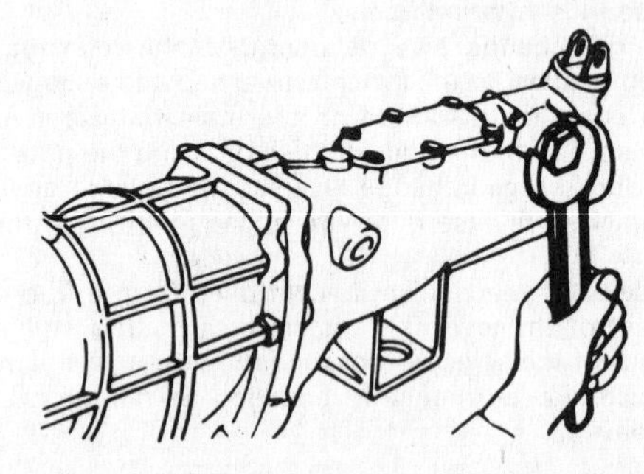

Fig. 6.5 Removing the return spring and plug assembly (Sec 3)

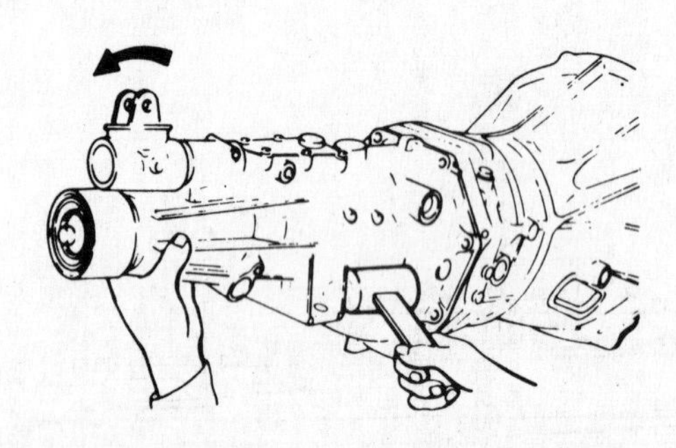

Fig. 6.6 Removing the rear extension housing (Sec 3)

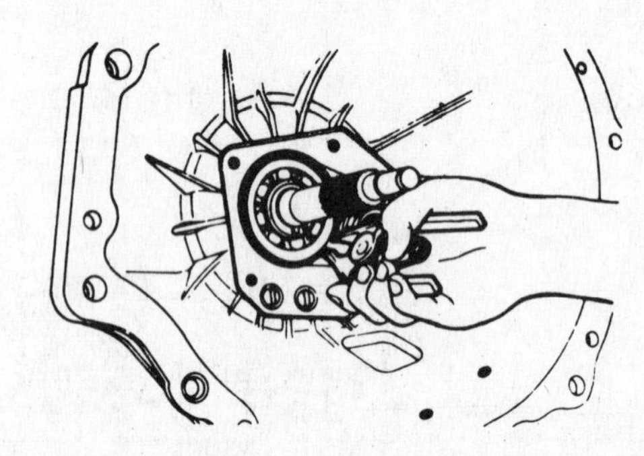

Fig. 6.7 Removing the input shaft bearing snap-ring (Sec 3)

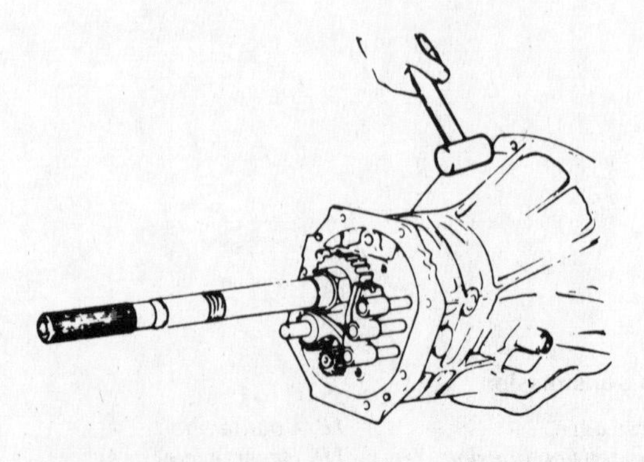

Fig. 6.8 Removing the one-piece transmission casing/clutch bell-housing (Sec 3)

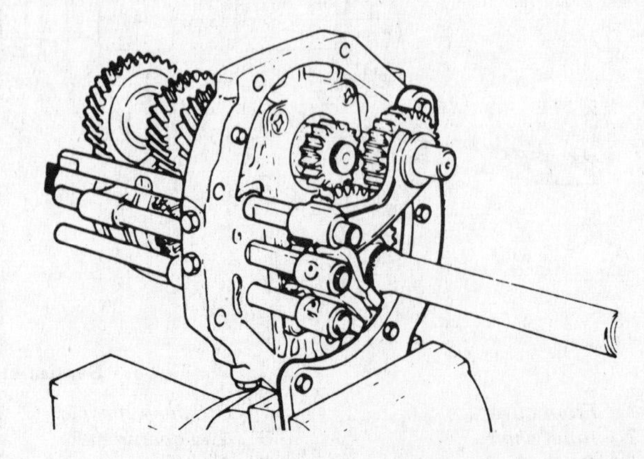

Fig. 6.9 The adaptor plate and gear assembly mounted on a support plate (Sec 3)

10 Unbolt and remove the clutch operating cylinder and tie it back out of the way. There is no need to disconnect the hydraulic line. (Further information can be found in Chapter 5).
11 Disconnect and remove the propeller shaft as described in Chapter 7.
12 Support the engine under the oil pan using a suitable jack and a block of wood as an insulator.
13 Place a second jack under the transmission.
14 Undo and remove the transmission mounting insulator securing nuts and the mounting-to-bodyframe bolts.
15 Remove the starter motor from the bellhousing.
16 Unscrew and remove the bolts which secure the clutch bellhousing to the engine crankcase.
17 Lower each of the two jacks simultaneously, until the trasnmission can be withdrawn to the rear and removed from beneath the vehicle. Do not allow the weight of the transmission to hang upon the input shaft while it is engaged with the splines of the clutch driven plate.
18 Installation is a reversal of removal but check that the clutch driven plate is centralized (see Chapter 5) and apply a smear of high melting point grease to the input shaft splines.
19 Check the clutch pedal free-travel and adjust if necessary (Chapter 5) and remember to fill the transmission with the correct grade and quantity of oil.

3 Transmission – dismantling into major assemblies

1 With the transmission removed, thoroughly clean the external surfaces.
2 Remove the rubber dust boot from the withdrawal lever aperture in the clutch bellhousing.
3 Remove the release bearing and hub together with the withdrawal lever (Chapter 5)
4 Unscrew the reversing lamp switch from the rear extension housing.
5 Unbolt the lockplate from the rear extension housing and remove the speedometer pinion and sleeve.
6 Remove the E-ring and stopper pin, also the return spring plug assembly, all from the rear extension housing.
7 Unscrew the securing bolts, and drive the rear extension housing from the main transmission casing, using a soft mallet.
8 Unscrew and remove the front cover retaining bolts; remove the front cover and extract the countershaft bearing shim and the input shaft bearing snap-ring.
9 Drive off the one-piece bellhousing/transmission casing from the adaptor plate.
10 Make up a suitable support plate and bolt it to the transmission adaptor plate and then secure the support plate in a vise.
11 Drive out the securing pins from each of the shift forks, using a suitable thin drift.
12 Unscrew and remove the three detent ballplugs (Fig. 6.11)
13 Withdraw the selector rods from the adaptor plate.
14 Catch the shift forks, and extract the balls and springs as the selector rods are withdrawn. The four smaller balls are the interlock balls.
15 Lock the gears and using a suitable two legged extractor, draw the front bearing from the countershaft.
16 Now is the time to check the gears for backlash, using a dial gauge. The backlash should be between 0.0020 and 0.0039 in (0.05 and 0.10 mm). Where this is exceeded, renew the driving and driven gears as a matched set. Now check the gear endfloat using a feeler blade. For first gear this should be between 0.0126 and 0.0154 in (0.32 and 0.39 mm), for second gear between 0.0047 and 0.0075 in (0.12 and 0.19 mm) and between 0.0051 and 0.0146 in (0.13 and 0.37 mm) for third gear. Reverse idler gear endfloat should be between 0.0020 and 0.0079 in (0.05 and 0.20 mm). Selective snaprings are available to provide the specified endfloat in the following ranges:

Mainshaft, countershaft and reverse idler gear:
0.055 in (1.4 mm)
0.059 in (1.5 mm)
0.063 in (1.6 mm)

Input shaft:
0.0709 in (1.80 mm)

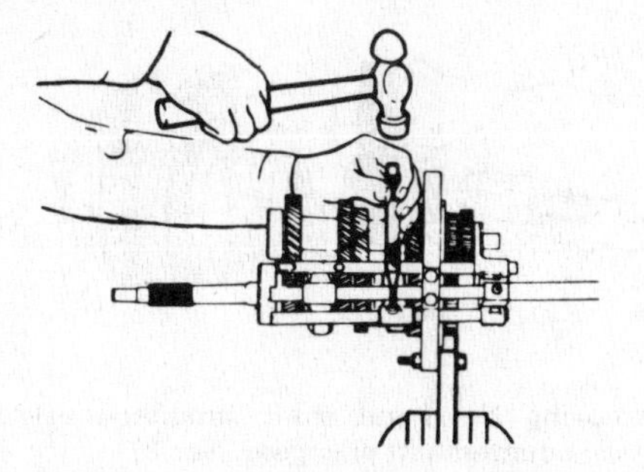

Fig. 6.10 Driving out the shift fork pins (Sec 3)

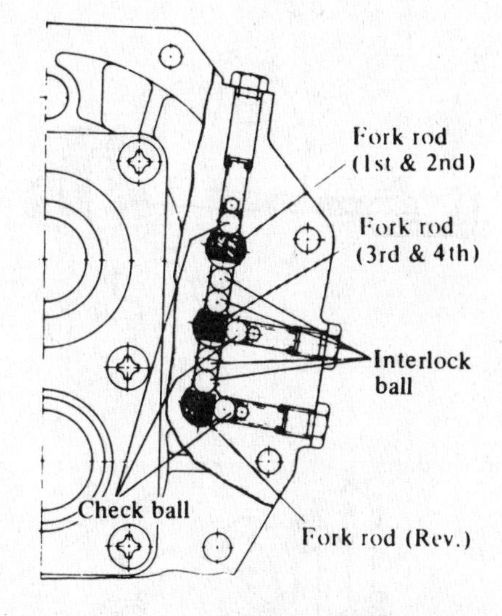

Fig. 6.11 Location of the detent ball plugs (Sec 3)

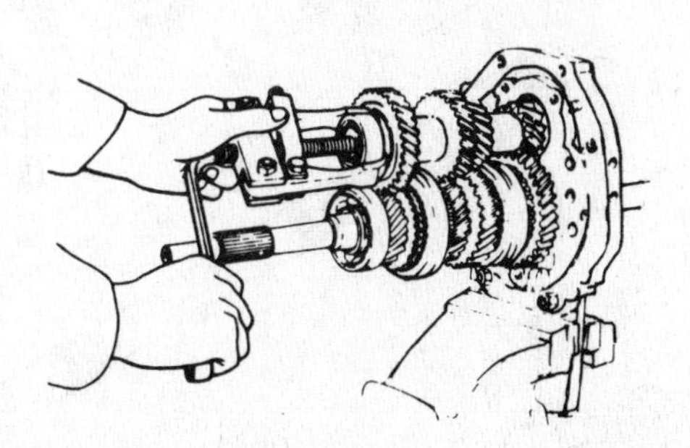

Fig. 6.12 Removing the countershaft front bearing (Sec 3)

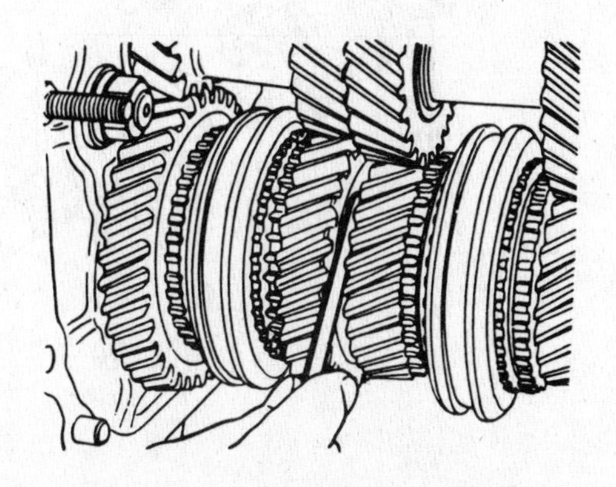

Fig. 6.13 Checking gear endfloat (Sec 3)

6

Fig. 6.14 Removing the input shaft simultaneously with the countershaft drivegear (Sec 3)

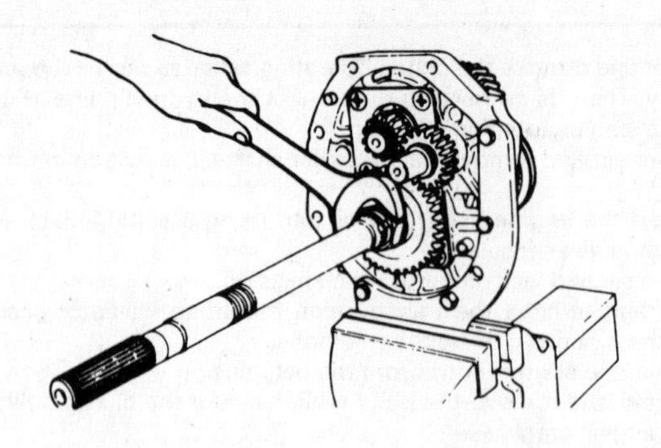

Fig. 6.15 Removing the mainshaft nut (Sec 3)

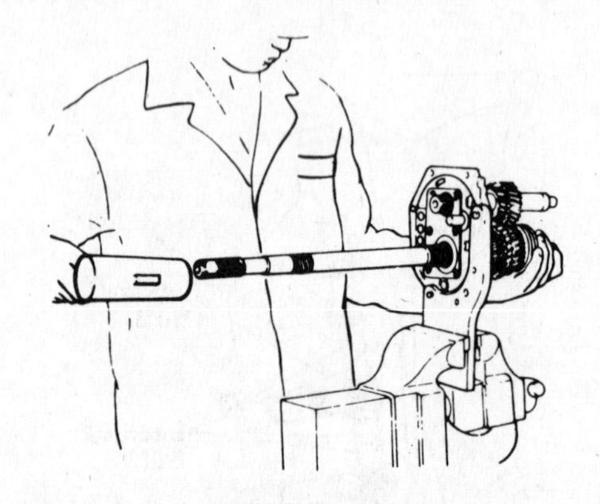

Fig. 6.16 Removing the mainshaft from the adaptor plate (Sec 3)

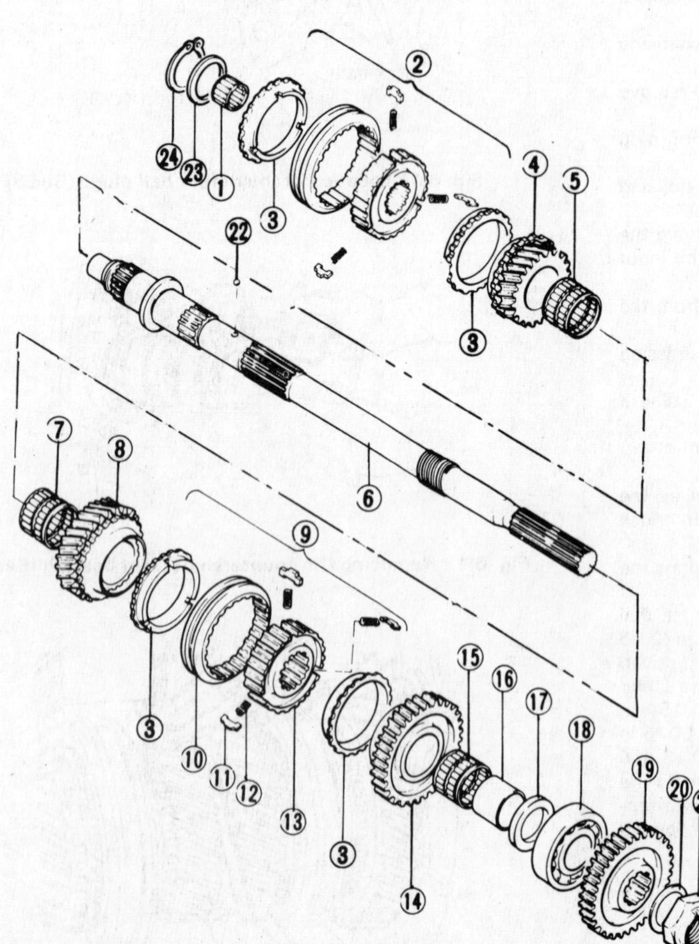

Fig. 6.17 The mainshaft assembly (Sec 4)

1 *Needle bearing*
2 *3rd/4th synchro unit*
3 *Baulk ring*
4 *3rd gearwheel*
5 *Needle bearing*
6 *Mainshaft*
7 *Needle bearing*
8 *2nd gearwheel*
9 *1st/2nd synchro unit*
10 *Synchro sleeve*
11 *Blocker bar*
12 *Spring*
13 *Synchro hub*
14 *1st gear*
15 *Needle bearing*
16 *Bush*
17 *Thrust washer*
18 *Mainshaft rear bearing*
19 *Reverse gear*
20 *Thrust washer*
21 *Mainshaft nut*
22 *Steel ball*
23 *Thrust washer*
24 *Snap-ring*

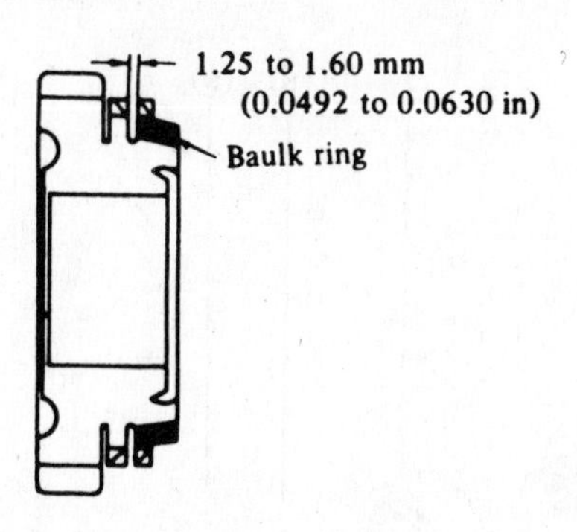

Fig. 6.18 Checking the gap between the baulk ring and synchro cone (Sec 4)

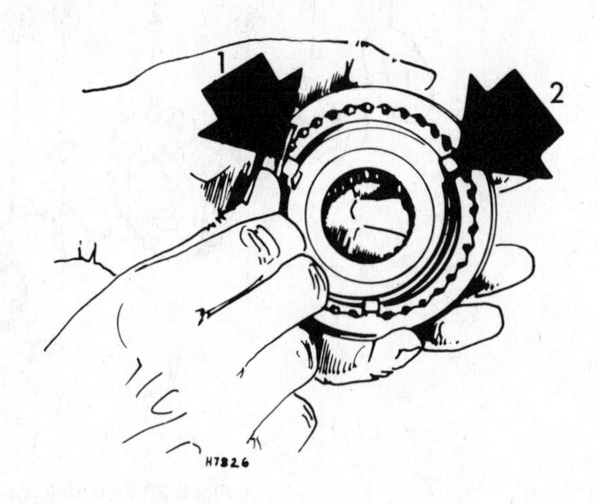

Fig. 6.19 Synchro unit spring (1) and blocker bar (2) (Sec 4)

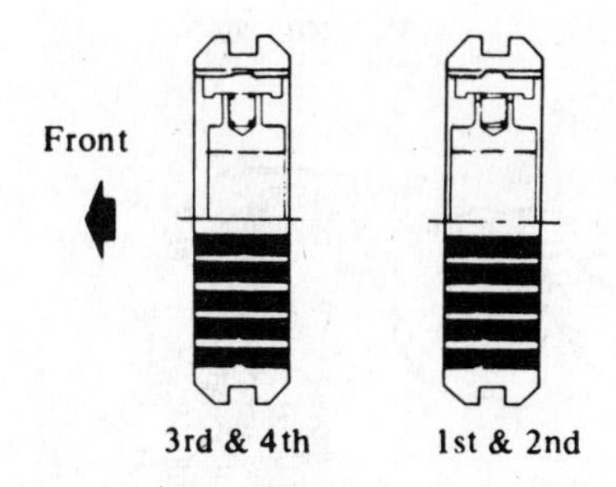

Fig. 6.20 Correct directional fitting of synchro units to mainshaft (Sec 4)

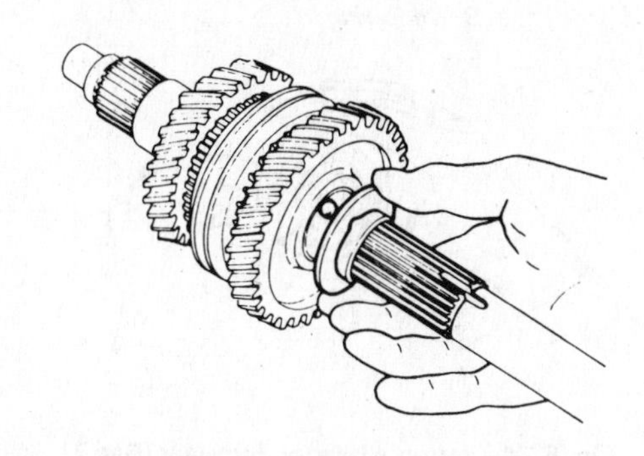

Fig. 6.21 Installing the mainshaft thrust washer (Sec 4)

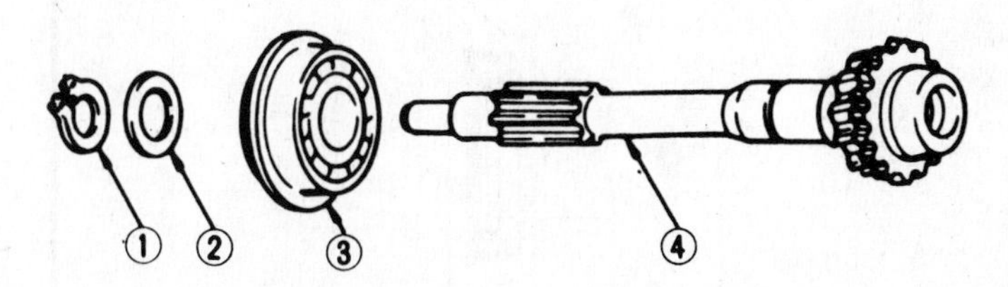

Fig. 6.22 The component parts of the input shaft (Sec 5)

1 Snap-ring
2 Spacer
3 Bearing
4 Input shaft

0.0736 in (1.87 mm)
0.0764 in (1.94 mm)
0.0791 in (2.01 mm)
0.0819 in (2.08 mm)
0.0846 in (2.15 mm)

17 Extract the now exposed snap-ring from the countershaft.
18 Withdraw the countershaft gear together with the input shaft. Take care not to drop the needle roller bearing which is located on the front of the mainshaft.
19 Extract the snap-ring from the front end of the mainshaft, followed by the thrust washer.
20 Withdraw 3rd/4th synchronizer unit, followed by 3rd gear.
21 Release the caulking on the mainshaft nut and then slacken it.
22 Remove the mainshaft nut, the thrust washer and reverse gear.
23 Extract the snap-ring from the rear end of the countershaft and remove the reverse idler gear.
24 Drive the mainshaft and countershaft assemblies simultaneously from the adaptor plate, using a soft-faced mallet.

4 Mainshaft – dismantling, servicing and reassembly

1 Carefully examine the gearwheels and shaft splines for chipping of the teeth or wear and then dismantle the gear train into its component parts, renewing any worn or damaged items.
2 Examine the shaft itself for scoring or grooving, also the splines for twist, taper or general wear.
3 Examine the synchromesh units for cracks or wear or general slackness in the assembly and renew if evident, particularly if there has been a history of noisy gearchange or where the synchromesh can be easily 'beaten'.
4 Press the baulk ring tight against the synchromesh cone and measure the gap between the two components. If it is less than specified, renew the components. (Fig. 6.18).
5 When reassembling the synchromesh unit ensure that the ends of the snap-ring on opposite sides of the unit do not engage in the same slot.
6 Commence assembly of the mainshaft by installing the 2nd gear needle bearing, 2nd gearwheel, the baulk ring followed by the 1st/2nd synchromesh unit, noting carefully the direction of installing the latter.
7 Now install the 1st gear baulk ring, 1st gear bush and the needle bearing followed by the 1st gearwheel and the steel ball (well greased) and the thrust washer.

5 Input shaft – bearing renewal

1 Remove the snap-ring and spacer.
2 Withdraw the bearing using a two legged extractor or a press. Once removed (by means of its outer track) discard the bearing.
3 Press the new bearing onto the shaft, applying pressure to the center track only. Snap-rings are available in a range of thicknesses as listed in Section 3 of this Chapter.

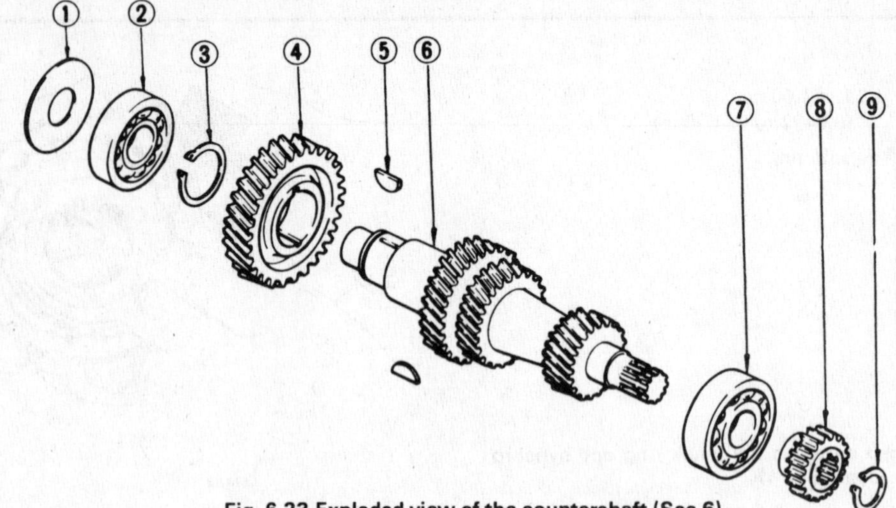

Fig. 6.23 Exploded view of the countershaft (Sec 6)

1 *Shim*
2 *Front bearing*
3 *Snap-ring*
4 *Drivegear*
5 *Woodruff key*
6 *Countershaft*
7 *Rear bearing*
8 *Reverse gear*
9 *Snap-ring*

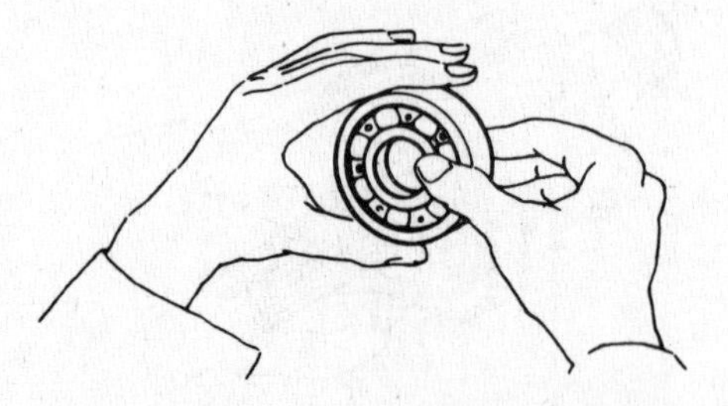

Fig. 6.24 Testing a bearing for wear (Sec 7)

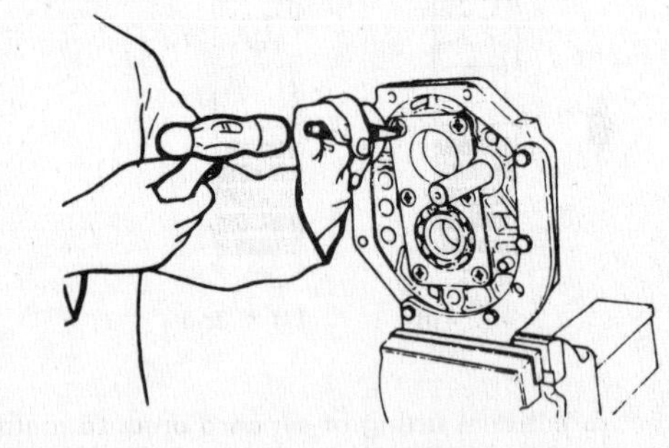

Fig. 6.25 Staking the bearing retainer plate screws (Sec 7)

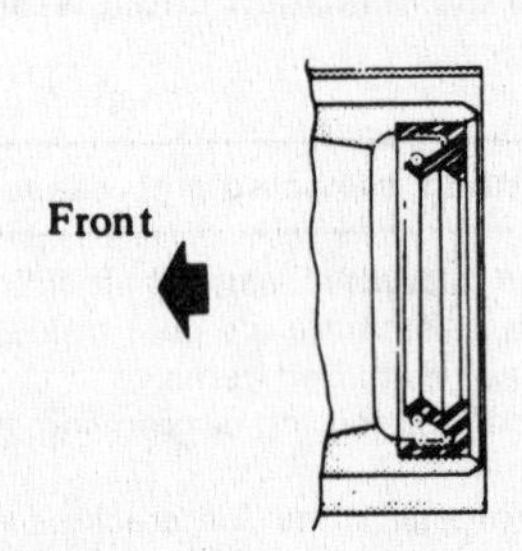

Fig. 6.26 Installation diagram for the rear extension oil seal (Sec 8)

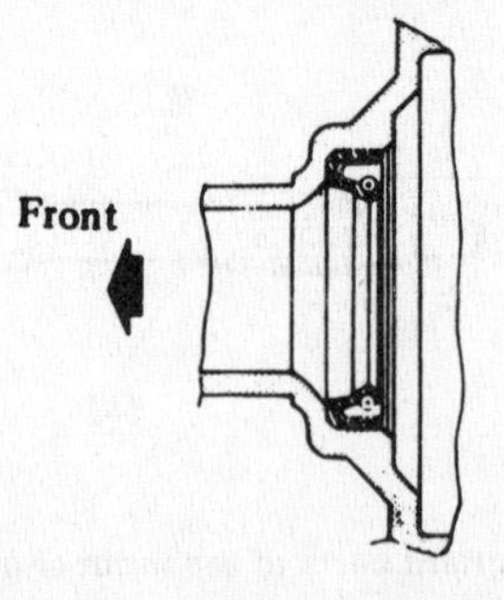

Fig. 6.27 Front cover oil seal installation diagram (Sec 8)

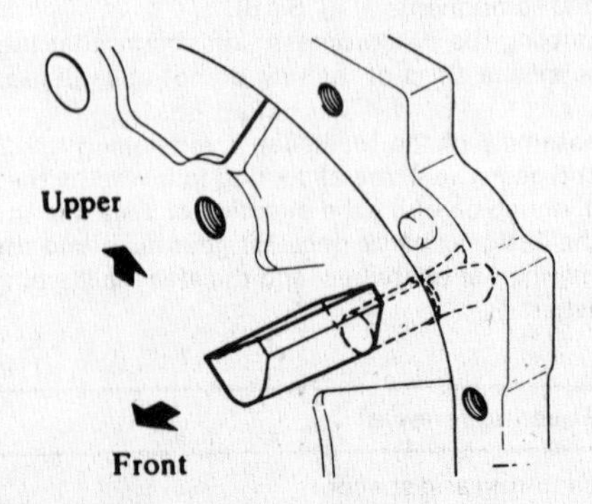

Fig. 6.28 Correct location of the oil trough and dowel (Sec 9)

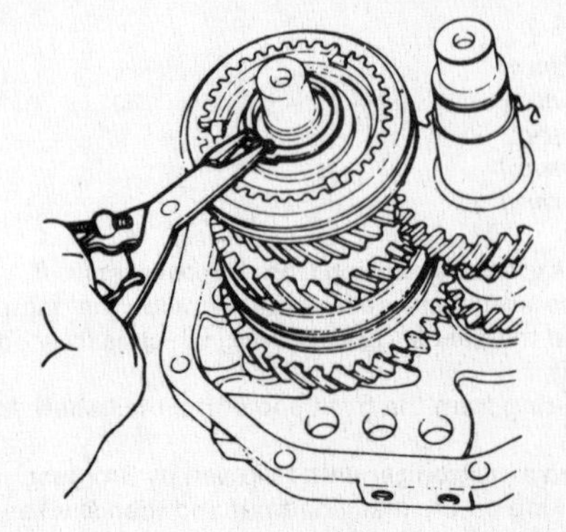

Fig. 6.29 Installing the snap-ring on the mainshaft assembly (Sec 9)

6 Countershaft – dismantling, servicing and reassembly

1 The countershaft front bearing was removed at the time of dismantling the transmission into major units (Section 3).
2 The countershaft rear bearing was left in position in the adaptor plate.
3 Withdraw the countershaft drive gear and extract the two Woodruff keys.
4 Check all components for wear, especially the gear teeth and shaft splines for chipping. Renew the Woodruff keys and the snap-ring.
5 Reassembly is a reversal of dismantling.

7 Mainshaft and countershaft adaptor plate bearings – renewal

1 Before commencing to reassemble the transmission, the mainshaft and countershaft adaptor plate bearings should be removed, examined and renewed if worn. To do this, unscrew the six screws which retain the bearing retainer plate to the adaptor plate. The use of an impact driver will probably be required for this operation.
2 With the bearing retainer plate removed, press the mainshaft and countershaft bearings from the adaptor plate. Apply pressure only to the outer tracks of the bearings.
3 Check the bearings for wear by first washing them in clean gasoline and drying in air from a tire pump. Spin them with the fingers, and if they are noisy or slack in operation, renew them.
4 Tap the countershaft bearing into position in the adaptor plate, after the mainshaft bearing and retaining plate have been installed.

8 Oil seals – renewal

1 Pry out the oil seal from the rear extension and drive in a new one, with the seal lips facing inwards.
2 Renew the speedometer pinion sleeve O-ring seal.
3 Renew the oil seal in the front cover by prying out the old one and driving in a new one, with a piece of tubing used as a drift.

9 Transmission – reassembly

1 Check that the dowel pin and oil trough are correctly positioned on the adaptor plate.
2 Tap the mainshaft bearing lightly and squarely into position in the adaptor plate.
3 Drive the reverse idler shaft into the adaptor plate so that $\frac{2}{3}$rds of its length is projecting rearwards. Ensure that the cutout in the shaft is positioned to receive the edge of the bearing retainer plate.
4 Install the bearing retainer plate and tighten the screws to the specified torque.
5 Tap the countershaft rear bearing into position in the adaptor plate.
6 Press the mainshaft assembly into position in the bearing in the adaptor plate. Support the rear of the bearing center track during this operation.
7 Press the countershaft assembly into position in the bearing in the adaptor plate. Again support the rear of the bearing center track during this operation.
8 Install the needle bearing, 3rd gear, baulk ring and the 3rd/4th synchromesh unit to the front of the mainshaft.
9 Install the thrust washer and a selective snap-ring and check for endfloat as described in Section 3.
10 Insert the needle pilot bearing in its recess at the end of the input shaft.
11 Mesh the countershaft drive-gear with the 4th gear on the input shaft. Push the drivegear and input shaft onto the countershaft and mainshaft simultaneously, but a piece of tubing will be needed to drive the countershaft gear into position while supporting the rear end of the countershaft.
12 Install a snap-ring to the front of the countershaft, checking the endfloat and selecting the snap-ring as described in Section 3.
13 Using a tubular drift, drive the front bearing onto the countershaft.
14 To the rear of the mainshaft, install the reverse gear, the plain washer and screw on the nut, finger-tight.

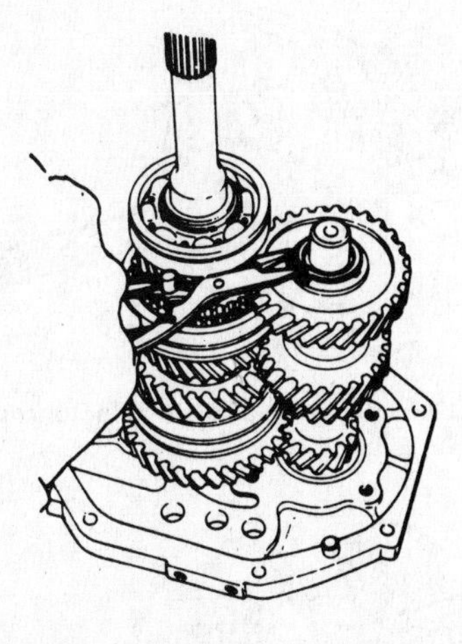

Fig. 6.30 Installing the countershaft snap-ring (Sec 9)

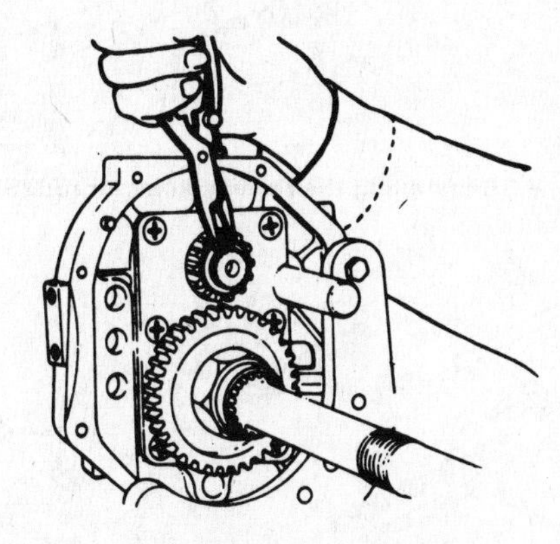

Fig. 6.31 Installing the reverse gear snap-ring to the countershaft (Sec 9)

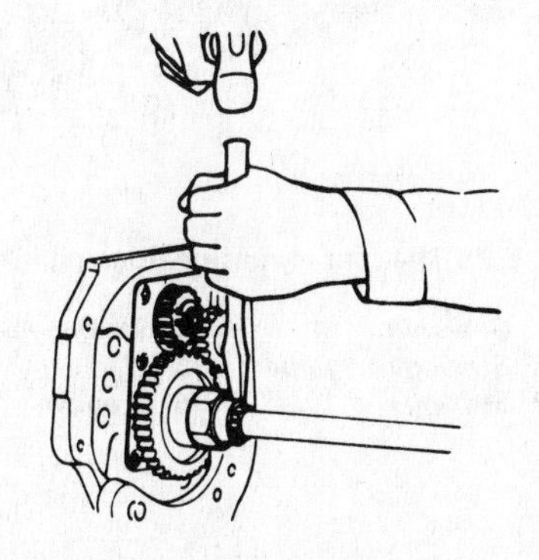

Fig. 6.32 Staking the mainshaft nut (Sec 9)

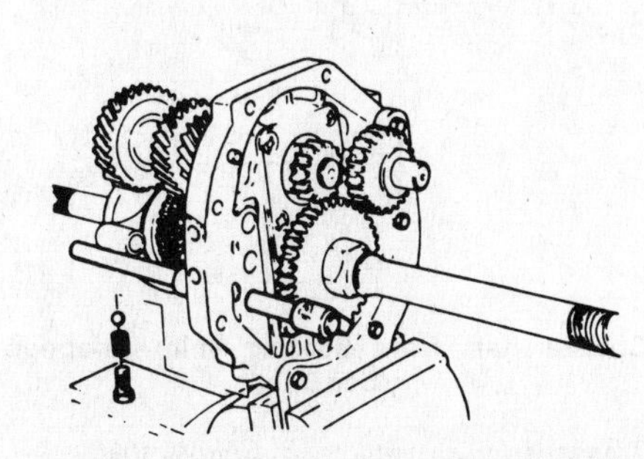

Fig. 6.33 Installing the 1st/2nd selector rod (Sec 9)

6

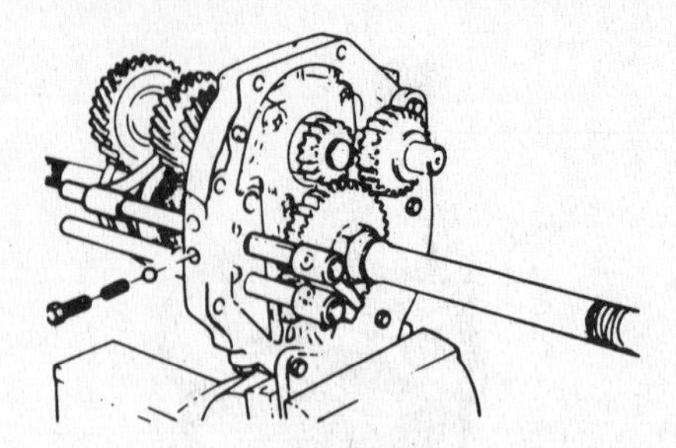

Fig. 6.34 Installing the 3rd/4th selector rod (Sec 9)

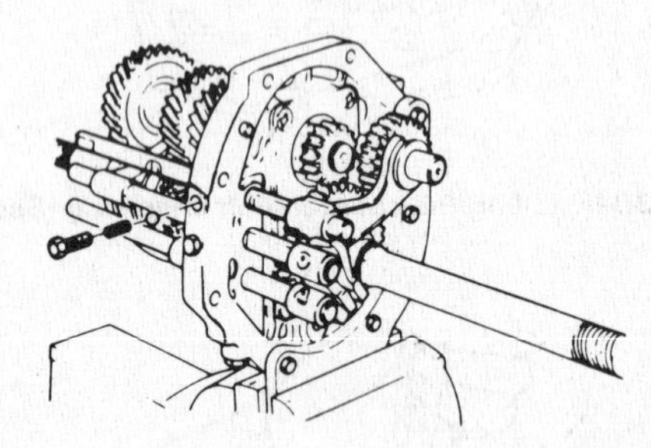

Fig. 6.35 Installing the reverse selector rod (Sec 9)

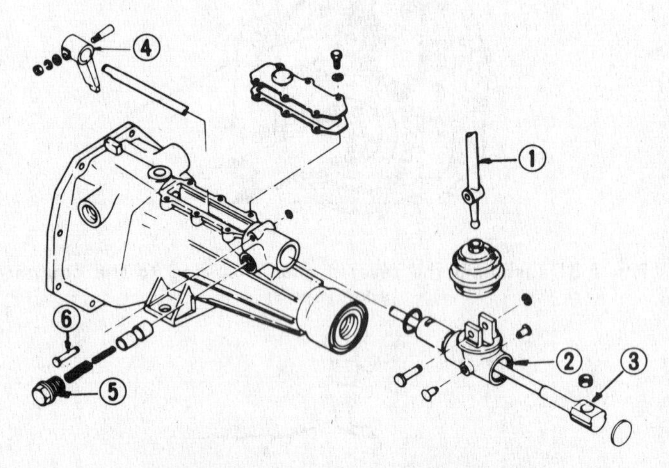

Fig. 6.36 The rear extension housing (Sec 9)

1 Gearshift *4 Striking lever*
2 Striking rod guide *5 Plug*
3 Striking rod *6 Stop pin*

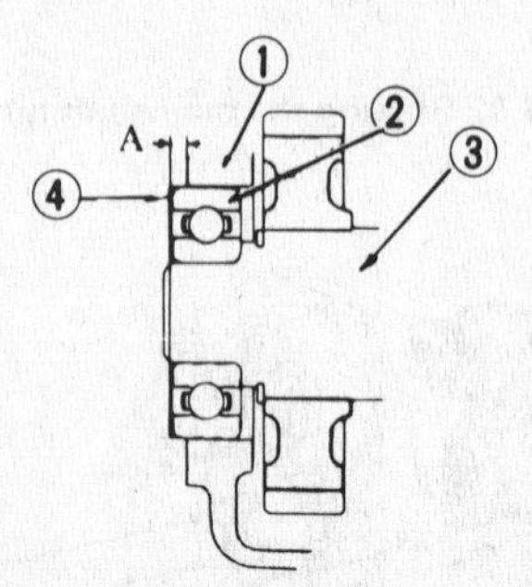

Fig. 6.37 Countershaft front bearing shim selection diagram (Sec 9)

1 Transmission casing *3 Countershaft*
2 Front bearing *4 Shim*

15 Install the reverse gear to the countershaft and use a snap-ring from the thicknesses listed in Section 3 to give minimum endfloat.
16 Install the reverse idler gear to the reverse idler shaft.
17 Tighten the mainshaft nut (after locking the gears) to the specified torque.
18 Stake the collar of the nut into the groove of the mainshaft.
19 Locate the 1st/2nd shift fork onto the 1st/2nd synchronizer unit, (the long end of the shift fork must be towards the countershaft). Now locate the 3rd/4th shift fork onto the 3rd/4th synchronizer unit, (the long end of the shift fork must be the opposite side to the 1st/2nd shift fork).
20 Slide the 1st/2nd selector rod through the adaptor plate and into the 1st/2nd shift fork; align the hole in the rod with the hole in the fork and drive in a new retaining pin.
21 Align the notch in the 1st/2nd selector rod with the detent (check) ball bore, then install the detent (check) ball, spring and screw in the detent ball plug. Apply a little thread sealant to the detent ball plug (Fig. 6.11).
22 Now invert the adaptor plate assembly (hold the 3rd/4th shift fork in position) so that the check ball plug assembled at paragraph 21, is lowermost. Drop two interlock balls into the 3rd/4th detent ball plug hole and, using a suitable thin probe, push them up against the 1st/2nd selector rod (if the adaptor plate is correctly positioned, the interlock balls will drop into position). Slide the 3rd/4th selector rod through the adaptor plate, ensuring that the interlock balls are held between this selector rod and the 1st/2nd selector rod, and into the 3rd/4th shift fork. Align the holes in the shift fork and selector rod, and drive in a new retaining pin. Now install a detent ball, spring and detent ball plug (with thread sealant applied) to the 3rd/4th detent ball plug bore. Ensure that the notch in the 3rd/4th selector rod is aligned with the detent ball plug bore before assembling the detent ball.
23 Drop two interlock balls into the remaining detent ball plug bore, ensuring that they locate against the 3rd/4th selector rod. Locate the reverse shift fork to the reverse idler gear and slide the reverse selector rod through the reverse shift fork and into the adaptor plate. Ensure that the two interlock balls are held in position between the 3rd/4th selector rod and the reverse selector rod, sliding the reverse selector rod into the adaptor plate until the notch in the selector rod aligns with the detent ball plug bore. Insert the detent ball, spring and detent ball plug as before. Drive in a new retaining pin to retain the reverse shift fork to the reverse selector rod.
24 Finally, tighten the three detent ball plugs to the specified torque.
25 Thoroughly oil the entire assembly and check to see that the selector rods operate correctly and smoothly.
26 Clean the mating faces of the adaptor plate and the transmission casing and apply gasket sealant to both surfaces.
27 Tap the transmission casing into position on the adaptor plate using a soft faced mallet, taking particular care that it engages correctly with the input shaft bearing and countershaft front bearing.
28 Fit the outer snap-ring to the input shaft bearing.
29 Clean the mating faces of the adaptor plate and rear extension housing and apply gasket sealant.
30 Arrange the shift forks in their neutral mode and then lower the rear extension housing onto the adaptor plate so that the striking lever engages correctly with the selector rods.
31 Fit the tie bolts which secure the sections of the transmission together and tighten them to the specified torque.
32 Measure the amount by which the countershaft front bearing stands proud of the transmission casing front face. Use feeler blades for this and then select the appropriate shims after reference to the following table:

Projection (A)	*Shim (thickness)*
0.1150/0.1185 in (2.92/3.01 mm)	*0.0236 in (0.6 mm)*
0.1189/0.1224 in (3.02/3.11 mm)	*0.0197 in (0.5 mm)*
0.1228/0.1264 in (3.12/3.21 mm)	*0.0157 in (0.4 mm)*
0.1268/0.1303 in (3.22/3.31 mm)	*0.0118 in (0.3 mm)*
0.1307/0.1343 in (3.32/3.41 mm)	*0.0079 in (0.2 mm)*
0.1346/0.1382 in (3.42/3.51 mm)	*0.0039 in (0.1 mm)*

33 Stick the shim in position using a dab of thick grease, then fit the front cover to the transmission casing (within the clutch bellhousing) complete with a new gasket and taking care not to damage the oil seal as it passes over the input shaft splines.
34 Tighten the securing bolts to the specified torque, making sure that the bolt threads are coated with gasket sealant to prevent oil seepage.

35 Install the speedometer pinion assembly to the rear extension housing.
36 Install the reversing lamp switch tightening it to the specified torque.
37 Install the release bearing and withdrawal lever within the clutch bellhousing (Chapter 5).
38 Check the security of the drain plug.

10 Fault diagnosis – manual transmission

Symptom	Reason/s
Weak or ineffective synchromesh	Synchronising cones worn, split or damaged Baulk ring synchromesh dogs worn or damaged
Jumps out of gear	Broken shift fork rod spring Transmission coupling dogs badly worn Selector fork rod groove badly worn
Excessive noise	Incorrect grade of oil in transmission or oil level too low Bush or needle roller bearings worn or damaged Gear teeth excessively worn or damaged Countershaft snap-rings worn allowing excessive endplay
Excessive difficulty in engaging gear	Clutch pedal adjustment incorrect

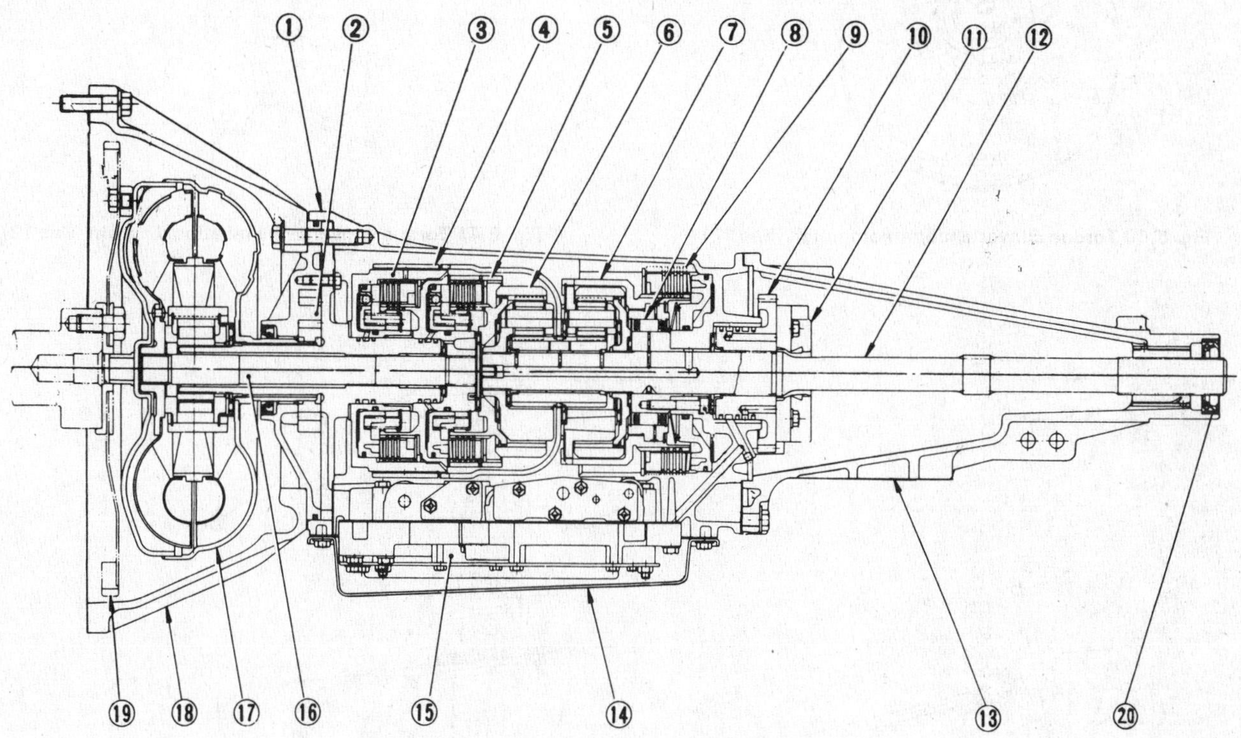

Fig. 6.38 Sectional view of the automatic transmission unit

1 Transmission housing
2 Oil pump
3 Front clutch
4 Brake band
5 Rear clutch
6 Front planetary gear
7 Rear planetary gear
8 One way clutch
9 Low/reverse brake
10 Oil distributor
11 Governor
12 Output shaft
13 Rear extension
14 Oil pan
15 Control valve
16 Input shaft
17 Torque converter
18 Converter housing
19 Drive plate
20 Rear extension oil seal

PART 2: AUTOMATIC TRANSMISSION

11 General description

The automatic transmission unit fitted to the Datsun 810 range is the type JATCO 3N71B .

The unit provides three forward ratios and one reverse. Changing of the forward gear ratios is completely automatic in relation to the vehicle speed and engine torque output and is dependent upon the vacuum pressure in the manifold and the vehicle road speed to actuate the gear change mechanism at the precise time.

The transmission has six selector positions:

P – Parking position which locks the output shaft to the interior wall of the transmission housing. This is a safety device for use

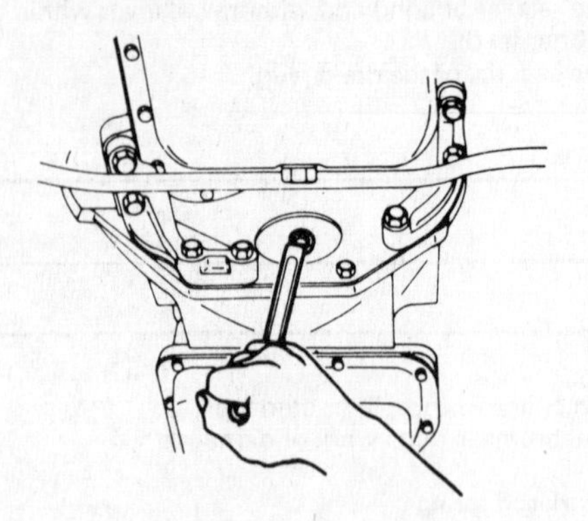
Fig. 6.39 Unscrewing a driveplate-to-torque converter bolt (Sec 12)

when the vehicle is parked on an incline. The engine may be started with 'P' selected and this position should always be selected when adjusting the engine while it is running. Never attempt to select 'P' when the vehicle is in motion.
R – Reverse gear.
N – Neutral. Select this position to start the engine or when idling in traffic for long periods.
D – Drive, for all normal motoring conditions.
2 – Locks the transmission in second gear for wet road conditions or steep hill climbing or descents. The engine can be over revved in this position.
1 – The selection of this ratio above road speeds of approximately 25mph (40kph) will engage second gear and as the speed drops below 25 mph (40kph) the transmission will lock into first gear. Provides maximum retardation on steep descents.

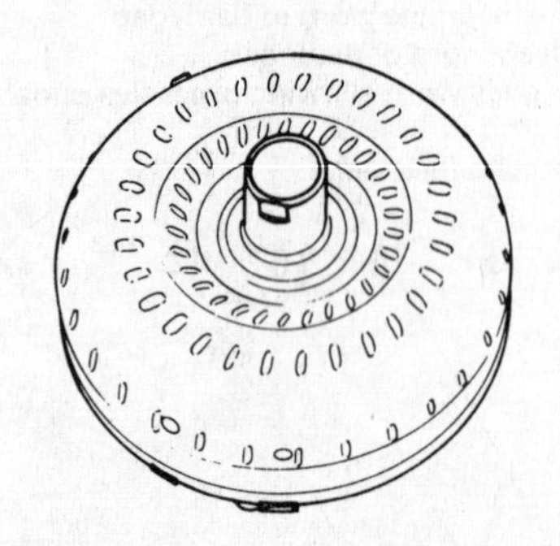
Fig. 6.40 Torque converter alignment notch (Sec 12)

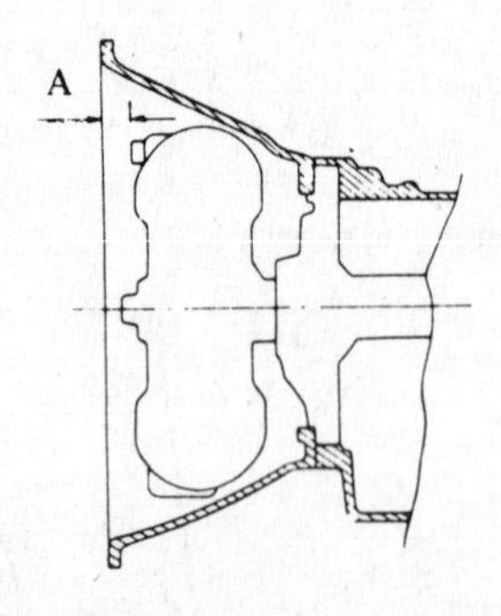

Fig. 6.41 Torque converter installation diagram (Sec 12)

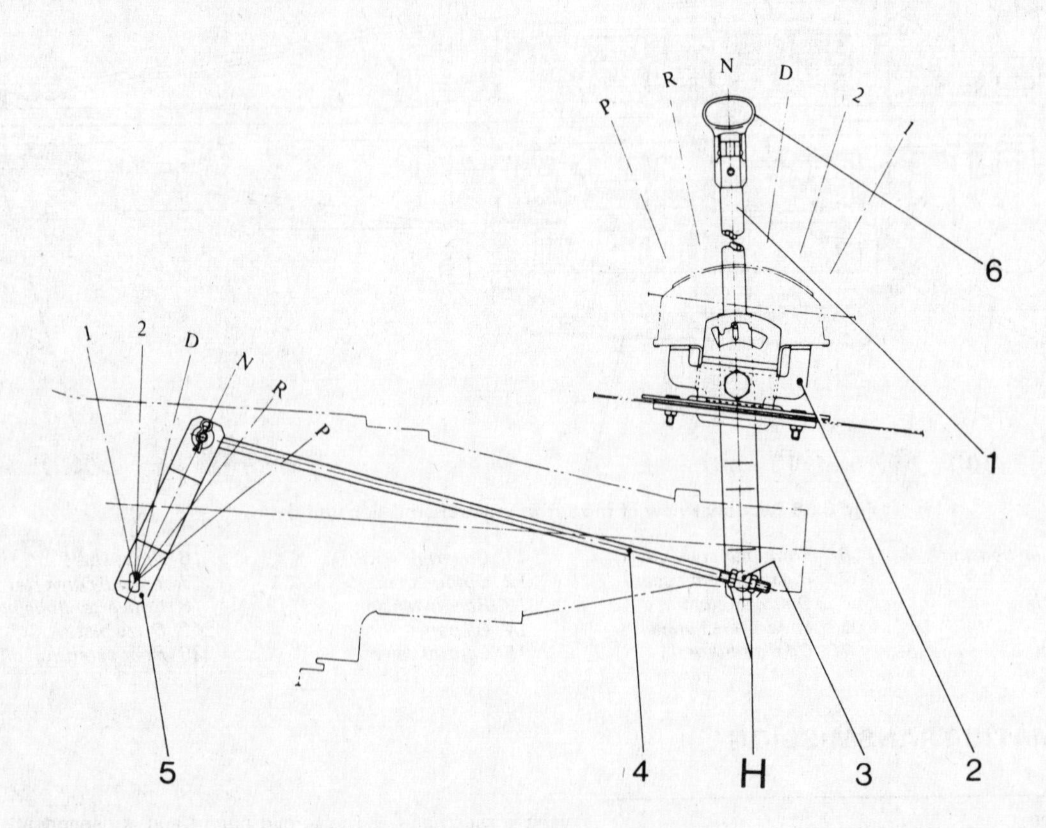

Fig. 6.42 The speed selector linkage (Sec 13)

1 Control lever
2 Bracket
3 Trunnion
4 Selector rod
5 Selector range lever
6 Control lever knob

Due to the complexity of the automatic transmission unit, any internal adjustment or servicing should be left to a main Datsun agent. The information given in this Chapter is therefore confined to those operations which are considered within the scope of the home mechanic. An automatic transmission should give many tens of thousands of miles service provided normal maintenance and adjustment is carried out. When the unit finally requires major overhaul, consideration should be given to exchanging the old transmission for a factory reconditioned one, the removal and installation being well within the capabilities of the home mechanic as described later in this Chapter. The hydraulic fluid does not require periodic draining or refilling but the fluid level must be regularly checked and maintained as described in the Routine Maintenance Section at the front of this manual.

Periodically clean the outside of the transmission housing as the accumulation of dirt and oil is liable to cause overheating of the unit under extreme conditions.

12 Automatic transmission – removal and installation

1 Removal of the engine and automatic transmission as a combined unit is described in Chapter 1 of this manual. Where it is decided to remove the transmission leaving the engine in position in the vehicle, proceed as follows:
2 Disconnect the battery ground lead.
3 Drain the fluid from the transmission unit, retaining it in a clean container if required for further use.
4 Jack the car to an adequate working height and support on stands or blocks.
5 Disconnect the exhaust downpipe and the accelerator linkage. On California vehicles, disconnect the catalyzer harness and remove the harness protector. The catalytic converter and shield should also be removed as described in Chapter 3.
6 Disconnect the wires from the starter inhibitor switch.
7 Disconnect the wire from the downshift solenoid.
8 Disconnect the vacuum pipe from the vacuum capsule which is located just forward of the downshift solenoid.
9 Separate the selector lever from the selector linkage.
10 Disconnect the speedometer drive cable from the rear extension housing.
11 Disconnect the fluid filler tube. Plug the opening.
12 Disconnect the fluid cooler tubes from the transmission casing and plug the openings.
13 Mark the edges of the propeller shaft rear driving flange and the pinion flange (for exact alignment or installation), remove the four retaining bolts and withdraw the propeller shaft from its connection with the transmission rear extension housing.
14 Support the engine oil pan with a jack and use a block of wood to prevent damage to the surface of the oil pan.
15 Remove the cover from the lower half of the torque converter housing. Mark the torque converter housing and drive plate in relation to each other for exact replacement.
16 Unscrew and remove the four bolts which secure the torque converter to the drive plate. Access to each of these bolts in turn is obtained by rotating the engine slowly, using a wrench on the crankshaft pulley bolt.
17 Unbolt and withdraw the starter motor.
18 Support the transmission with a jack (preferably a trolley type).
19 Detach the rear transmission mounting from the transmission housing and the vehicle body frame.
20 Unscrew and remove the transmission-to-engine securing bolts.
21 Lower the two jacks sufficiently to allow the transmission unit to be withdrawn from below and to the rear of the vehicle. The help of an assistant will probably be required due to the weight of the unit.
22 Installation is a reversal of removal but should the torque converter have been separated from the main assembly, ensure that the notch on the converter is correctly aligned with the corresponding one on the oil pump. To check that the torque converter has been correctly installed, the dimension A should exceed 0.846 in (21.5 mm) (Figs. 6.40 and 6.41).
23 Tighten all bolts to the specified torque settings and refill the unit with the correct grade and quantity of fluid.
24 Check the operation of the inhibitor switch and the selector linkage and adjust if necessary as described later in this Chapter.

13 Selector linkage – adjustment

1 Working beneath the vehicle, slacken the adjuster nuts H (Fig. 6.42).
2 Set the control lever (1) and the selector lever on the transmission to the 'N' position.
3 Turn the adjuster nuts until both levers are in the centers of their detents, without any tendency for the selector rod to push or pull one lever against the other, in an effort to override the selector position.
4 Check the operation of the linkage in all speed selector positions. Worn pivots or elongated holes in the lever will cause incorrect operation and they should be renewed.

14 Selector linkage – removal and installation

1 Remove the two small screws which secure the knob to the speed selector lever. Remove the knob.
2 Remove the console from the transmission tunnel.
3 Unbolt the selector lever bracket and the lever on the side of the transmission and withdraw the complete selector linkage.
4 Installation is a reversal of removal but adjust the linkage as described in the preceding Section.

15 Kickdown switch and downshift solenoid – checking

1 If the kickdown facility fails to operate or operates at an incorrect change point, first check the security of the switch on the accelerator pedal arm and the wiring between the switch and the solenoid.
2 Turn the ignition key so that the ignition and oil pressure lamps illuminate but without operating the starter motor. Depress the accelerator pedal fully and as the switch actuates, a distinct click should be heard from the solenoid. Where this is absent, drain approximately 3 US pt (1.5 liters) of fluid from the transmission unit, unscrew the solenoid and install a new one. Replenish the transmission fluid.

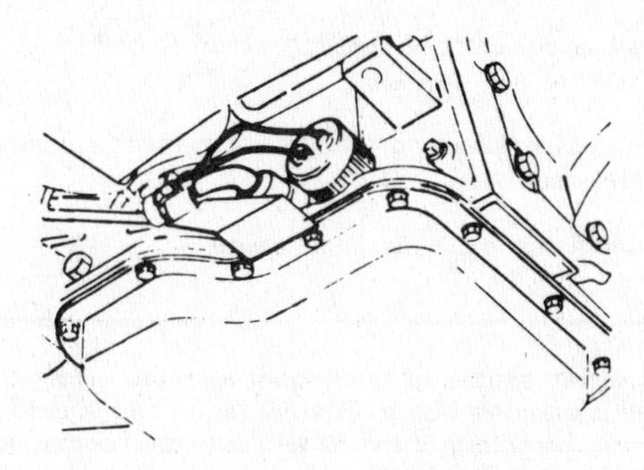

Fig. 6.43 Location of the downshift solenoid (Sec 15)

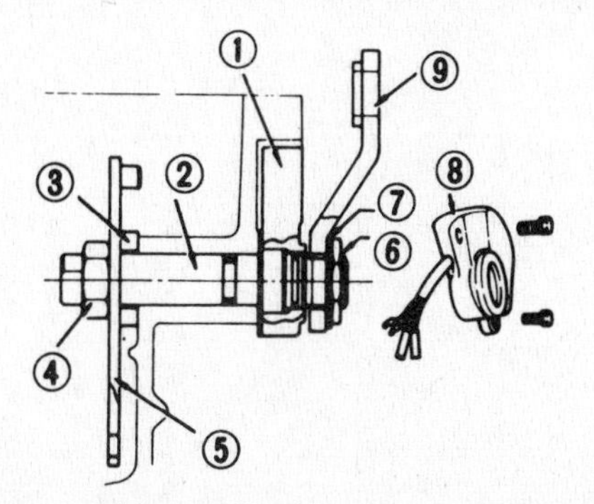

Fig. 6.44 Starter inhibitor and back-up lamp switch (Sec 16)

1 Switch	*4 Nut*	*7 Washer*
2 Shaft	*5 Plate*	*8 Switch (detached)*
3 Washer	*6 Nut*	*9 Range select lever*

16 Starter inhibitor and reverse lamp switch – testing and adjustment

1 Check that the starter motor operates only in 'N' and 'P' and the reversing lamps illuminate only with the selector lever in 'R'.
2 Any deviation from this arrangement should be rectified by adjustment, first having checked the correct setting of the selector linkage.
3 Detach the range selector lever from the selector rod which connects it to the hand control. Now move the range selector lever to the 'N' position, (slot in shaft vertical).
4 Connect an ohmmeter to the black and yellow wires of the inhibitor switch. With the ignition switch on, the meter should indicate continuity of circuit when the range select lever is within 3 degrees (either side) of the 'N' and 'P' positions.
5 Repeat the test with the meter connected to the red and black wires and the range lever in 'R'.
6 Where the switch requires adjusting to provide the correct moment of contact in the three selector positions, move the range level to 'N' and then remove the retaining nut, the two inhibitor switch securing bolts and the screw located below the switch.
7 Align the hole, from which the screw was removed, with the pinhole in the manual shaft. A thin rod or piece of wire may be used to do this. Holding this alignment, install the inhibitor switch securing bolts and tighten them. Remove the alignment rod and install the screw.
8 Install the remaining switch components and test for correct operation as previously described. If the test procedure does not prove positive, renew the switch.

17 Rear extension oil seal – renewal

1 After a considerable mileage, leakage may occur from the seal which surrounds the shaft at the rear end of the automatic transmission extension housing. This leakage will be evident from the state of the underbody and from the reduction in the level of the hydraulic fluid.
2 Remove the propeller shaft as described in Chapter 7.
3 Taking care not to damage the splined output shaft and the alloy housing, pry the old seal from its location. Drive in the new one using a tubular drift.
4 Should the seal be very tight in its recess, then remove the transmission housing securing bolts.
5 Pull the extension housing straight off over the output shaft and governor assembly.
6 Using a suitable drift applied from the interior of the rear extension housing, remove the old oil seal. At the same time check the bush and renew it if it is scored or worn.
7 Installation is a reversal of removal, but always use a new gasket between the rear extension and main housing, and tighten the securing bolts to the specified torque.

18 Fault diagnosis – automatic transmission

In addition to the information given in this Chapter, reference should be made to Chapter 3 for the servicing and maintenance of the emission control equipment fitted to models equipped with automatic transmission.

Symptom	Reason/s
Engine will not start in 'N' or 'P'	Faulty starter or ignition circuit Incorrect linkage adjustment Incorrectly installed inhibitor switch
Engine starts in selector positions other than 'N' or 'P'	Incorrect linkage adjustment Incorrectly installed inhibitor switch
Severe bump when selecting 'D' or 'R' and excessive creep when handlebar released	Idle speed too high Vacuum circuit leaking
Poor acceleration and low maximum speed	Incorrect oil level Incorrect linkage adjustment

The most likely causes of faulty operation are incorrect oil level and linkage adjustment. Any other malfunction of the automatic transmission unit must be due to internal faults and should be rectified by your Datsun dealer. An indication of a major internal fault may be gained from the colour of the oil which under normal conditions should be transparent red. If it becomes discolored or black then burned clutch or brake bands must be suspected.

Chapter 7 Propeller shaft

Refer to Chapter 13 for specifications and information applicable to 1980 through 1984 models.

Contents

Specifications

Type	Tubular steel with three universal joints, center support bearing and sliding sleeve at front end	
Universal joint axial play	Less than 0·0008 in (0·02 mm)	
Torque wrench settings	**lbf ft**	**kgf m**
Propeller shaft-to-companion flange bolts (axle end)	17 to 24	2·4 to 3·3
Center flange-to-flange yoke (rear shaft) bolts	17 to 24	2·4 to 3·3
Center flange nut	145 to 174	20·0 to 24·0
Center bearing bracket fixing nuts	5·71 to 8	0·79 to 1·06
Center bearing bracket-to-body bolts	37 to 50	5·1 to 6·9

1 General description

The Datsun 810 uses a two-piece propeller shaft, which is supported by a center bearing mounted to the chassis frame. It incorporates three universal joints; one at the transmission end, one behind the center bearing and one at the axle companion flange. The universal joints are of the sealed-type; therefore no maintenance can be carried out on them.

The propeller shaft is finely balanced during manufacture and, in the case of worn universal joints, a replacement propeller shaft will have to be purchased. When removing the propeller shaft to service the center bearing, the yoke-to-companion flange relationships must be maintained. This is best ensured by marking each joint before removal.

2 Universal joints – testing for wear

1 Wear in the needle roller bearings is characterized by vibration in the transmission, 'clonks' on taking up the drive and, in extreme cases, metallic squeaking and ultimately grating and shrieking sounds as the bearings break up.

2 It is easy to check if the needle roller bearings are worn with the propeller shaft in position, by trying to turn the shaft with one hand, the other hand holding the rear axle flange when the rear universal joint is being checked, and the front half coupling when the front universal joint is being checked. Any movement between the propeller shaft and the front half couplings, and round the rear half couplings, is indicative of considerable wear.

3 If wear is evident, a new propeller shaft will have to be purchased, as it is not possible to repair the universal joints.

4 A final test for wear, is to attempt to lift the shaft and note any movement between the yokes of the joints.

3 Propeller shaft – removal and installation

1 Jack-up the rear of the car, or position the rear of the car over a pit.

2 If the rear of the car is jacked-up, supplement the jack with support blocks so that danger is minimized should the jack collapse.

3 If the rear wheels are off the ground, place the car in gear and apply the handbrake to ensure that the propeller shaft does not turn when an attempt is made to loosen the four bolts securing the propeller shaft to the rear axle.

4 The propeller shaft is carefully balanced to fine limits and it is important that it is installed in exactly the same position it was in prior to removal. Scratch marks on the propeller shaft to axle companion flange; also mark the center flange-to-propeller shaft joint, to ensure accurate mating when the time comes for installation.

5 Unscrew and remove the four bolts and spring washers which hold the flange on the propeller shaft to the flange on the rear axle.

6 Unscrew and remove the bolts which secure the bearing carrier to the body frame, at the center of the propeller shaft.

7 Slightly push the shaft forward to separate the two flanges, then lower the end of the shaft and pull it rearward to disengage it from the transmission mainshaft splines.

8 Place a can or tray under the rear of the transmission extension to catch any oil which may leak past the oil seal when the propeller shaft is removed.

9 Installation is a reversal of removal but ensure that the rear flange marks are in alignment and tighten all securing bolts to the specified torque. Check the oil level in the transmission unit and top-up if necessary.

4 Center bearing – dismantling and reassembly

1 With the propeller shaft removed as already described, disconnect the rear section of the shaft from the front section by unbolting the flanges which are located just to the rear of the center bearing, (do not forget to mark,the flange alignment).

2 Relieve the caulking on the locknut which is now exposed, and unscrew and remove the locknut. This nut will be very tight and the best method of holding the shaft still while it is unscrewed is to pass two old bolts through two of the flange holes and secure them in a vise. Alternatively, a special flange securing wrench can be used.

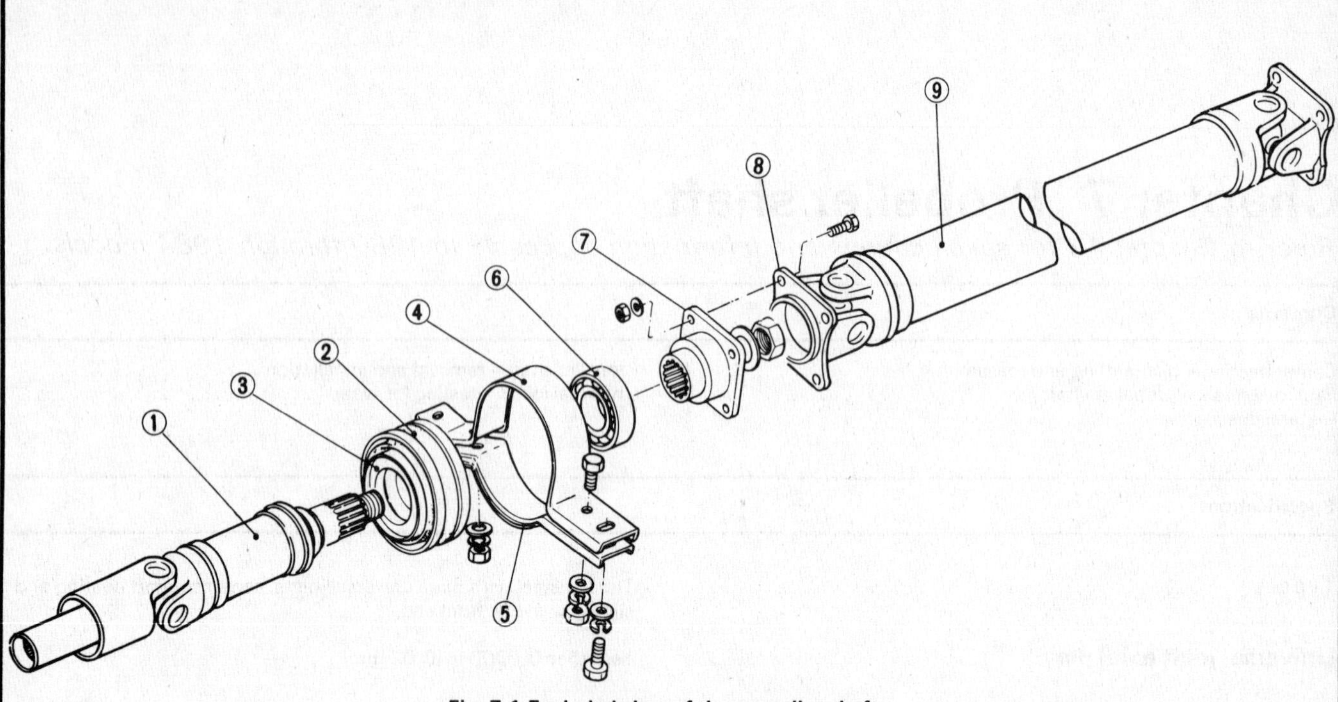

Fig. 7.1 Exploded view of the propeller shaft

1 Front shaft
2 Cushion rubber
3 Bearing insulator
4 Center bearing bracket
5 Center bearing support
6 Center bearing
7 Center flange
8 Flange yoke
9 Rear shaft

Fig. 7.2 The propeller shaft rear flange securing bolts arrowed

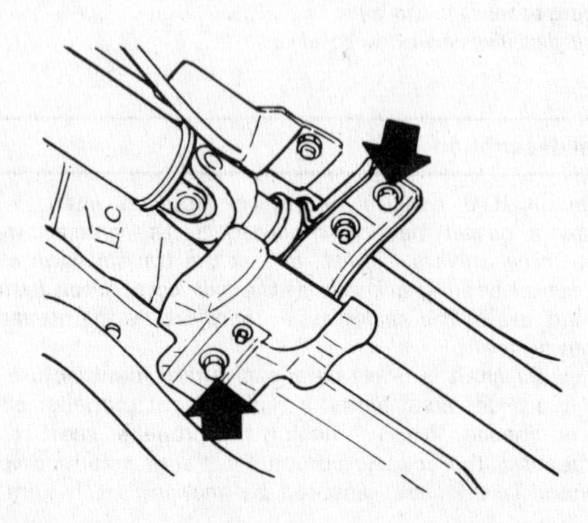

Fig. 7.3 The center bearing mounting bracket bolts arrowed

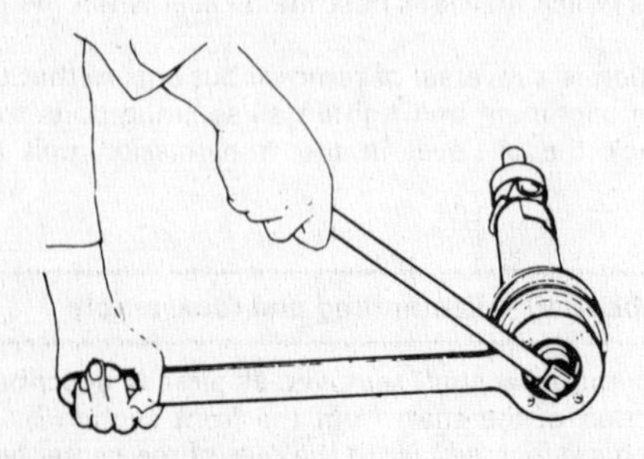

Fig. 7.4 Removing the center flange locknut with a securing wrench

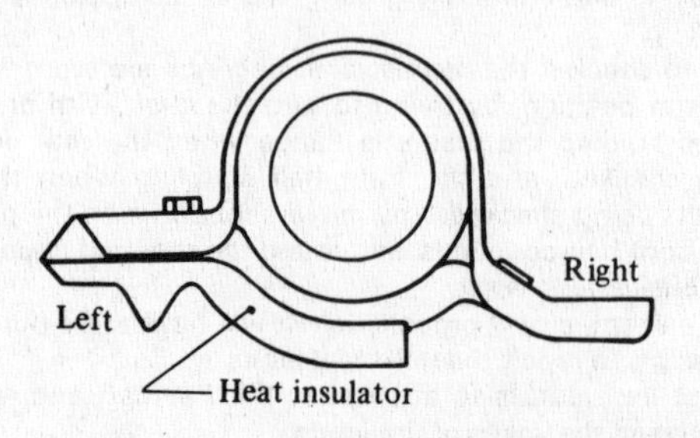

Fig. 7.5 Correct installation of the center bearing bracket

3 Mark the relative positions of the center bearing assembly to the clamp/support, and the bearing and clamp/support to the propeller shaft itself.
4 Unbolt and remove the bearing clamp/support and, using a suitable extractor, draw off the center bearing complete with insulator.
5 Commence reassembly by inserting the center bearing into its insulator. If a new bearing is being installed, do not lubricate it as it is of a grease-sealed type.
6 Install the bearing/insulator and center flange to the shaft and align the match marks previously made.
7 When installing the bearing assembly and the center flange, considerable pressure will be needed to ensure that the bearing assembly and center flange are fully seated. If a suitable press is not available, provided the shaft is suitably supported, and cannot slip or be damaged in any way, a piece of tube with a suitable diameter, can be placed against the center flange and a series of carefully aimed blows with a hammer should seat the assembly.
8 Install the washer and locknut, and tighten to the specified torque. Stake the nut into the shaft groove using a punch.
9 Reconnect the front shaft to the rear shaft, tightening the securing bolts to the specified torque.
10 Install the clamp/support to the center bearing, aligning it correctly, and tighten the bolts to the specified torque.

5 Fault diagnosis – propeller shaft

Symptom	Reason/s
Vibration when car is running on road	Out of balance shaft Wear in splined shaft Loose flange securing bolts Worn universal joint bearings Worn center bearing

7

Chapter 8 Rear axle

Refer to Chapter 13 for specifications and information applicable to 1980 through 1984 models.

Contents

Specifications

Sedan models

Differential type	Hypoid final drive and differential unit mounted on underside of bodyframe
Type reference number	R180
Final drive ratio	3·700 : 1
Driveshaft type	Open, two universal joints and sliding sleeve
Oil capacity	$2\frac{1}{8}$ US pints (1 liter)

Station Wagon models

Differential type	Hypoid final drive and differential unit; semi-floating rear axle
Type reference number	H190
Final drive ratio	3·700 : 1
Oil capacity	$2\frac{1}{8}$ US pints (1 liter)

Torque wrench settings	**lbf ft**	**kgf m**
Sedan		
Pinion nut (see text)	123 to 145	17 to 20
Driveshaft yoke bolt	23 to 31	3·2 to 4·3
Driveshaft outer flange bolt	36 to 43	5 to 6
Rear hub bearing locknut (see text)	180 to 240	25 to 33
Shock absorber lower end bolts	43 to 50	6 to 8
Front differential mounting bolts	43 to 58	6 to 8
Differential mounting bar-to-body nut	61 to 83	8·5 to 11·5
Suspension cross-member-to-body nut	58 to 80	8 to 11
Station Wagon		
Pinion nut (see text)	101 to 123	14 to 17
Differential carrier retaining nuts	14 to 18	2·0 to 2·5
U-bolt nut	43 to 47	6·0 to 6·5
Shock absorber lower end nut	26 to 35	3·6 to 4·8
Brake backplate nut	16 to 20	2·2 to 2·7

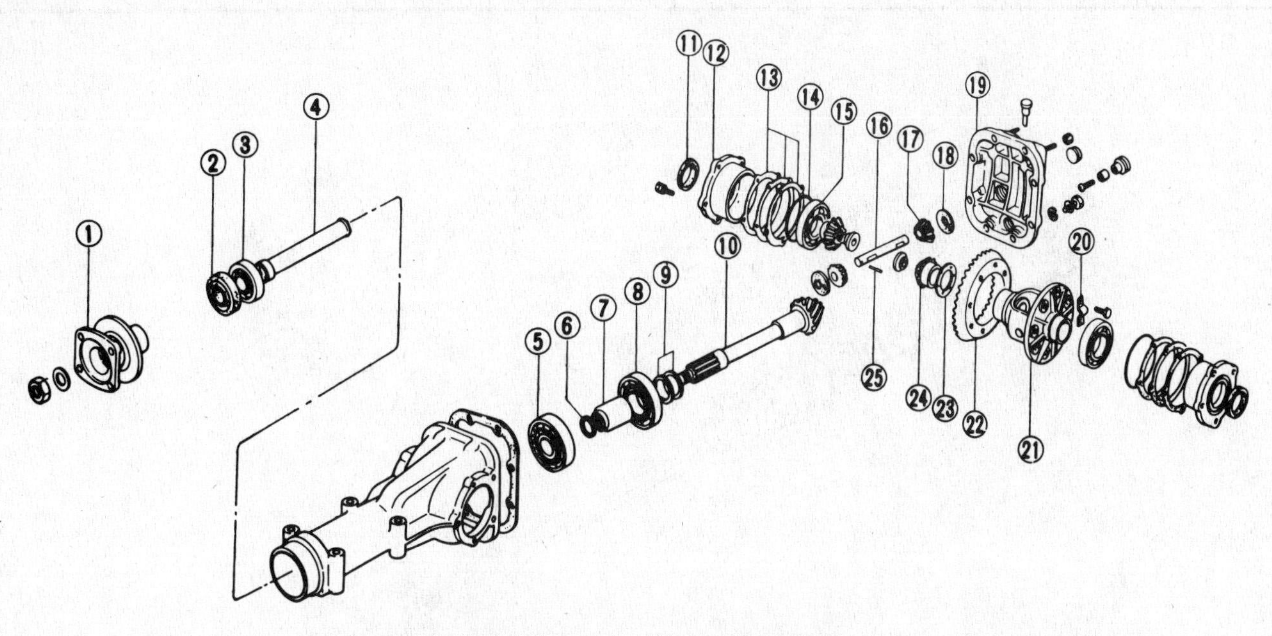

Fig. 8.1 Exploded view of the differential unit (Sedan)

1 *Companion flange*
2 *Oil seal*
3 *Bearing*
4 *Spacer*
5 *Bearing*
6 *Adjusting washer*
7 *Adjusting spacer*
8 *Bearing*
9 *Adjusting washer*
10 *Drive pinion*
11 *Oil seal*
12 *Retainer flange*
13 *Adjusting shim*
14 *O-ring*
15 *Side bearing*
16 *Pinion shaft*
17 *Pinion gear*
18 *Thrust washer*
19 *Rear cover*
20 *Lock strap*
21 *Differential gear case*
22 *Crownwheel*
23 *Thrust washer*
24 *Side gear*
25 *Lock pin*

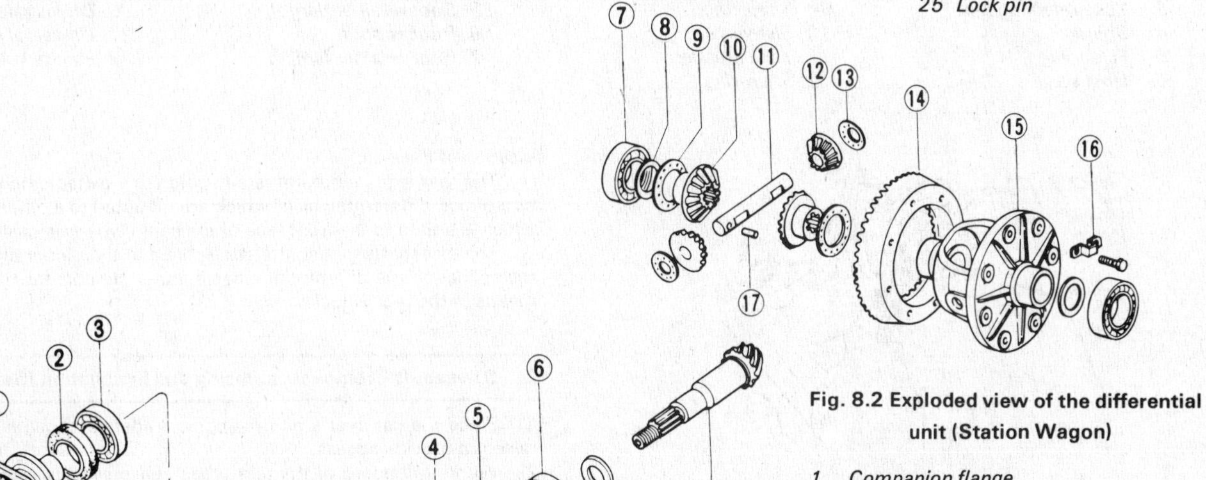

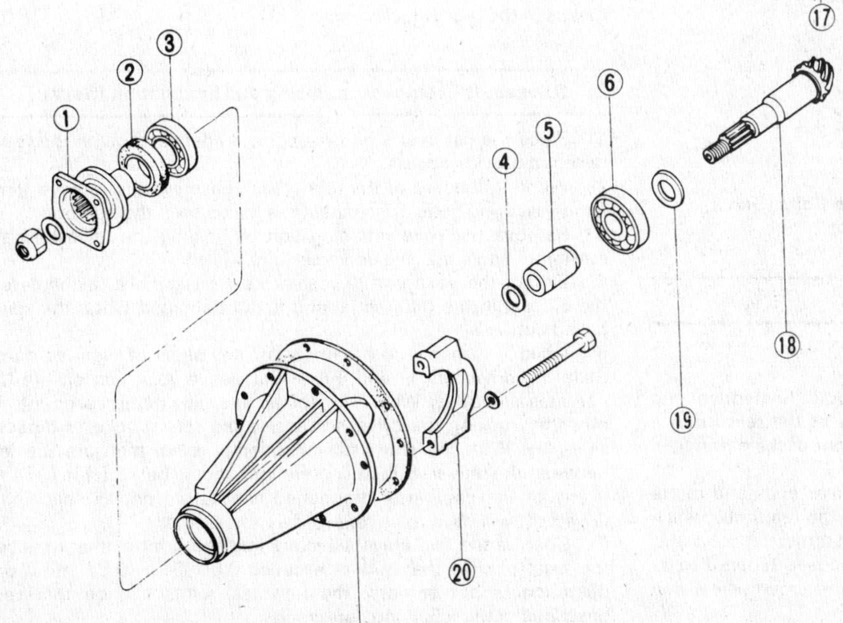

Fig. 8.2 Exploded view of the differential unit (Station Wagon)

1 *Companion flange*
2 *Oil seal*
3 *Pinion front bearing*
4 *Adjusting washer*
5 *Adjusting spacer*
6 *Pinion rear bearing*
7 *Bearing*
8 *Adjusting washer*
9 *Thrust washer*
10 *Differential gear*
11 *Pinion shaft*
12 *Pinion gear*
13 *Thrust washer*
14 *Crownwheel*
15 *Differential gear case*
16 *Lock strap*
17 *Lock pin*
18 *Drive pinion*
19 *Adjusting washer*
20 *Bearing cap*
21 *Differential housing*

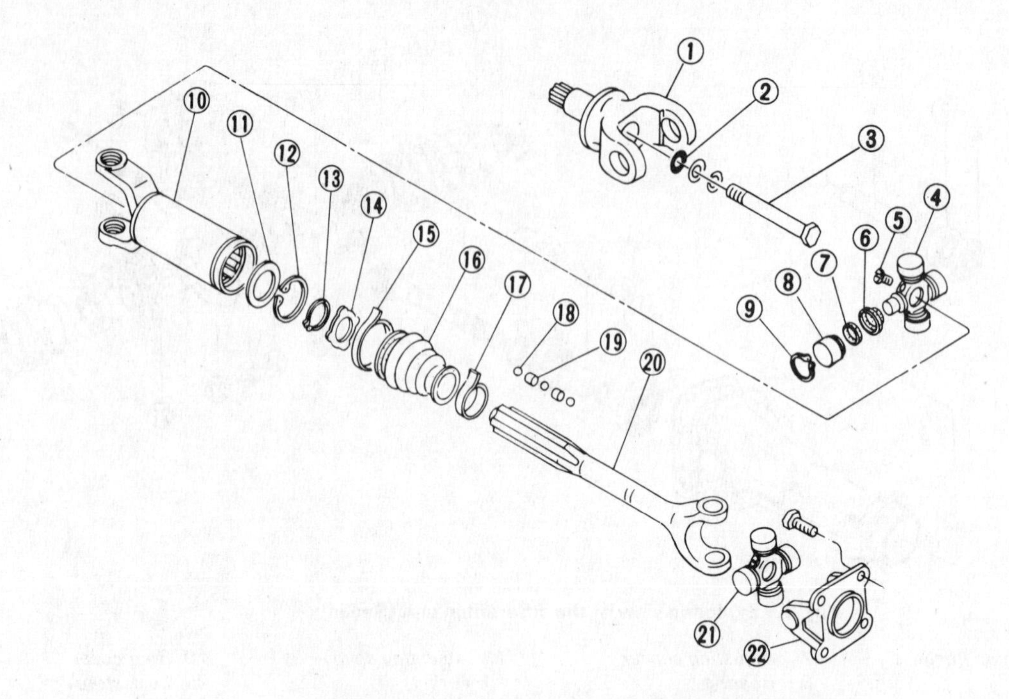

Fig. 8.3 An exploded view of a driveshaft (Sedan) (Sec 2)

1 *Side yoke*
2 *O-ring*
3 *Yoke retaining bolt*
4 *Spider*
5 *Filler plug*
6 *Dust seal*
7 *Oil seal*
8 *Bearing race assembly*
9 *Snap-ring*
10 *Sleeve yoke*
11 *Yoke stopper*
12 *Snap-ring*
13 *Driveshaft snap-ring*
14 *Driveshaft stopper*
15 *Boot retainer (large)*
16 *Rubber boot*
17 *Boot retainer (small)*
18 *Ball*
19 *Spacer*
20 *Driveshaft*
21 *Universal joint*
22 *Flange yoke*

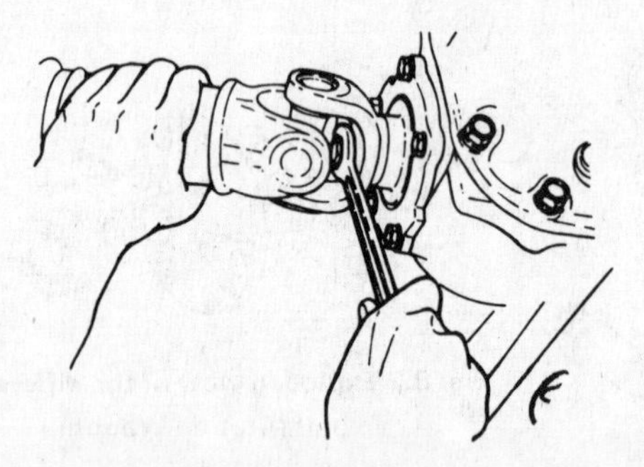

Fig. 8.4 Removing the inner yoke retaining bolts (Sec 2)

1 General description

Sedan models

The main rear axle component is the hypoid final drive and differential unit, which is fixed to the bodyframe at the rear using a cross-bracket located in rubber mountings. The front of the differential unit is mounted to the suspension cross-member.

The driveshafts, which are splined at their inner ends and rotate through two universal joints, transfer the drive to the rear hubs which are mounted on the trailing ends of the suspension arms.

The crownwheel and pinion each run on opposed tapered roller bearings, the bearing preload and meshing of the crownwheel and pinion being controlled by shims.

Station Wagon

The rear axle is semi-floating and is held in place by semi-elliptic springs. These springs provide the necessary lateral and longitudinal location of the axle.

The rear axle incorporates a hypoid crownwheel and pinion, and a two-pinion differential, all of which are mounted to a differential carrier which is bolted to the front face of the banjo type axle casing.

The axleshafts (halfshafts) are splined at their inner ends to fit into the splines in the differential wheels; outer support for the shaft is by means of the rear wheel bearing.

2 Driveshaft – removal, servicing and installation (Sedan)

1 Place the car over a pit or support it adequately with the rear end raised to provide access.

2 Working inboard of the rear wheel, unscrew and remove the four flange securing bolts. Disconnect the flange from the rear hub.

3 Remove the yoke retaining bolt by holding the driveshaft at the maximum angle that the universal joint will allow.

4 Extract the yoke and driveshaft as an assembly, taking care that the oil seal in the differential unit is not damaged when the spline is pulled out.

5 Visually inspect the assembly for any signs of wear or damage. Slide the driveshaft in and out of the sleeve yoke and ensure that it operates smoothly. When the splines are fully compressed into each other try rocking the driveshaft inside the sleeve yoke to detect any side-play. With the shaft still fully compressed, measure the length between the universal joint centers, this should be 13.54 in (344 mm). If any of the previously mentioned checks are not as specified, the driveshaft will have to be renewed as an assembly.

6 Operate the two universal joints and check for a smooth action. If the axial play in the spiders exceeds 0.0008 in (0.02 mm), or the operation is not smooth, the universal joints can be renewed as described in the following paragraphs.

7 Extract the snap-rings from the four journals of the spider, then, using a soft-faced mallet, tap the yoke to remove the bearing races. With the bearing races removed, it is quite simple to remove the spider from the yoke. To reassemble the universal joints, reverse the

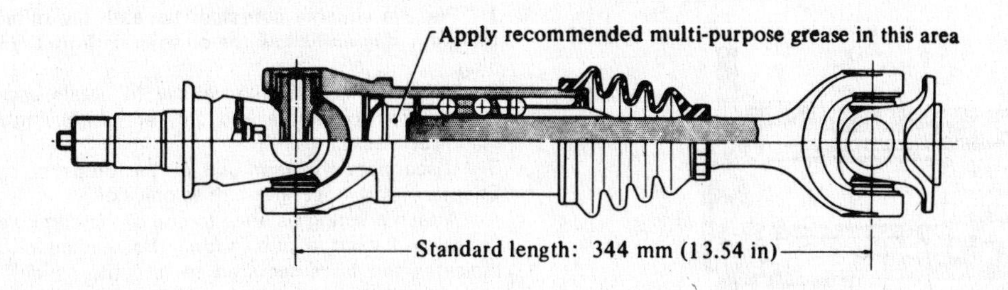

Fig. 8.5 The correct length of the driveshaft (Sec 2)

dismantling procedure.

8 If the driveshaft assembly has to be dismantled (to renew a rubber boot) remove the two boot retainers, peel back the larger diameter of the boot and extract the snap-ring from the sleeve yoke. Remove the yoke stopper and carefully draw out the driveshaft from the sleeve yoke taking care not to loose the balls and spacers.

9 Before commencing reassembly, thoroughly clean all the parts in gasoline; slide the new rubber boot over the driveshaft and align the two yokes so that they are on the same center line. Assemble the yoke stopper and snap-ring to the driveshaft. Fit the steel balls and spacers, after greasing their running groove and the area inside the sleeve yoke, and slide the driveshaft into the sleeve yoke. Now locate the stopper into the sleeve yoke and install the snap-ring.

10 Manipulate the rubber boot into its correct position and secure with the boot retainers.

11 Installation of the driveshaft is a reversal of removal but ensure that the bolts are tightened to the specified torque. When installing the yoke retaining bolt, always use a new O-ring.

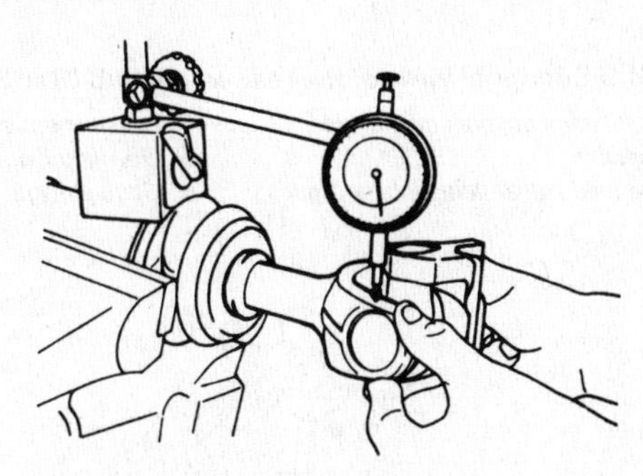

Fig. 8.6 Checking a universal joint for wear (Sec 2)

3 Rear axle stub, bearing and oil seal – dismantling and reassembly (Sedan)

1 Jack-up the rear of the vehicle and remove the roadwheel. Support the bodyframe and the suspension securely.

2 Unscrew and remove the four bolts which secure the driveshaft flange at the wheel hub; move the driveshaft to one side out of the way.

3 Temporarily install the roadwheel and lower the jacks. Apply the parking brake fully and then unscrew the wheel bearing locknut. This nut is very tight and will require leverage from a socket having an operating arm extension of from two to three feet (0.60 to 0.90 meters).

4 Jack-up the vehicle again, remove the brake drum and using a slide hammer extract the rear axle shaft stub. This will come out complete with outer bearing.

5 Remove the inner flange, distance piece and the bearing washer.

6 Using a tubular drift, drive out the inner bearing and oil seal.

7 Examine the condition of the outer bearing/oil seal. If the seal face is cracked or the bearing is rough or noisy in operation, it must be pressed off and a new one installed so that the side of the bearing with the seal will be facing the roadwheel.

8 Drive in the inner bearing and new oil seal.

9 Pack general purpose grease into the bearings and into the space between them, and then install the distance piece. Should the ends of the distance piece be deformed it should be renewed after reference to the grading mark stamped on it (this is repeated on the bearing housing).

10 Insert the axle stub and the inner flange, taking care not to damage the oil seal.

11 Fit the thrust washer and bearing locknut. Tighten the locknut to 180 lbf ft (25 kgf m) after having temporarily installed the roadwheel and brakedrum and lowered the car to the ground. Raise the car again and remove the roadwheel. Attach a spring balance to a roadwheel stud and check the point at which the hub will rotate. This should be at a reading on the spring balance of 2.6 lb (1.2 kg) or less. Where this condition is not met, tighten the locknut progressively up to 240 lbf ft (33 kgf m) testing each increase in torque tightening until the preload is correct. When the preload is correctly set, the endfloat of the rear axleshaft stub should not exceed 0.012 in (0.3 mm).

12 Securely stake the bearing locknut, install the driveshaft, brake drum and roadwheel, and lower the vehicle.

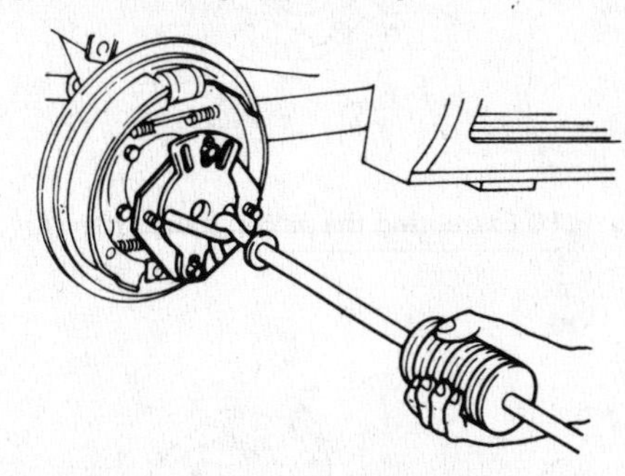

Fig. 8.7 Removing the rear axleshaft stub using a slide hammer (Sec 3)

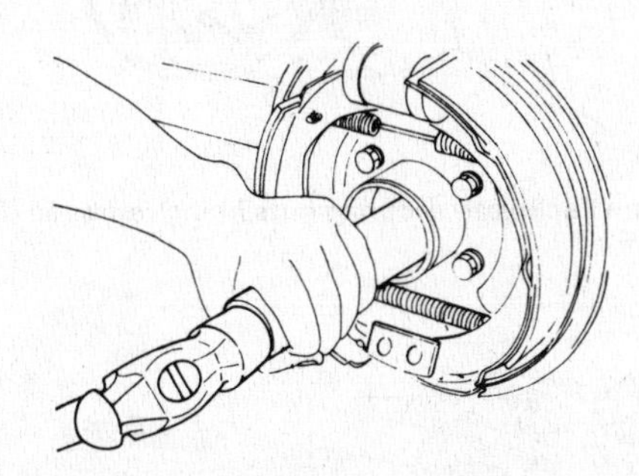

Fig. 8.8 Driving out the inner hub bearing and oil seal using a tubular drift (Sec 3)

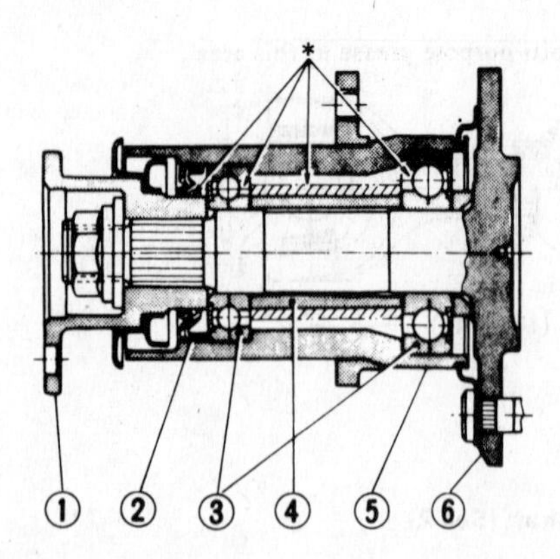

Fig. 8.9 Sectional view of the rear wheel hub (Sec 3)

1 Driveshaft companion flange
2 Oil seal
3 Inner and outer wheel bearings
4 Distance piece
5 Bearing housing
6 Stub flange

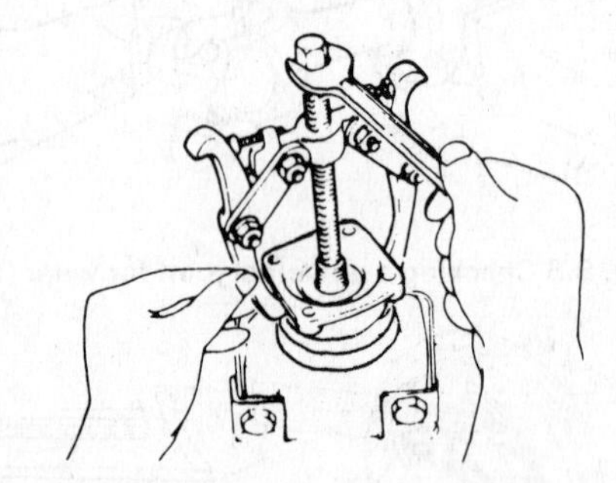

Fig. 8.10 Extracting the pinion flange (Sec 4)

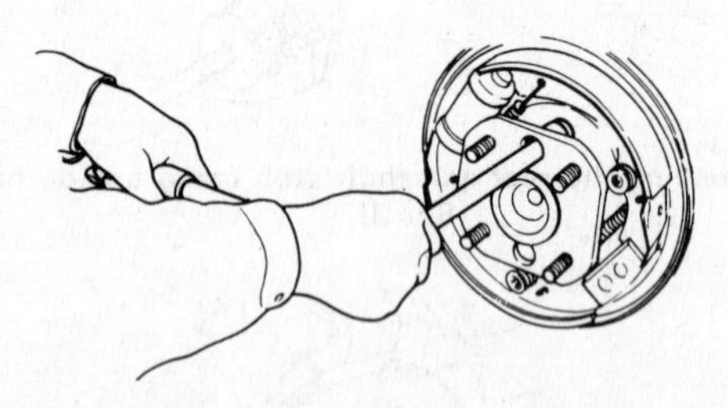

Fig. 8.11 The backplate securing nuts being removed (Sec 8)

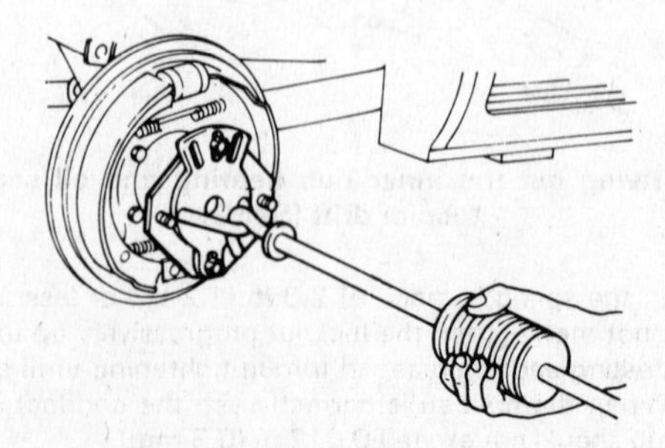

Fig. 8.12 Using a slide hammer to extract the axleshaft (Sec 8)

4 Pinion oil seal – renewal (Sedan)

1 Place a suitable container beneath the differential then, remove the drain plug and allow the oil to drain from the unit. Install the drain plug.
2 Raise the rear of the vehicle to obtain access to the unit, and support the bodyframe and differential housing securely on suitable stands or blocks.
3 Disconnect the rear end of the propeller shaft as described in Chapter 7 and move the shaft to one side.
4 Attach a spring balance to one of the bolt holes in the pinion drive flange and exert a pull, noting the reading on the balance (which indicates the force required to turn the pinion). This is the bearing preload and should be between 7.1 and 8.8 lbf (3.2 and 4.0 kgf).
5 Hold the pinion flange by placing two 2 in (50 mm) long bolts through two opposite holes. Bolting them up tight, undo the self-locking nut whilst holding a large screwdriver or tire lever between the two bolts as a lever.
6 Using a suitable extractor, withdraw the pinion flange from the differential unit.
7 Remove the defective oil seal by driving in one side of it and levering it out.
8 Install the new oil seal first having greased the mating surfaces of the seal and the axle housing. The lips of the oil seal must face inwards. Using a suitable piece of tubing, carefully drive the new oil seal into the axle housing recess until the face of the seal is flush with the housing. Make sure that the end of the pinion is not knocked during this operation.
9 Install the pinion flange and thrust washer, and screw on the pinion nut.
10 Again holding the pinion flange still with the screwdriver or tire lever, tighten the pinion nut to 123 lbf ft (17.0 kgf m).
11 Check the bearing preload by the spring balance method and if necessary tighten the nut by increments in the torque wrench setting of 5lbf ft (0.7 kgf m) until the reading on the spring balance matches that which applied before dismantling.
12 Install the propeller shaft and lower the vehicle to the ground.
13 Refill the differential unit with the correct grade and quantity of oil.

5 Differential side bearing oil seals – renewal (Sedan)

1 Jack-up the rear of the vehicle sufficiently to gain access to the differential unit. Firmly support the bodyframe and the differential unit with suitable blocks or stands.
2 Disconnect the driveshaft flange from the wheel hub by unscrewing and removing the four bolts.
3 Unscrew and remove the single center bolt which retains the inner universal joint yoke to the differential unit.
4 Withdraw the driveshaft complete with splined inner joint yoke.
5 Pry out the defective oil seal and drive in the new one, using a suitable piece of tubing as a drift.
6 Reassembly is a reversal of removal and dismantling, but tighten the flange bolts and the center yoke bolt to the specified torque.

6 Differential unit – servicing general (Sedan)

1 Due to the need for special gauges and tools, it is not recommended that the unit is dismantled.
2 Operations should be limited to the renewal of oil seals which are the only components likely to require renewal after a high mileage.
3 A full description is given to enable the unit to be removed from the vehicle, so that it can be renewed on an exchange basis or taken to a specialist repairer.

7 Differential unit – removal and installation (Sedan)

1 Jack-up the rear of the vehicle and support the bodyframe on suitable stands or blocks.
2 Place jack (preferably a trolley type) beneath the differential unit.
3 Disconnect the propeller shaft from the differential pinion drive flange (Chapter 7).
4 Disconnect the driveshafts and remove them from the vehicle, as

described in Section 2.
5 Raise the jack beneath the differential unit to take its weight.
6 Unscrew and remove the two nuts which hold the differential rear cross-member to the bodyframe.
7 Unscrew and remove the four bolts that secure the front of the differential unit to the suspension cross-member. The differential unit can now be carefully lowered and removed from the underside of the vehicle.
8 Installation is a reversal of removal but ensure that all nuts and bolts are tightened to the specified torque.

8 Axleshaft, bearing and oil seal – removal and installation (Station Wagon)

1 Chock the front wheels, jack-up the rear of the vehicle and place on firmly based axle stands. Remove the rear wheel.
2 Refer to Chapter 9. Remove the brake drum, disconnect the pipe to the wheel cylinder and the parking brake linkage from the backplate.
3 Undo and remove the four nuts that secure the backplate to the axle casing. There are two holes in the axleshaft flange to allow for access.
4 The axleshaft assembly may now be withdrawn from the axle casing. If tight, it will be necessary to use a slide hammer or tire levers to ease the wheel bearing from the axle casing.
5 Remove the bearing collar by splitting with a chisel; a new one will be required on reassembly!
6 Should a new bearing be required this job should be left to the local Datsun garage. They are installed under a force of between 4 and 5 tons, and a press is necessary.
7 Reassembling and installation of the axleshaft assembly is the reverse sequence to removal. The following additional points should be noted:

(a) Make sure the wheel bearing is well packed with grease.
(b) Pack the lip of the oil seal with a little grease.
(c) Top-up the rear axle oil level.
(d) Bleed the brake hydraulic system as described in Chapter 9.

9 Pinion oil seal – renewal (Station Wagon)

1 The procedure for installing a new pinion oil seal is basically similar to that described for the Sedan model, which is described in Section 4.
2 The pinion nut should be tightened to a torque of 94 lbf ft (13.0 kgf m), at which the bearing preload (force required to turn the pinion flange) should be 7.1 to 9.0 lbf (3.2 to 4.1 kgf). If this is not so, progressively tighten the pinion nut until the correct preload is obtained. If the torque applied to the pinion nut exceeds 130 lbf ft (18.0 kgf m), renew the pinion nut and repeat this process.

10 Differential assembly – servicing general (Station Wagon)

The overhaul of the rear axle differential unit is not within the scope of the home mechanic, due to the specialized gauges and tools which are required. Where the unit requires servicing or repair, due to wear or excessive noise, it is most economical to exchange it for a factory-reconditioned assembly.

11 Differential carrier – removal and installation (Station Wagon)

1 If it is wished to exchange the differential unit for a reconditioned one, first remove the axleshafts as described in Section 8.
2 Disconnect the propeller shaft and remove it from the vehicle, as described in Chapter 7.
3 Drain the oil from the rear axle.
4 Unscrew, evenly and in opposite sequence, the nuts from the differential unit securing studs. Pull the differential unit from the axle casing.
5 Scrape all traces of old gasket from the mating surfaces of the axle casing. Position a new gasket on the axle casing having first smeared it with jointing compound.
6 Install the differential carrier so that the pinion assembly is at the lowest point. Tighten the securing nuts to the specified torque.
7 Install the axle shafts and the propeller shaft.
8 Fill the axle to the correct level with the correct grade of oil.

12 Rear axle – removal and installation (Station Wagon)

1 Chock the front wheels, jack-up the rear of the vehicle and place on suitable support blocks or stands. Remove the rear wheels.
2 Refer to Chapter 9 and disconnect the parking brake linkages and the hydraulic brake hose.
3 Drain the oil from the unit
4 Refer to Chapter 7, and disconnect the propeller shaft from the rear axle pinion flange.
5 Disconnect the shock absorbers at their lower ends and remove the nuts that retain the U-bolts in position.
6 Now place a jack beneath the axle casing to support its weight. Remove the U-bolts from either side of the axle, then raise the axle a little with the jack. With assistance from a second person, the axle assembly can now be removed from the vehicle by passing it over the leaf springs and out through the side of the vehicle.
7 Installation of the rear axle is the reverse of removal. The following additional points should be noted:

(a) Final tightening of the U-bolts and the shock absorbers should be carried out when the vehicle is lowered to the ground.
(b) It will be necessary to bleed the brake hydraulic system as described in Chapter 9.

13 Fault diagnosis – rear axle

Symptom	Reason/s
Sedan	
Noise on drive, coasting or overrun	Shortage of oil Incorrect crownwheel-to-pinion mesh Worn pinion bearings Worn side bearings Loose bearing cap bolts
Noise on turn	Differential side gears worn, damaged or tight
Knock on taking up drive or during gearchange	Excessive crownwheel-to-pinion backlash Worn gears Worn axleshaft splines Pinion bearing preload too low Loose securing bolts or nuts within unit Loose roadwheel nuts or elongated wheel nut holes
Oil leakage	Defective gaskets or oil seals

Station Wagon

Symptom	Reason/s
Noise on drive, coasting or overrun	Shortage of oil Incorrect crownwheel-to-pinion mesh Worn pinion bearings
Noise on turn	Worn differential gears
Knock on taking up drive or during gearchange	Excessive crownwheel-to-pinion backlash Worn gears Worn axleshaft splines Worn propeller shaft joints Loose pinion flange bolts
Oil leakage	Defective gaskets or oil seals

Chapter 9 Braking system

Refer to Chapter 13 for specifications and information applicable to 1980 through 1984 models.

Contents

Specifications

System type .... Front disc, rear drum, dual hydraulic circuit with servo assistance. Parking brake mechanically-operated on rear wheels

Disc brake

Minimum permissible disc thickness	0·413 in (10·5 mm)
Maximum permissible disc runout	0·006 in (0·15 mm)
Minimum brake pad thickness	0.080 in (2.0 mm)

Drum brake

Type	Leading and trailing brake shoes with fixed wheel cylinder
Maximum permissible drum diameter	9·060 in (230 mm)
Maximum permissible drum runout	0·0008 in (0·020 mm)
Minimum permissible lining thickness	0·060 in (1·5 mm)

Master cylinder

Allowable clearance between cylinder and piston	0·006 in (0·15 mm)

Brake pedal

Free height	7·090 in (180·0 mm)
Free play at pedal pad	0·040 to 0·20 in (1·0 to 5·0 mm)
Depressed height	More than 2·95 in (75·0 mm)

Torque wrench settings	**lbf ft**	**kgf m**
Master cylinder-to-servo unit nuts	6 to 8	0·8 to 1·1
Caliper-to-stub axle fixing bolts	53 to 72	7·3 to 9·9
Disc-to-hub bolts	28 to 38	3·9 to 5·3
Disc backplate/shield bolts	2 to 3	0·34 to 0·44
Drum backplate retaining bolts (Sedan)	20 to 27	2·7 to 3·7
Drum backplate retaining bolts (Station Wagon)	16 to 20	2·2 to 2·7
Wheel cylinder mounting bolts	4 to 6	0·6 to 0·8
Vacuum servo mounting nuts	6 to 8	0·8 to 1·1
Vacuum servo pushrod locknut	12 to 16	1·6 to 2·2
Vacuum servo pushrod adjusting nut	12 to 16	1·6 to 2·2

1 General description

The braking system used on the Datsun 810 range of vehicles is of the four-wheel servo-assisted, hydraulic type with discs at the front and drum brakes at the rear.

A mechanically-operated parking brake operates on the rear wheels only. All brakes are self-adjusting.

If the hydraulic fluid in the master cylinder reservoir falls to an excessively low level, a switch, incorporated within the fluid reservoir, actuates a panel-mounted warning light. This warning light also operates when the parking brake is operated.

The hydraulic circuit incorporates a pressure regulating valve to prevent the rear brakes locking under heavy brake application. The pressure regulating valve completely separates the front and rear brake fluid lines from the tandem master cylinder and, in the event of a leakage in either system, automatically proportions the pressure delivery to each brake line.

2 Front disc pads – inspection, removal and installation

1 Jack-up the front of the car and remove the roadwheel.
2 Inspect the thickness of the friction material. If it is 0.080 in (2.0

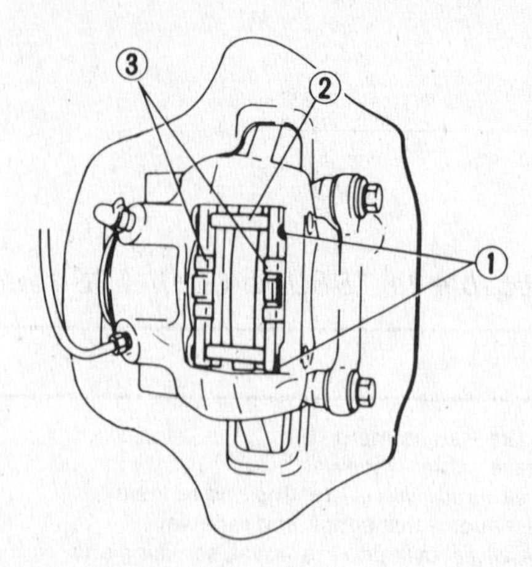

Fig. 9.1 A view of the caliper assembly (Sec 2)

1 Clips
2 Pad retaining pins
3 Anti-squeal springs

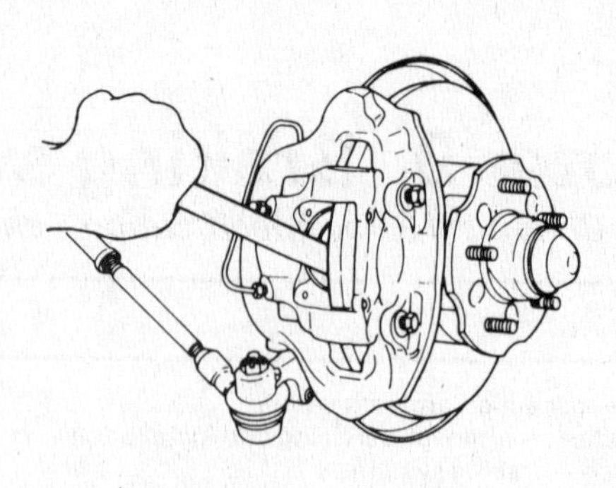

Fig. 9.2 Depressing the caliper pistons (Sec 2)

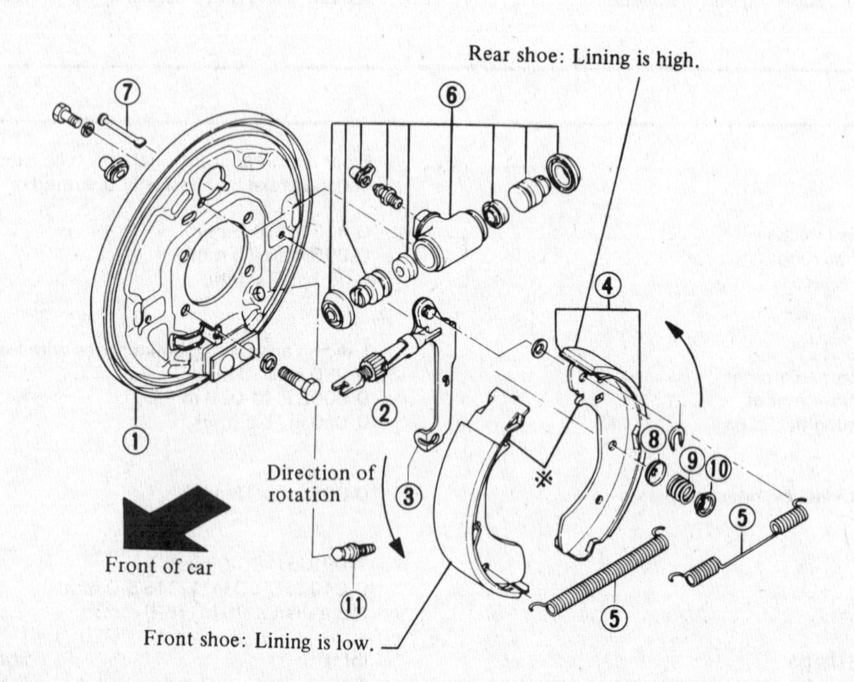

Fig. 9.3 Exploded view of the left-hand side brake assembly (Sedan) (Sec 3)

1 Backplate
2 Knurled wheel and adjuster assembly
3 Parking brake lever
4 Brake shoe
5 Return springs
6 Wheel cylinder
7 Shoe steady post
8 Spring seat
9 Spring
10 Cup washer
11 Stopper assembly
**Adjuster location slots*

mm) or less, renew the pads on both front wheels.
3 Remove the clips from the ends of the retaining pins.
4 Withdraw the retaining pins and springs.
5 Using a pair of pliers, pull out each of the disc pads together with the anti-squeal shims located at the rear face of the pad backing plate.
6 Brush out any accumulated dust from within the caliper body and from the area of the piston seal.
7 Using a flat bar or piece of wood, depress the piston fully and squarely into its cylinder bore in order to accomodate the new thicker pads. This action will cause the level of fluid in the master cylinder to rise, so either syphon some fluid from the reservoir before commencing the operation or alternatively open the bleed nipple on the caliper unit and allow the surplus fluid to be ejected.
8 Insert the new disc pads complete with anti-squeal shims (arrow in direction of disc forward rotation).
9 Install the springs, retaining pins and retaining pin clips.
10 Check that the bleed nipple is closed and then depress the brake pedal several times to bring the pad into contact with the disc.
11 Check and top-up if necessary the fluid level in the brake master cylinder.
12 Repeat the operations on the opposite wheel.

3 Rear brake shoes – inspection and renewal

1 Jack-up the rear of the car, chock the front wheels and ensure that the car is safely supported.
2 Remove the roadwheel followed by the brakedrum. Should it be stuck tight, lightly tap the drum outwards using a block of wood as an insulator between the drum and the face of the hammer.

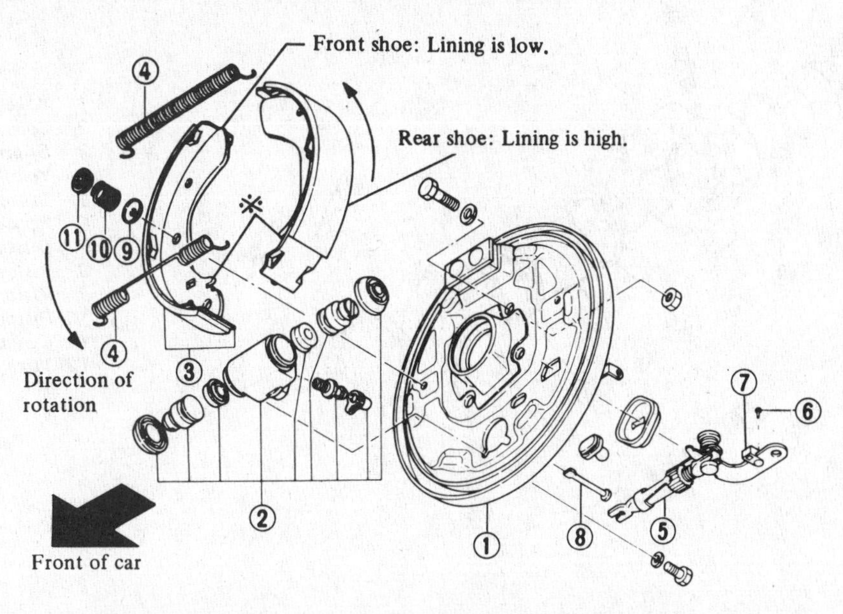

Fig. 9.4 Exploded view of the left-hand side brake assembly (Station Wagon) (Sec 3)

1 Backplate
2 Wheel cylinder
3 Brake shoe
4 Return springs
5 Adjuster assembly
6 Stopper pin
7 Stopper
8 Shoe steady post
9 Spring seat
10 Spring
11 Cup washer
**Adjuster location slots*

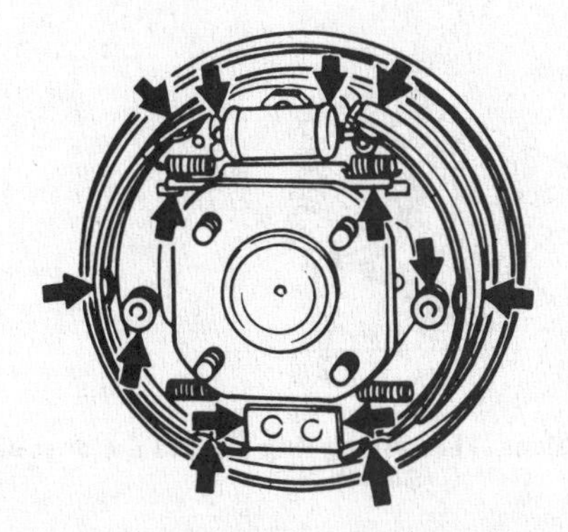

Fig. 9.5 The grease points arrowed (Sec 3)

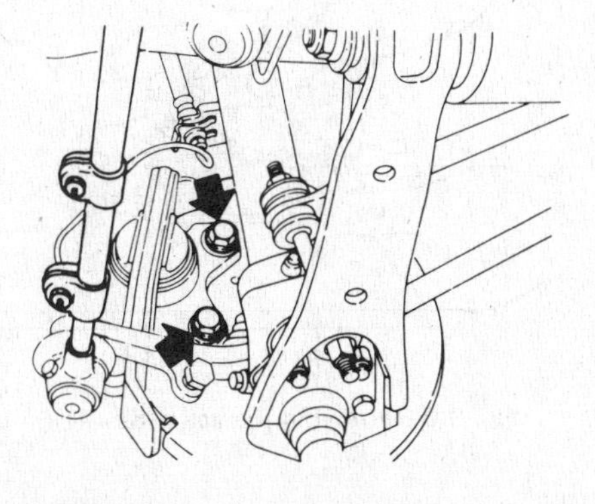

Fig. 9.6 The disc caliper securing bolts (Sec 4)

3 Brush away all accumulated dust from the surface of the linings and the interior of the drum. Inspect the linings for wear and if they are worn down to (or very near) the level of the securing rivets, the shoes must be renewed. With bonded type linings, if they are worn down to 0.060 in (1.5 mm) then renew the shoes. It is not worth attempting to reline shoes yourself but rather exchange them for factory-reconditioned ones.

4 Remove the cup washers from the shoe steady posts. To do this, grip the edges of the cup washer with a pair of pliers and depress it against the coil spring, then turn the washer through 90° and release it. The T-shaped head of the steady post will now pass through the slot in the cup washer, and the washer and spring can be withdrawn, followed by the spring seat.

5 Using a screwdriver, lever one brake shoe from engagement with the slot in the anchor block and, at the other end of the brake shoe, the knurled adjuster. (If the adjustment is fully extended on the lever assembly, it can easily be retracted by lifting the ratchet plate and rotating the knurled adjuster to close the assembly). Pull the shoe forward slightly and then release it so that it moves towards the wheel center. The opposite shoe can be detached in a similar manner.

6 Before separating the shoes and return springs, note the holes in the shoe webs in which the return springs engage, also the position of the shoes with regard to leading and trailing ends.

7 Before installing the new shoes, clean the brake backplate and apply a smear of high melting point grease to the sliding surfaces on the backplate, also the slots of the anchor block and wheel cylinder.

8 Ensure that the adjuster assembly is fully retracted and apply a smear of grease to the threads.

9 Lay out the new brake shoes on a bench and engage the return springs in their correct holes.

10 Maintaining tension on the return springs by prying the shoes apart, first engage one shoe with the anchor block and wheel cylinder then pull the opposite shoe over and into engagement with the anchor block and adjuster screw slot.

11 Install the shoe steady posts, springs and cup washers.

12 Turn the knurled adjusting wheel and expand the shoes until the brake drum will just pass over them.

13 Fit the drum and apply the parking brake then release it several times until the knurled adjuster wheel ceases to click over any more notches, indicating that full shoe adjustment has been obtained.

14 Install the roadwheel and lower the car to the ground. Repeat the operations on the opposite wheel.

4 Disc brake caliper – removal, servicing and installation

1 Refer to Section 2, and remove the disc pads.

2 Remove the cap from the front circuit reservoir and place a piece

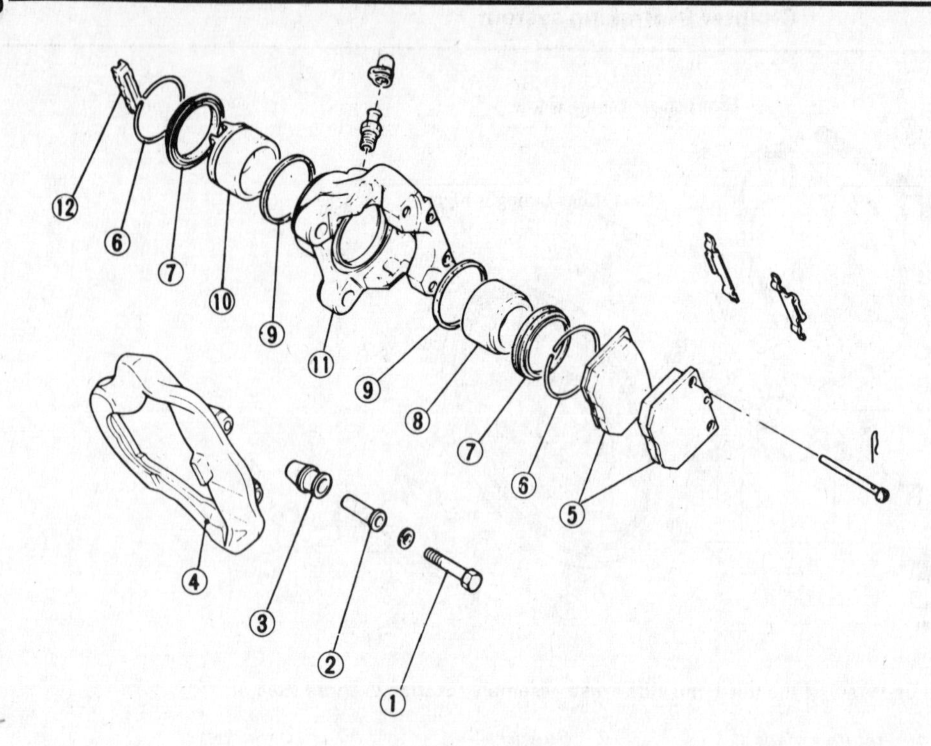

1 Yoke retaining bolt
2 Collar
3 Gripper
4 Yoke
5 Brake pads
6 Retaining rings
7 Dust seal
8 Piston B
9 Piston seal
10 Piston A
11 Cylinder body
12 Yoke holder

Fig. 9.7 An exploded view of the caliper assembly (Sec 4)

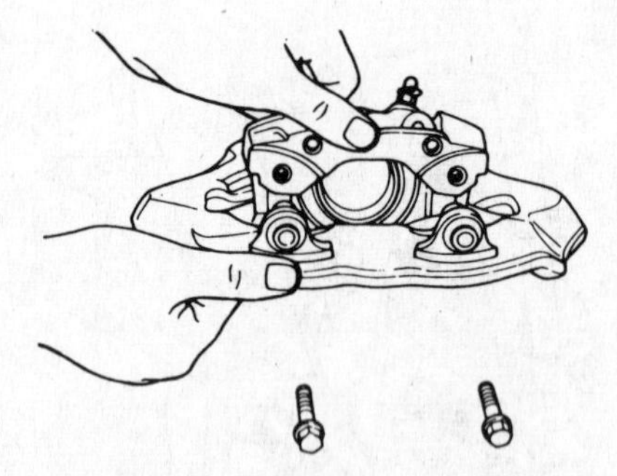

Fig. 9.8 Removing the yoke (Sec 4)

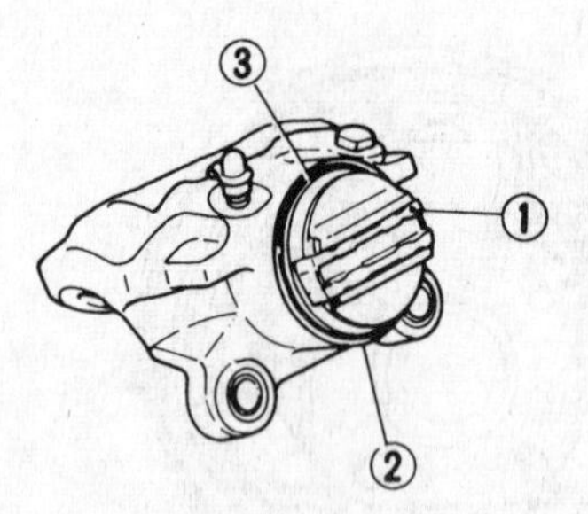

Fig. 9.9 The yoke holder (1) retaining ring (2) and the dust seal (3) (Sec 4)

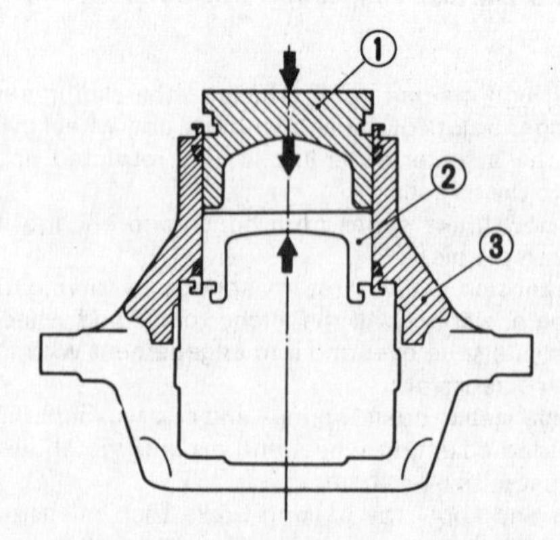

Fig. 9.10 Correct installation of the pistons (Sec 4)

1 Piston A
2 Piston B
3 Cylinder body

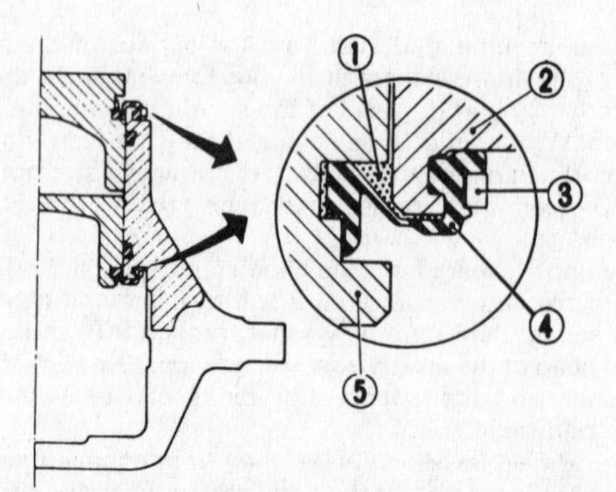

Fig. 9.11 Dust seal greasing and installation diagram (Sec 4)

1 Grease
2 Cylinder body
3 Retaining ring
4 Dust seal
5 Piston

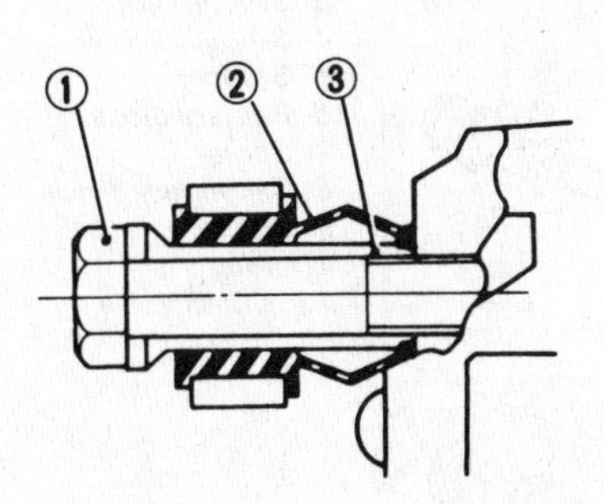

Fig. 9.12 Installing the gripper (Sec 4)

1 Retaining bolt
2 Gripper
3 Collar

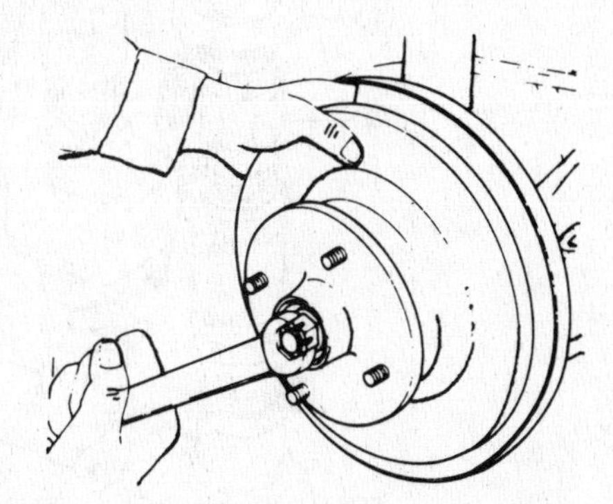

Fig. 9.13 Removing a hub nut (Sec 5)

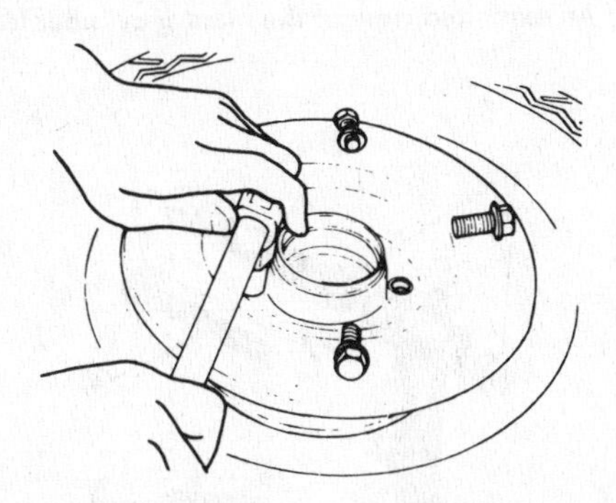

Fig. 9.14 Removing the disc to hub setscrews (Sec 5)

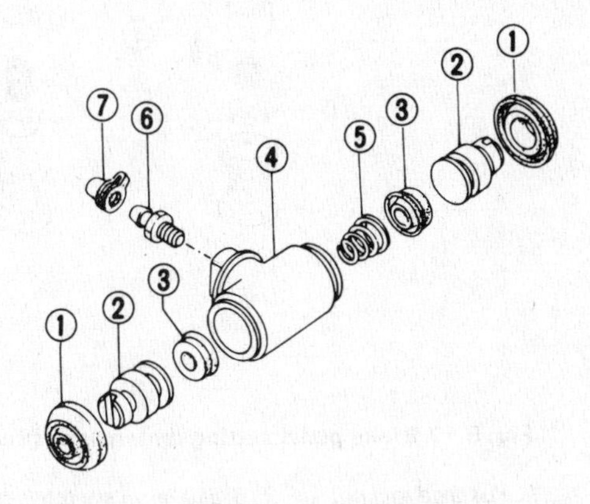

Fig. 9.15 The component parts of the wheel cylinder (Sec 6)

1 Dust cover
2 Piston
3 Piston cup seal
4 Cylinder body
5 Spring
6 Bleed valve
7 Bleed valve cap

of polythene sheeting over the reservoir opening and secure it with a tightly fitting rubber band. This will create a vacuum and prevent loss of fluid when the hydraulic pipe is disconnected.

3 Disconnect the flexible brake hose at the support bracket on the suspension leg.

4 Unscrew and remove the two bolts which secure the caliper unit to the stub axle carrier and lift the caliper away.

5 Clean away all external dust and dirt, ensuring that none enters the open end of the brake pipe.

6 Commence dismantling by unscrewing and removing the two bolts that secure the yoke to the cylinder body. Remove the yoke from the cylinder body.

7 Remove the yoke holder from the piston.

8 Prise out the retaining ring and remove the dust excluder from each piston.

9 Apply air from a tire pump to the end of the flexible hose and eject the pistons to a point where they can be withdrawn with the fingers. Mark the pistons to ensure that they are reassembled in the correct order.

10 At this stage, examine the surfaces of the pistons and cylinder bore. If they are scored or any 'bright' wear areas are evident, renew the caliper unit complete.

11 Where the components are in good order, wash them in clean brake fluid. Discard all rubber seals and obtain a repair kit. When extracting the piston seal from the groove in the cylinder bore, take great care not to scratch or score the bore.

12 Install the new piston seals, using fingers only to manipulate them into position in the cylinder bore grooves.

13 Dip each piston into clean brake fluid and enter it squarely into the cylinder bore. Piston A should be inserted from one side of the cylinder bore, and piston B should be inserted from the other side. Do not push the pistons too far into the bore. Piston A should be installed so that its yoke groove coincides with the groove on the cylinder.

14 Apply a smear of brake seal grease to the dust seals and manipulate them into position. Install the retaining rings.

15 Install the yoke holder to the piston A.

16 If the grippers were removed, install them using soap solution to aid assembly then drive in the collars.

17 Drive the yoke into the yoke holder on the end of the piston. Use a soft-faced hammer to do this.

18 Install the yoke to the cylinder body and install the retaining bolts, tightening them to 12 to 15 lbf ft (1.6 to 2.1kgf m).

19 Installation of the caliper assembly is a reversal of removal. Tighten the caliper securing bolts to a torque of 53 to 72 lbf ft (7.3 to 9.9 kgf m) and bleed the hydraulic system on completion as described in Section 13.

5 Front brake disc – inspection and servicing

1 Jack-up the front of the car and remove the roadwheel.

2 Examine the surface of the disc for deep scoring or grooving. Light scoring is normal and should be ignored.

3 The discs can be skimmed professionally, provided that their thickness will not be reduced below 0.413 in (10.5 mm).

4 Disc run-out (buckle or out-of-true) must not exceed 0.006 in (0.15 mm). This can only be satisfactorily measured with a dial gauge applied at the center of the disc pad contact area. Where the disc is

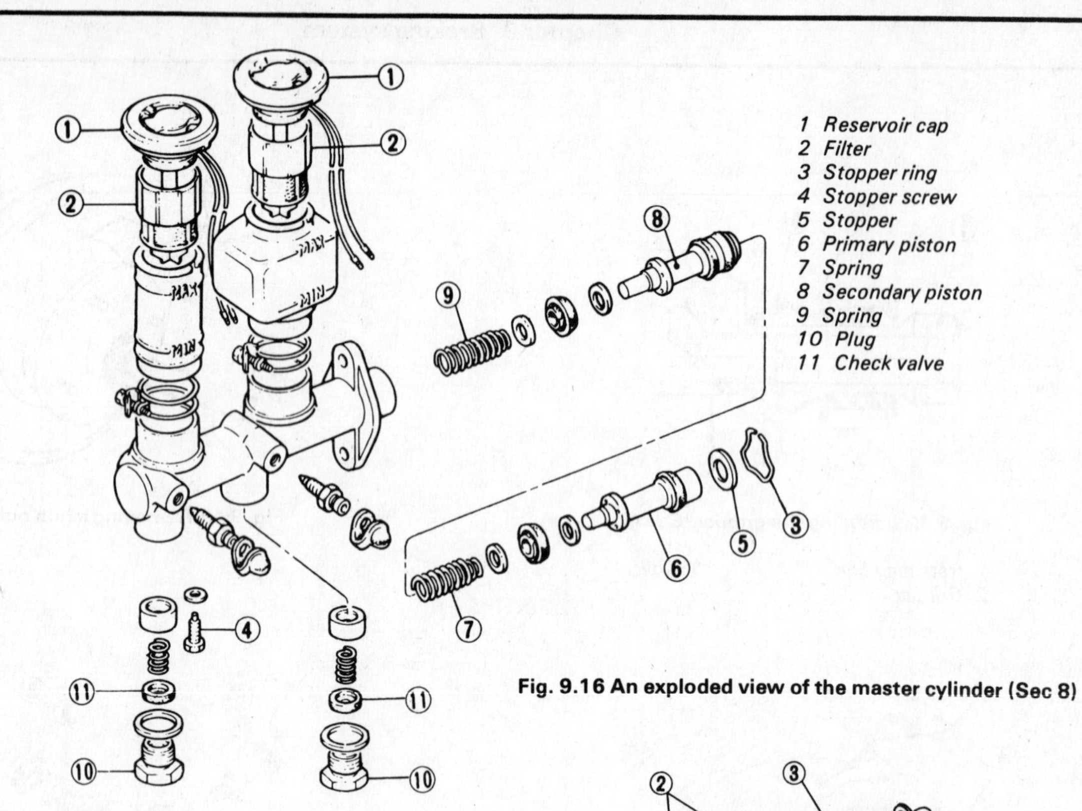

Fig. 9.16 An exploded view of the master cylinder (Sec 8)

Fig. 9.17 Brake pedal setting dimensions (Sec 9)

1 Pushrod locknut *3 Brake lamp switch locknut*
2 Brake lamp switch

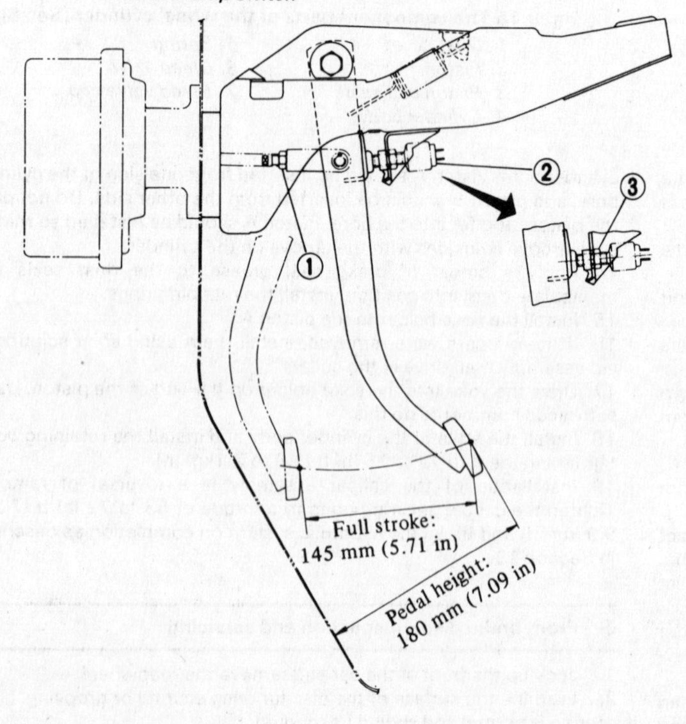

Fig. 9.18 The component parts of the brake pedal (Sec 9)

1 Brake pedal *4 Fulcrum bolt*
2 Pedal bush *5 Return spring*
3 Sleeve

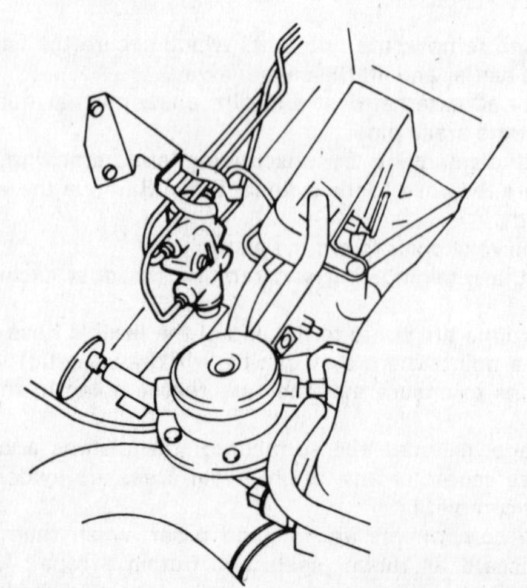

Fig. 9.19 The location of the pressure regulating valve (Sec 10)

out-of-true, renew it.
5 To remove the brake disc, first remove the caliper (Section 4) then pry off the grease cap from the end of the front hub, remove the cotter pin and unscrew the castellated nut.
6 Pull the hub/disc assembly towards you and catch the thrust washer and outer bearing.
7 Withdraw the hub/disc assembly.
8 Unscrew and remove the four setscrews which secure the brake disc to the hub. It will prevent the disc rotating if the assembly is dropped into a roadwheel used as a support and mounting.
9 If required, the backplate shield may be unbolted from the suspension leg.
10 Reassembly is a reversal of dismantling but tighten the disc bolts to a torque of 35 lbf ft (4.8 kgf m) and the backplate shield bolts to only 3 lbf ft (0.4 kgf m). The front hub should be adjusted as described in Chapter 11. Bleed the hydraulic system (Section 13).

6 Rear brake wheel cylinder – removal, servicing and installation

1 Remove the brake shoes as described in Section 3.
2 Remove the cap from the rear circuit reservoir of the master cylinder; place a sheet of polythene over the reservoir opening and secure it with a tightly fitting rubber band. This will create a vacuum and prevent loss of fluid when the hydraulic pipe is disconnected.
3 Disconnect the hydraulic fluid pipe by unscrewing the union on the wheel cylinder.
4 Unscrew and remove the bolts that retain the wheel cylinder to the backplate. Lift away the wheel cylinder.
5 Peel off the rubber dust covers at either end of the assembly and remove the two piston cup assemblies, followed by the spring.
6 Examine the surfaces of the pistons and cylinder bore. If they are scored or show evidence of 'bright' wear areas, renew the wheel cylinder as an assembly. Where these components are in good order, discard the piston cup seals and obtain a repair kit.
7 Manipulate the new seals into position using the fingers only. To aid this assembly, the pistons and seals should be dipped into clean brake fluid.
8 Insert the spring (narrow coil first) into the cylinder and then dip the piston assemblies into clean brake fluid, before entering them into the cylinder bores. Install the dust covers.
9 Installation of the wheel cylinder is a reversal of removal but the hydraulic system will need bleeding as described in Section 13.

7 Brake drum – inspection and renovation

1 Inspect the interior friction surfaces of the rear brake drum. If they are deeply grooved or scored, the drum must either be renewed or turned on a lathe, provided that the inner diameter of the drum will not exceed 9.060 in (230.0 mm). Ovality of the drum must also be corrected by one of these methods.

8 Master cylinder – removal, servicing and installation

1 The master cylinder is mounted on the front face of the vacuum servo unit on the engine compartment firewall.
2 Disconnect the fluid level warning switch wires from each reservoir.
3 Disconnect the brake pipes from the master cylinder and allow the fluid to drain into a suitable container.
4 Unscrew and remove the flange securing nuts and remove the master cylinder from the vacuum servo unit.
5 Commence dismantling by removing the two reservoir caps and filters, and draining out the hydraulic fluid.
6 Using a small screwdriver, pry out the stopper ring from the open end of the body. Remove the stopper.
7 Undo and remove the stopper screw, then extract the primary piston assembly and spring, followed by the secondary piston and spring.
8 Unscrew and remove the two plugs which house the check valves, remove the check valves and their springs.
9 Examine the surfaces of the pistons and cylinder bores. If they are scored or any 'bright' wear areas are evident, renew the master cylinder complete.
10 Where the components are in good order, wash them thoroughly in clean brake fluid and obtain a repair kit.
11 Reassembly is a reversal of dismantling, but dip each component in clean brake fluid before assembly, and observe absolute cleanliness to avoid any dirt entering the unit.
12 Installation of the master cylinder is a reversal of removal. Tighten the flange mounting nuts to a torque of 6 to 8 lbf ft (0.8 to 1.1 kgf m), and then bleed the hydraulic system (Section 13).
13 Check the pedal height and adjust if necessary, as described in Section 9.

9 Foot-brake pedal – adjustment, removal and installation

Adjustment

1 Loosen the brake lamp switch locknut and, by rotating the switch in the required direction, obtain a toe-board-to pedal pad dimension of 7.090 in (180.0 mm). Once this dimension has been achieved, tighten the locknut.
2 With the dimension set as specified in paragraph 1, loosen the locknut on the pushrod between the pedal and the vacuum servo pushrod.
3 Screw the pushrod in the required direction until the pedal free-play is between 0.040 and 0.20 in (1.0 and 5.0 mm). Once this dimension has been achieved, tighten the pushrod locknut.
4 Recheck the dimensions previously set when, if all is well in the brake hydraulic system, the pedal depressed height from the toe-board to the pedal pad should be 2.95 in (75.0 mm) or greater.

Removal and installation

5 To remove the brake pedal, detach the return spring and extract the cotter pin from the clevis pin. Pull out the clevis pin and separate the servo unit push rod from the brake pedal.
6 Unscrew and remove the fulcrum bolt (which has a left-hand thread and should be turned clockwise). Remove the brake pedal complete with bushes and sleeve.
7 Installation is a reversal of removal but apply a smear of grease to fulcrum bolt and sleeve, and also to the clevis pin.

10 Pressure regulating valve – testing and renewal

1 The purpose of the pressure regulating valve is to prevent the rear brakes locking in advance of the front brakes.
2 To test the valve, drive the car in a straight line at 31 mph (50khp) and apply the brakes hard to lock the wheels. The skid marks for the front wheels should be longer than those for the rear, indicating that the front wheels locked first.
3 Where the test proves the valve to be faulty, renew it as an assembly.
4 The pressure regulating valve is bolted to the inner wing and once the four hydraulic pipes have been disconnected it is quite simply unbolted.
5 When installing a new pressure regulator valve, observe any flow marks before reconnecting the hydraulic pipes.
6 Bleed the hydraulic system as described in Section 13.

11 Flexible hoses – inspection, removal and installation

1 Inspect the condition of the flexible hydraulic hoses. If they are swollen, perished or chafed, they must be renewed.
2 To remove a flexible hose, hold the flats on its end-fitting in an open-ended wrench and unscrew the union nut which couples it to the rigid brake line.
3 Disconnect the flexible hose from the rigid line and support bracket, then unscrew the hose from the caliper or wheel cylinder circuit as the case may be.
4 Installation is a reversal of removal. The flexible hoses may be twisted not more than one quarter turn in either direction if necessary to provide a 'set' to ensure that they do not rub or chafe against any adjacent component.
5 Bleed the hydraulic system on completion (Seition 13).

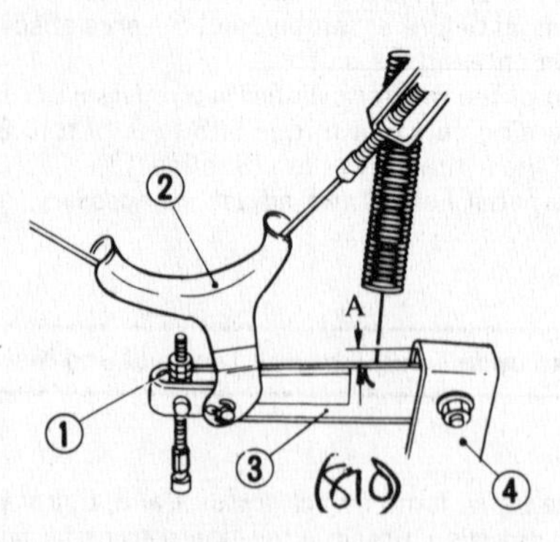

Fig. 9.20 The parking brake center lever assembly (Sedan) (Sec 14)

1 Adjusting nut and locknut
2 Equalizer
3 Center lever
4 Center lever bracket

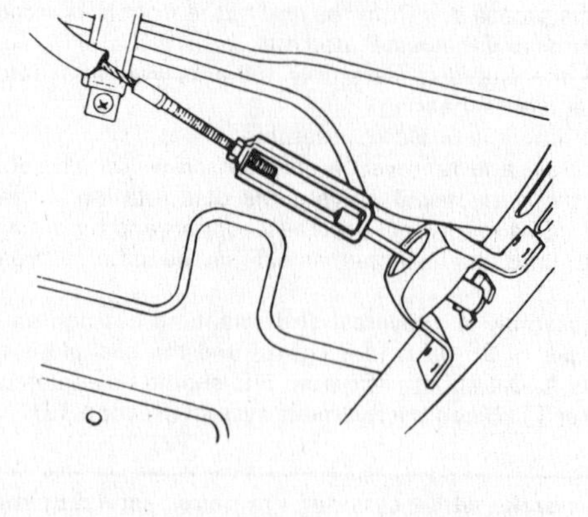

Fig. 9.21 The rear cable adjuster (Sedan) (Sec 4)

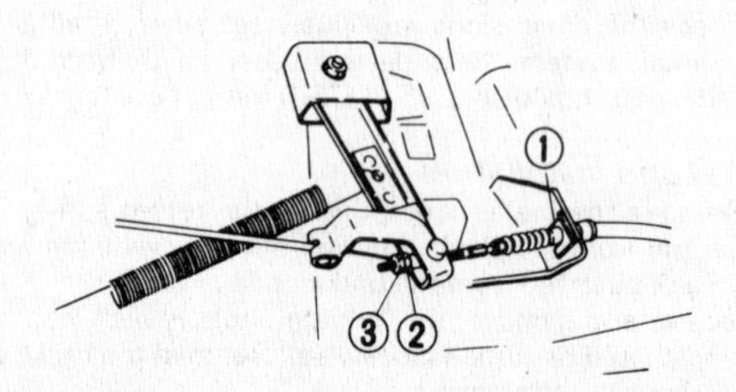

Fig. 9.22 The parking brake center lever assembly (Station Wagon) (Sec 14)

1 Lockplate
2 Adjuster nut
3 Locknut

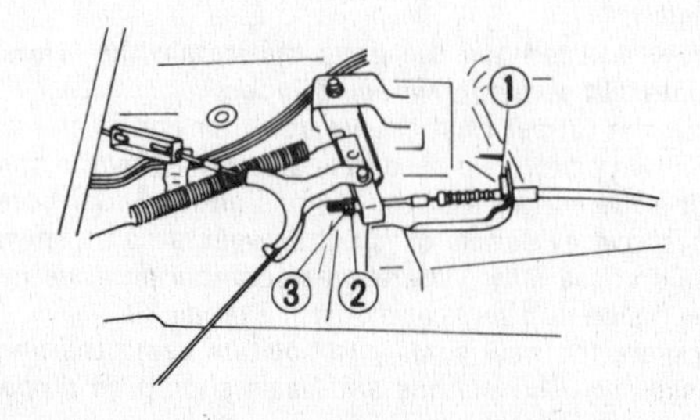

Fig. 9.23 The center lever assembly (Sedan) (Sec 15)

1 Lockplate
2 Adjuster nut
3 Locknut

12 Rigid brake lines – inspection, removal and installation

1 At regular intervals wipe the steel brake pipes clean and examine them for signs of rust or denting caused by flying stones.
2 Examine the securing clips for signs of corrosion. Bend the tongues of the clips if necessary to ensure that they hold the brake pipes securely without letting them rattle or vibrate.
3 Check that the pipes are not touching any adjacent components or rubbing against any part of the vehicle. Where this is observed, bend the pipe gently away to clear.
4 Although the pipes are galvanized, any section of pipe may become rusty through chafing and should be renewed. Brake pipes are available to the correct length and fitted with end unions from most Datsun dealers and can be made to pattern by many accessory suppliers. When installing the new pipes use the old pipes as a guide to bending and do not make any bends sharper than is necessary.
5 The system will of course have to be bled when the circuit has been reconnected (Section 13).

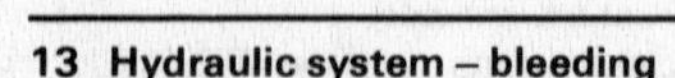

13 Hydraulic system – bleeding

1 Removal of all the air from the hydraulic system is essential to the correct working of the braking system. Before undertaking this, examine the fluid reservoir cap to ensure that both vent holes, one on top and the second underneath but not in line, are clear; check the fluid level and top up if required.
2 Check all brake line unions and connections for possible seepage, and at the same time check the condition of the rubber hoses, which may be perished.
3 If the condition of the wheel cylinders is in doubt, check for possible signs of fluid leakage.
4 If there is any possibility of incorrect fluid having been put into the system, drain all the fluid out and flush through with methylated spirit. Renew all piston seals and cups since these will be affected and could possibly fail under pressure.
5 Gather together a clean jar, a length of tubing which fits tightly over the bleed nipples, and a tin of the correct type of brake fluid.
6 Depress the brake pedal several times in order to destroy the vacuum in the servo system.
7 As the front and rear circuits are independent, it will be obvious that only one circuit need be bled if only one hydraulic line has been disconnected.
8 Clean dirt from around the bleed nipples on the master cylinder and bleed one or both as necessary. To do this, push one end of the rubber tube onto the nipple and immerse the other end in a little brake fluid contained in the jar. Keep the open end of the tube submerged throughout the operation.
9 Open the bleed nipple about one half of a turn and then have an assistant depress the brake pedal fully. The foot should then be removed quickly from the pedal so that it returns unobstructed. Pause and then repeat the operation until no more air bubbles can be seen emerging from the end of the bleed tube which is submerged in the jar. Tighten the bleed nipple (do not force it) when the pedal is held in the fully depressed position.
10 Now bleed the calipers and wheel operating cylinders working on the principle of bleeding the unit first which is furthest from the master cylinder then the next furthest away, and so on, in sequence.
11 It is vital that the reservoir supplying the circuit which is being bled, is kept topped-up throughout the operation with clean hydraulic fluid which has been stored in an airtight container and has remained unshaken for the previous 24 hours. *Always discard fluid which is expelled into the jar – never use it for topping-up.*

14 Parking brake – adjustment

1 The parking brake is normally adjusted automatically when the clearance between the brake shoe linings and the drum is reduced by action of the drum internal adjuster mechanism. However, when the parking brake cables have stretched, or after dismantling and reassembling the linkage or the rear brake assembly, set the parking brake cables in the following manner.

Sedan

2 Jack-up the rear end of the vehicle and firmly support on suitable blocks or stands. For added safety, chock the front wheels.
3 Ensure that the parking brake lever is in the fully off position.
4 Working beneath the vehicle at the center lever, loosen the locknut and turn the front cable adjusting nut in the required direction, until dimension A (Fig. 9.20) is approximately 0.280 in (7.0 mm). When this dimension is obtained tighten the locknut.
5 The parking brake is designed to operate when the parking brake lever has passed over 5 or 6 ratchet notches. To adjust the rear cable, pull the parking brake lever over 4 notches only.
6 Now, working at the rear cable adjuster, loosen the locknut and turn the adjuster in the required direction to remove any slack in the rear cable. With an assistant rotating each rear roadwheel in turn, continue adjusting as previously mentioned until the roadwheels just start to bind. At this point, firmly tighten the adjuster locknut. When the parking brake lever is pulled over the 5th or 6th ratchet notch, the rear wheels should be firmly locked. The vehicle can now be lowered to the ground.

Station Wagon

7 Any adjustment required on the Station Wagon parking brake can only be carried out at the front cable, there being no provision for rear cable adjustment.
8 Jack-up the rear end of the vehicle and firmly support on suitable blocks or stands. For added safety, chock the front wheels.
9 Pull on the parking brake lever over 4 notches of the ratchet only.
10 Working beneath the vehicle at the center lever, loosen the locknut and adjust the front cable adjusting nut to remove any slack from the rear cable. With an assistant rotating each rear wheel in turn, continue adjusting the rear cable as previously mentioned until the roadwheels just start to bind. At this point, firmly tighten the adjuster locknut. When the parking brake lever is pulled over the 5th or 6th ratchet notch, the rear wheels should be firmly locked.
11 If all is satisfactory the vehicle can be lowered to the ground.

15 Parking brake cable – renewal

Front cable

1 Working inside the vehicle, remove the center console and peel back the carpet in the vicinity of the parking brake lever assembly.
2 Disconnect the terminals from the warning light switch.
3 Unscrew and remove the bolts that secure the parking brake lever assembly to the floor.
4 Working beneath the vehicle, unscrew and remove the locknut and adjusting nut from the center lever.
5 Detach the front cable from the center lever by prying out the cable lockplate.
6 With the front cable now disconnected from the center lever, pull out the rubber grommet in the floor and remove the parking brake lever assembly together with the front cable from inside the vehicle.
7 Remove the front cable from the parking brake lever assembly by removing the cotter pin and the clevis pin.
8 Reassembly and installation is the reversal of removal and dismantling, but apply a smear of grease to all sliding surfaces.

Rear cable (Sedan)

9 Jack-up the rear of the vehicle and remove the roadwheel and the brake drum.
10 Unscrew and remove the stopper and fastener as an assembly from the brake backplate.
11 Disengage the parking brake cable from the adjuster lever. Repeat this operation on the other wheel.
12 Working beneath the vehicle, detach the return spring from the center lever.
13 Disconnect the cable from the rear adjuster.
14 Remove the cable guides from either side of the suspension.
15 Remove the rear cable.
16 Installation of the cable is a reversal of removal but apply a smear of grease to all sliding surfaces of the assembly, and adjust the cable as described in Section 14.

Rear cable (Station Wagon)

17 Jack-up the rear of the vehicle and support it firmly on stands or blocks. Disconnect the rear cable at the center lever.
18 Unhook the center lever return spring and the cross-rod return spring at the axle casing.
19 Remove the clevis pins at each cross-rod connection to the brake lever at each backplate.
20 Remove the clevis pin from the balance lever, and the rear cable, together with the cross-rods can be removed from the vehicle.
21 Installation is a reversal of removal but apply a smear of grease to all sliding surfaces and adjust the cable as described in Section 14.

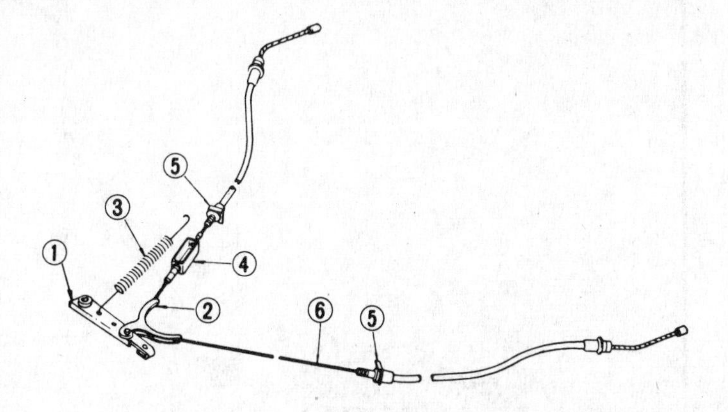

Fig. 9.24 The rear brake cable assembly (Sedan) (Sec 15)

1 Center lever
2 Equalizer
3 Spring
4 Cable adjuster
5 Cable guide
6 Rear cable

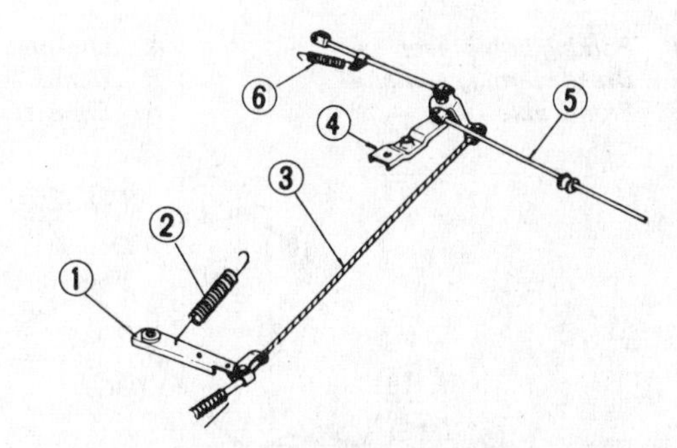

Fig. 9.25 The rear brake cable assembly (Station Wagon) (Sec 15)

1 Center lever
2 Spring
3 Rear cable
4 Swing arm
5 Cross-rod
6 Return spring

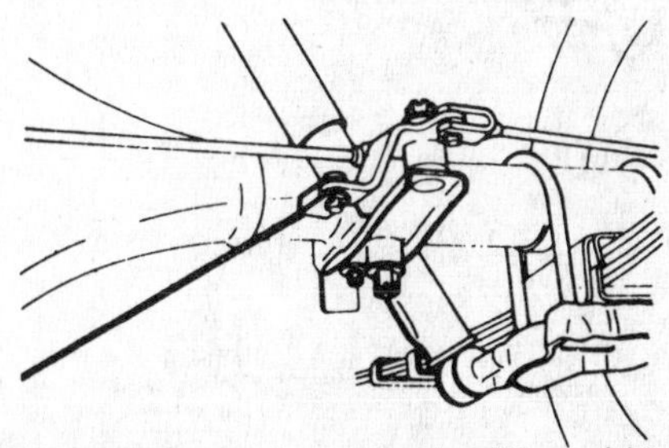

Fig. 9.26 The balance lever (Station Wagon) (Sec 15)

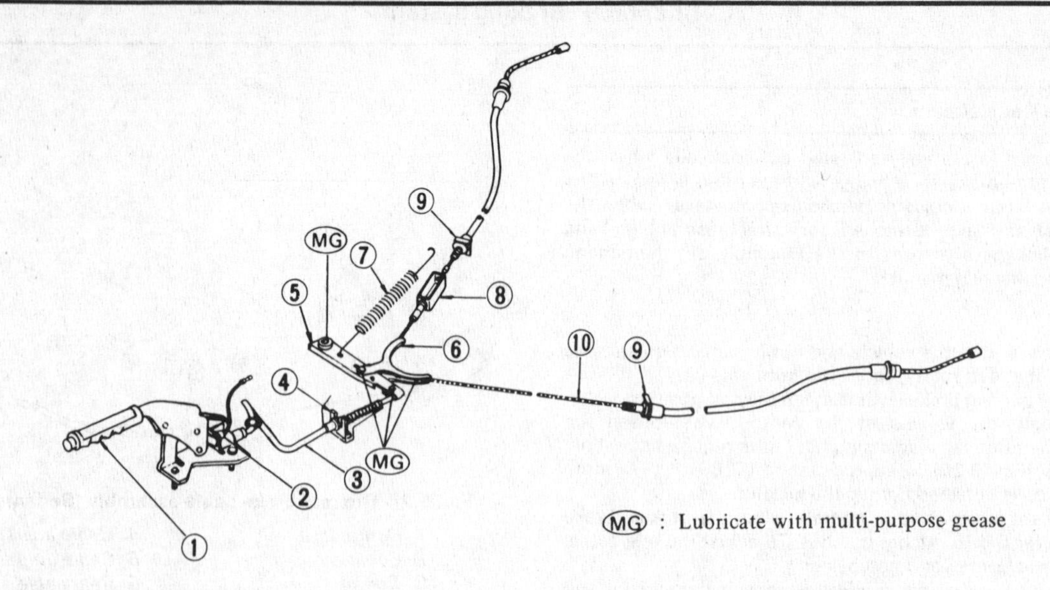

Fig. 9.27 The component parts of the parking brake assembly (Sedan) (Sec 15)

1 Parking brake lever
2 Brake warning switch
3 Front cable
4 Lockplate
5 Center lever
6 Equalizer
7 Spring
8 Rear cable adjuster
9 Cable guides
10 Rear cable

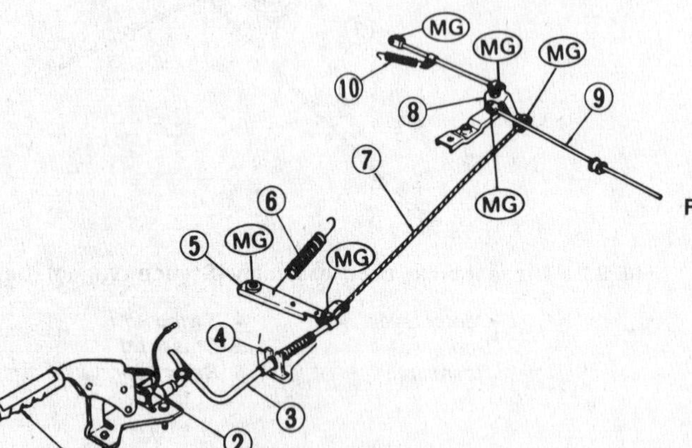

Fig. 9.28 The component parts of the parking brake assembly (Station Wagon) (Sec 15)

1 Parking brake lever
2 Brake warning switch
3 Front cable
4 Lockplate
5 Center lever
6 Spring
7 Rear cable
8 Swing arm
9 Cross-rod
10 Spring

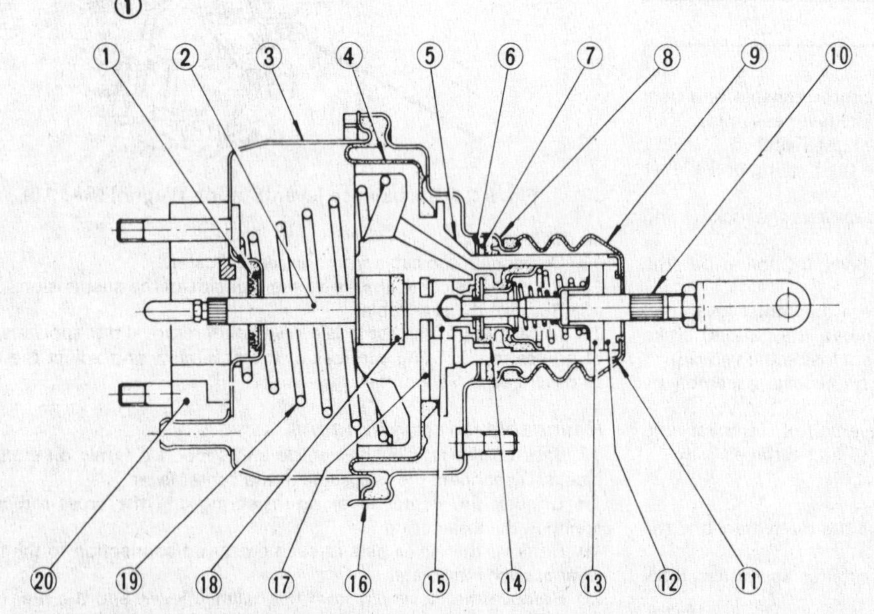

1 Plate and seal assembly
2 Pushrod
3 Front shell
4 Diaphragm
5 Diaphragm plate and valve body
6 Retainer
7 Bearing
8 Valve body seal
9 Valve body cover
10 Valve operating rod
11 Silencer
12 Silencer filter (felt)
13 Silencer filter (rubber)
14 Poppet assembly
15 Plunger assembly (valve operating rod)
16 Rear shell
17 Valve plunger stop key
18 Reaction disc
19 Diaphragm return spring
20 Flange

Fig. 9.29 Sectional view of the brake vacuum servo unit (Sec 16)

16 Vacuum servo unit – description and testing

1 A vacuum servo unit is fitted into the brake hydraulic circuit in series with the master cylinder, to provide assistance to the driver when the brake pedal is depressed. This reduces the effort required by the driver to operate the brakes under all braking conditions.
2 The unit operates by vacuum obtained from the induction manifold and comprises basically a booster diaphragm and non-return valve. The servo unit and hydraulic master cylinder are connected together so that the servo unit piston rod acts as the master cylinder pushrod. The driver's braking effort is transmitted through another pushrod to the servo unit piston and its built-in control system. The servo unit piston does not fit tightly into the cylinder, but has a rolling diaphragm, so assuring an air-tight seal between the two parts. The forward chamber is held under vacuum conditions created in the inlet manifold of the engine and, during periods when the brake pedal is not in use, the controls open a passage to the rear chamber so placing it under vacuum conditions as well. When the brake pedal is depressed, the vacuum passage to the rear chamber is cut off and the chamber opened to atmospheric pressure. The consequent differential pressure pushes the servo piston forward in the vacuum chamber and operates the main pushrod to the master cylinder.
3 The controls are designed so that assistance is given under all conditions and, when the brakes are not required, vacuum in the rear chamber is established when the brake pedal is released. All air from the atmosphere entering the rear chamber is passed through a small air filter.
4 Under normal operating conditions the vacuum servo unit is very reliable and does not require overhaul except at very high mileages. In this case it is far better to obtain a service-exchange unit, rather than repair the original unit. Servicing procedures are however described for those wishing to carry out this work.
5 It is emphasised, that the servo unit assists in reducing the braking effort required at the foot pedal and in the event of its failure, the hydraulic braking system is in no way affected except that the need for higher pedal pressures will be noticed.
6 To test the efficiency of the vacuum servo unit, a vacuum gauge should be connected in the line between the non-return valve and the servo unit.
7 Start the engine and run it until the vacuum gauge shows a reading of 19.69 in Hg (500 mm Hg). Switch off the engine and wait 15 seconds without touching the brake pedal. The pressure should not drop by more than 0.98 in Hg (25.0 mm Hg). If it does, then this will probably be due to one of the following faults:

(a) Insecure or split connecting pipes.
(b) Faulty or worn pushrod seals.
(c) Faulty sealing between main valve body and seal.
(d) Faulty sealing of main valve plunger seat.

8 Repeat the operation but this time wait 15 seconds with the brake pedal fully applied. Again, the pressure should not drop by more than 0.98 in Hg (25.0 mm Hg). If it does, it will probably be due to one of the following faults:

(a) Damaged diaphragm.
(b) Loose or displaced reaction disc.
(c) Faulty non-return valve or leaking connections.
(d) Poppet assembly not seating to provide air-tight seal.

9 Even where the servo unit is performing satisfactorily, it is recommended that it is overhauled at two-yearly intervals.

17 Vacuum servo unit – removal and installation

1 Disconnect the pushrod from the arm of the brake pedal by withdrawing the cotter pin and clevis pin.
2 Disconnect the brake pipes from the master cylinder.
3 Remove the master cylinder from the servo unit (See Section 8).
4 Disconnect the vacuum pipe from the servo unit.
5 Unscrew and remove the four securing nuts from the mounting studs of the vacuum servo unit and withdraw the unit from the engine compartment rear firewall.
6 Installation is a reversal of removal but bleed the hydraulic system

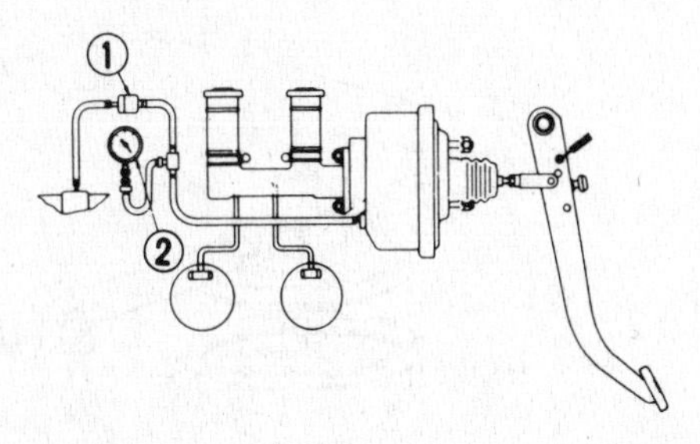

Fig. 9.30 Testing the servo unit using a vacuum gauge (Sec 16)

1 Non-return valve *2 Vacuum gauge*

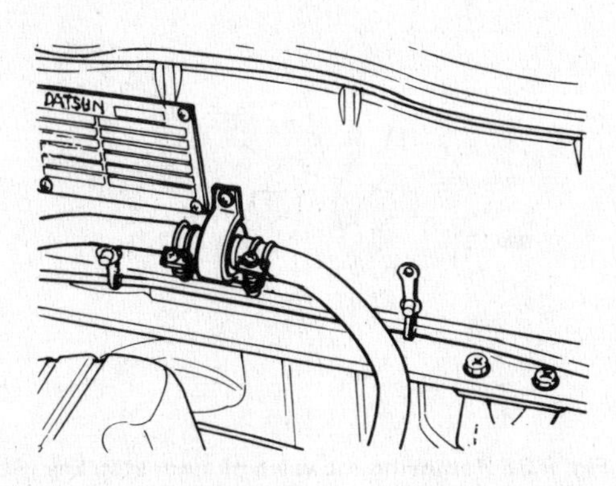

Fig. 9.31 Location of the brake servo unit non-return valve (Sec 16)

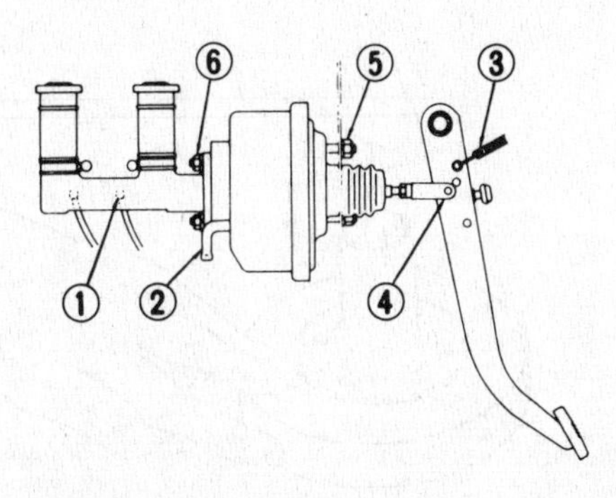

Fig. 9.32 Installation diagram of servo unit, master cylinder and brake pedal (Sec 17)

1 Master cylinder
2 Servo unit
3 Pedal return spring
4 Pushrod clevis
5 Servo retaining nut
6 Master cylinder retaining nut

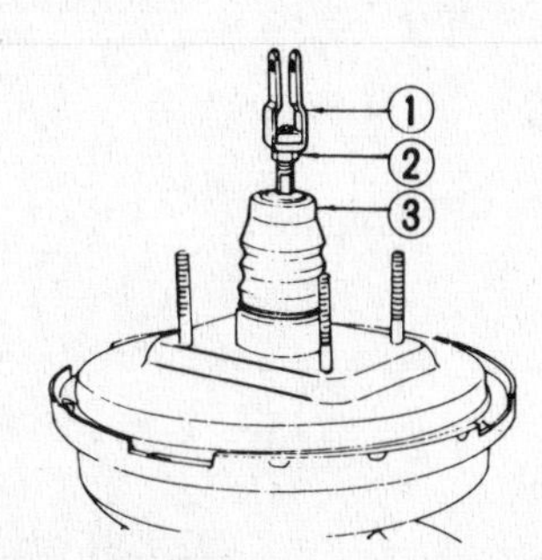

Fig. 9.33 Dismantling clevis fork (1) locknut (2) and the valve protective cover (3) from the servo unit (Sec 18)

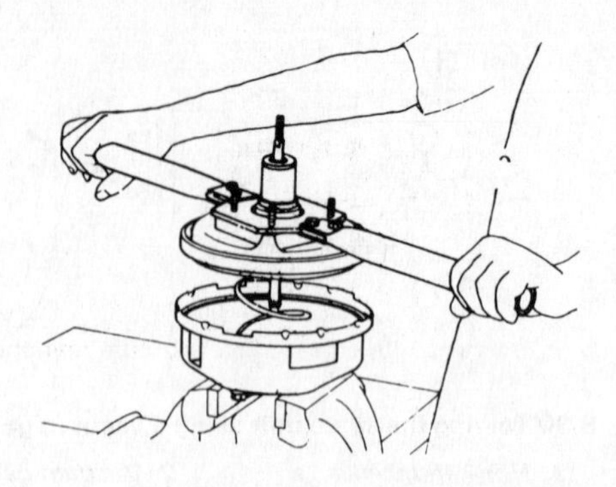

Fig. 9.34 Releasing the rear shell of the vacuum servo unit (Sec 18)

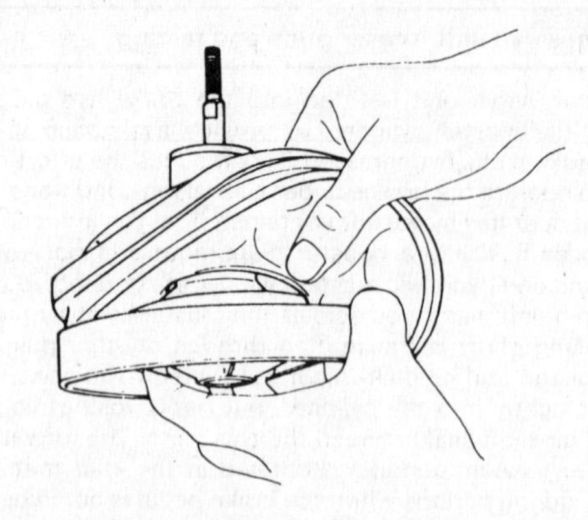

Fig. 9.35 Removing the servo unit diaphragm (Sec 18)

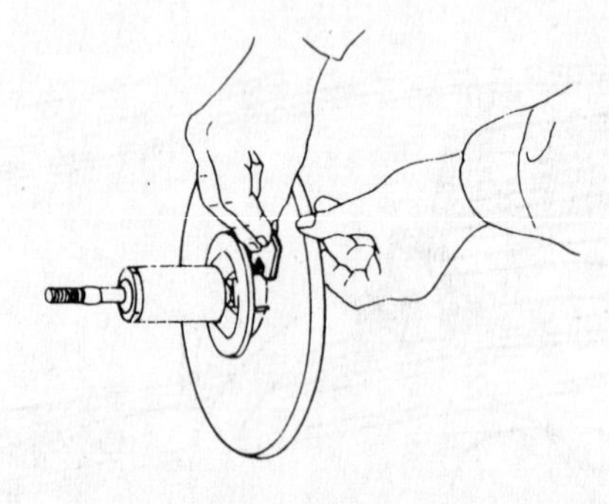

Fig. 9.36 Removing the valve plunger stop key (Sec 18)

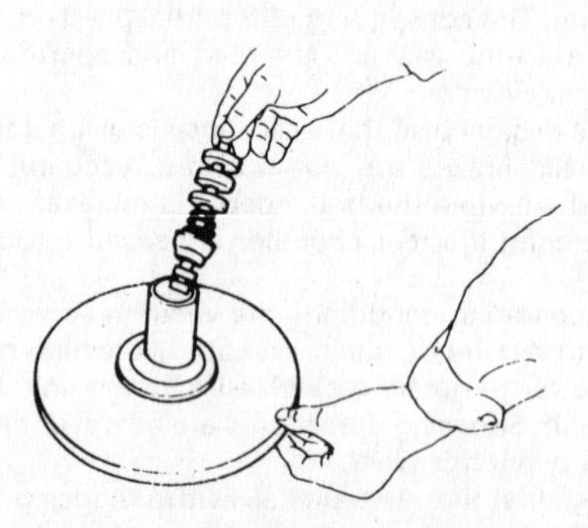

Fig. 9.37 Removing the air silencer and valve operating rod assembly (Sec 18)

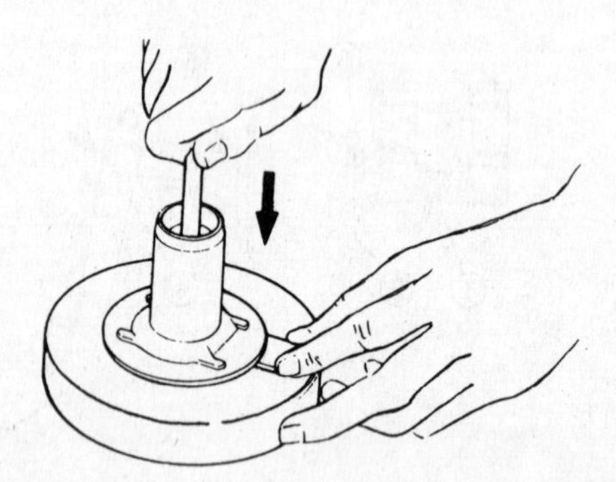

Fig. 9.38 Installing the valve operating rod and stop key (Sec 18)

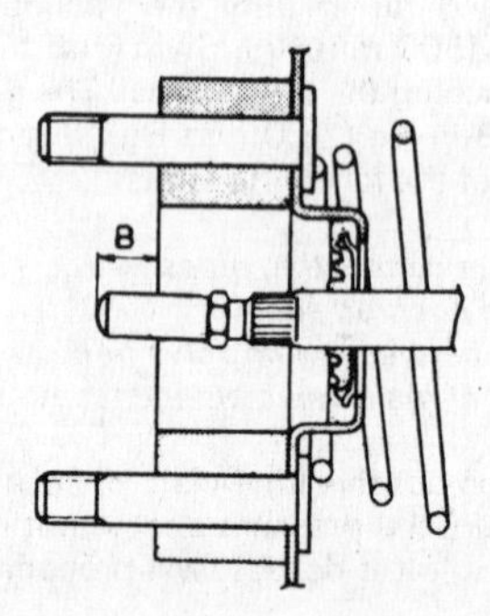

Fig. 9.39 Pushrod setting length B (Sec 18)

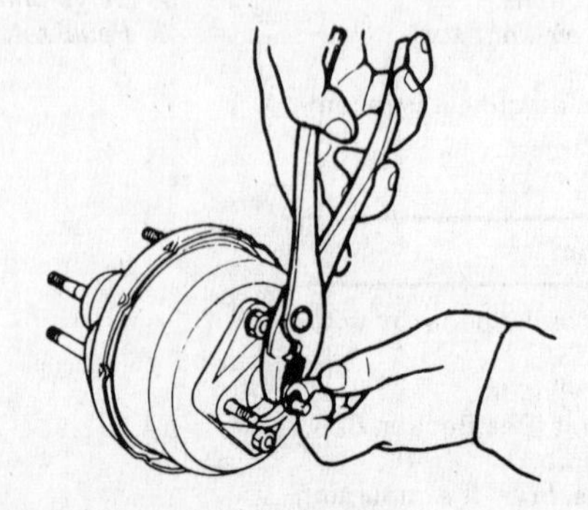

Fig. 9.40 Adjusting the pushrod of the vacuum servo unit (Sec 18)

on completion (See Section 13).

18 Vacuum servo unit – servicing

1 Remove all external dust and accumulated dirt.
2 Make alignment marks on the front and rear shells to facilitate reassembly.
3 Secure the master cylinder mounting studs and flange in the jaws of a vise but take care not to damage the threads of the studs or distort them.
4 Remove the clevis fork (1) locknut (2) and the valve protective cover (3) (Fig. 9.33).
5 A tool will now have to be made up which can be slipped over the mounting studs of the servo unit to enable the rear shell to be slightly depressed and then released by turning it in a counter-clockwise direction.
6 Remove the pushrod from the diaphragm plate.
7 Detach the valve body and the diaphragm from the rear shell.
8 Use a screwdriver to detach the retainer, bearing and seal from the rear shell.
9 Pull the diaphragm from its groove in the diaphragm plate.
10 Tap off the air silencer retainer carefully using a hammer and screwdriver applied at several points around its edge.
11 Depress the valve operating rod so that the keyhole opening faces downwards and then shake until the valve plunger stop key is ejected.
12 Withdraw the valve operating rod assembly and air silencer filter from the valve body and diaphragm plate.
13 Push out the reaction disc from the direction of the valve body.
14 Unscrew and remove the nuts which secure the master cylinder mounting flange to the front shell.
15 Examine all components for wear or deterioration and renew as required. If it is a routine overhaul, use the components supplied as a repair kit.
16 Reassembly is a reversal of dismantling but apply silicone grease sparingly to the seal at its contact face with the rear shell, to the lip of the poppet, to both faces of the reaction disc, to the edges of the diaphragm, to the plate and seal assembly and to the end of the pushrod. A pack of specified grease is supplied in the kit provided for repair purposes.
17 Insert the valve operating rod, holding it vertically and depressing it so that the stop key can be slid into position from the side.
18 Install the retainer to the rear shell using a piece of tubing as a drift.
19 Position the servo unit so that the pushrod (at the master cylinder mounting flange end) is vertical and then adjust its length (locknut and sleeve) so that the measurement between the face of the mounting flange and the end of the pushrod (dimension B in Fig. 9.39) is between 0.394 and 0.413 in (10.0 to 10.5 mm).

19 Fault diagnosis – braking system

Symptom	Reason/s
Pedal travels almost to toe-board before brakes operate	Brake fluid level too low Caliper leaking Master cylinder leaking (bubbles in master cylinder fluid) Brake flexible hose leaking Brake line fractured Brake system unions loose Pad or shoe linings over 75% worn
Brake pedal feels springy	New linings not yet bedded-in Brake discs or drums badly worn or cracked Master cylinder securing nuts loose
Brake pedal feels spongy and soggy	Caliper or wheel cylinder leaking Master cylinder leaking (bubbles in master cylinder reservoir) Brake pipe line or flexible hose leaking Unions in brake system loose
Excessive effort required to brake car	Pad or shoe linings badly worn New pads or shoes recently installed – not yet bedded-in Harder linings installed than standard causing increase in pedal pressure Linings and brake drums contaminated with oil, grease or hydraulic fluid Servo unit inoperative or faulty One half of brake hydraulic circuit not operating
Brakes uneven and pulling to one side	Linings and discs or drums contaminated with oil, grease or hydraulic fluid Tire pressures unequal Radial ply tires installed at one end of the car only Brake caliper loose Brake pads or shoes installed incorrectly Different type of linings installed at each wheel Anchorages for front suspension or rear suspension loose Brake discs or drums badly worn, cracked or distorted
Brakes tend to bind, drag or lock-on	Air in hydraulic system Wheel cylinders seized Parking brake cables too tight

Chapter 10 Electrical system

Refer to Chapter 13 for specifications and information applicable to 1980 through 1984 models.

Contents

Specifications

System type
12 volt negative ground

Alternator

Type:	
1977 models	Hitachi LT160–39
1978 models	Hitachi LR160–42
Rating	12 volt, 60 amp
Minimum brush length:	
1977 models	0·31 in (7·5 mm)
1978 models	0·28 in (7·0 mm)
Brush spring tension	9 to 12 ozf (255 to 345 gf)
Minimum slip ring diameter	1·18 in (30·0 mm)
Regulating voltage (1978 models)	14·4 to 15V

Voltage regulator unit (1977 models)

Type	TL1Z–82C
Regulator:	
Regulating voltage (fully charged battery)	14·3 to 15·3 volts at 20°C (68°F)
Coil resistance	10·5 ohms at 20°C (68°F)
Core gap	0·024 to 0·039 in (0·6 to 1·0 mm)
Points gap	0·014 to 0·018 in (0·35 to 0·45 mm)
Cut-out:	
Release voltage	4·2 to 5·2 volts at N terminal
Coil resistance	37·8 ohms at 20°C (68°F)
Core gap	0·031 to 0·039 in (0·8 to 1·0 mm)
Points gap	0·016 to 0·024 in (0·4 to 0·6 mm)

Starter motor

Type:	
1977 models, manual transmission	Hitachi S114–173B, pre-engaged
1977 models, automatic transmission	Hitachi S114–182B, pre-engaged
1978 models	Hitachi S114–254, reduction gear type

Minimum brush length:	
1977 models	0·47 in (12·0 mm)
1978 models	0·43 in (11·0 mm)
Brush spring tension:	
1977 models	3·1 to 4·0 lbf (1·4 to 1·8 kgf)
1978 models	3·5 to 4·4 lbf (1·6 to 2·0 kgf)
Maximum armature bearing clearance	0·008 in (0·2 mm)

Bulbs	**Wattage**
Headlamp outer (sealed beam)	37·5/50
Headlamp inner (sealed beam)	50
Turn signal light	27
Side marker light	8
Front parking light	8
Rear stop/tail light	27/8
Back-up light	27
License plate light	10 or 8
Room (interior) light	10
Instrument panel illuminating lights	3·4
Warning light cluster bulbs	3·4
High beam pilot light	3·4
Turn signal pilot light	3·4
Ashtray light	3·4
Glovebox light	3·4
Cigarette lighter illumination bulb	3·4
Hazard warning light	1·4
Defogger switch light	1·4
Key lamp	1·4
Trunk compartment light (Sedan)	5
Luggage compartment light (Station Wagon)	10

1 General description

1 The electrical system is a 12 volt, negative ground type comprising an alternator and its associated regulator, the battery, starting motor, the ancillaries such as lighting equipment, windshield wipers etc, and the associated wiring and protective circuits.
2 The battery provides starting power and a reserve of energy should the loading of the system exceed the alternator output.
3 When installing electrical accessories to your vehicle it is important to ensure that they are suitable for negative ground vehicles (most vehicles are negative ground, but accessories are still available for positive ground systems). Equipment which incorporates semiconductor devices may well be permanently damaged if they are not suited to your particular system.
4 If the battery is to be boost-charged from an external charging source, it is important to disconnect the battery ground cable in order to protect the semi-conductor devices in the alternator and possibly any electrical accessories which may have been installed. Similarly, when any electric (arc) welding or power tools are used the same precaution should be taken.

2 Battery – removal and installation

1 The battery is located within the engine compartment at the front of the right-hand fender inner wall.
2 Disconnect the battery negative (ground) terminal.
3 Then remove the positive terminal, and remove the nuts from the battery clamp. Remove the battery clamps.
4 The battery can now be removed from the vehicle, but take care to keep it upright to avoid spilling electrolyte on the paintwork.
5 Installation is a reversal of removal procedure but when reconnecting the terminals, clean off any white deposits present and smear on a little petroleum jelly.

3 Battery – maintenance and inspection

1 Keep the top of the battery clean by wiping away dirt and moisture.
2 Remove the plugs or lid from the cells and check that the electrolyte level is just above the separator plates. If the level has fallen, add only distilled water until the electrolyte level is just above the separator plates.
3 As well as keeping the terminals clean and covered with petroleum jelly, the top of the battery, and especially the top of the cells, should be kept clean and dry. This helps prevent corrosion and ensures that the battery does not become partially discharged by leakage through dampness and dirt.
4 Once every three months, remove the battery and inspect the battery securing bolts, the battery clamp plate, tray and battery leads for corrosion (white fluffy deposits on the metal which are brittle to touch). If any corrosion is found, clean off the deposits with ammonia or soda and paint over the clean metal with an anti-rust/anti-acid paint.
5 At the same time inspect the battery case for cracks. If a crack is found, clean and plug it with one of the proprietary compounds marketed for this purpose. If leakage through the crack has been excessive then it will be necessary to refill the appropriate cell with fresh electrolyte as detailed later. Cracks are frequently caused to the top of battery cases by pouring in distilled water in the middle of winter *after* instead of *before* a run. This gives the water no chance to mix with the electrolyte and so the former freezes and splits the battery case.
6 If topping-up the battery becomes excessive and the case has been inspected for cracks that could cause leakage, but none are found, the battery is being over-charged and the alternator or regulator may be at fault.
7 With the battery on the bench at the three monthly interval check, measure its specific gravity with a hydrometer to determine the state of charge and condition of the electrolyte. There should be very little variation between the different cells and if a variation in excess of 0·25 is present it will be due to either:

(a) Loss of electrolyte from the battery at some time caused by spillage or a leak, resulting in a drop in the specific gravity of electrolyte when the deficiency was replaced with distilled water instead of fresh electrolyte.

(b) An internal short-circuit caused by buckling of the plates or a similar malady pointing to the likelihood of total battery failure in the near future.

8 The specific gravity of the electrolyte for fully charged conditions at the electrolyte temperature indicated, is listed in Table A. The specific gravity of a fully discharged battery at different temperatures

of the electrolyte is given in Table B.

Table A
1.268 at 100°F or 38°C electrolyte temperature
1.272 at 90°F or 32°C electrolyte temperature
1.276 at 80°F or 27°C electrolyte temperature
1.280 at 70°F or 21°C electrolyte temperature
1.284 at 60°F or 16°C electrolyte temperature
1.288 at 50°F or 10°C electrolyte temperature
1.292 at 40°F or 4°C electrolyte temperature
1.296 at 30°F or –1.5°C electrolyte temperature

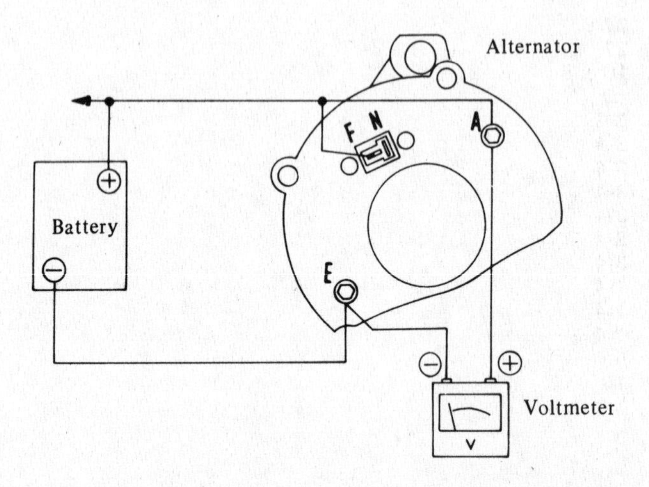

Fig. 10.1 Alternator test circuit (Sec 8)

Table B
1.098 at 100°F or 38°C electrolyte temperature
1.102 at 90°F or 32°C electrolyte temperature
1.106 at 80°F or 27°C electrolyte temperature
1.110 at 70°F or 21°C electrolyte temperature
1.114 at 60°F or 16°C electrolyte temperature
1.118 at 50°F or 10°C electrolyte temperature
1.122 at 40°F or 4°C electrolyte temperature
1.126 at 30°F or –1.5°C electrolyte temperature

4 Battery – electrolyte replenishment

1 If the battery is in a fully charged state and one of the cells maintains a specific gravity reading which is 0·25 or more lower than the others, and a check of each cell has been made with a suitable meter to check for short-circuits (a four to seven second test should give a steady reading of between 1·2 to 1·8 volts), then it is likely that electrolyte has been lost from the cell with the low reading at some time.
2 Top-up the cell with a solution of 1 part sulphuric acid to 2·5 parts of water. If the cell is already fully topped-up draw some electrolyte out of it with a pipette.
3 Continue to top-up the cell with freshly-made electrolyte and then recharge the battery and check the hydrometer readings. **CAUTION** *When mixing the sulphuric acid and water, never add water to the sulphuric acid. Always pour the acid slowly onto the water in a glass container. If water is added to sulphuric acid it will react violently.*

5 Battery charging

1 In winter time when heavy demand is placed upon the battery, such as when starting from cold, and much electrical equipment is continually in use, it is a good idea to occasionally have the battery fully charged from an external source at the rate of 3·5 to 4 amps (see Section 7).

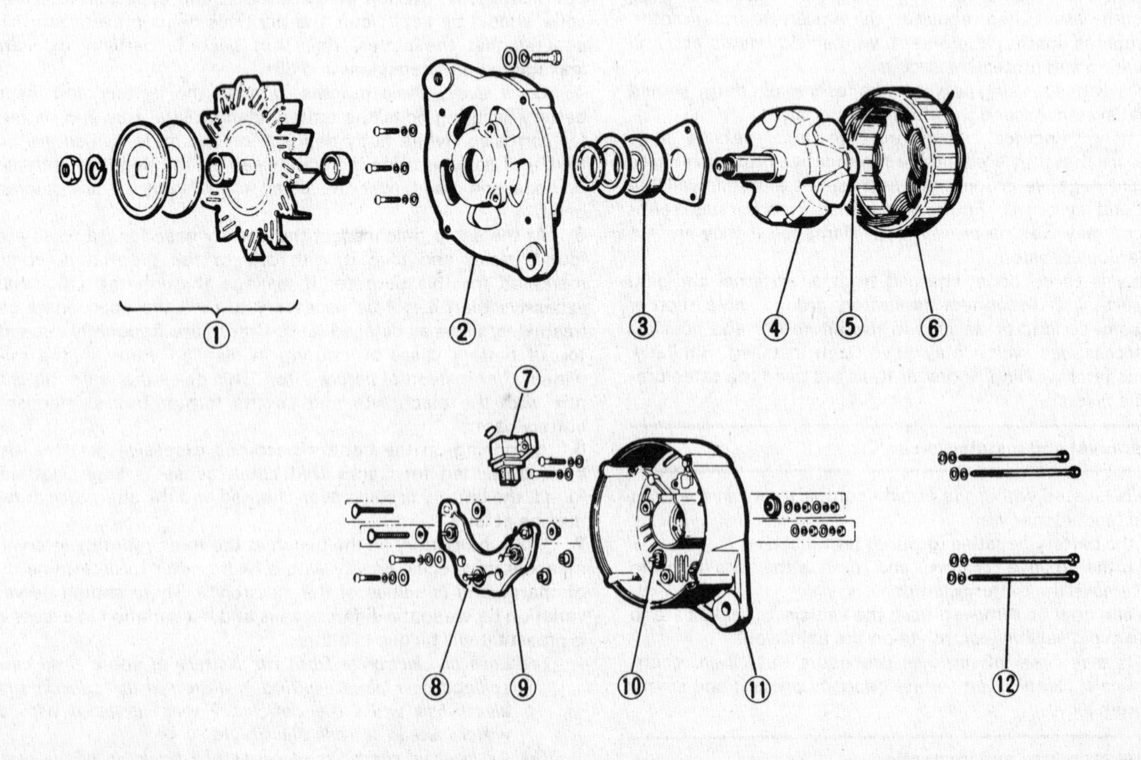

Fig. 10.2 An exploded view of the early-type alternator (Sec 10)

1 Pulley assembly	*4 Rotor*	*7 Brush assembly*	*10 Negative diode*
2 Front cover	*5 Rear bearing*	*8 Positive diode*	*11 Rear cover*
3 Front bearing	*6 Stator assembly*	*9 Diode mounting plate*	*12 Through-bolts*

2 Continue to charge the battery at this rate until no further rise in specific gravity is noted over a four hour period.
3 Alternatively, a trickle charger at the rate of 1·5 amps can be safely used overnight.
4 Specially rapid 'boost' charges which are claimed to restore the power of the battery in 1 to 2 hours are most dangerous as they can cause serious damage to the battery plates.

6 Alternator – general description and maintenance

1 Briefly, the alternator comprises a rotor and stator. Current is generated in the coils of the stator as soon as the rotor revolves. This current is three-phase alternating which is then rectified by positive and negative silicon diodes and the level of voltage required to maintain the battery charge is controlled by a regulator unit.
2 Maintenance consists of occasionally wiping away any oil or dirt which may have accumulated on the outside of the unit.
3 No lubrication is required as the bearings are grease-sealed for life.
4 Check the drivebelt tension periodically to ensure that its specified deflection is correctly maintained as described in Chapter 1, Section 43.

7 Alternator – special precautions

Take extreme care when making circuit connections to a vehicle which has an alternator and observe the following:

When making connections to the alternator from a battery, always match correct polarity. Before using electric arc welding equipment to repair any part of the vehicle, disconnect the connector from the alternator, and disconnect the battery ground lead.

Never start the car with a battery charger connected.

Always disconnect both battery leads before using a mains charger.

If boost starting from another battery, always connect in parallel using heavy duty cables.

8 Alternator – testing in position in the car

1 Where a faulty alternator is suspected, first ensure that the battery is fully charged, if necessary from an outside source.
2 Obtain a 0 to 30V dc voltmeter.
3 Disconnect the leads from the alternator terminals, and link together the alternator A and F terminals.
4 Connect a test probe from the voltmeter positive terminal to the A terminal of the alternator. Connect the voltmeter negative terminal to ground and check that the voltmeter indicates battery voltage (12 volts).
5 Switch the headlamps to main beam.
6 Start the engine and gradually increase its speed to approximately 1100 rev/min and check the reading on the voltmeter. If it registers over 12·5 volts then the alternator is in good condition; if it registers below 12·5 volts then the alternator is faulty and must be removed and repaired.

9 Alternator – removal and installation

1 Disconnect the battery ground cable.
2 Disconnect the cable connectors from the rear of the alternator.
3 Loosen the alternator adjusting bolt and mounting bolt.
4 Push the unit towards the engine block sufficiently to enable the drivebelt to be slipped off the alternator pulley.
5 Remove the adjusting bolt and mounting bolt, then lift away the alternator.
6 Installation is a reversal of the removal procedure, but ensure that connections are correctly made and that the drivebelt is correctly adjusted.

10 Alternator – dismantling, servicing and reassembly

1 Unscrew and remove the four through-bolts.
2 Separate the front cover and rotor assembly from the rear cover and stator assembly, by lightly tapping the front cover with a soft-faced mallet.
3 Holding the outer diameter of the rotor in protected vise jaws, unscrew and remove the pulley and fan retaining nut. Remove the spring washer and pry off the pulley, followed by the fan. Be sure to recover the spacers that are either side of the fan.
4 With the pulley and fan assembly removed from the rotor shaft, three set screws will now be visible. These screws retain the front cover to the bearing retainer plate. Unscrew and remove them.
5 Using a soft-faced mallet, carefully tap the rotor out of the front cover. The bearing will remain on the rotor shaft.
6 The bearings may be removed from the rotor shaft, if necessary, by extracting them with a puller.
7 Disconnect the stator coil-to-diode terminal wires with a soldering iron.
8 Unscrew and remove the two screws retaining the brush assembly to the inside of the rear cover. The stator coil together with the brush holder (and IC regulator on later models), can now be removed from the rear cover.
9 Carefully, using nippers, remove the protective sleeve over the soldered joint between the brush holder assembly and the stator coil wire. The brush assembly can then be removed by unsoldering the joint. **Note**: *On later models with the transistorized integral regulator, carefully detach the wires from the diode terminal using a soldering iron; then loosen the retaining screws and disconnect the brush unit and IC regulator.*
10 Working inside the rear cover, unscrew and remove the two diode plate securing screws and the A and E terminals, and lift away the diode plate.
11 On the earlier types, six diodes are used in the alternator; three are positive and are pressed into the diode mounting plate; three are negative and are pressed into the rear cover. On later models, the alternator also has three sub-diodes which are soldered onto the brush unit. To carry out continuity tests on the diodes, which will require the use of an ohmmeter, connect the negative probe of the ohmmeter to the rear

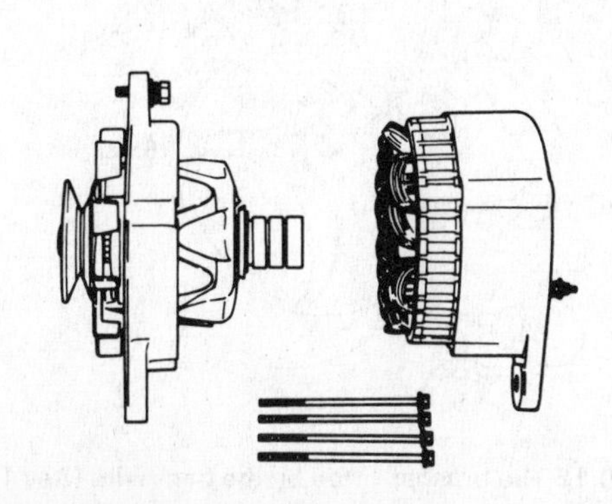

Fig. 10.3 The front and rear covers separated (Sec 10)

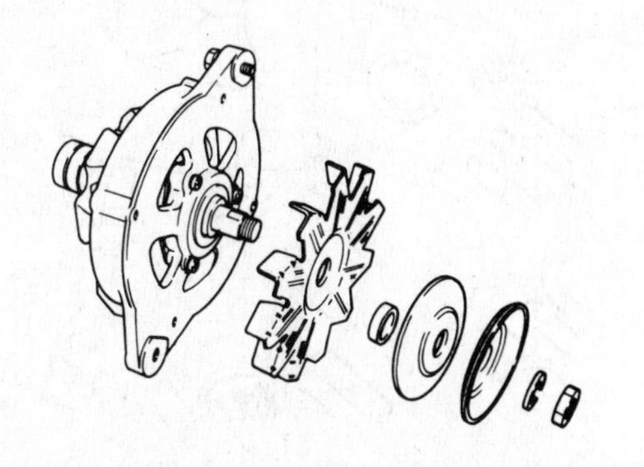

Fig. 10.4 Removing the pulley and fan (Sec 10)

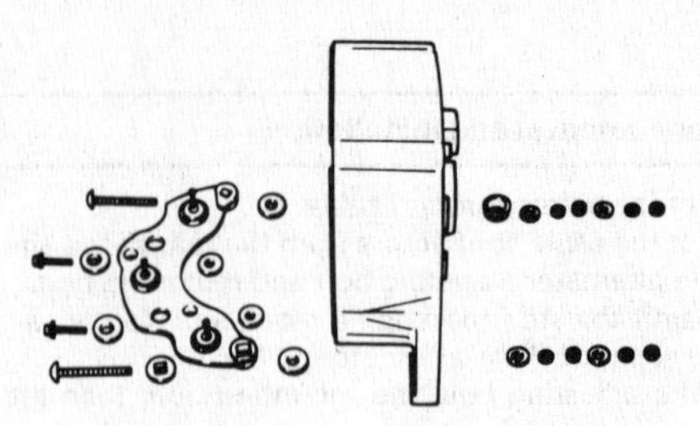

Fig. 10.5 The diode mounting plate removed from the rear cover (Sec 10)

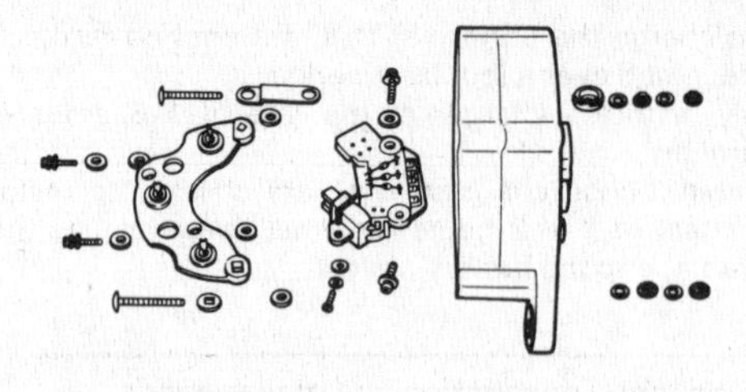

Fig. 10.6 The diode mounting plate and regulator/brush unit removed from the rear cover (later models) (Sec 10)

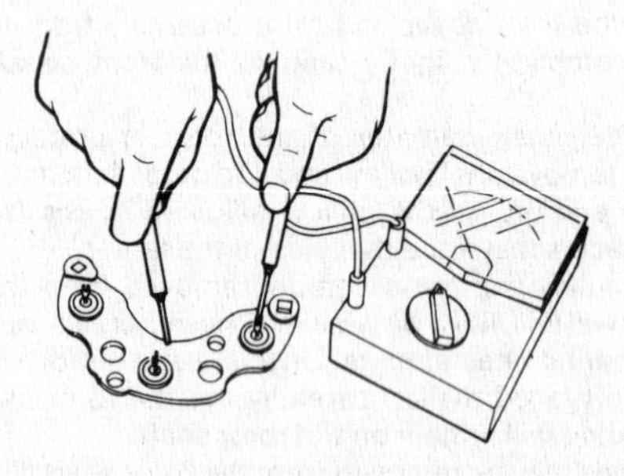

Fig. 10.7 Testing a positive diode for continuity (Sec 10)

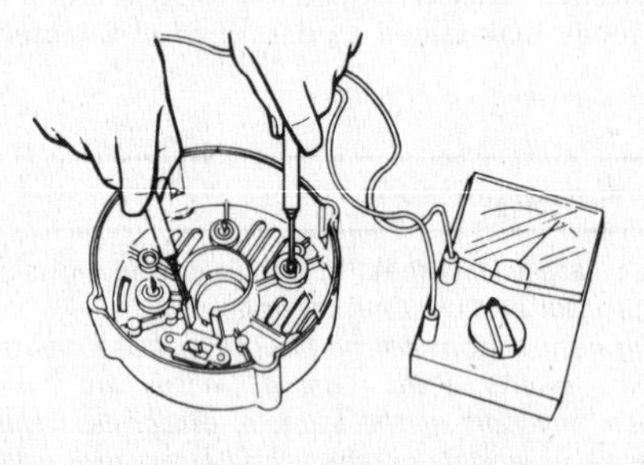

Fig. 10.8 Testing a negative diode for continuity (Sec 10)

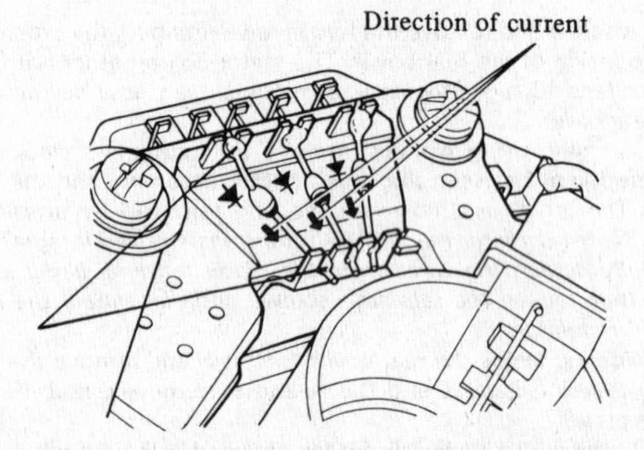

Fig. 10.9 The auxiliary diodes (sub-diodes) on the late model alternator (Sec 10)

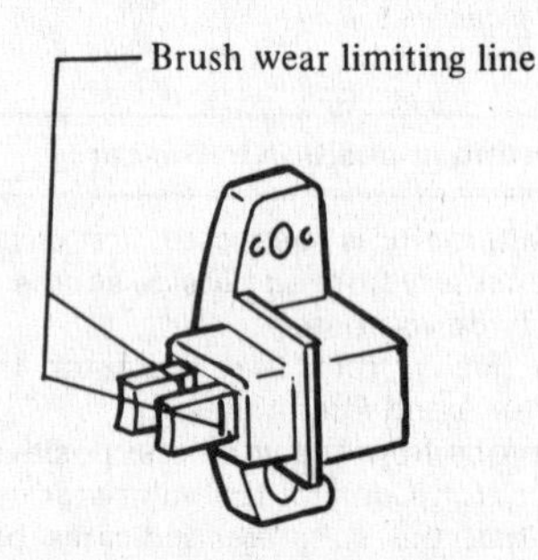

Fig. 10.10 The brush wear limit lines (Sec 10)

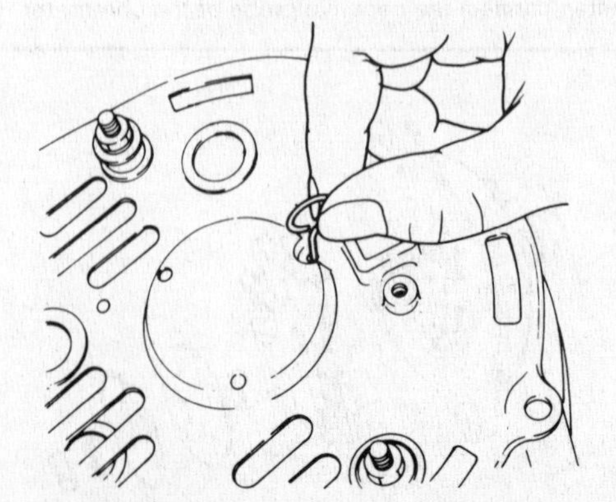

Fig. 10.11 Using a piece of bent wire to lift the brushes prior to joining the two halves of the alternator together (Sec 10)

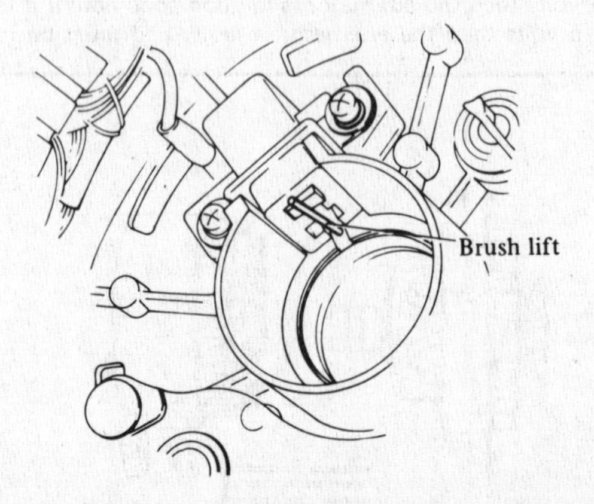

Fig. 10.12 The brushes lifted by the bent wire (Sec 10)

cover case and the positive probe to a diode terminal; the ohmmeter should show continuity. Now connect the negative probe to the diode terminal and the positive probe to the rear case; the ohmmeter should not show continuity. Repeat this test on the other two diodes. If any of the diodes in the rear case fail this test, they must be renewed. If any of the sub-diodes are to be renewed, pinch the diode lead wire with a pair of pliers when unsoldering, which will help prevent any heat being transferred from the iron to the diode. To check the positive diodes on the diode mounting plate, connect the negative probe of the ohmmeter to the mounting plate and the positive probe to a diode terminal; the ohmmeter should not show continuity. Now connect the negative probe to the diode terminal and the positive probe to the mounting plate; the ohmmeter should now show continuity. Repeat this test at the other two diodes. If any diode fails this test it must be pressed out and a new one pressed in.

12 Using an ohmmeter, test the rotor coils for continuity. Apply the probes of the meter to the slip rings, when the meter should show continuity. If there is no reading, there is probably a break in the field coils and the rotor must be renewed.

13 Next test the rotor for insulation breakdown by placing the probes, one on the slip ring and one on the rotor core. If the meter shows any reading, then the slip ring or rotor coil is grounding, probably due to faulty insulation.

14 Check the stator for continuity by connecting the meter probes to the stator coil wires. If no reading is indicated on the meter, the coil wiring is open-circuit. Now connect one probe to the stator core and touch each of the stator wires with the other probe. As each wire is touched by the probe there should be no reading on the meter. If there is a reading on the meter, it indicates that the coil is grounding to the stator coil.

15 Check the condition of the brush assembly. If the brushes have worn down to the wear limit line, the brush assembly will have to be renewed complete. Using a suitable spring balance, test the tension of the brush springs when the front face of the brush is 0·079 in (2·0 mm) from the front face of the brush holder. If the tension is not as specified, renew the brush assembly.

16 Reassembly is a reversal of dismantling but the following points should be noted:

(a) When soldering each stator coil lead to its diode, carry out the operation as quickly as possible, localising the heat.
(b) When installing the diode A terminal, ensure that the insulating bush is correctly installed.
(c) A length of 'heat and shrink' sleeve must be placed over the soldered joint between the brush holder assembly and the stator wire.
(d) Before installing the front cover assembly to the rear cover assembly, use a piece of bent wire inserted through the cover hole, to lift the brushes whilst the rear bearing passes through the bore and the slip rings are in position. Once the two halves of the assembly are together, the wire can be removed.

11 Voltage regulator and cutout – description, testing and adjustment

1 The voltage cutout/regulator is situated on the relay mounting bracket within the engine compartment on early models. On later models it is integral with the alternator.

2 The earlier electro-mechanical type employs two sets of adjustable contact points to make or break the charging circuit at a regulated amount depending on the alternator load and voltage requirements. This type can be checked and adjusted (see below from paragraph 5 onwards).

3 The 1978 models are fitted with a regulator (IC regulator) having integrated circuits and transistors which replace the contact points. This type of regulator is shown in Fig. 10.16 and if it is suspected of malfunction, have it checked out by your Datsun dealer. It cannot be repaired and if faulty must be renewed.

4 The electro-mechanical voltage regulator controls the output from the alternator depending upon the state of the battery and the demands of the vehicle electrical equipment and it ensures that the battery is not overcharged. The cutout is virtually an automatic switch which completes the charging circuit as soon as the alternator starts to rotate and isolates it when the engine stops so that the battery cannot be discharged to ground through the alternator. One visual indication

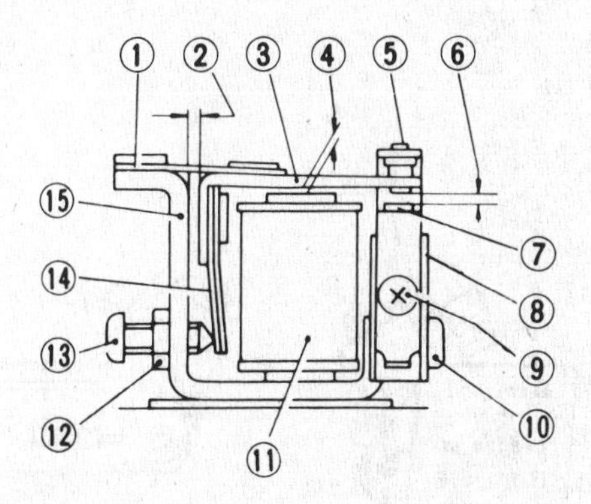

Fig. 10.13 Sectional view of the voltage regulator (Sec 11)

1 Connecting spring
2 Yoke gap
3 Armature
4 Core gap
5 Low speed contact
6 Points gap
7 High speed contact
8 Contact set
9 Screw
10 Screw
11 Coil
12 Locknut
13 Adjusting screw
14 Adjuster spring
15 Yoke

Fig. 10.14 Sectional view of the cutout (Sec 11)

1 Points gap
2 Charge relay contact
3 Core gap
4 Armature
5 Connecting spring
6 Yoke gap
7 Yoke
8 Adjusting spring
9 Adjusting screw
10 Locknut
11 Coil
12 Screw
13 Screw
14 Contact set
15 Voltage regulator contact

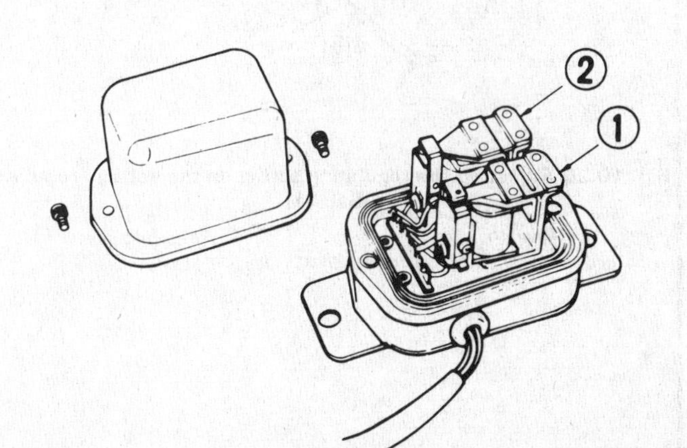

Fig. 10.15 The electro-mechanical cutout (1) and voltage regulator (2) (Sec 11)

10

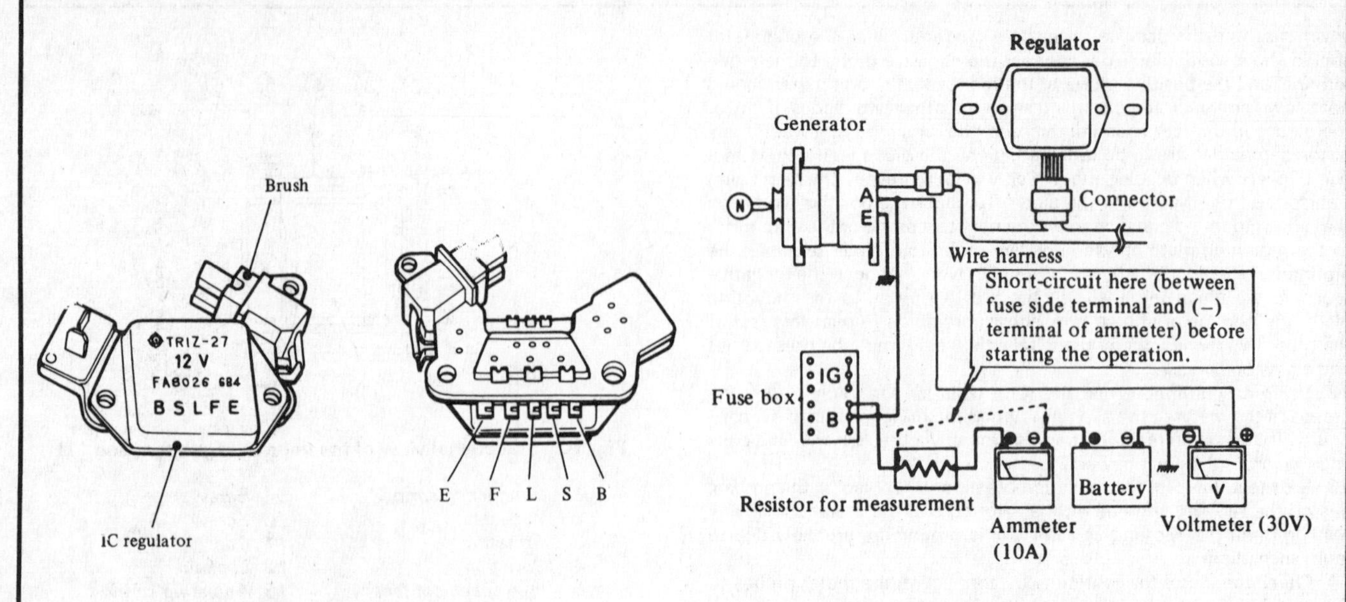

Fig. 10.16 The later type IC regulator

Fig. 10.17 Circuit for testing the voltage regulator in the car (Sec 11)

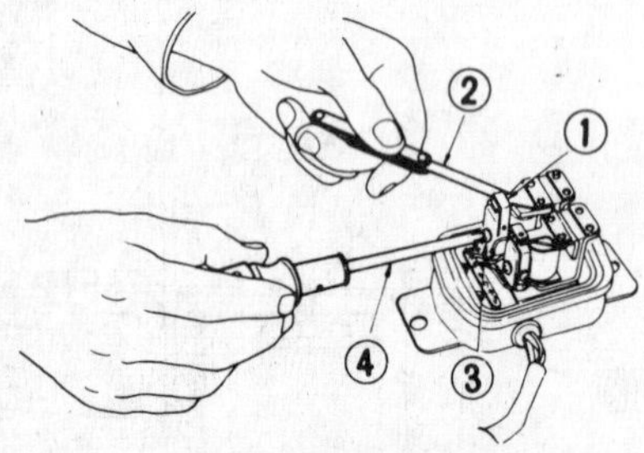

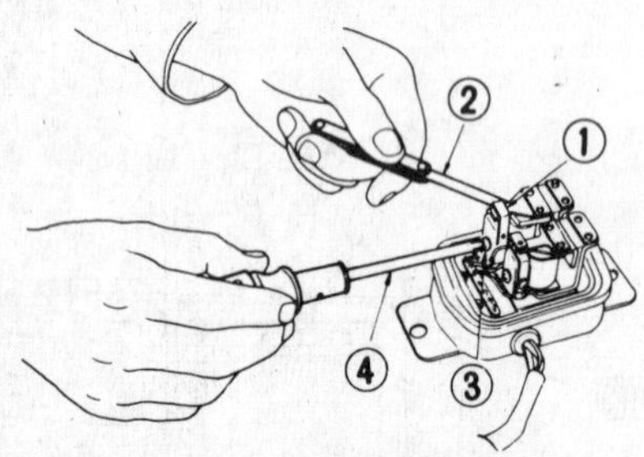

Fig. 10.18 Checking and adjusting the voltage regulator core gap (Sec 11)

1 Contacts
2 Feeler blade
3 Adjusting screw
4 Crosshead type screwdriver

Fig. 10.19 Checking and adjusting the voltage regulator points gap (Sec 11)

1 Upper contact
2 Adjusting screw
3 Crosshead type screwdriver
4 Feeler blade

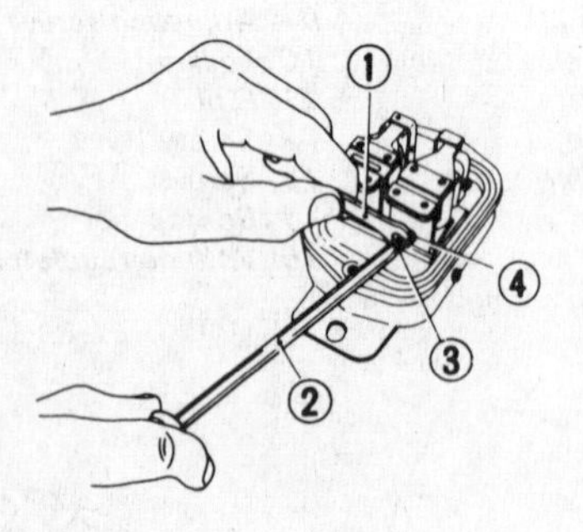

Fig. 10.20 Adjusting the regulating screw on the voltage regulator (Sec 11)

1 Wrench
2 Crosshead type screwdriver
3 Adjusting screw
4 Locknut

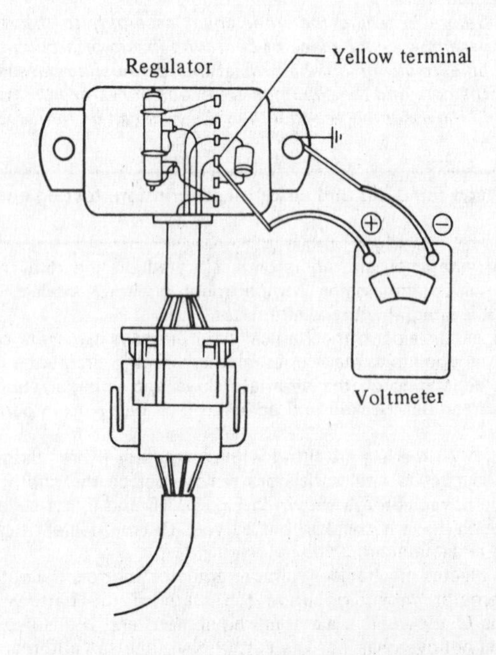

Fig. 10.21 Test circuit for the cutout (Sec 11)

of the correct functioning of the cutout is the ignition warning lamp. When the lamp is out, the system is charging. **Note**: *Before testing, check that the alternator drivebelt is not broken or slack and that all electrical leads are secure.*
5 Test the regulator voltage with the unit still installed in the vehicle. Carry out the testing with the engine compartment cold and complete the test within one minute to prevent the regulator heating up and affecting the specified voltage readings.
6 Establish the ambient temperature within the engine compartment, turn off all vehicle electrical equipment and ensure that the battery is in a fully-charged state. Connect a good quality ammeter, voltmeter and resistor as shown.
7 Start the engine and immediately detach the short circuit wire. Increase the engine speed to 2500 rev/min and check the voltmeter reading according to the pre-determined ambient temperature table below.
8 If the voltage does not conform to that specified, continue to run the engine at 2500 rev/min for several minutes and then with the engine idling check that the ammeter reads below 5 amps. If the reading is above this, the battery is not fully charged and must be removed for charging as otherwise accurate testing cannot be carried out.

Ambient temperature		*Rated regulating voltage*
°C	*(°F)*	*(V)*
–10	*(14)*	*14.75 to 15.74*
0	*(32)*	*14.60 to 15.60*
10	*(50)*	*14.45 to 15.45*
20	*(68)*	*14.30 to 15.30*
30	*(86)*	*14.15 to 15.15*
40	*(104)*	*14.00 to 15.00*

9 Switch off the engine, remove the cover from the voltage regulator and inspect the surfaces of the contacts. If these are rough or pitted, clean them by drawing a strip of fine emery cloth between them.
10 Using feeler gauges, check and adjust the core gap if necessary, to between 0·024 and 0·040 in (0·6 and 1·0 mm).
11 Check and adjust the contact point gap if necessary, to between 0·014 and 0·018 in (0·35 and 0·45 mm) (Fig. 10.18).
12 By now the voltage regulator will have cooled down so that the previous test may be repeated. If the voltage/temperature is still not compatible, switch off the engine and adjust the regulator screw. Do this by loosening the locknut and turning the screw clockwise to increase the voltage reading and anti-clockwise to reduce it (Fig. 10.19).
13 Turn the adjuster screw only fractionally before retesting the voltage charging rate gain with the unit cold. Finally tighten the locknut.
14 If the cutout is operating incorrectly, first check the alternator drivebelt and the ignition warning lamp bulb. Connect the positive terminal of a moving coil voltmeter to the N socket of the regulator connecting plug and the voltmeter negative terminal to ground as shown (Fig. 10.20).
15 Start the engine and let it idle. Check the voltmeter reading. If the reading is 0 volts check for continuity between the N terminals of the regulator unit and the alternator. If the reading is below 5·2 volts and the ignition warning lamp remains on, check and adjust the core gap to between 0·032 and 0·040 in (0·8 and 1·0 mm) and the points gap to 0·016 and 0·024 in (0·4 and 0·6 mm). Remember that this time the adjustments are carried out to the cutout *not the voltage regulator* although the procedure is similar.
16 If the reading is over 5·2 volts with the ignition warning lamp on the core and points gap are correctly set, the complete regulator unit must be renewed.
17 The cutout is operating correctly if the voltmeter shows a reading of more than 5·2 volts (ignition lamp out).

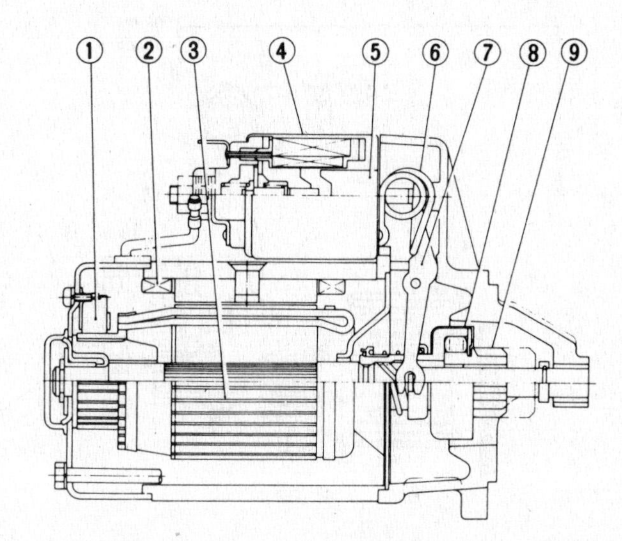

Fig. 10.22 Sectional view of the starter motor – 1977 models (Sec 13)

1 Brush
2 Field coil
3 Armature
4 Solenoid
5 Plunger
6 Torsion spring
7 Shift lever
8 Clutch mechanism
9 Pinion

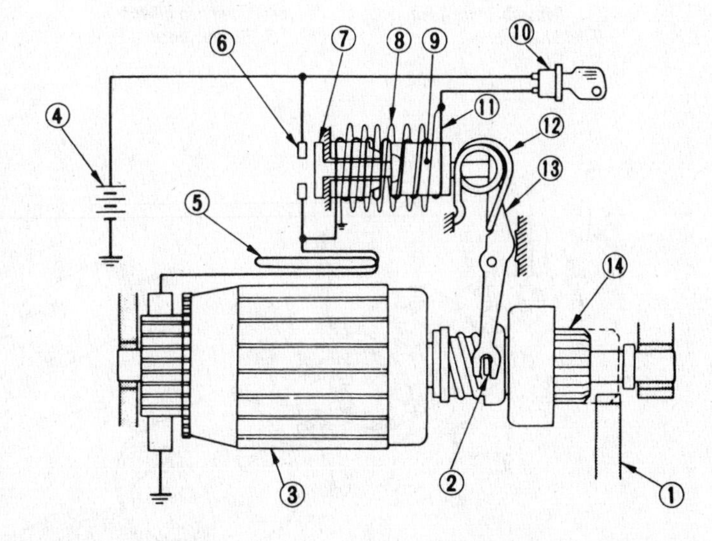

Fig. 10.23 Starter motor circuit – 1977 models (Sec 13)

1 Starter ring gear
2 Shift fork
3 Armature
4 Battery
5 Field coil
6 Stationary contact
7 Movable contact
8 Shunt coil
9 Plunger
10 Ignition switch
11 Series coil
12 Torsion spring
13 Shift lever
14 Pinion

12 Starter motor – general description

Two types of starter motor have been used. On early models (1977) the starter motor incorporates a solenoid mounted on top of the starter motor body. When the ignition switch is operated, the solenoid moves the starter drive pinion, through the medium of the shift lever, into engagement with the flywheel or driveplate starter ring gear. As the solenoid reaches the end of its stroke and with the pinion by now fully engaged with the flywheel ring gear, the main fixed and moving contacts close and engage the starter motor to rotate the engine.

This fractional pre-engagement of the starter drive does much to reduce the wear on the flywheel ring gear associated with inertia type starter motors.

On later models (1978 on), a reduction-gear type starter motor is used. Whilst the solenoid and starter motor functions are basically the same as the earlier type, a reduction gear is situated between the

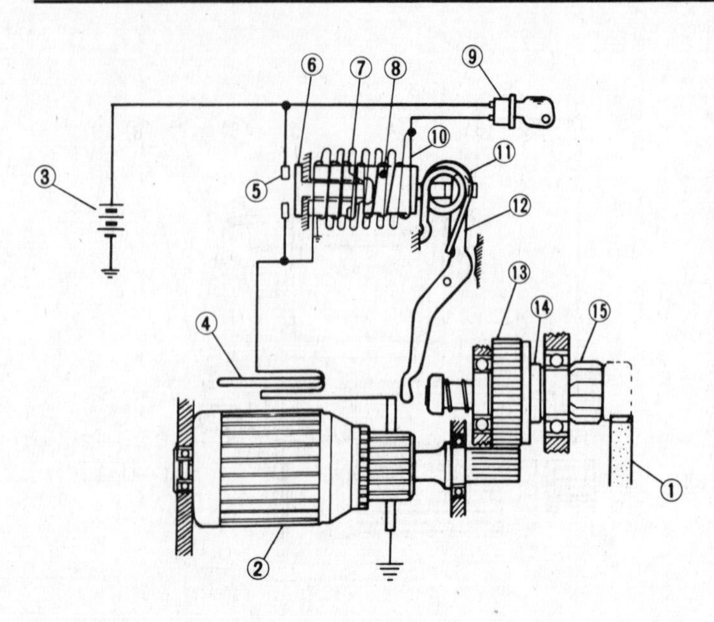

Fig. 10.24 Starter motor circuit showing reduction gear on 1978 models (Sec 13)

1 Ring gear
2 Armature
3 Battery
4 Field coil
5 Stationary contact
6 Movable contact
7 Shunt coil
8 Plunger
9 Ignition switch
10 Series coil
11 Torsion spring
12 Shift lever
13 Reduction gear
14 Overrun clutch
15 Pinion gear

pinion gear and armature. The armature-to-pinion gear actuation is operated by this reduction gear which, whilst reducing the speed of rotation at the flywheel, increases the actual rotational torque. This starter motor is shown in Fig. 10.24.

13 Starter motor – removal and installation

1 Disconnect the cable from the battery ground terminal.
2 Disconnect the black and yellow wire from the S terminal on the solenoid and the black cable from the B terminal also on the end cover of the solenoid.
3 Unscrew and remove the two starter motor securing bolts, pull the starter forward, tilt it slightly to clear the motor shaft support from the flywheel ring gear and withdraw it.
4 Installation is a reversal of removal.

14 Starter motor – dismantling, servicing and reassembly

Early type – dismantling

1 Disconnect the lead from the 'M' terminal of the solenoid.
2 Remove the two solenoid securing screws and withdraw the solenoid from the starter motor.
3 At the rear end of the starter motor, remove the dust cover and extract the E-ring from the groove in the armature shaft. With the E-

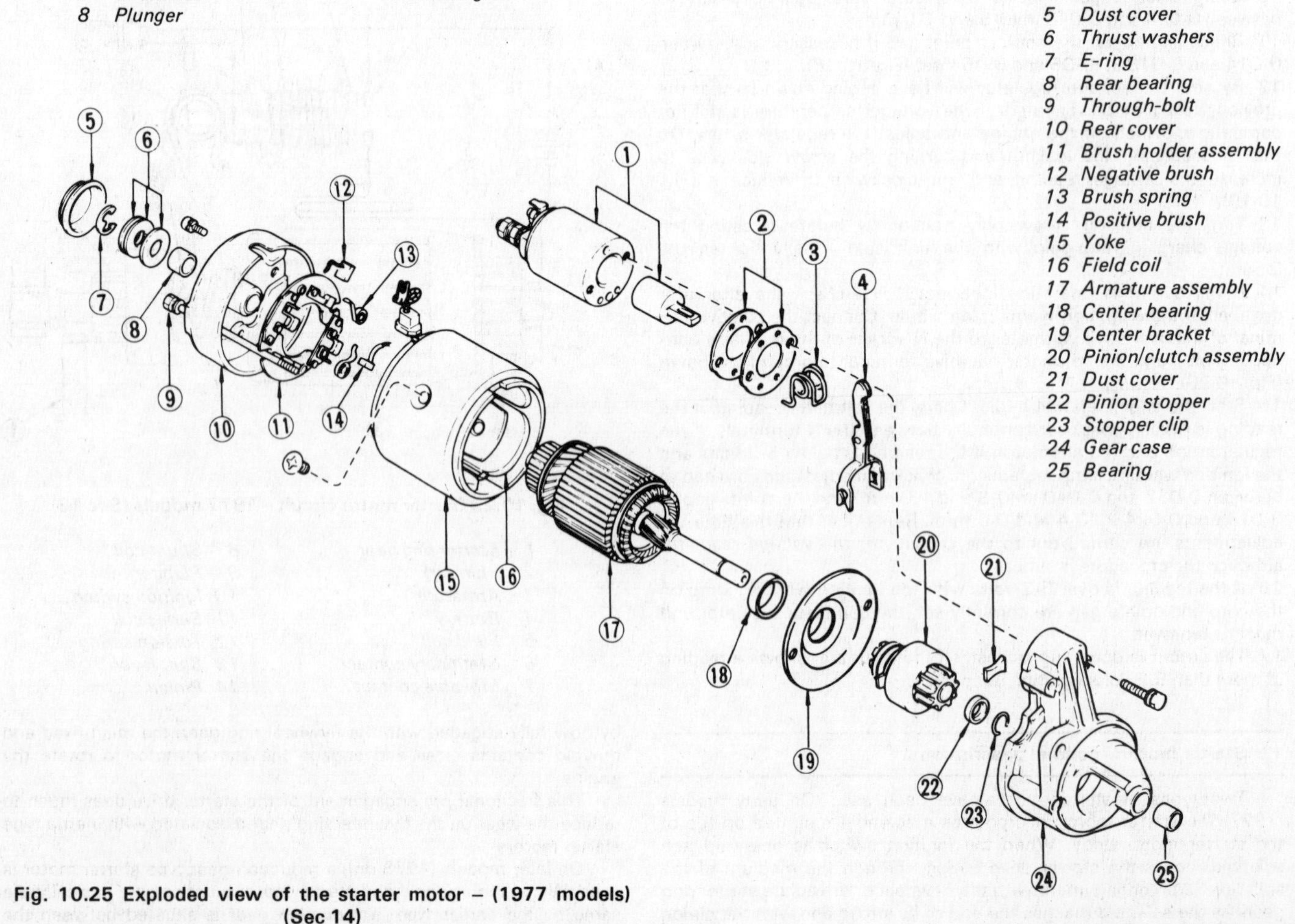

1 Solenoid assembly
2 Adjusting washers
3 Torsion spring
4 Shift lever
5 Dust cover
6 Thrust washers
7 E-ring
8 Rear bearing
9 Through-bolt
10 Rear cover
11 Brush holder assembly
12 Negative brush
13 Brush spring
14 Positive brush
15 Yoke
16 Field coil
17 Armature assembly
18 Center bearing
19 Center bracket
20 Pinion/clutch assembly
21 Dust cover
22 Pinion stopper
23 Stopper clip
24 Gear case
25 Bearing

Fig. 10.25 Exploded view of the starter motor – (1977 models) (Sec 14)

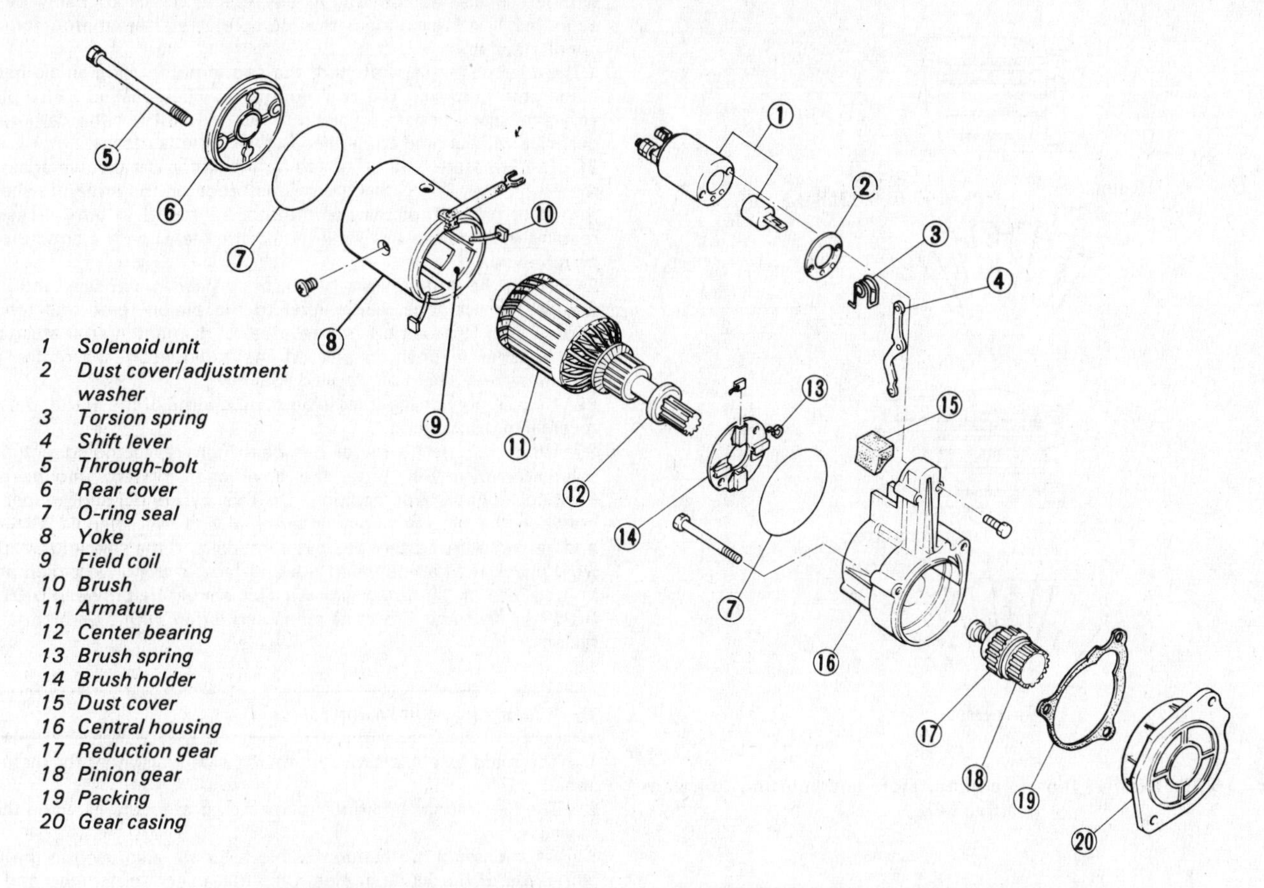

Fig. 10.26 Exploded view of starter motor on 1978 models – reduction gear type (Sec 14)

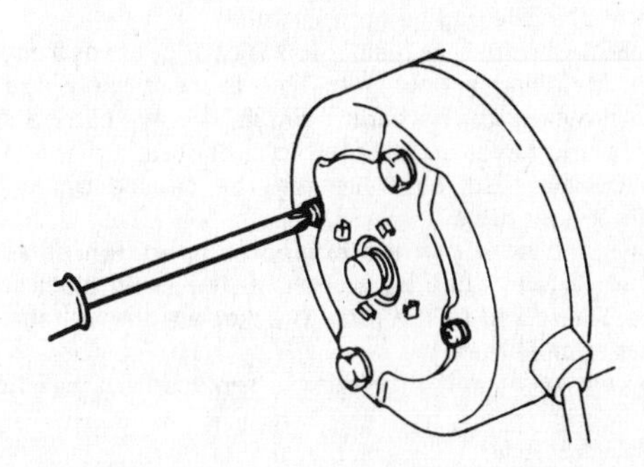

Fig. 10.27 Removing the brush holder setscrews (Sec 14)

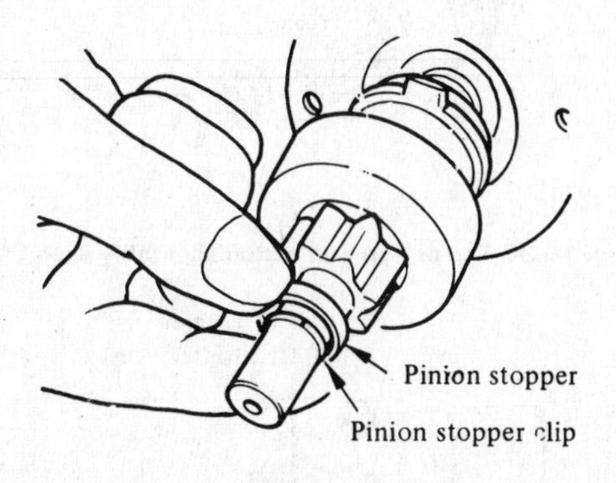

Fig. 10.28 Removing the pinion stopper (Sec 14)

ring removed, take off the thrust washer(s).

4 Unscrew and remove the two brush holder retaining screws.

5 Unscrew and remove the two through-bolts and lift away the rear cover.

6 Using a piece of wire shaped into a hook, pull away each brush spring and pull out the brushes. Lift away the brush holder assembly.

7 Drive off the yoke from the gearcase by tapping it lightly with a soft-faced mallet.

8 Withdraw the armature assembly and shift lever.

9 Using a piece of tubing, drive the pinion stopper down the shaft to expose the pinion stopper clip. Remove the stopper clip, and then from the armature shaft remove the stopper followed by the pinion and clutch assembly.

Reduction gear type – dismantling

10 Detach the lead from the 'M' terminal of the solenoid.

11 Unscrew the two solenoid screws and withdraw the solenoid from the starter motor. Detach the torsion spring from the solenoid.

12 Unscrew and extract the rear cover through-bolts and remove the cover. Pry the rear cover free using a screwdriver blade or similar tool but take care not to damage the packing.

13 The yoke, armature and brush holder can now be removed from the center housing. Take care not to damage the brushes.

14 Using a piece of hooked wire, pull each brush holder spring back and extract each brush. Note that the positive brush is insulated from its holder, and has its lead connected to the field coil.

15 To separate the center housing from the gear case, unscrew and remove the retaining bolts. Finally remove the pinion gear.

Servicing and reassembly

16 Check the brushes for wear. If their length is less than 0·472 in (12·0 mm) renew them. Using a suitable spring balance, test the tension of the brush springs which should be 3·5 lbf (1·6 kgf). Renew the springs if the tension is less than 3·1 lbf (1·4 kgf).

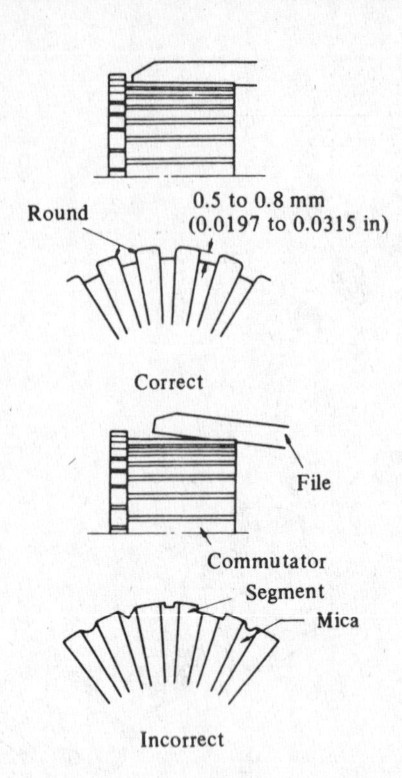

Fig. 10.29 Starter motor commutator undercutting diagram (Sec 14)

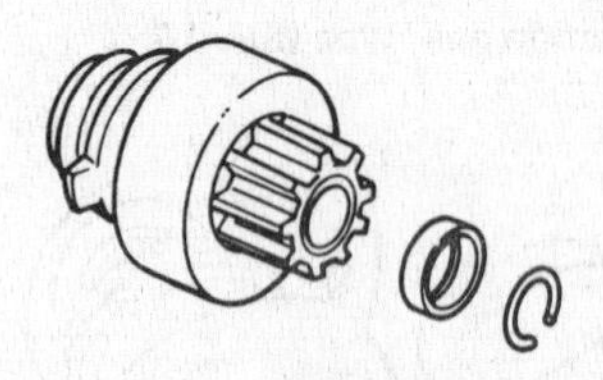

Fig. 10.30 The pinion and clutch assembly (Sec 14)

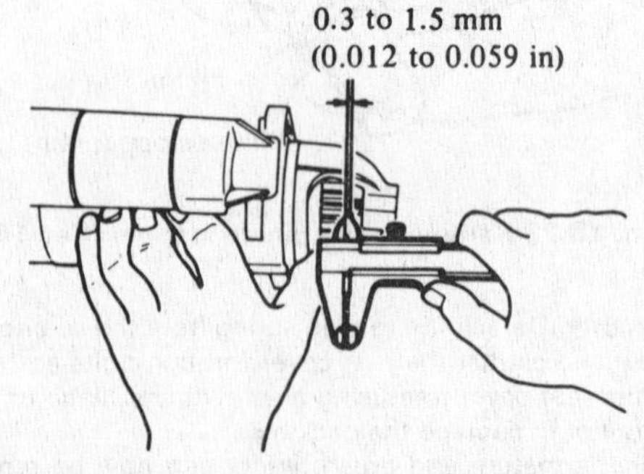

Fig. 10.31 Checking the starter motor pinion-to-thrust washer gap (Sec 14)

17 If an ohmmeter is available, test the field coil for continuity. To do this, connect one probe of the meter to the field coil positive terminal and the other to the positive brush holder. If no reading is indicated then the field coil circuit has a break in it.

18 Connect one probe of the meter to the field coil positive lead and the other one to the yoke. If there is little or no resistance then the field coil is grounded due to a breakdown in insulation. When this fault is discovered, the field coils should be renewed by an automotive electrician as it is very difficult to remove the field coil securing screws without special equipment. In any event, it will probably be more economical to exchange the complete starter motor for a re-conditioned unit.

19 Undercut the separators of the commutator using an old hacksaw blade ground to suit. The commutator may be polished with a piece of very fine glass paper – never use emery cloth as the carborundum particles will become embedded in the copper surfaces.

20 The armature may be tested for insulation breakdown again using the ohmmeter. To do this, place one probe on the armature shaft and the other on each of the commutator segments in turn. If there is a reading indicated at any time during the test then the armature must be renewed.

21 Wash the components of the drive gear in kerosene and inspect for wear or damage, particularly to the pinion teeth and renew as appropriate. Reassembly is a reversal of dismantling but stake a new stop washer in position and oil the sliding surfaces of the pinion assembly with a light oil, applied sparingly.

22 Reassembly of the remaining components of the starter motor is a reversal of dismantling.

23 When the starter motor has been fully reassembled, actuate the solenoid which will throw the drive gear forward into its normal flywheel engagement position. Do this by connecting jumper leads between the battery negative terminal and the solenoid M terminal and between the battery positive terminal and the solenoid S terminal. Now check the gap between the end-face of the drive pinion and the mating face of the thrust washer. This should be between 0·012 and 0·059 in (0·3 and 1·5 mm) measured either with a vernier gauge or feelers.

15 Fuses, fusible links and relays

1 The main fusebox is located on the side wall under the instrument panel.

2 The fuse ratings and the circuits they protect are listed on the fuse box cover.

3 As additional protection, fusible links are also used in the wiring harnesses of the ignition, electronic fuel injection, lighting and alternator systems. The fusible link for the electronic fuel injection system is connected between the battery positive terminal and the fuel injection harness. The fusible link for the remaining circuits is located on the relay bracket within the engine compartment.

4 In the event of a fuse or fusible link blowing, always establish the cause before installing a new one. This is most likely due to faulty insulation somewhere in the wiring circuit. Always carry a spare fuse for each rating and never be tempted to substitute a piece of wire or a nail for the correct fuse, as a fire may be caused or, at least, the electrical component ruined.

5 The relay bracket within the engine compartment is so installed that a number of relays can be located in the same place for ease of maintenance. Also fixed to the relay bracket are the voltage regulator and the wiper amplifier unit.

6 In the event of a fault in a relay a replacement will have to be fitted.

16 Headlamps – removal and installation

1 Disconnect the battery ground cable.

2 Remove the headlamp finisher, which is retained by two screws at the top edge and two screws at the lower edge.

3 Loosen and remove the three headlamp retaining ring screws, and lift away the retaining ring. (Ensure that the aim adjusting screws are not disturbed). The retaining ring should be turned slightly clockwise, then the ring and headlamp can be removed.

4 Disconnect the connector from the rear of the lens unit.

5 The new lens unit should be installed in the reverse sequence of removal. Ensure that the *TOP* mark on the lens is positioned correctly.

17 Headlamp aim – adjustment

1 Headlamp aim adjusting screws are provided on each light unit but it is not advisable to adjust the aim haphazardly, since not only may this be illegal, but it is not very easy to do accurately.

2 A vertical and horizontal aim screw is used on each light unit, and

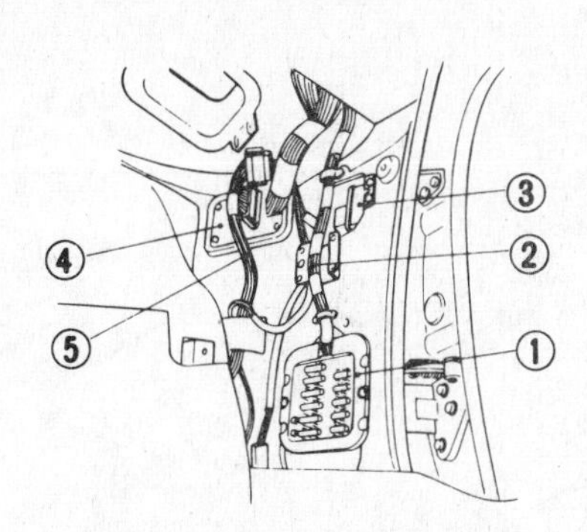

Fig. 10.32 Location of the fuse box (Sec 15)

1 *Fuse block*
2 *Main harness*
3 *Ignition relay*
4 *Junction box*
5 *Body harness*

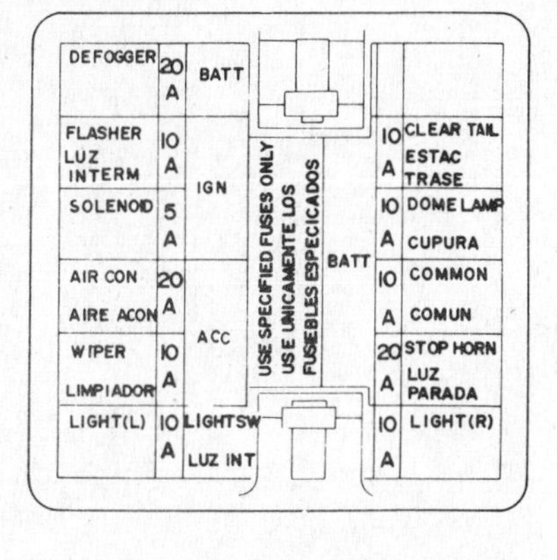

Fig. 10.33 The fuse box cover (Sec 15)

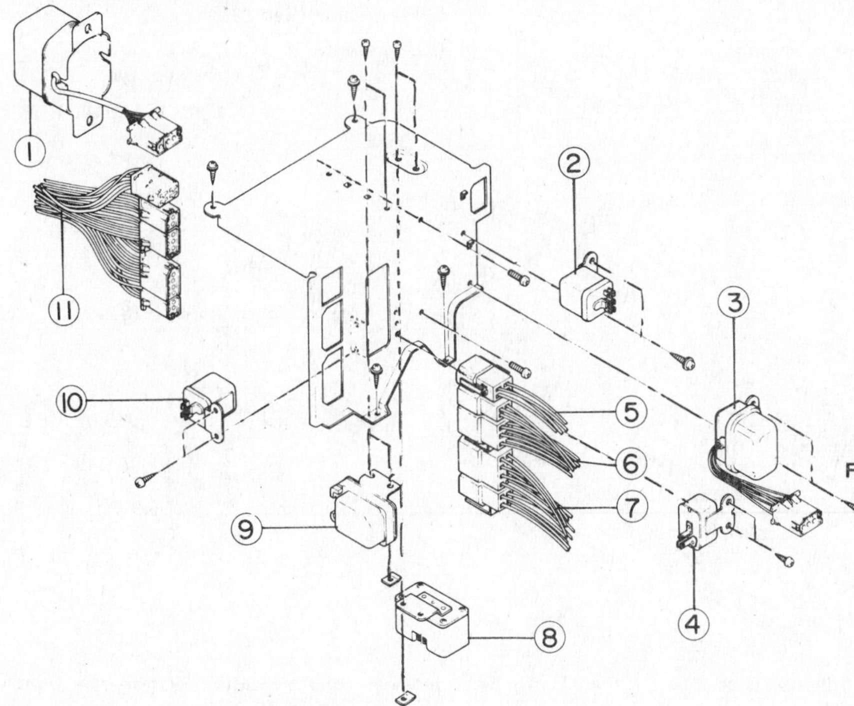

1 *Voltage regulator*
2 *Starter inhibitor relay*
3 *Auto-choke relay*
4 *Horn relay*
5 *Fusible link*
6 *Engine harness*
7 *Engine harness*
8 *Headlamp sensor*
9 *Wiper amplifier*
10 *Lighting relay*
11 *Main harness*

Fig. 10.34 The relay bracket (Sec 15)

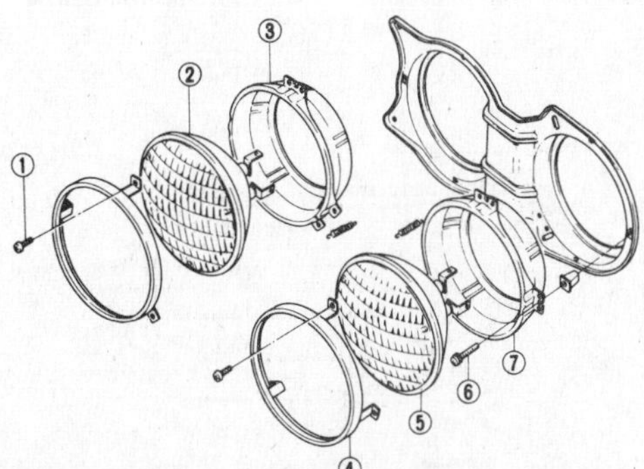

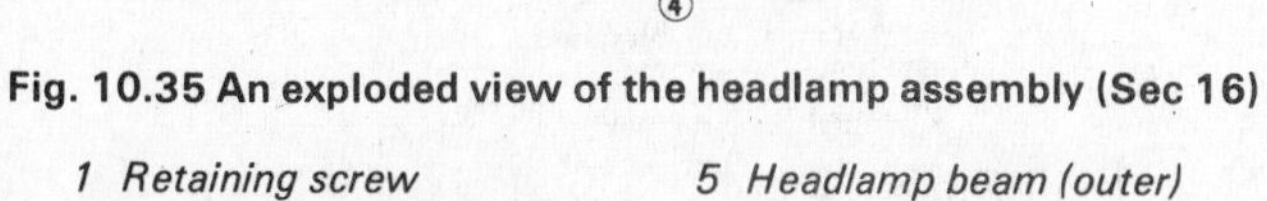

Fig. 10.35 An exploded view of the headlamp assembly (Sec 16)

1 *Retaining screw*
2 *Headlamp beam (inner)*
3 *Mounting ring (inner)*
4 *Retaining ring (outer)*
5 *Headlamp beam (outer)*
6 *Aim adjusting screw*
7 *Mounting ring (outer)*

Vertical adjustment
Horizontal adjustment

Fig. 10.36 The headlamp aim adjusting screws (Sec 17)

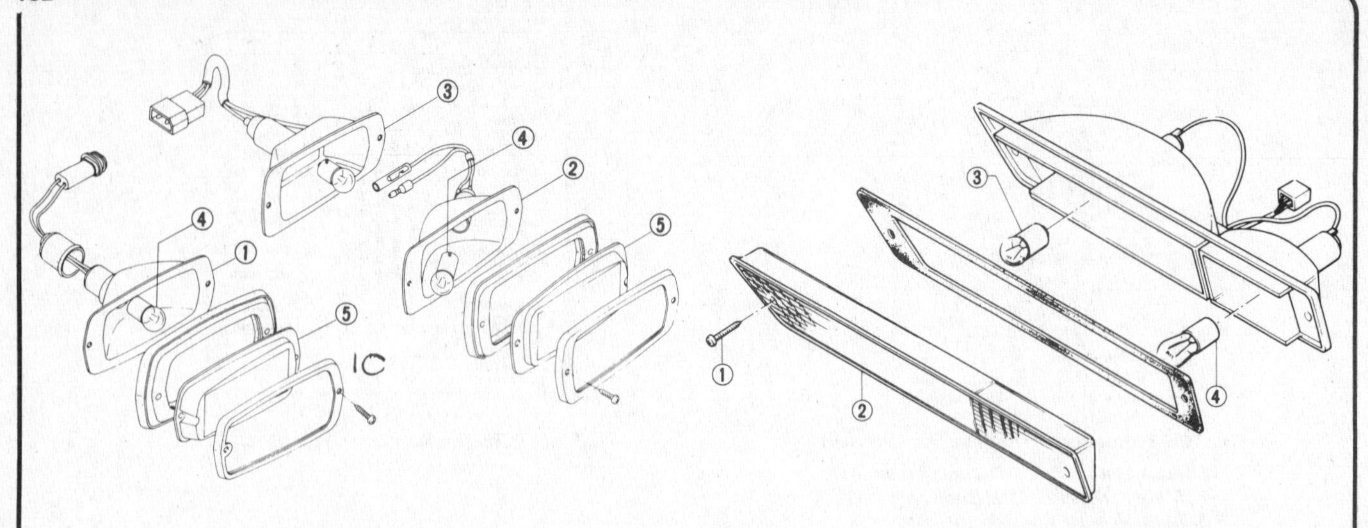

Fig. 10.37 Exploded view of the side marker lamp assemblies (Sec 18)

1 *Front lamp body (Sedan and Station Wagon)*
2 *Rear lamp body (Sedan)*
3 *Rear lamp body (Station Wagon)*
4 *Bulb*
5 *Lens*

Fig. 10.38 Exploded view of the front parking and turn signal lamp unit (Sec 19)

1 *Retaining screws*
2 *Lens*
3 *Turn signal bulb*
4 *Parking bulb*

Fig. 10.39 Exploded view of the Sedan rear light unit (Sec 20)

1 *Backplate*
2 *Stop and tail bulb*
3 *Turn signal bulb*
4 *Stop and tail bulb*
5 *Back-up bulb*
6 *Lens*

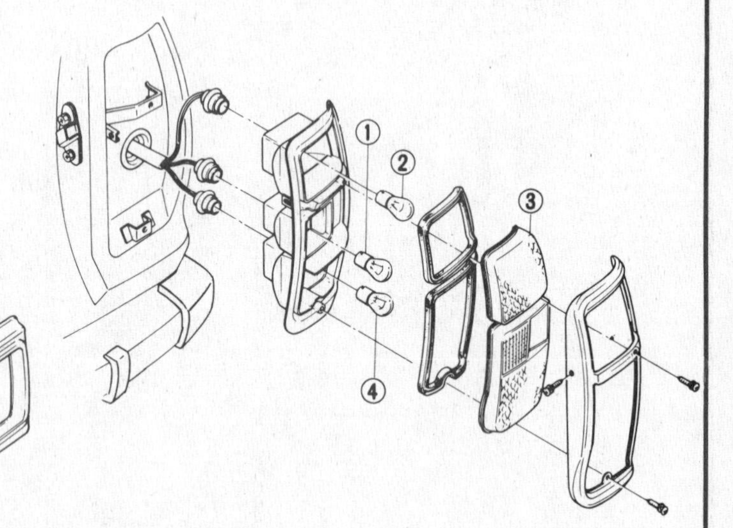

Fig. 10.40 Exploded view of the Station Wagon rear light unit (Sec 20)

1 *Back-up bulb*
2 *Turn signal bulb*
3 *Lens*
4 *Stop and tail bulb*

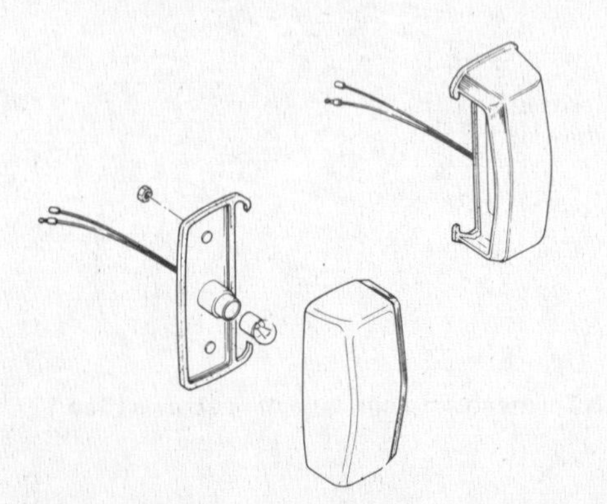

Fig. 10.41 The Sedan license plate lamp unit (Sec 21)

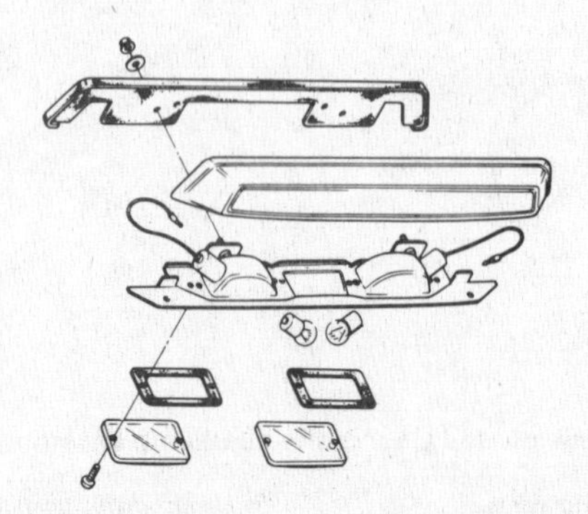

Fig. 10.42 The Station Wagon license plate lamp assembly (Sec 21)

access to them can be gained without removing the finisher.

18 Side marker lamps

1 The side marker lamps are installed at the front and rear. The construction of the rear lamp body varies slightly between Sedan and Station Wagon. However, bulb renewal is identical in both cases.
2 Disconnect the battery ground cable.
3 Remove the two lens retaining screws and lift away the lens.
4 The bulb is of the bayonet type and can be removed with a counter-clockwise twist after pressing in slightly.
5 Installation is the reverse of removal but ensure that the lens gasket is correctly installed.

19 Front parking and turn signal lamps

1 Disconnect the battery ground cable.
2 Undo and remove the two lens retaining screws and lift away the lens and gasket.
3 Press in and, at the same time, turn the bulb counter-clockwise and remove it.
4 Installation is a reversal of removal.
5 To remove the lamp assembly, after the lens has been removed, simply pry it from its aperture. (The lens retaining screws also secure the lamp assembly). Once the lamp assembly is clear of the car, disconnect the lead connector at the rear of the unit.

20 Tail, rear turn and back-up signal lamps

Sedan

1 To gain access to a bulb, open the trunk and remove the two nuts and washers securing the lamp unit lens. From the outside of the car, withdraw the lens retainer, gasket and lens.
2 The faulty bulb is removed by pressing it in and turning it counter-clockwise.
3 To remove the lamp unit, first disconnect the battery ground cable.
4 Working inside the trunk, unscrew and remove the seven flange nuts securing the unit to the rear panel.
5 Disconnect the harness connector from the rear of the lamp unit and remove the unit from the outside of the car.
6 To install a bulb, or the lamp unit itself, reverse the removal sequence.

Station Wagon

7 Bulb renewal in this unit can be carried out after removing the three lens retaining screws and the lens retainer, lens and gasket.
8 To remove the lamp unit, first disconnect the battery ground cable.
9 Remove the three screws securing the lamp assembly to the fender.
10 Remove the bulb sockets from the lamp housing and lift away the lamp unit.
11 Installation is a reversal of removal.

21 License plate lamp

Sedan

1 To renew the license plate bulb, open the trunk lid and pull the bulb socket from the unit.
2 Press the bulb inwards and, at the same time, turn it counter-clockwise.
3 Installation is a reversal of the removal procedure.
4 To remove the lamp unit, first disconnect the battery ground cable.
5 Working inside the trunk lid, pull out the bulb socket; undo and remove the two nuts securing the lamp unit to the trunk lid. Lift the unit from the trunk lid.
6 To install the unit, reverse the removal procedure.

Station Wagon

7 To renew a bulb in this unit, remove the two screws securing the lens to the lamp assembly. Lift away the lens and gasket.
8 Press on the bulb, turning it counter-clockwise to remove it.
9 When installing the gasket and lens ensure that the gasket is correctly installed.
10 To remove the lamp unit, first disconnect the battery ground cable.
11 Remove the finishing trim from the tailgate.
12 Unscrew and remove the two nuts securing the lamp unit to the tailgate.
13 Disconnect the lead wires at the connector and remove the lamp unit.
14 To install the lamp unit, reverse the removal procedure.

22 Room (interior) lamp

1 To gain access to the room lamp bulb, pry the front end of the light unit from the roof panel.
2 Install the bulb and press the unit back into the roof panel.
3 To remove the lamp unit, first disconnect the battery ground cable.
4 Pry the unit from the roof panel aperture until the wires can be disconnected.
5 Installation of the room lamp is a reversal of the removal procedure.

23 Trunk and luggage compartment lamps

Trunk lamp (Sedan)

1 The trunk lamp bulb is removed by pressing it inwards and, at the same time, turning it counter-clockwise.

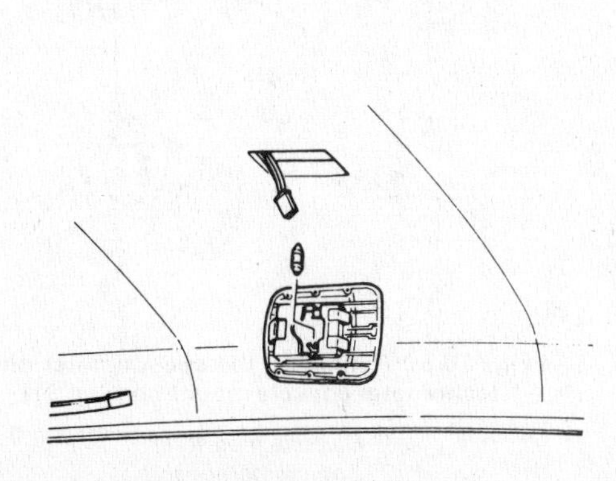

Fig. 10.43 The room (interior) lamp unit (Sec 22)

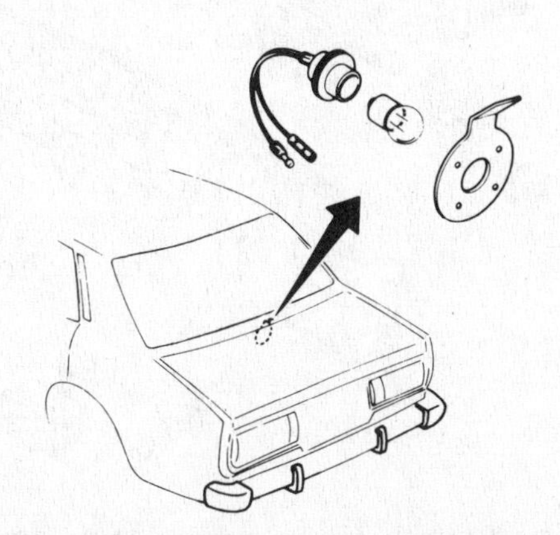

Fig. 10.44 The trunk compartment lamp (Sedan) (Sec 23)

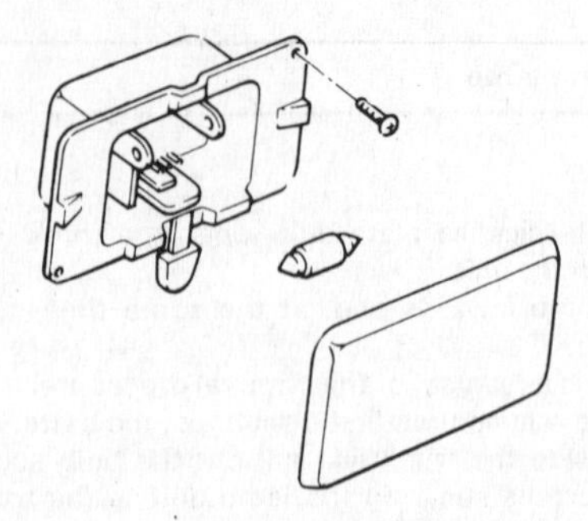
Fig. 10.45 The luggage compartment lamp (Station Wagon) (Sec 23)

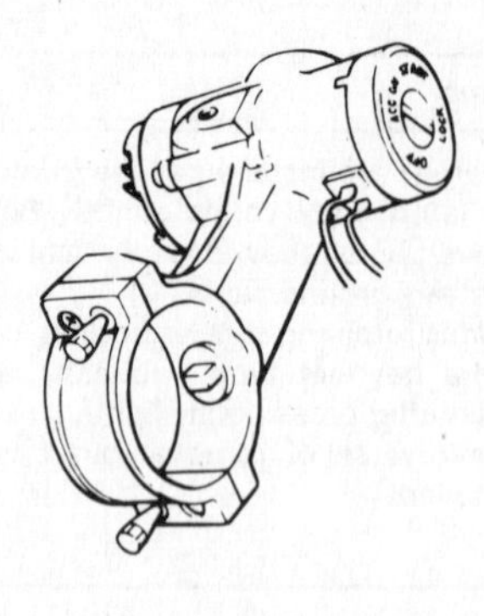
Fig. 10.46 The ignition switch (Sec 24)

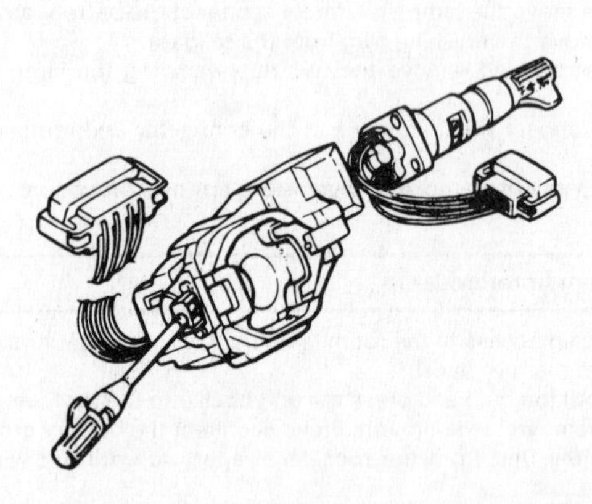
Fig. 10.47 The combination switch assembly (Sec 25)

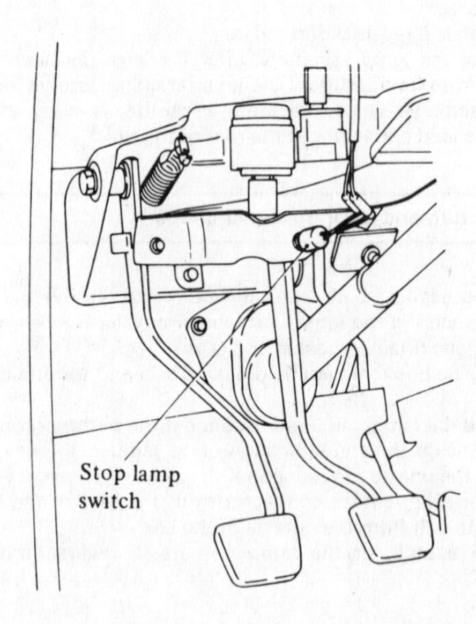

Fig. 10.48 The stop lamp switch (Sec 26)

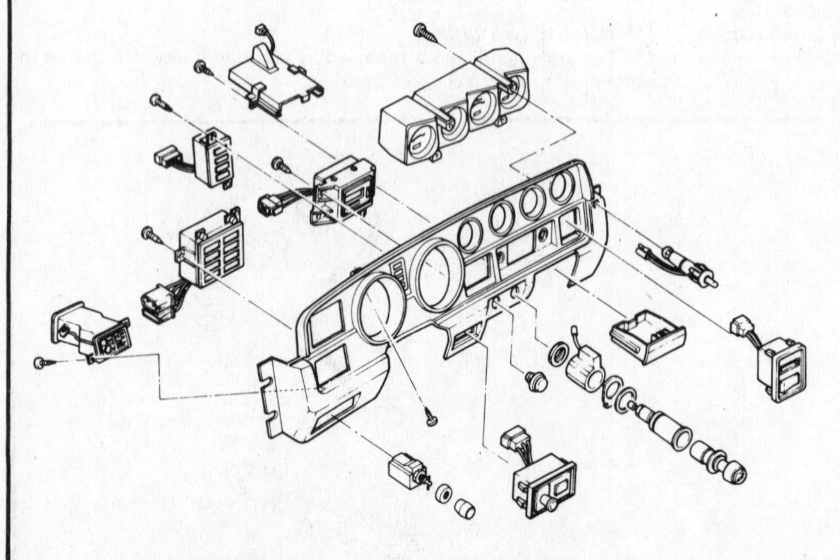
Fig. 10.49. Removing the cluster lid (Sec 30)

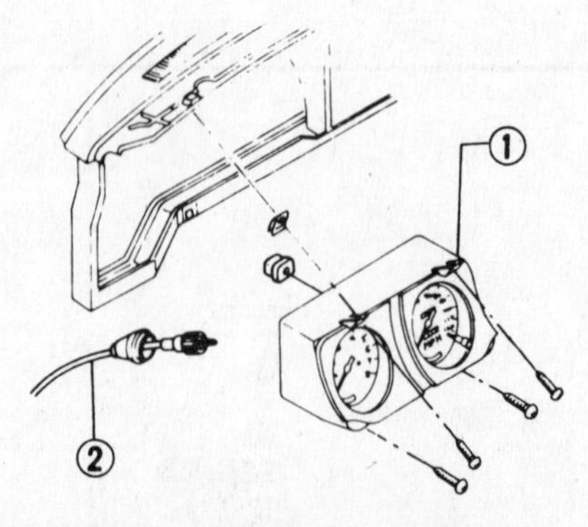

Fig. 10.50 Removing the speedometer and tachometer console mounting (Sec 31)

1 Console mounting
2 Speedometer cable

Luggage compartment lamp (Station Wagon)

2 Remove the lamp unit lens by gently squeezing it inwards and pulling it from the lamp unit.
3 The bulb is simply pulled from its contacts.
4 To remove the lamp unit, first disconnect the battery ground cable.
5 Remove the lamp lens as previously mentioned.
6 Undo and remove the two screws securing the unit to the roof.
7 Remove the unit and disconnect the wire connector; lift the unit away from the roof.
8 Installation of the lamp unit is the reverse of removal.

24 Ignition switch – removal and installation

1 Disconnect the battery ground cable.
2 Remove the screws that retain the two halves of the steering column cover.
3 Disconnect the harness connector.
4 Remove the small screw retaining the switch body to the steering lock. The switch assembly can now be taken out.
5 Install the ignition switch in the reverse order of removal.

25 Combination switch – removal and installation

1 Disconnect the battery ground cable.
2 Remove the screws from the rear side of the steering wheel and lift away the horn pad, disconnecting the horn wire.
3 Remove the steering wheel as described in Chapter 11.
4 Remove the screws that retain the two halves of the steering column cover in position. Lift away the steering column cover.
5 Disconnect the combination switch harness connector.
6 Loosen the two retaining screws and remove the switch from the steering column.
7 Installation of the combination switch is a reversal of the removal procedure.

26 Stop lamp switch – removal and installation

1 The stop lamp switch is located at the upper end of the brake pedal mounted on a cross-plate.
2 To remove the switch, disconnect the pedal return spring, remove the locknut and unscrew the switch from the cross-plate. Disconnect the harness wires.
3 Installation of the stop lamp switch is a reversal of the removal procedure but the brake pedal height will have to be checked and adjusted as described in Chapter 9.

27 Hazard warning lamp switch – removal and installation

1 Disconnect the battery ground cable.
2 Using a small screwdriver, carefully pry the switch from its aperture.
3 Pull the switch out far enough to enable the two harness connectors to be pulled apart.
4 Installation of the switch is a reversal of the removal procedure.

28 Rear window defogger switch – removal and installation

1 Disconnect the battery ground cable.
2 Using a small screwdriver, carefully pry the switch from its aperture.
3 Disconnect the wiring harness at the back of the switch and remove it.
4 Installation of the switch is a reversal of the removal procedure.

29 Rheostat (instrument panel dimmer) control – removal and installation

1 Disconnect the battery ground cable.
2 Pull off the knob from the control unit.
3 Loosen and remove the ring nut that secures the unit.
4 Disconnect the wires to the control unit, then remove it from behind the cluster lid.
5 Installation of the rheostat control is a reversal of the removal procedure.

30 Cluster lid – removal and installation

1 Disconnect the battery ground cable.
2 Pull off the knobs from the radio controls, and loosen and remove the retaining nuts.
3 Remove the ashtray.
4 Loosen and remove the screws securing the two halves of the steering column cover in position. Remove the steering column cover.
5 Working beneath the instrument panel, first identify, then disconnect the wire harnesses to the switches that will be removed with the cluster lid.
6 Remove the cluster lid retaining screws and lift the cluster lid out.
7 Installation of the cluster lid is a reversal of removal but ensure that the harness connections are correctly made.

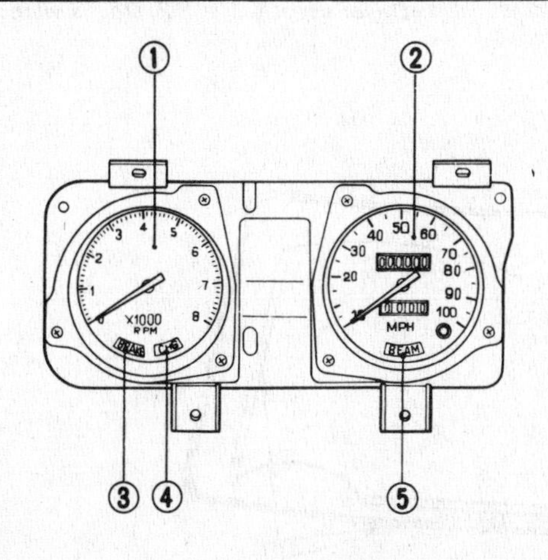

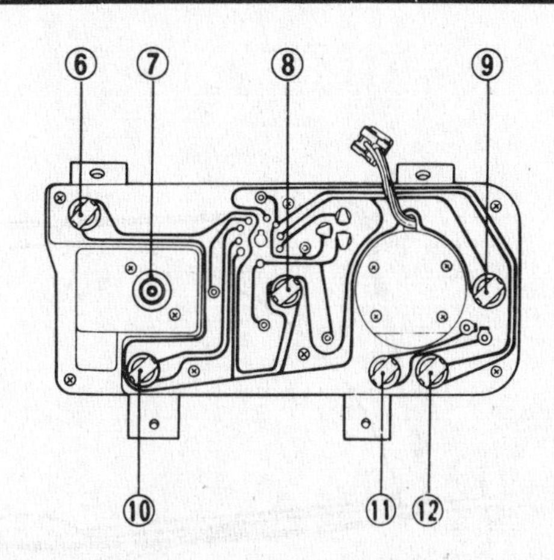

Fig. 10.51 Front and rear views of the speedometer and tachometer unit (Sec 31)

1 *Tachometer*
2 *Speedometer*
3 *Brake warning lamp*
4 *Charge warning lamp*
5 *Main beam lamp*
6 *Illumination lamp*
7 *Speedometer connection*
8 *Illumination lamp*
9 *Illumination lamp*
10 *Main beam lamp*
11 *Charge warning lamp*
12 *Brake warning lamp*

10

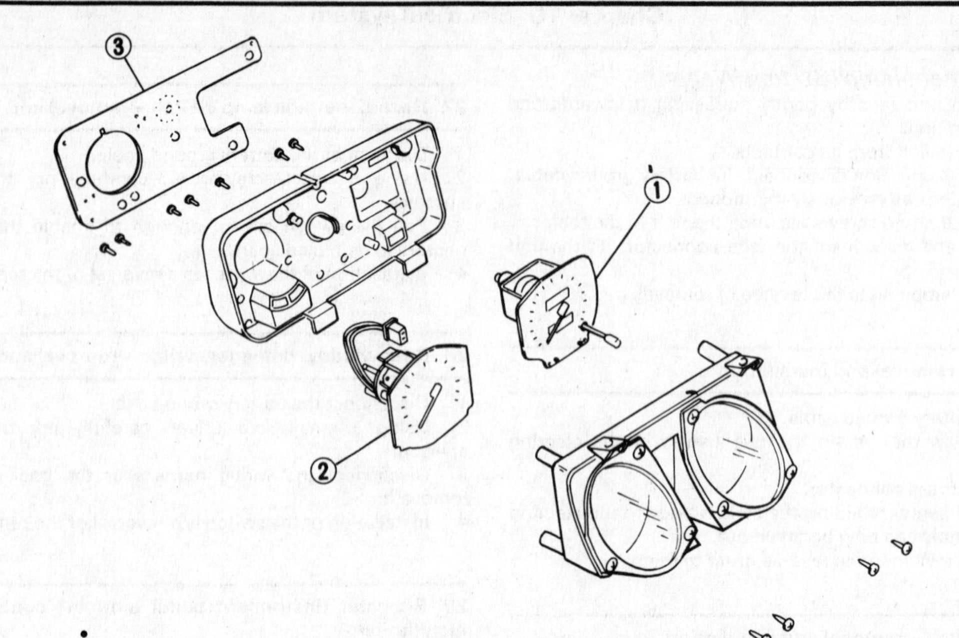

Fig. 10.52 The mounting console dismantled (Sec 31)

1 Speedometer *2 Tachometer* *3 Printed circuit board*

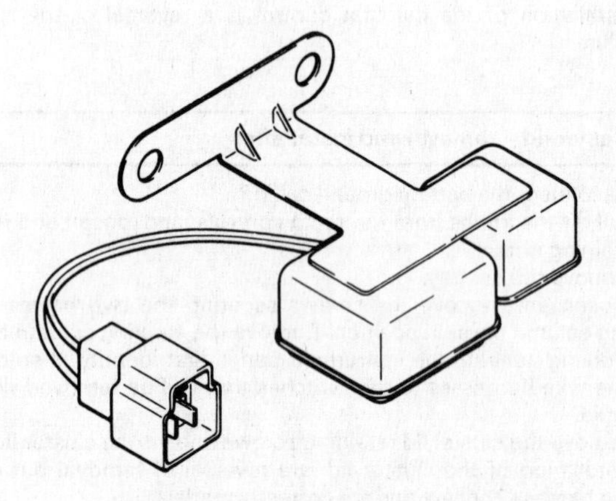

Fig. 10.53 The warning buzzer (Sec 32)

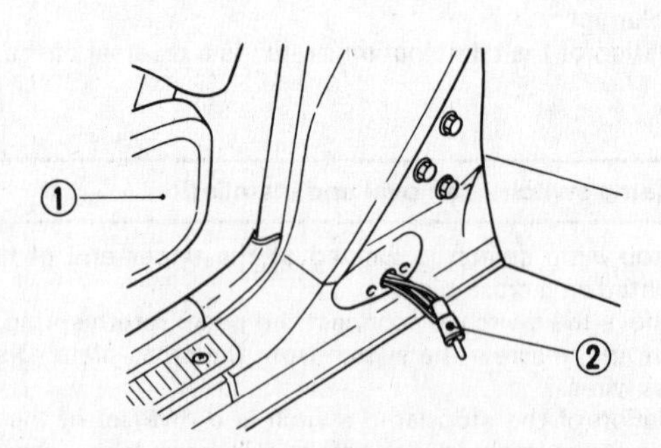

Fig. 10.54 Location of the door switch (Sec 32)

1 Driver's seat *2 Door switch*

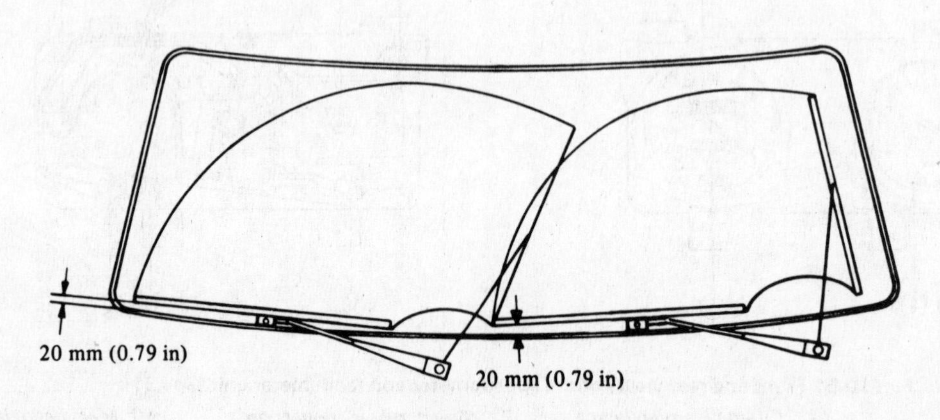

Fig. 10.55 Windshield wiper arm setting diagram (Sec 34)

31 Gauges (general) – removal and installation

Speedometer and tachometer

1 To remove either of these units the mounting console will first have to be removed.
2 Disconnect the battery ground cable.
3 Remove the cluster lid as described in Section 30.
4 Working behind the gauges, disconnect the multi-pole and the two-pole connectors.
5 From the rear of the speedometer, disconnect the speedometer cable.
6 Loosen and remove the four screws that secure the mounting console in position and lift it away from the instrument panel.
7 If any of the bulbs associated with the speedometer or tachometer need renewing, this can be carried out at the rear of the units. (The bulbs can be removed by reaching up behind the units).
8 To remove the gauges from the mounting console, first remove the screws that retain the printed circuit board to the back of the unit. Remove the printed circuit board.
9 Loosen and remove the screws that retain the speedometer or tachometer to the mounting console. The required meter can then be removed from the console box.
10 Reassembly and installation of the meters is a reversal of the removal and dismantling procedure.

Oil, fuel, water temperature gauges and voltmeter

11 These four gauges are mounted in one unit, and to remove them the cluster lid must first be removed as described in Section 30.
12 With the cluster lid removed the complete gauge unit can be unscrewed from the back of the cluster lid.
13 If only an illumination bulb is to be renewed there is no need to remove the gauge unit from the cluster lid.
14 Removal of any of the gauges will require removing the printed circuit board from the back of the assembly, and is a similar process to that described for the speedometer and tachometer.

32 Warning units – removal and installation

Warning lamp clusters

1 The combined warning lamp clusters, which are mounted on the instrument panel, monitor the various systems used on the car. Trouble in any of the systems is indicated by these lamps.
2 Access to the two warning lamp clusters is gained after removing the cluster lid, which is described in Section 30.

Warning buzzer

3 The warning buzzer, which operates as an anti-theft device, 'seat belt fasten' warning and 'open door' warning, is located behind the cluster lid.

Horn

4 The two horns are located behind the front grille. To remove them, first remove the front grille as described in Chapter 12.
5 Disconnect the wire connection behind the horn.
6 Remove the horn retaining bolt and lift away the horn.
7 To install the horn, reverse the removal procedure.

Door switch

8 The door switch, which illuminates the room lamp and actuates the warning buzzer when the driver's door is opened, is located on the lower end of the left-hand door pillar.
9 To remove the switch, first disconnect the battery ground cable.
10 Pry out the rubber grommet from around the switch.
11 Withdraw the switch from the door pillar and disconnect the wires from the harness.
12 The door switch can be installed following the reverse of the removal procedure.

Warning switches general

13 In addition to the units previously mentioned, warning switches are fitted to the caps of the brake fluid reservoir, the windshield washer tank, the battery and to the fuel level sensor. These switches are quite easily removed after the caps have been removed. With regard to the fuel level sensor, this can be removed after access to the fuel tank has been gained.
14 The seat belt switch is an integral part of the floor-mounted portion of the belt assembly and, in the event of its failure, must be renewed as an assembly.
15 All the previously mentioned switches are fed through harnesses, via relays, to the warning lamp clusters.

33 Windshield wipers and washers – description

1 The windshield wiper consists of a wiper motor unit, link mechanism, wiper arms, blades and an intermittent amplifier.
2 The motor incorporates an auto-stop device and operates the wipers in three different stages: intermittent, low speed, and high speed.
3 The electrically operated windshield washer consists of a reservoir tank (with built-in motor and pump), washer nozzles and vinyl tubes used to connect the components.
4 The tailgate on the Station Wagon is equipped with a similar wiper motor and washer components, but the motor is of the single-speed type. The washer pump, whilst being very similar to that used on the front washers, is housed in a flexible-type bag in the rear trim of the vehicle.

34 Wiper arm – removal and installation

1 Before removing a wiper arm, turn the windshield wiper switch on and off to ensure the arms are in their normal parked position parallel with the bottom of the screen, and the outer tips 0·79 in (20·0 mm) from the sealing rubber.
2 To remove the arms, pivot the arm back, slacken the arm securing nut and detach the arm from the spindle.
3 When installing the arm, place it so that it is in its correct relative parked position and tighten the arm securing nut.

35 Wiper motor and linkage – removal and installation

1 Remove the wiper arm blade as described in the preceding Section.
2 Raise the hood and disconnect the wiper motor lead connector plug.
3 Loosen and remove the screws that retain the cowl top grille in position. Lift away the grille.
4 The wiper motor is attached to the fire-wall by three bolts which must be removed, leaving the motor attached to the connecting balljoint.
5 Remove the balljoint connecting the motor shaft to the linkage.
6 Lift away the motor and withdraw the wiper linkage from the top grille aperture.
7 Installation of the wiper motor and linkage is a reversal of removal but ensure that the wiper arms are correctly installed to obtain the correct sweeping zone.

36 Wiper motor and linkage – servicing

1 It is not recommended that the wiper motor is overhauled beyond renewing the brushes if they are badly worn. Always check the fuses in the circuit before dismantling.
2 Check for wear in the linkage and renew any worn bushes. Slackness in these components can cause the wiper blades to strike or ride over the windshield frame surround at the ends of the arc of travel of the arm.
3 When reassembling the linkage, apply a little grease to the pivots and moving parts of the assembly.

37 Intermittent amplifier

1 The intermittent amplifier, which is located on the engine compartment relay mounting bracket, controls the intermittent operation of the wipers.

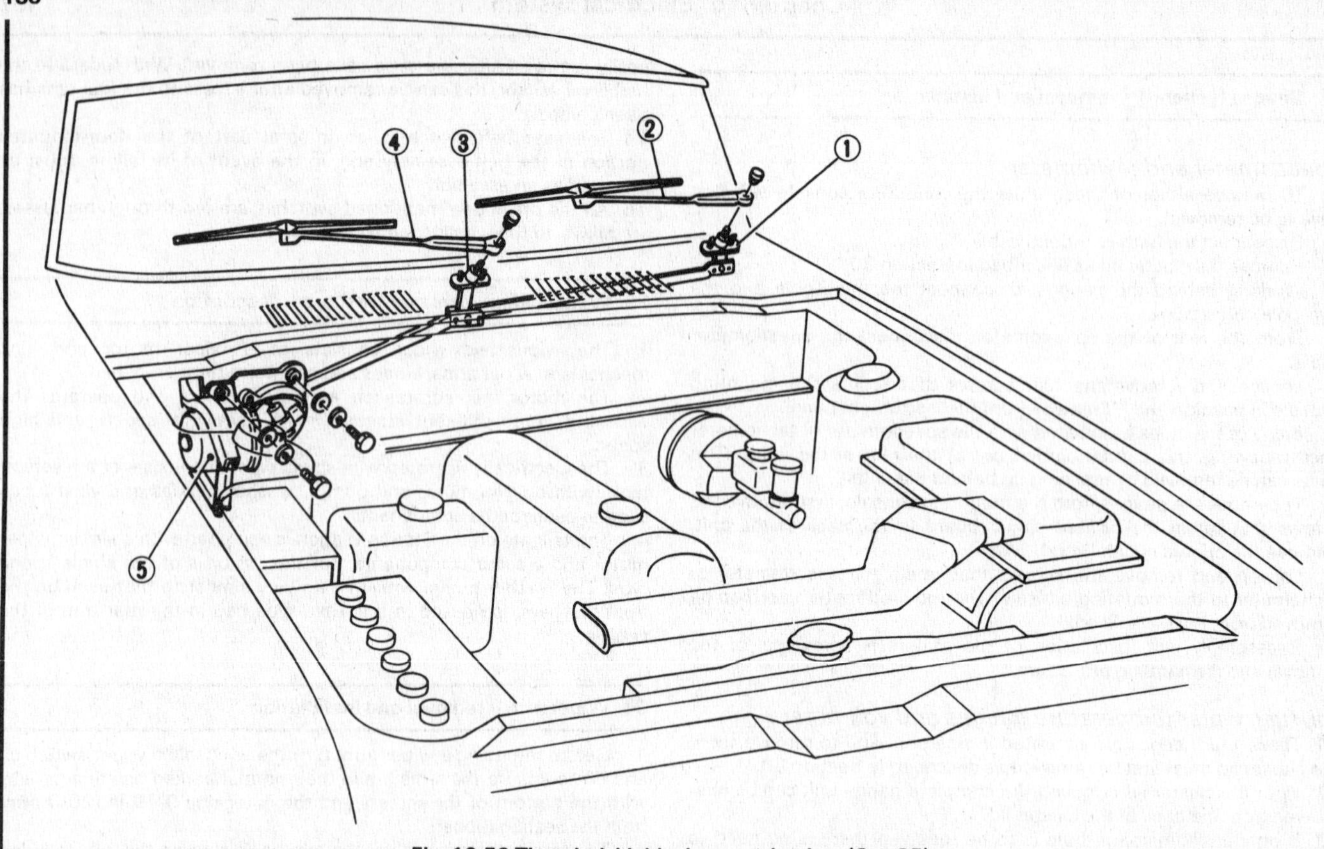

Fig. 10.56 The windshield wiper mechanism (Sec 35)

1 *Left-hand pivot*
2 *Left-hand wiper arm*
3 *Right-hand pivot*
4 *Right-hand wiper arm*
5 *Wiper motor*

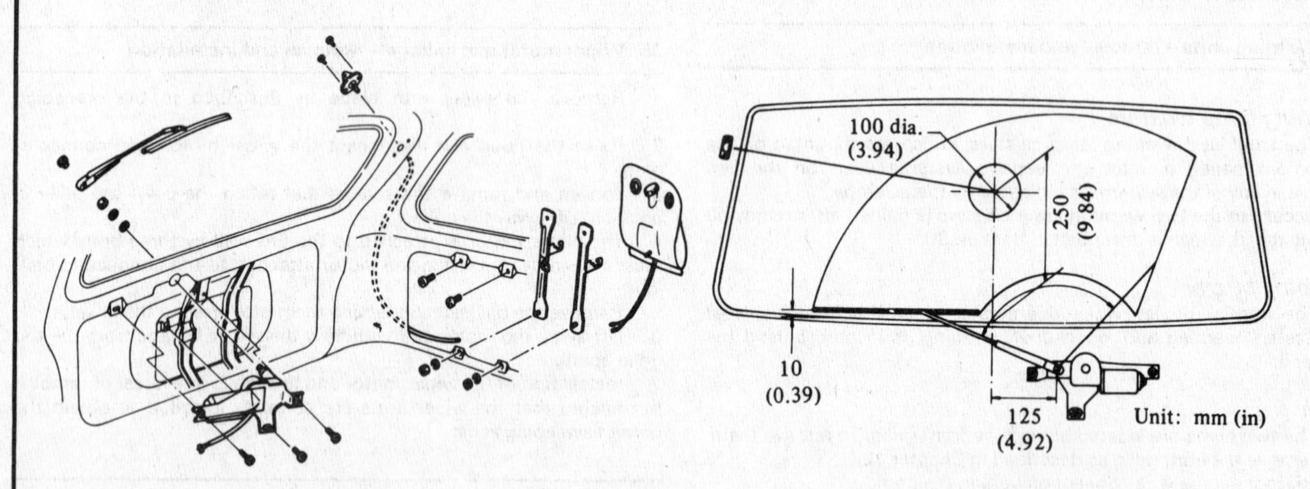

Fig. 10.57 The rear wiper and washer assemblies (Station Wagon) (Sec 39)

Fig. 10.58 Rear wiper blade installation diagram (Station Wagon) (Sec 39)

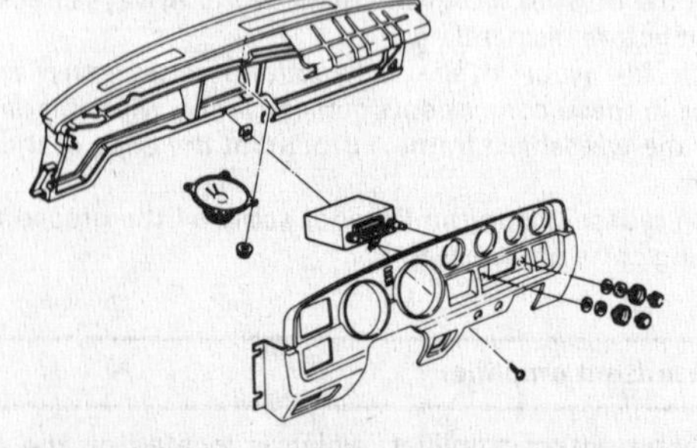

Fig. 10.59 Removing the radio receiver (Sec 43)

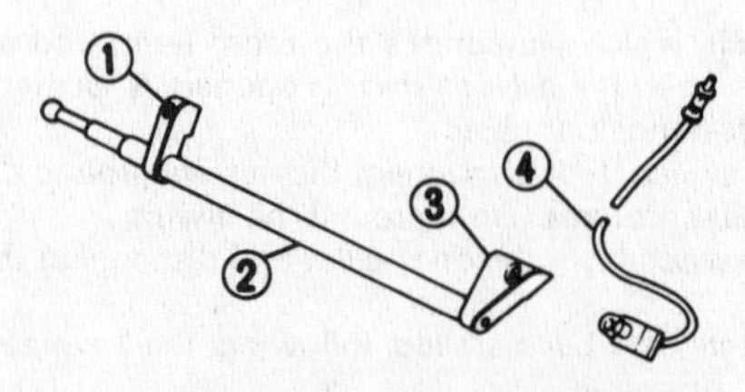

Fig 10.60 The antenna (Sec 43)

1 *Support*
2 *Antenna*
3 *Lower joint*
4 *Antenna cable*

2 In the unlikely event of its failure, it is not possible to repair the amplifier and it must be renewed as an assembly.

38 Wiper switch – removal and installation

1 Disconnect the battery ground cable.
2 Loosen and remove the screws that secure the two halves of the steering column cover in position. Lift away the steering column cover.
3 Remove the wiper switch from the combination switch by removing the retaining screws and disconnecting the wire harness.
4 Installation of the wiper switch is a reversal of removal.

39 Rear windshield wiper motor (Station Wagon) – removal and installation

1 Disconnect the battery ground cable.
2 Remove the wiper blade arm as described in Section 34.
3 Open the tailgate and carefully remove the trim panel.
4 Disconnect the wiper motor wire connectors.
5 Loosen and remove the three motor attaching bolts and lift away the motor.
6 When installing the motor, follow the reverse sequence to removal.
7 Finally, with the motor installed, position the wiper arm assembly to the motor spindle in such a manner that there is 0·390 (10·0 mm) clearance between the bottom edge of the tailgate glass and the wiper arm.

40 Rear wiper switch – removal and installation

1 The rear wiper switch also incorporates the washer operating switch. To remove the switch, proceed as described in the following paragraphs.
2 Disconnect the battery ground cable.
3 Refer to Section 30, and remove the cluster lid.
4 Disconnect the harness connector from the switch.
5 Remove the switch knob by pushing it downwards and twisting it, to align the slots inside the knob with the pins on the switch shaft; the knob can then be pulled off the shaft.
6 Loosen and remove the nut that secures the switch to the cluster lid. The switch can then be removed from the cluster lid.
7 Install the switch in the reverse sequence to the removal procedure.

41 Windshield washers – servicing

1 Normally the windshield washers fitted to both the Sedan and the Station Wagon tailgate require no maintenance other than keeping the reservoirs topped-up with water, to which a little windshield cleaning fluid has been added.
2 If the front washer nozzles are incorrectly aligned to provide satisfactory coverage of the glass, they can be adjusted by using a pair of long nosed pliers inserted through the grille.
3 If the washer reservoirs or electric pump have to be renewed individually, they may be separated after disconnecting the connecting pipes and leads.
4 With regard to the front washer reservoir, when reconnecting the pump to the base of the reservoir, warm the reservoir by immersing it in hot water and use a solution of soapy water to lubricate the neck of the pump opening in the reservoir.

42 Clock – removal and installation

1 If it is only necessary to renew the illumination bulb in the clock, this can be carried out by reaching up behind the instrument panel. The socket, together with the illumination bulb, is removed by turning the socket counter-clockwise. With the socket and bulb removed from the clock, it is very easy to install the bulb.
2 To remove the clock from the instrument panel, first disconnect the battery ground cable.
3 Remove the cluster lid as described in Section 30.
4 Disconnect the clock harness connector.
5 Loosen and remove the retaining screws and lift away the clock.
6 Installation of the clock is a reversal of removal.

43 Radio – removal and installation

Radio receiver

1 Disconnect the battery ground cable.
2 Refer to Section 30, and remove the cluster lid.
3 Remove the radio bracket-to-instrument panel screw.
4 Disconnect the speaker wires at the connector.
5 At the harness connector disconnect the power supply to the receiver.
6 Pull out the antenna feeder at the receiver.
7 Lift the receiver from the instrument panel.
8 Installation of the radio receiver is a reversal of removal.

Speakers

9 The radio is equipped with three speakers, one of which is installed behind the instrument panel. The Sedan rear speakers are installed on the rear parcel shelf and in the Station Wagon they are behind the side trim panels.
10 Removal of the rear speakers can be carried out after removing the parcel shelf trim (Sedan) or the side trim panels (Station Wagon).
11 To remove the front speaker will necessitate removal of the cluster lid (Section 30).

Antenna

12 The antenna is attached to the front pillar and is easily removed after disconnecting the cable from the receiver and removing the antenna retaining screws.

44 Fault diagnosis – electrical system

Symptom	Reason/s
Starter fails to turn engine	Battery discharged Battery defective internally Battery terminal leads loose or ground lead not securely attached to body Loose or broken connections in starter motor circuit Starter motor solenoid faulty Starter motor pinion jammed in mesh with flywheel gear ring Starter brushes badly worn, sticking, or brush wires loose Commutator dirty, worn or burnt Starter motor armature faulty Field coils grounded
Starter turns engine very slowly	Battery in discharged condition

Symptom	Reason/s
	Starter brushes badly worn, sticking or brush wires loose Loose wires in starter motor circuit
Starter spins but does not turn engine	Pinion or flywheel gear teeth broken or worn
Starter motor noisy or excessively rough engagement	Pinion or flywheel gear teeth broken or worn Starter motor retaining bolts loose
Battery will not hold charge for more than a few days	Battery defective internally Electrolyte level too low or electrolyte too weak due to leakage Plate separators no longer fully effective Battery plates severely sulphated Alternator drivebelt slipping Battery terminal connections loose or corroded Alternator not charging Short-circuit causing continual battery drain Regulator not working correctly
Ignition light fails to go out, battery runs flat in a few days	Alternator drivebelt loose and slipping or broken Alternator brushes worn, sticking, broken or dirty Alternator brush springs weak or broken Internal fault in alternator Regulator incorrectly set Cut-out incorrectly set Open circuit in wiring of cut-out and regulator unit

Failure of individual electrical equipment to function correctly is dealt with alphabetically, item by item, under the headings listed below

Horn

Symptom	Reason/s
Horn operates all the time	Horn pad either grounded or stuck down Horn cable to horn push grounded
Horn fails to operate	Blown fuse Cable or cable connection loose, broken or disconnected Horn has an internal fault

Lights

Symptom	Reason/s
Lights do not come on	If engine not running, battery discharged Wire connections loose, disconnected or broken Light switch shorting or otherwise faulty
Lights come on but fade out	If engine not running, battery discharged
Lights give very poor illumination	Lamp glasses dirty Lamps badly out of adjustment
Lights work erratically – flashing on and off, especially over bumps	Battery terminals or ground connection loose Lights not grounding properly Contacts in light switch faulty

Wipers

Symptom	Reason/s
Wiper motor fails to work	Blown fuse Wire connections loose, disconnected or broken Brushes badly worn Armature worn or faulty Field coils faulty
Wiper motor works very slowly and takes excessive current	Commutator dirty, greasy or burnt Armature bearings dirty or unaligned Armature badly worn or faulty
Wiper motor works slowly and takes little current	Brushes badly worn Commutator dirty, greasy or burnt Armature badly worn or faulty
Wiper motor works but wiper blades remain static	Wiper motor gearbox parts badly worn

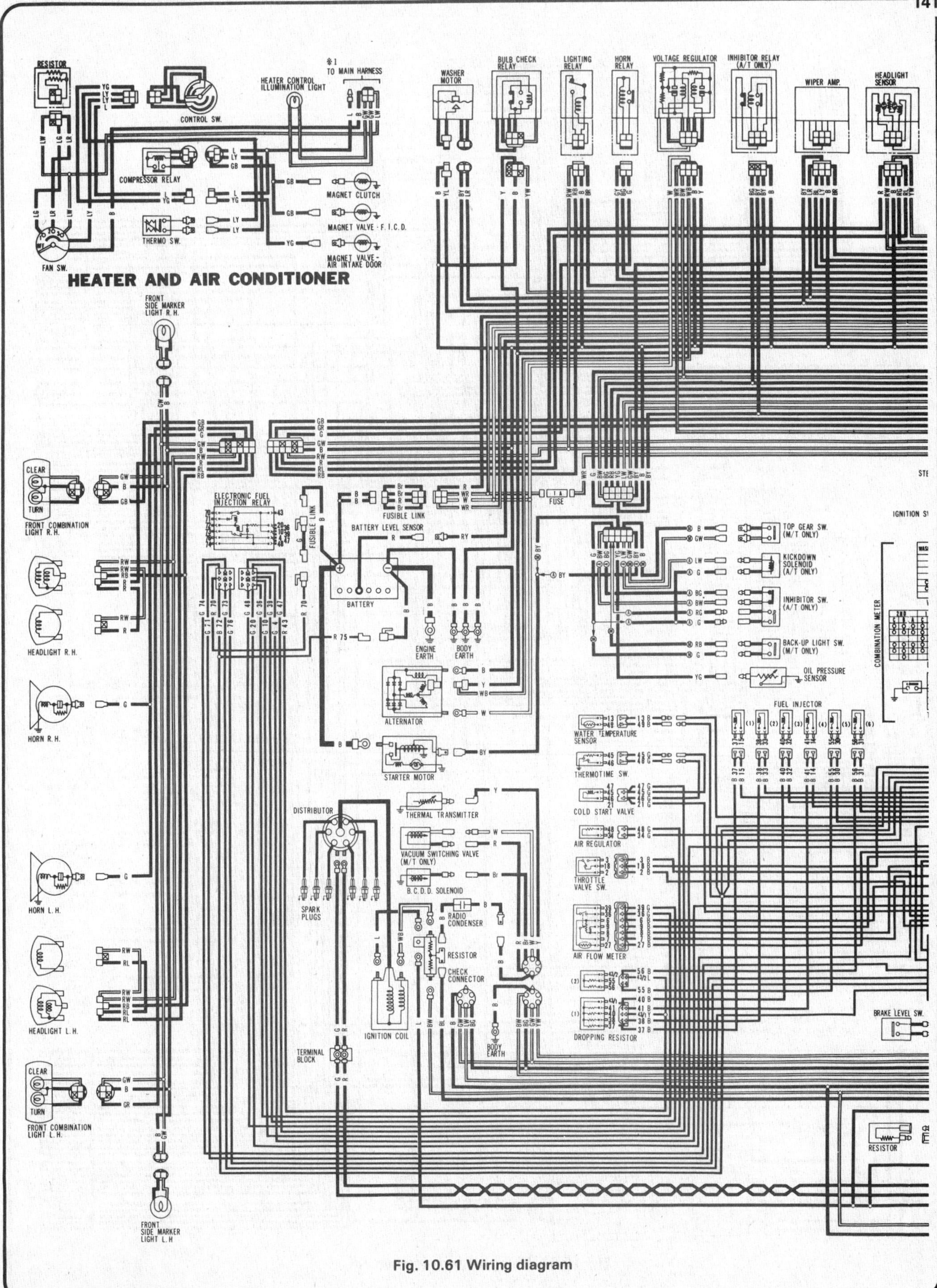

Fig. 10.61 Wiring diagram

10

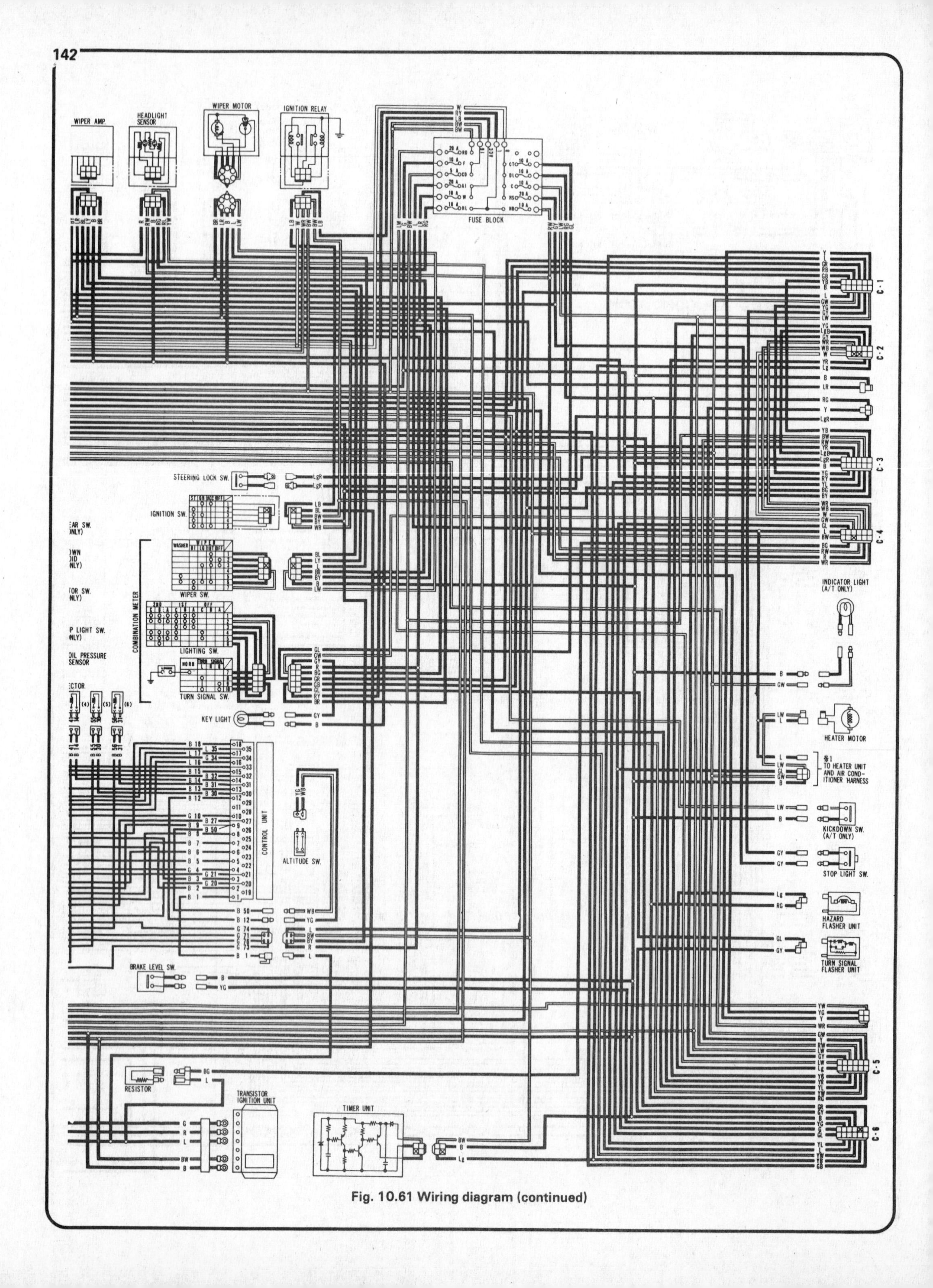

Fig. 10.61 Wiring diagram (continued)

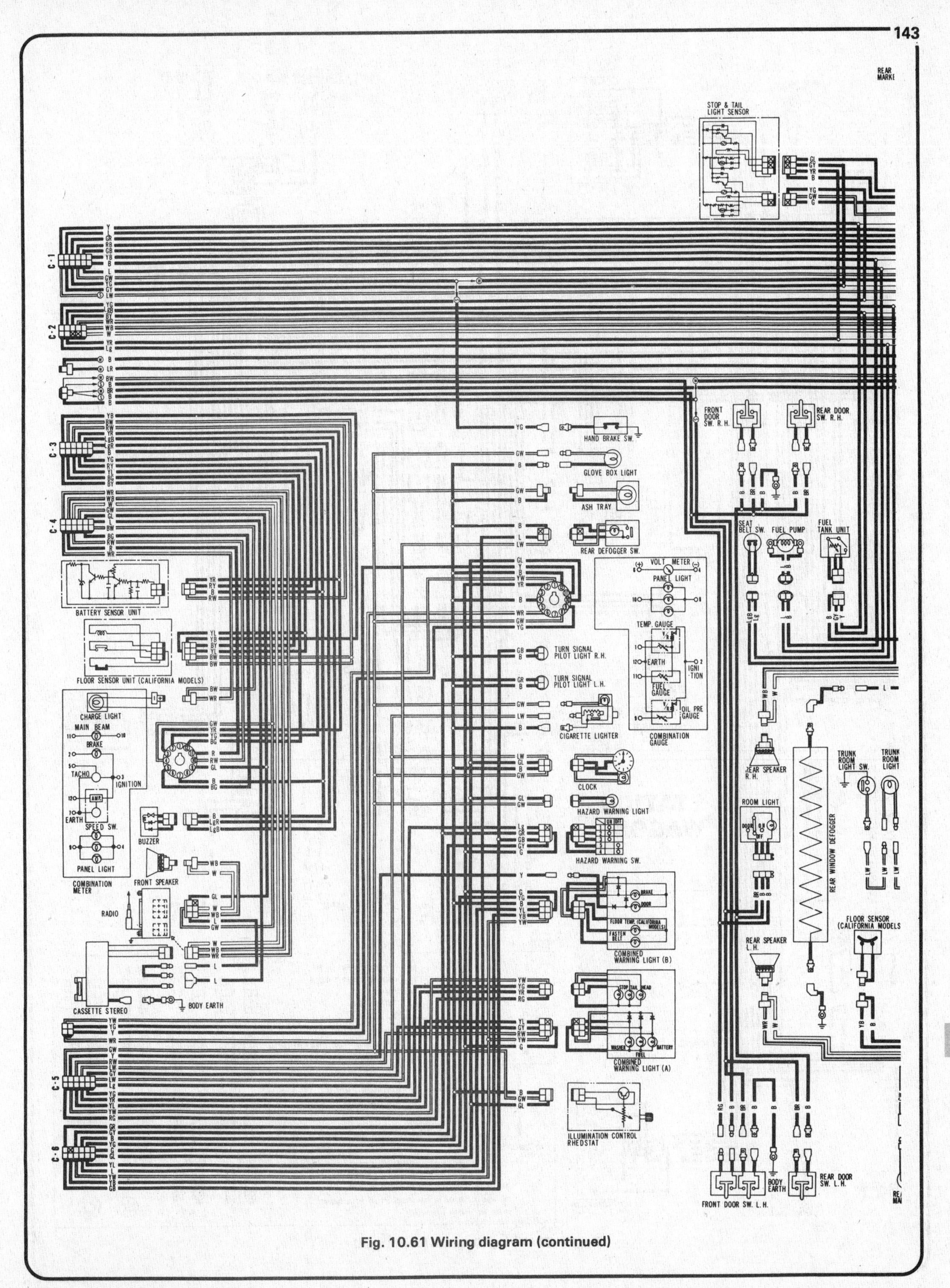

Fig. 10.61 Wiring diagram (continued)

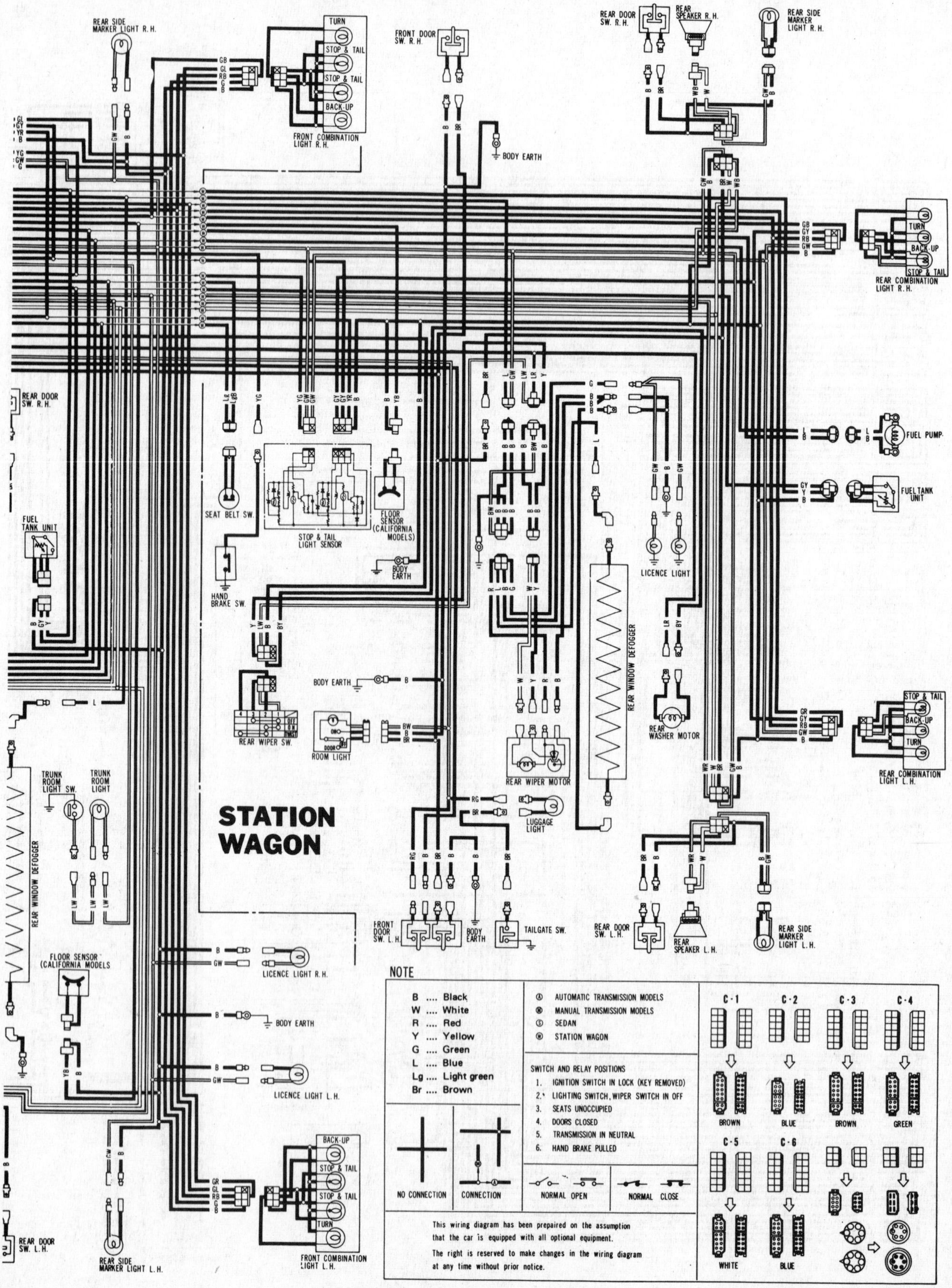

STATION WAGON

REAR SIDE MARKER LIGHT R.H.
FRONT COMBINATION LIGHT R.H.
TURN
STOP & TAIL
BACK-UP
FRONT DOOR SW. R.H.
BODY EARTH
REAR DOOR SW. R.H.
REAR SPEAKER R.H.
REAR COMBINATION LIGHT R.H.
FUEL PUMP
FUEL TANK UNIT
SEAT BELT SW.
STOP & TAIL LIGHT SENSOR
FLOOR SENSOR (CALIFORNIA MODELS)
HAND BRAKE SW.
LICENCE LIGHT
REAR WINDOW DEFOGGER
REAR WIPER SW.
ROOM LIGHT
REAR WIPER MOTOR
REAR WASHER MOTOR
REAR COMBINATION LIGHT L.H.
TRUNK ROOM LIGHT SW.
TRUNK ROOM LIGHT
LUGGAGE LIGHT
FRONT DOOR SW. L.H.
TAILGATE SW.
REAR DOOR SW. L.H.
REAR SPEAKER L.H.
REAR SIDE MARKER LIGHT L.H.
LICENCE LIGHT R.H.
LICENCE LIGHT L.H.
REAR DOOR SW. L.H.
REAR SIDE MARKER LIGHT L.H.
FRONT COMBINATION LIGHT L.H.

NOTE

B .... Black
W .... White
R .... Red
Y .... Yellow
G .... Green
L .... Blue
Lg .... Light green
Br .... Brown

Ⓐ AUTOMATIC TRANSMISSION MODELS
Ⓜ MANUAL TRANSMISSION MODELS
Ⓢ SEDAN
Ⓦ STATION WAGON

SWITCH AND RELAY POSITIONS
1. IGNITION SWITCH IN LOCK (KEY REMOVED)
2. LIGHTING SWITCH, WIPER SWITCH IN OFF
3. SEATS UNOCCUPIED
4. DOORS CLOSED
5. TRANSMISSION IN NEUTRAL
6. HAND BRAKE PULLED

NO CONNECTION
CONNECTION
NORMAL OPEN
NORMAL CLOSE

C-1 BROWN
C-2 BLUE
C-3 BROWN
C-4 GREEN
C-5 WHITE
C-6 BLUE

This wiring diagram has been prepaired on the assumption that the car is equipped with all optional equipment.
The right is reserved to make changes in the wiring diagram at any time without prior notice.

Fig. 10.61 Wiring diagram (continued)

Chapter 11 Suspension and steering

Refer to Chapter 13 for specifications and information applicable to 1980 through 1984 models.

Contents

Specifications

Front suspension

Type ..... Macpherson strut with stabilizer bar

Coil spring free length:
Standard models ..... 16.752 in (425.5 mm)
Air conditioning models ..... 16.949 in (430.5 mm)

Rear suspension (Sedan)

Type ..... Independent with coil springs and suspension arm. Double-acting hydraulic shock absorber.

Coil spring free length ..... 14.17 in (360 mm)

Rear wheel alignment
Toe-in ..... 0.16 to 0.55 in (4.0 to 14.0 mm)
Camber ..... 27′ to 1°57′

Rear suspension (Station Wagon)

Type ..... Semi-elliptic leaf springs with double-acting hydraulic shock absorbers

Steering

Type ..... Worm and nut, recirculating ball (power-assisted on air-conditioned models); collapsible steering column

Number of turns of steering wheel (lock to lock):
Manual steering 3.9
Power steering 3.2

Minimum turning radius ..... 17.4 ft (5.3 m)

Steering gear oil capacity:
Manual steering ..... 5/8 US pint 0.29 liters
Power steering ..... 1 1/2 US quarts 1.4 liters

Steering angles (vehicle unladen):
Camber ..... 0 to 1°30′

Caster	1°10′ to 2°40′
Steering axis inclination	7°10′ to 8°40′
Toe-in	0 to 0.08 in (0 to 2.0 mm)

Wheels and tires

Wheel type and size	Pressed steel, 5J–14
Tire type and size	Steel-braced radial ply, tubeless. 185/70HR-14
Tire pressures (maximum load of 750 lbs)	28 lbf/in² (2.0 kgf/cm²) front and rear

Torque wrench settings	**lbf ft**	**kgf m**
Front suspension		
Stabilizer bracket-to-bodyframe nuts	20 to 27	2.7 to 3.7
Stabilizer bracket–to–bodyframe bolts	20 to 27	2.7 to 3.7
Radius rod securing nut	33 to 40	4.5 to 5.5
Radius rod-to-transverse link bolts	33 to 40	4.5 to 5.5
Transverse link pin nut	58 to 80	8 to 11
Lower balljoint-to-transverse link bolts	33 to 40	4.5 to 5.5
Balljoint-to-knuckle arm nut	71 to 88	9.8 to 12.2
Strut assembly-to-knuckle arm bolts	53 to 72	7.3 to 9.9
Strut assembly-to-bodyframe nuts	18 to 25	2.5 to 3.5
Piston rod self-locking nut	43 to 54	6.0 to 7.5
Gland packing nut	51 to 94	7.0 to 13.0
Track rod balljoint nut	40 to 72	5.5 to 10
Suspension member-to-bodyframe bolt	20 to 27	2.7 to 3.7
Suspension member-to-bodyframe nut	37 to 50	5.1 to 6.9
Suspension member-to-engine mounts	12 to 16	1.6 to 2.2
Rear suspension (Sedan)		
Rear wheel bearing locknut	181 to 239	25 to 33
Shock absorber mounting insulator-to-body nut	19 to 29	2.6 to 4.0
Shock absorber lower end fixing bolt	43 to 58	6 to 8
Rear suspension member mounting locknut	58 to 80	8 to 11
Suspension arm pin nut	65 to 80	9 to 11
Rear suspension (Station Wagon)		
Shock absorber upper end (bracket bolts)	7 to 9	0.9 to 1.2
Shock absorber lower end nut	26 to 35	3.6 to 4.8
Leaf spring U-bolt nut	43 to 47	6.0 to 6.5
Spring shackle nuts	43 to 47	6.0 to 6.5
Front mounting pin nuts	43 to 47	6.0 to 6.5
Bumper rubber fixing nut	7 to 9	0.9 to 1.2
Steering		
Steering wheel nut	27 to 38	3.8 to 5.2
Steering column clamp bolts	9 to 13	1.3 to 1.8
Steering column jacket bracket-to-dash panel screws	2.5 to 3.3	0.35 to 0.45
Rubber coupling to wormshaft bolt:		
Manual steering	29 to 36	4.0 to 5.0
Power steering	24 to 28	3.3 to 3.9
Manual steering gear		
Pitman arm nut	94 to 108	13 to 15
Steering box-to-bodyframe bolts	38 to 46	5.3 to 6.3
Rear cover bolts	11 to 18	1.5 to 2.5
Sector shaft cover bolts	11 to 18	1.5 to 2.5
Sector shaft adjusting screw locknut	12 to 18	1.7 to 2.5
Power steering gear		
Oil pump installing bolts	14 to 19	1.9 to 2.6
Pitman arm nut	94 to 108	13 to 15
Steering box-to-bodyframe	38 to 46	5.3 to 6.3
Sector shaft adjusting screw locknut	21 to 25	2.9 to 3.5
Hose attachments-to-oil pump	22 to 36	3 to 5
Hose attachments-to-steering box	36 to 51	5 to 7
Bleed screw	5 to 7	0.7 to 0.9
Sector cover bolt	20 to 24	2.7 to 3.3
Rear housing bolt	20 to 24	2.7 to 3.3
Steering linkage		
Idler body-to-bodyframe bolts	23 to 31	3.2 to 4.3
Balljoint taper pin nuts	40 to 72	5.5 to 10

Track-rod end locknuts	8 to 12	1.1 to 1.7
Idler shaft nut	40 to 51	5.5 to 7.0
Wheels		
Securing nut	58 to 65	8 to 9

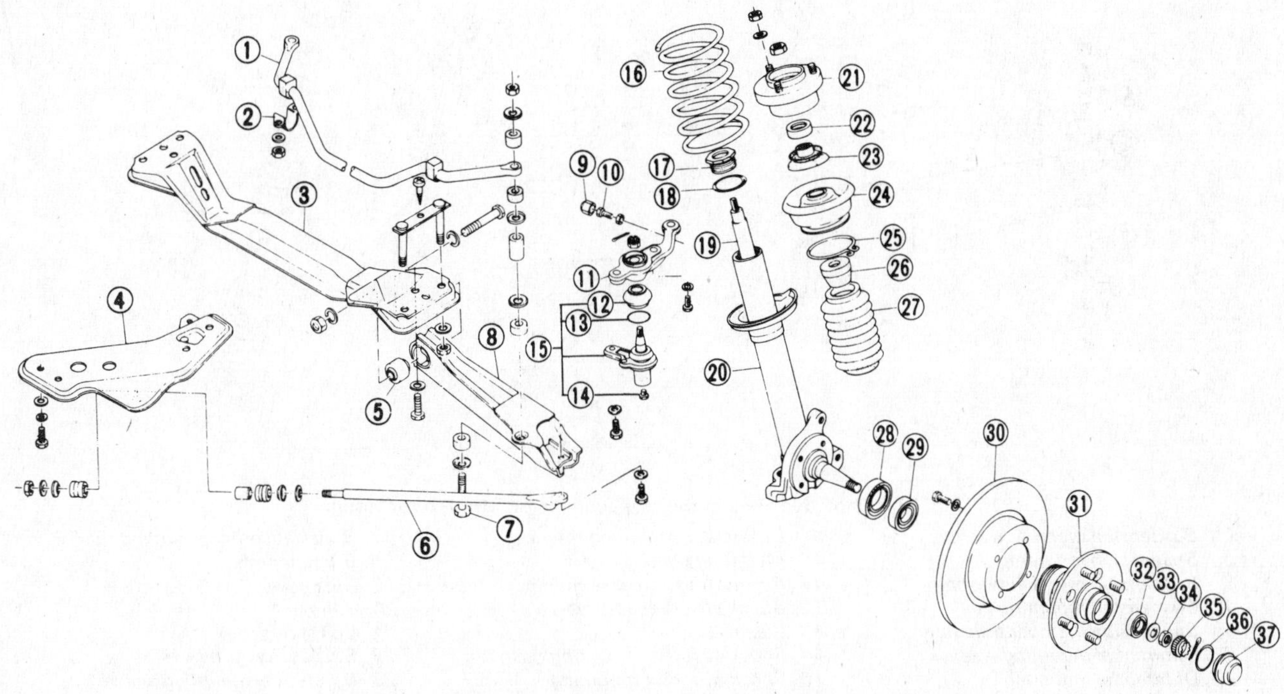

Fig. 11.1 Exploded view of the front axle and suspension assembly

1 Stabilizer bar
2 Stabilizer mounting bracket
3 Suspension crossmember
4 Radius rod bracket
5 Transverse link bush
6 Radius rod
7 Transverse link-to-stabilizer bar bolt
8 Transverse link
9 Cap
10 Steering stopper bolt
11 Knuckle arm
12 Dust cover
13 Retaining ring
14 Filler plug
15 Lower balljoint assembly
16 Coil spring
17 Gland packing nut
18 O-ring
19 Shock absorber
20 Strut assembly
21 Strut mounting insulator
22 Strut mounting bearing
23 Dust seal
24 Spring upper seat
25 Snap-ring
26 Bumper
27 Dust cover
28 Grease seal
29 Inner wheel bearing
30 Disc brake
31 Wheel hub
32 Outer wheel bearing
33 Washer
34 Wheel bearing nut
35 Adjusting cup washer
36 O-ring
37 Hub cap

1 General description

The front suspension on all models is of the independent Macpherson strut type. Each strut is secured at its upper end to rubber-mounted thrust bearings, while the lower end is secured to the lower balljoint located at the outer end of the transverse link, which allows the strut assembly to swivel. The transverse links are assembled to the suspension member through rubber bushings to avoid metal-to-metal contact. The entire suspension is designed to cater for thrust in all directions. Sideways thrust is controlled by a stabilizer bar and transverse links. Fore-and-aft thrust is controlled by the radius rods, whilst vertical thrust is controlled by the struts.

The strut mounting points are non-adjustable (set during production) and determine the castor, camber and steering axis. Each strut contains an integral direct-acting hydraulic damper; coil springs are mounted externally.

At the lower end of the strut, a stub axle carries the hub and disc brake assembly. The left and right-hand suspension struts are not interchangeable.

The rear suspension on the Sedan model is also of independent type, the layout comprising a crossmember assembly, suspension arm assemblies, and coil springs through which are passed the shock absorbers.

The rear suspension fitted to the Station Wagon is of the semi-elliptic leaf spring and shock absorber type. The rear axle assembly is located to the leaf springs by U-bolts which effectively clamp the axle case to them. The leaf springs are mounted to the body underframe, beneath the rear floor pan and the luggage compartment floor, using rubber-insulated pins and shackles. The shock absorbers, which are mounted between the body underframe and the lower spring seats, are also mounted in rubber bushes.

The steering system is of worm and nut recirculating ball type incorporating an idler, center track-rod and track-rod ends. As a factory option, power-assisted steering is available.

The steering column used on the manual and power steering models is of the collapsible type and great care is needed, when handling the column, to ensure that no undue stress is applied in an axial direction.

2 Maintenance and inspection

1 Regularly inspect the condition of all rubber dust excluders and balljoint covers for splits and deterioration. Remove them as necessary after reference to the appropriate Section of this Chapter.
2 Check the security of the track-rod end locknuts, also the ball-pin nuts.
3 Examine the condition of all suspension link rubber bushes. If they are perished, or permit movement of the adjacent components due to

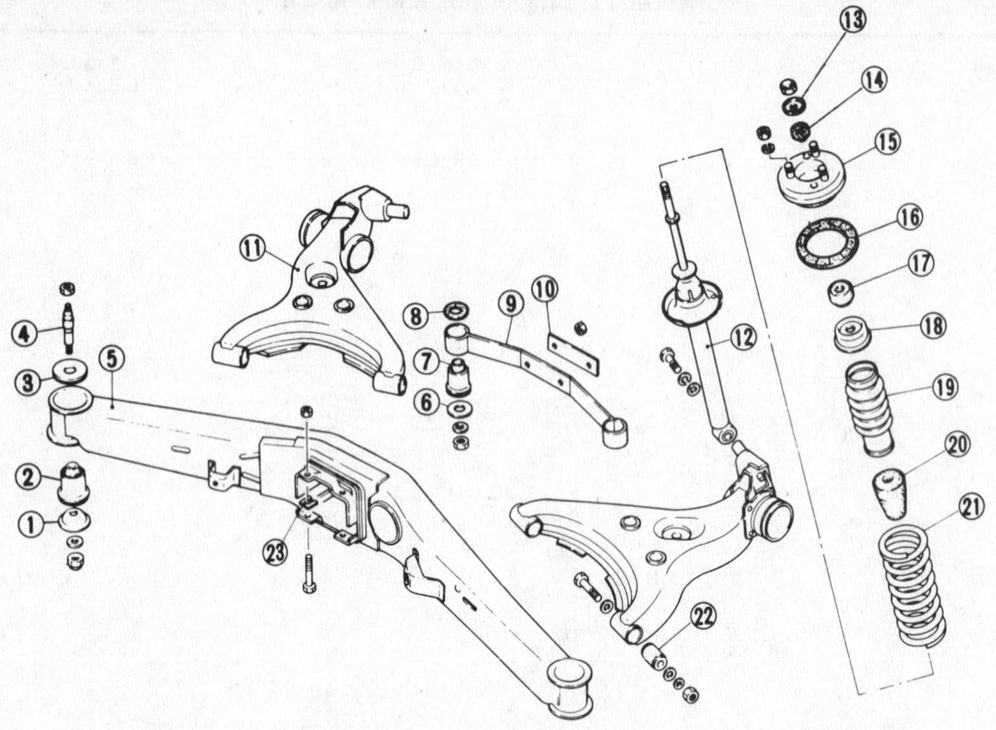

Fig. 11.2 Exploded view of the rear suspension fitted to the Sedan

1 Suspension member washer
2 Suspension mounting
3 Suspension member washer
4 Suspension mounting bolt
5 Suspension member assembly
6 Differential mounting washer
7 Differential mounting
8 Differential mounting washer
9 Differential mounting member
10 Mounting plate
11 Suspension arm assembly
12 Shock absorber assembly
13 Washer
14 Shock absorber mounting bush A
15 Shock absorber mounting
16 Spring seat rubber
17 Shock absorber mounting bush B
18 Bumper cover
19 Dust cover
20 Bumper
21 Coil spring
22 Suspension arm bush
23 Differential mounting spacer

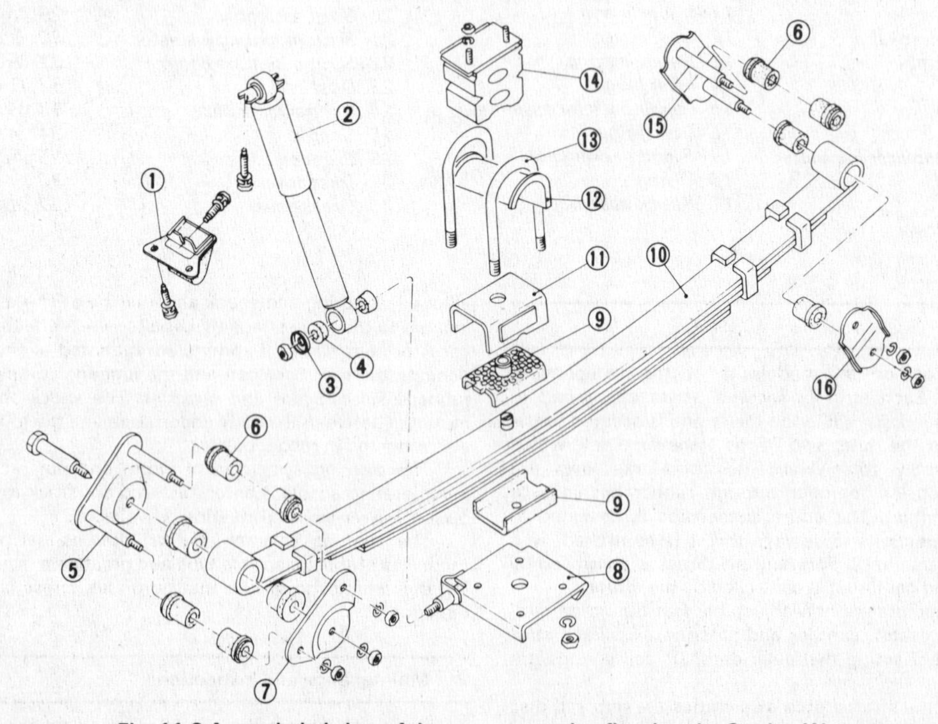

Fig. 11.3 An exploded view of the rear suspension fitted to the Station Wagon

1 Torque arrester
2 Shock absorber assembly
3 Washer
4 Bush
5 Front pin assembly
6 Bush
7 Front pin outer plate
8 Lower spring seat
9 Spring seating pad
10 Leaf spring assembly
11 Location plate
12 Axleplate
13 U-bolt
14 Rubber bumper
15 Shackle pin assembly
16 Bush

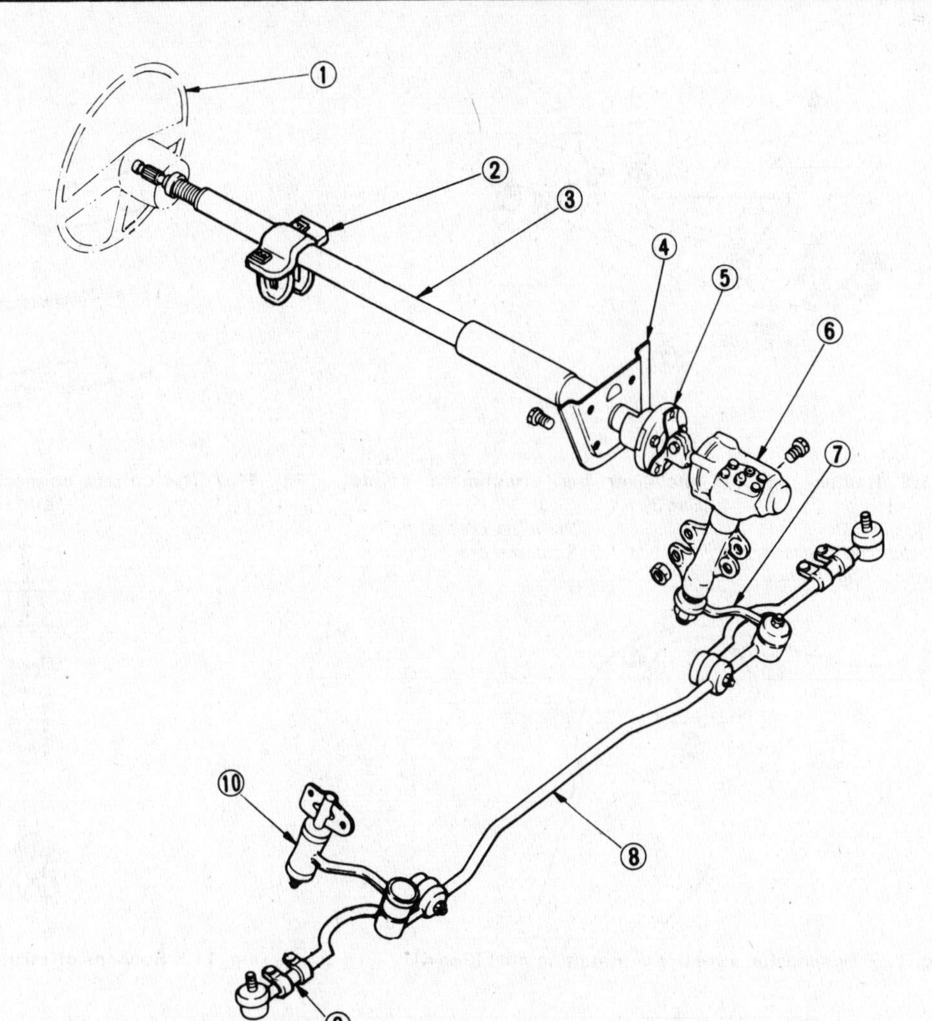

Fig. 11.4 The steering system

1 *Steering wheel*
2 *Column clamp*
3 *Steering column*
4 *Flange*
5 *Rubber coupling*
6 *Steering box*
7 *Pitman arm*
8 *Center track rod*
9 *Outer track rod*
10 *Idler*

elongation of their holes, then they must be renewed.
4 Every 30 000 miles (48 000 km) remove the plugs from the balljoints, screw in a grease nipple and recharge the balljoint with multi-purpose grease.
5 At a similar mileage interval, dismantle the front hubs, clean out the old grease and repack with fresh grease, as described in Section 8 of this Chapter.
6 On Sedan models, the rear hubs will require dismantling and repacking with grease at 48 000 miles (77 000 km) service intervals, as described in Chapter 8, Section 3.
7 On Station Wagons, no maintenance of the rear suspension is required except to periodically inspect the rubber bushes for wear and to check the security of the attachment bolts.

3 Front stabilizer and radius rod – removal and installation

1 Jack-up the front of the vehicle and support it securely on stands.
2 Remove the roadwheel.
3 Remove the under-engine splashboard.
4 Refer to Fig. 11.6, and unscrew and remove the radius rod nut (1) securing the radius rod to the bracket.
5 Unscrew and remove the bolts (2) securing the radius rod to the transverse link. Lift away the radius rod.
6 Unscrew and remove the nut (3) securing the stabilizer bar to the connecting rod on the transverse link.
7 Unscrew and remove the bolts (4) and the nuts (5) which secure the stabilizer bracket in position. The stabilizer can now be removed from one side of the vehicle. To remove the stabilizer bar completely, repeat these operations at the other side of the vehicle.

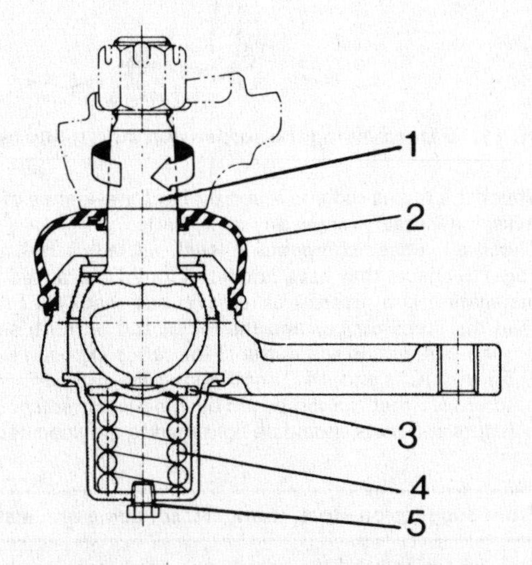

Fig. 11.5 Sectional view of a balljoint (Sec 2)

1 *Taper pin*
2 *Dust seal*
3 *Spring seat*
4 *Spring*
5 *Grease plug*

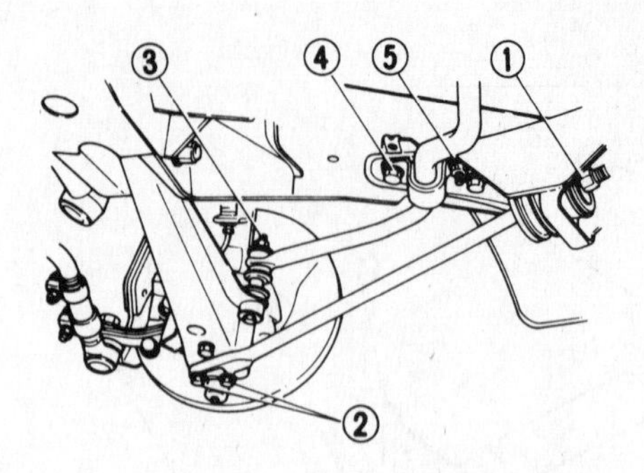

Fig. 11.6 Radius rod and stabilizer bar attachment points (Sec 3)

1 End nut
2 Radius rod-to-transverse link bolts
3 Connecting rod nut
4 Stabilizer bracket bolt
5 Stabilizer bracket nut

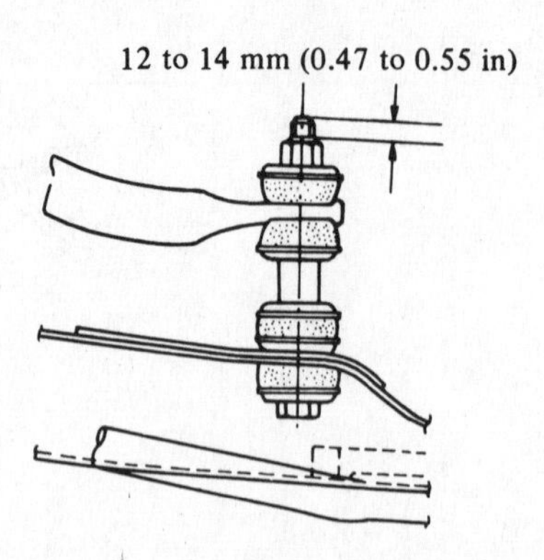

Fig. 11.7 The correct connecting rod protrusion above the nut surface (Sec 3)

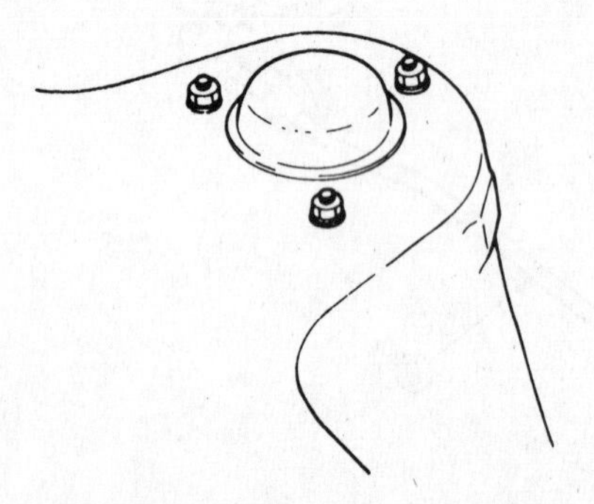

Fig. 11.8 Suspension strut upper mounting nuts (Sec 4)

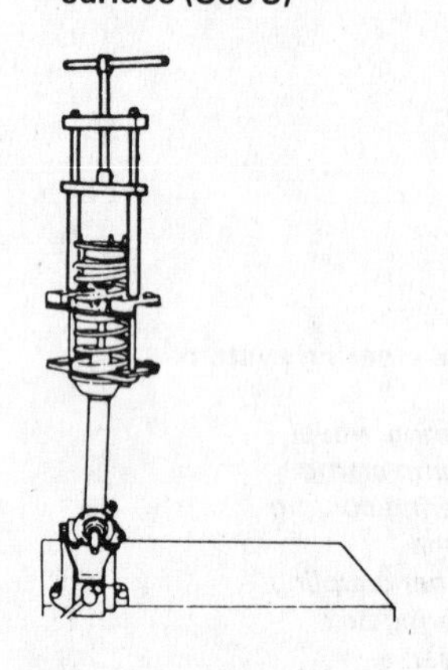

Fig. 11.9 Suspension strut coil spring compressed (Sec 4)

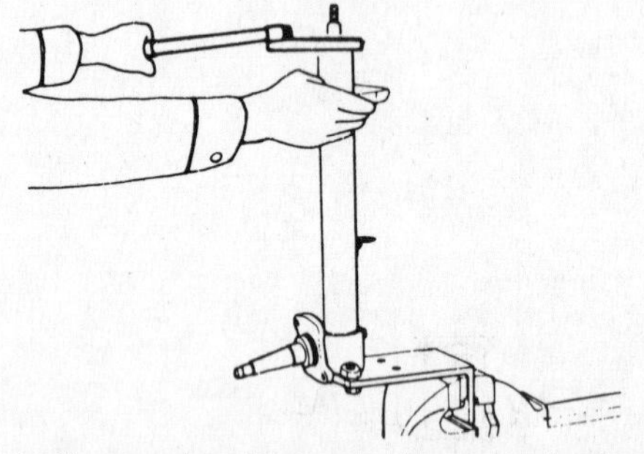

Fig. 11.10 Unscrewing the suspension strut gland nut (Sec 4)

8 Check the radius rod and stabilizer bar for evidence of deformation or cracks; if necessary renew any faulty part.
9 Check all rubber components (such as radius rod and stabilizer bushings) to ensure they have not deteriorated or cracked.
10 Installation is a reversal of removal but check that the clearance between the stabilizer bar and the radius rod on both sides is equal. Tighten the connecting rod nut until the rod is exposed between 0.47 and 0.55 in ((12.0 and 14.0 mm) above the top surface of the nut. Check to ensure that the radius rod bushing is correctly centered in its seat. All nuts and bolts should be tightened to the specified torque.

4 Front suspension strut – removal, servicing and installation

1 Jack-up the front of the vehicle and firmly support the bodyframe on suitable stands.
2 Disconnect the flexible brake hose from the rigid hydraulic pipe at the support bracket, as described in Chapter 9.
3 Remove the disc caliper unit also, as described in Chapter 9.

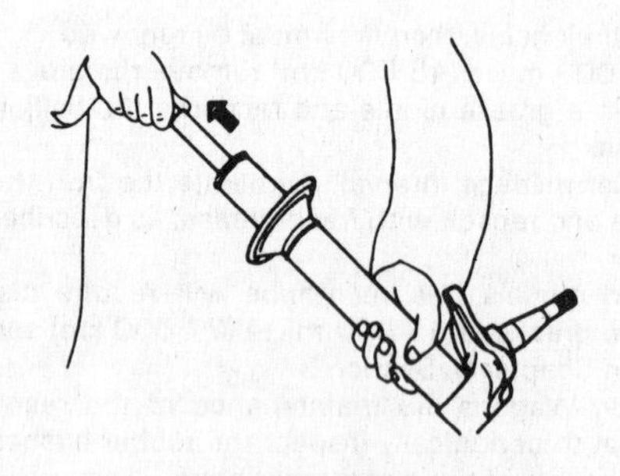

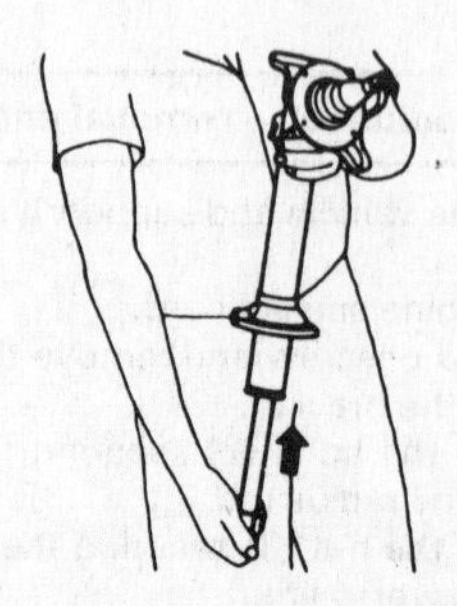

Fig. 11.11 Bleeding air from a suspension strut prior to installation (Sec 4)

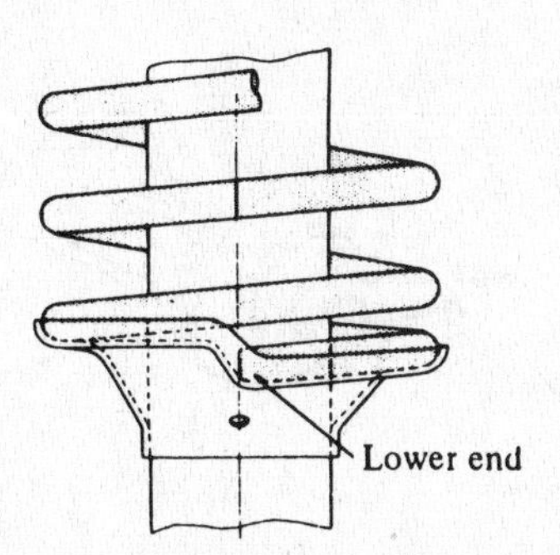

Fig. 11.12 Front suspension coil spring installation diagram (Sec 4)

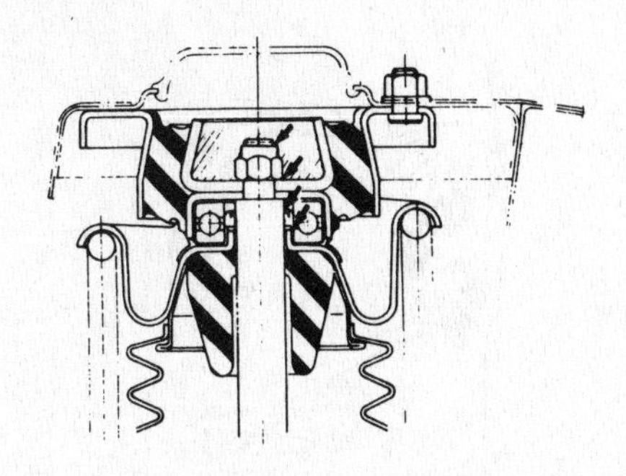

Fig. 11.13 Sectional view of suspension strut upper mounting. Lubrication points (arrowed)

4 Unscrew and remove the bolts securing the knuckle arm to the lower strut.
5 Detach the knuckle arm from the bottom of the strut by forcing the transverse link downwards with a suitable pry bar.
6 Support the strut assembly with a jack or a suitable stand, and unscrew and remove the three nuts within the hood compartment. The strut assembly can now be removed from beneath the fender.
7 With the strut assembly removed from the vehicle, proceed to remove the disc and hub assembly and the backplate.
8 A coil spring compressor must now be used to compress the spring until the top mounting assembly can be turned by hand.
9 Unscrew and remove the self-locking nut from the top face of the mounting. The mounting must be held stationary to do this and the way to do it is to pass a long bar between two of the mounting studs. Protect the threads of the studs to prevent damage.
10 Withdraw the top mounting insulator, the bearing and the upper spring seat.
11 The coil spring complete with compressor may now be withdrawn from the suspension strut. If the original spring is to be re-installed, there is no need to remove the compressor.
12 Inspect the suspension strut for oil leakage from around the piston gland. Hold the strut vertically, and fully extend and depress the piston rod several times. Unless there is a definite resistance with smooth operation in both directions, the unit must be renewed.
13 It is not recommended that the strut is dismantled due to the special tools required and the difficulty of obtaining individual components. Either renew the complete strut or obtain one of the sealed cartridge type repair units which are available. If the latter course is adopted then the upper gland nut and O-ring will have to be unscrewed, the internal components withdrawn from the strut outer casing and the new cartridge installed in accordance with the maker's instructions.
14 Before installing the suspension strut, bleed it of any air which may have accumulated while it has been stored in a horizontal position. This is carried out in exactly the same manner as described for testing the unit earlier in this Section. Check all mounting components for wear.
15 Locate the coil spring (in its compressed state) on the suspension strut so that its lower coil locates correctly into the spring pan (Fig. 11.12).
16 Install the upper spring seat, the bearing (packed with multi-purpose grease) and the insulator.
17 Screw on the self-locking nut, tightening it to the specified torque.
18 Release the coil spring compressor gently and then install the suspension leg, disc caliper and hub in the reverse manner to removal. Tighten all nuts and bolts to the specified torque settings. Adjust the front hub as described in Section 8. Reconnect the hydraulic line and bleed the brakes (Chapter 9).

5 Front coil spring – removal and installation

1 Removal of a coil spring using a compressor is described in the preceding Section. An alternative method which can be used when the original spring is to be re-installed, is to use clips.
2 Have an assistant sit on the top of the front fender to compress the front suspension leg and spring. Use a minimum of three clips to engage over three or four coils and then secure them in position with a strap or large worm drive clip.
3 Once the suspension strut is removed, the spring can be detached after the upper mounting has been withdrawn.
4 Installation is a reversal of removal. Remove the strap and clips after the weight of an assistant has again compressed the spring.

6 Transverse link and lower balljoint – removal, servicing and installation

1 Jack-up the front of the vehicle and firmly support it on suitable stands. Remove the appropriate roadwheels.
2 Remove the under-engine splashboard.
3 Disconnect the track-rod end balljoint (see Section 15) from the knuckle arm.
4 Disconnect the knuckle arm from the base of the suspension strut assembly (two bolts).
5 Remove the radius rod and stabilizer bar, as described in Section 3.
6 Loosen and remove the nut, and drive out the pivot pin attaching the transverse link to the suspension crossmember.
7 The transverse link can now be removed together with the suspension balljoint and knuckle arm.
8 With the transverse link removed, unbolt and remove the balljoint from the link.
9 Check the balljoint for wear by gripping the taper pin, and pushing and pulling it to test for seat movement. Now move it sideways in several different directions. There must be a stiff resistance during the complete arc of movement. If the rubber dust cover is split or has perished, renew it. The balljoints should be lubricated as described in Section 2.
10 Examine the rubber-bonded type bush of the transverse link. If this has perished or is deformed it must be renewed using a press and this is a job best left to your Datsum dealer who will have the necessary removal and installation tools.
11 Installation is a reversal of removal but tighten all nuts and bolts to the specified torque wrench settings. Tighten the transverse link pivot bolt only when the car has been lowered to the ground and is loaded with (or the equivalent of) two people in the front seats.

7 Suspension crossmember – removal and installation

1 Jack-up the front of the vehicle and firmly support the body sideframe members on suitable stands.
2 Remove the roadwheels.
3 Remove the under-engine splashboard.
4 Disconnect the transverse links, as described in Section 6.
5 Using a suitable hoist, raise the engine just enough to take its weight off the front mountings.
6 Loosen and remove the engine mounting-to-crossmember nuts.
7 Unbolt the crossmember from the chassis at either side of the

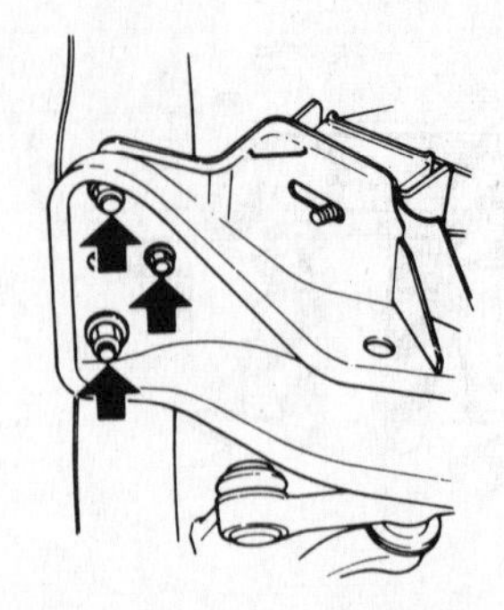

Fig. 11.14 The suspension member-to-chassis securing nuts arrowed (Sec 7)

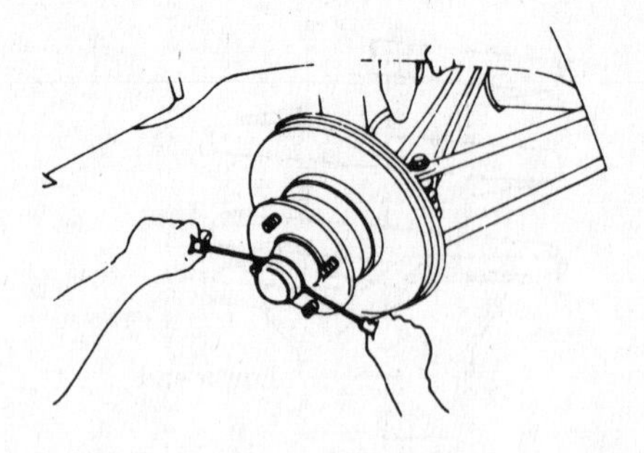

Fig. 11.15 Removing a front hub dust cap (Sec 8)

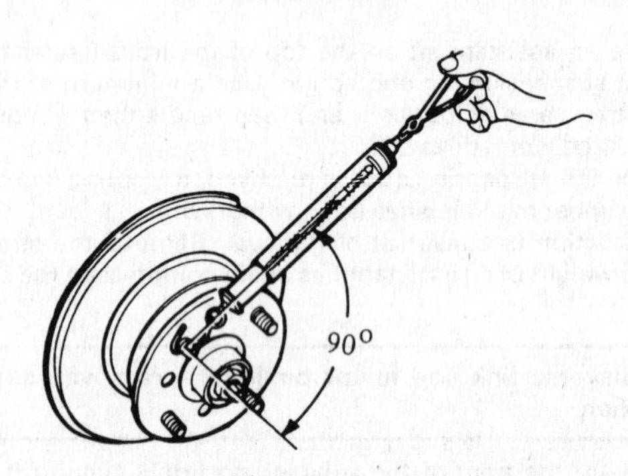

Fig. 11.16 Measuring front hub bearing preload (Sec 8)

Fig. 11.17 Hub adjusting cup washer and cotter pin (Sec 8)

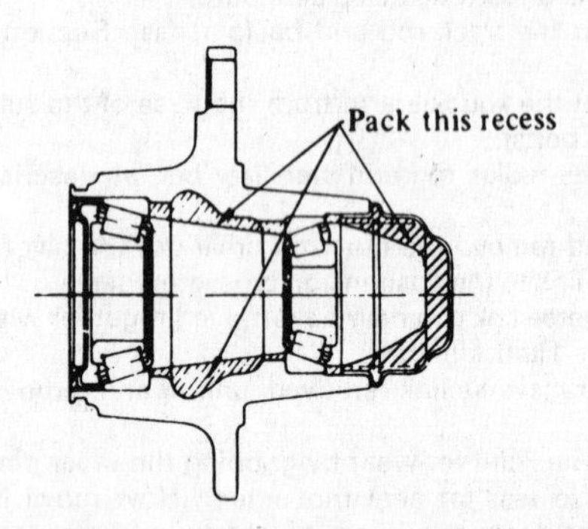

Fig. 11.18 Front wheel bearing greasing diagram (Sec 8)

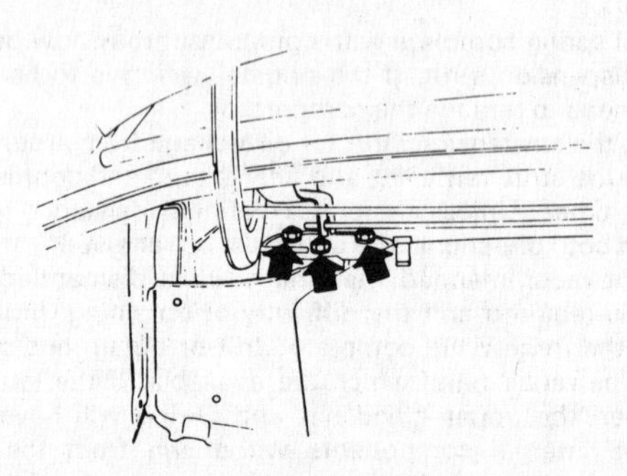

Fig. 11.19 The shock absorber mounting point inside the trunk compartment (Sec 9)

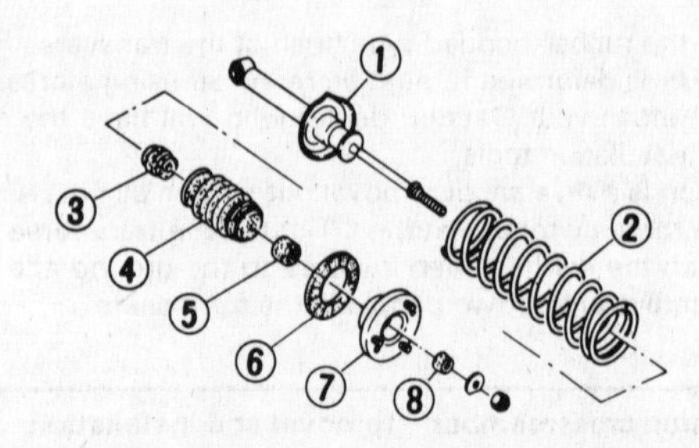

Fig. 11.20 Component parts of the coil spring and shock absorber assembly (Sec 9)

1 Shock absorber
2 Coil spring
3 Bump stop
4 Bumper cover and dust cover assembly
5 Inner bushing
6 Spring seat rubber
7 Shock absorber mounting
8 Outer bushing

vehicle and lower the suspension crossmember to the floor.

8 Check the crossmember for evidence of cracking and, if necessary, obtain a replacement.

9 Installation of the suspension crossmember is a reversal of the removal procedure, but ensure that the nuts and bolts are tightened to the specified torque.

8 Front hub bearings – adjustment, lubrication and renewal

1 Adjustment of the front hub bearings must be carried out if they develop any excessive play, whenever the hub components have been dismantled and reassembled or the bearings are repacked with grease. Every 3000 miles (4800 km) jack-up the front of the car and grip the top and bottom of the roadwheel. Rock it backwards and forwards and if there is a perceptible amount of play, check the adjustment as described in the following paragraphs.

2 Remove the roadwheel and withdraw the disc pads (Chapter 9).

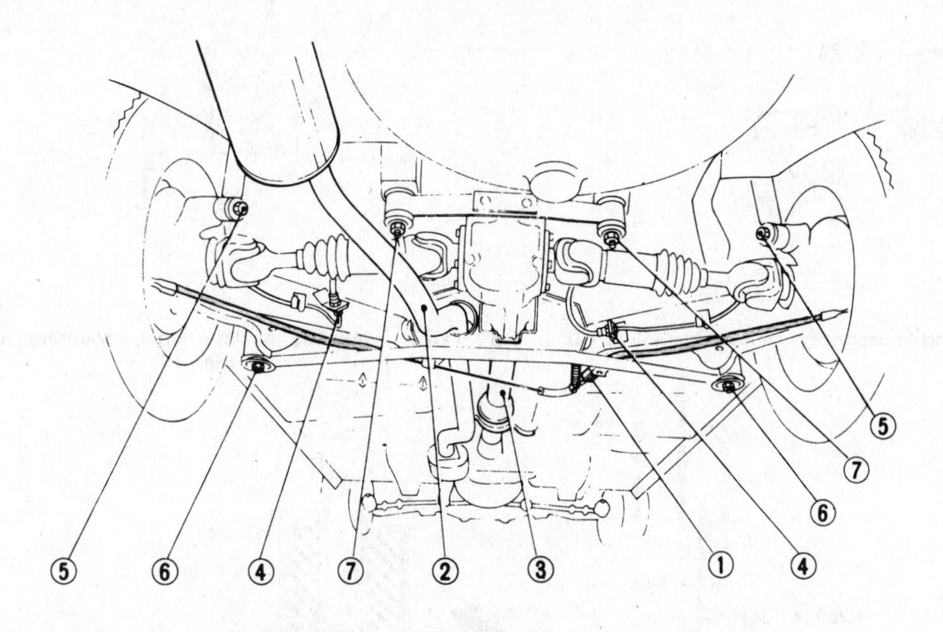

Fig. 11.21 Rear axle and suspension assembly disconnection points (Sec 10)

1 Parking brake cable locknut
2 Exhaust pipe and muffler
3 Propeller shaft
4 Rear brake hoses
5 Shock absorber mounting bolts
6 Suspension member nuts
7 Differential mounting member nuts

3 Remove the dust cap from the end of the hub, using two screwdrivers as levers or by tapping it off with a small cold chisel and hammer.
4 Remove the cotter pin from the castellated nut.
5 Using a torque wrench, tighten the hub nut to 22 lbf ft (3.0 kgf m) at the same time rotating the hub in its forward direction.
6 Unscrew the hub nut 60° and then install a new cotter pin and bend over the ends. If the cotter pin in the stub axle does not correspond with the slot in the castellated nut, move the nut in the loosening direction.
7 Using a suitable spring balance attached to one of the wheel hub studs, measure the force required to move the wheel hub. This should be between 0.4 and 1.8 lbf (0.2 and 0.8 kgf) with the original grease seal installed; or between 1.5 and 3.3 lbf (0.7 and 1.5 kgf) with a new grease seal installed.
8 To renew or lubricate the hub bearings, remove the dust cap, cotter pin and nut as previously described. Unbolt the brake caliper from the stub axle carrier and disconnect the hydraulic fluid pipe at the suspension leg bracket. Plug the fluid line or seal the master cylinder reservoir cap.
9 Withdraw the hub/disc assembly from the stub axle catching the outer bearing and thrust-washer.
10 Where the bearings are only to be repacked with grease, wipe all the old grease from the bearings and from the space inside the hub between the two bearings. Examine the condition of the grease seal and if there is evidence of leakage of grease, renew it by levering out the old one and driving in a new one with a piece of tubing used as a drift.
11 Repack the intervening space between the wheel bearings with fresh grease and also fill the dust cap not more than quarter full.
12 Where the bearings are slack and defy adjustment, they are worn and must be renewed. Use a drift to drive out the inner and outer bearing tracks and drive in the new one using a piece of tubing for the purpose. The grease seal will of course have to be renewed whenever the inner bearing track is removed and installed. If both front hubs are being serviced at the same time, do not mix the new bearings up but keep them in their packets until required as the roller cages and outer tracks are matched in production.
13 Reassembly of the front hub is a reversal of removal and dismantling but adjust the bearings as described earlier in this Section and bleed the brake hydraulic system once the caliper unit has been reconnected.

9 Rear coil spring and shock absorber assembly (Sedan) – removal, servicing and installation

1 Jack-up the rear of the vehicle and support the bodyframe sidemembers securely on suitable stands. The front wheels should also be chocked.
2 Working inside the trunk compartment, loosen and remove the three nuts securing the shock absorber upper mounting to the bodyframe.
3 Working beneath vehicle, disconnect the bottom of the shock absorber from the suspension arm (one bolt).
4 Remove the shock absorber and coil spring assembly from beneath the vehicle.
5 To separate the shock absorber from the coil spring a spring compressor will be needed.
6 Mark the relative positions of the upper shock absorber mounting and the lower end pin to ensure correct reassembly.
7 With the assembly held securely in a vise, compress the spring until the upper mounting can be turned by hand.
8 Unscrew and remove the self-locking nut from the top face of the mounting. The mounting must be held stationary to do this and the way to do it is to pass a long bar between two of the mounting studs. Protect the threads of the studs to prevent damage.
9 Withdraw the top mounting insulator together with the bushing and washer. Remove the spring seat rubber. The compressor and coil spring can now be removed from the assembly.
10 Remove the inner bushing, bumper and dust cover assembly, and the bumper stop from the assembly.
11 With the coil spring removed from the spring compressor, visually check the coil for any cracks or deformation. Measure the free length and compare the measurement with that given in the specifications.
12 Inspect the shock absorber for oil leakage. Hold the assembly vertically, and fully extend and depress the piston rod several times. Unless there is a definite resistance with smooth operation in both directions, the unit must be renewed.
13 Check all rubber components for wear, cracks, damage or deformation, renewing where necessary.
14 Reassembly of the coil spring and shock absorber assembly is a reversal of removal but ensure that the coil spring is correctly located in the lower spring seat (Fig. 11.12).
15 When installing the assembly to the vehicle, the upper end of the

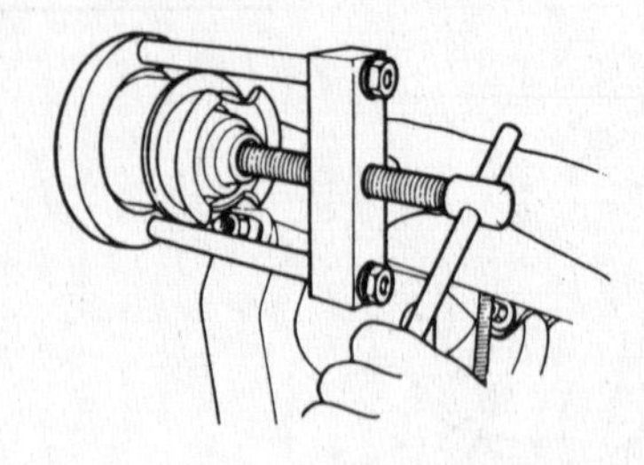

Fig. 11.22 Removing a suspension member bush (Sec 10)

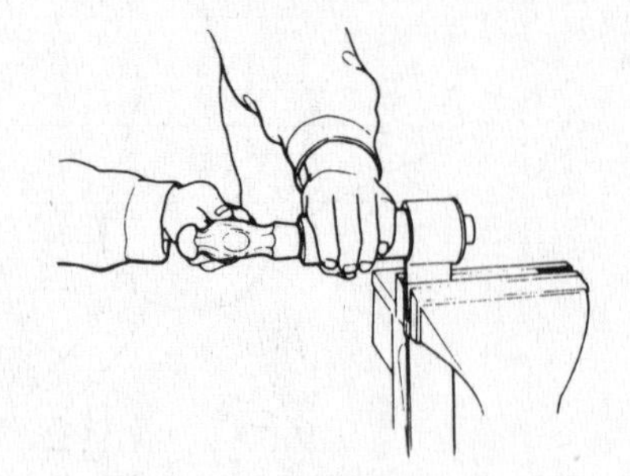

Fig. 11.23 Removing a differential mounting member bush (Sec 10)

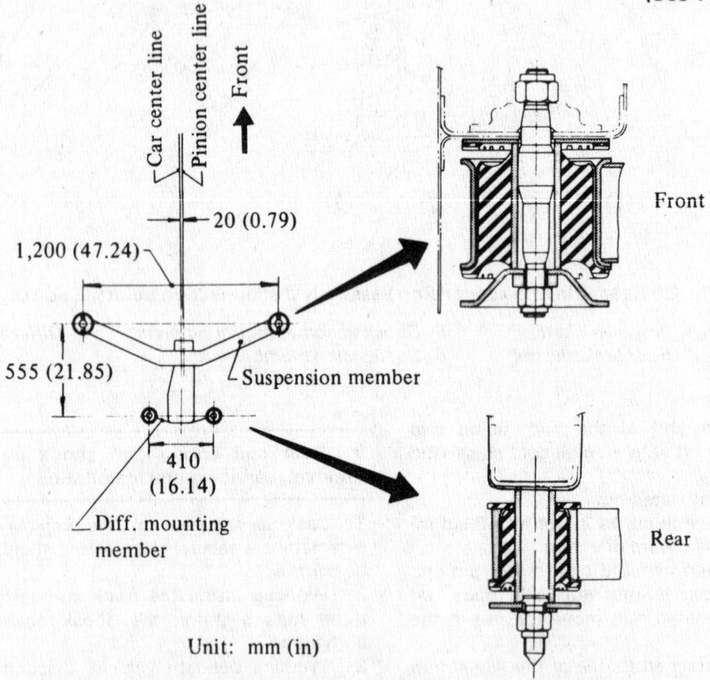

Fig. 11.24 Rear suspension assembly bush installation diagram (Sec 10)

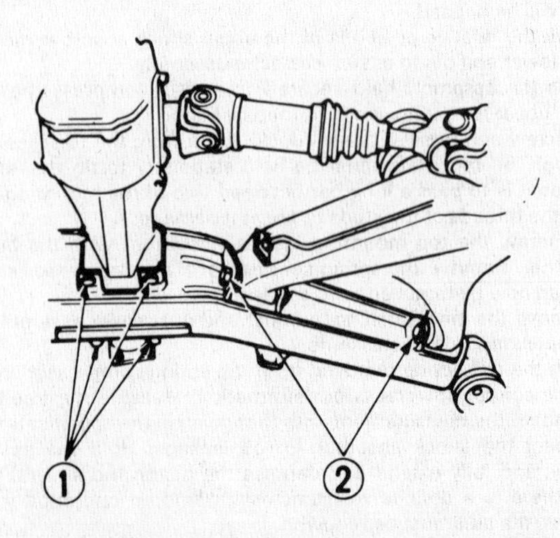

Fig. 11.25 Removing the suspension member (Sec 11)

1 Differential case-to-suspension member bolts
2 Suspension arm pivot pins

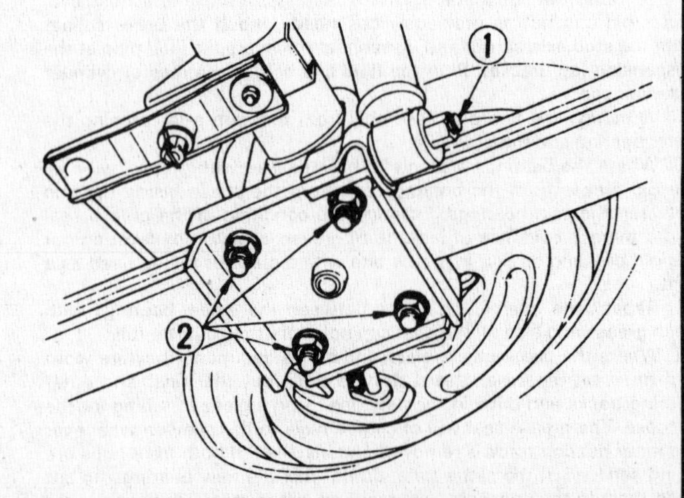

Fig. 11.26 The shock absorber lower mounting nut (1) and the U-bolt retaining nuts (2) (Sec 4)

assembly should be installed first. Tighten all nuts and bolts to the specified torque.

10 Rear axle and suspension assembly (Sedan) – removal and installation

1 Jack-up the rear end of the vehicle and support it on firmly based stands. Check the front wheels. Remove the rear wheels.
2 Refer to Chapter 3, and remove the rear exhaust pipe and muffler.
3 Remove the propeller shaft as described in Chapter 7.
4 Disconnect the brake flexible hoses from the rigid hydraulic lines on both sides of the vehicle. Before disconnecting the hoses, place a sheet of polythene over the mouth of the hydraulic master cylinder, to create a vacuum and prevent excessive loss of the hydraulic fluid. Ensure that the disconnected hoses and pipes are suitably covered to prevent dirt entering the system. Disconnect the parking brake linkage.
5 Using a trolley jack, support the differential housing in such a manner that when the assembly is lowered it will not tilt to one side or slip off the jack.
6 Disconnect each shock absorber from its lower suspension arm mounting point.
7 Loosen and remove the nut at each side of the suspension member.
8 Loosen and remove the two nuts securing the rear differential mounting member to the body underframe.
9 Very carefully lower the jack, ensuring that the assembly does not tilt to one side or slip off. Once the assembly has been lowered enough to clear the underframe it can be withdrawn from beneath the vehicle.
10 With the rear suspension removed, examine all parts for wear or damage. Particular attention should be given to the bushes in the suspension arms. Also check the condition of the rubber bushes in the suspension member and the differential mounting member. Any of these components, if worn, can result in noise and vibration.
11 To renew the suspension member bushes, will require the use of a suitable bush extractor. The old bush is simply forced out and a new one installed in the same way. Ensure that the bushes are correctly installed (Fig. 11.24), and inserted from the underside of the member.
12 Removal of a differential mounting member bush can be achieved by driving out the old bush with a piece of suitable tube; the new bush can be installed using the same tube and driving the bush squarely into position. Ensure that the new bush is correctly assembled (Fig. 11.24) and inserted from the underside of the member. Before installing the assembly, ensure that the suspension member and the differential mounting member are correctly aligned (Fig. 11.24).
13 Commence installation by jacking the assembly up into position to locate on the four mounting studs (two suspension member studs and two differential mounting member studs). Secure the assembly by installing the four securing nuts and tightening them to the specified torque.
14 Once the shock absorbers have been connected, the jack can be removed. The rest of the installation procedure is a reversal of removal but the rear brake hydraulic lines should be bled as described in Chapter 9.

11 Suspension member (Sedan) – removal and installation

1 Renewal of the suspension member bushes is described in Section 10. This can be carried out without removing the suspension member from the front of the differential case and the suspension arms. However, if the suspension member is to be renewed proceed as described in the following paragraphs.
2 Refer to Section 10, and remove the rear axle and suspension assembly from the vehicle.
3 Disconnect the front end of the differential case from the suspension member by removing the four securing bolts.
4 Disconnect each suspension arm by removing the nuts and driving out the two pivot pins. The suspension member can now be removed from the assembly.
5 Installation of the suspension member is a reversal of removal, but before tightening the suspension arm pivot pins the vehicle must be lowered to the ground. This will ensure that the rubber bushes are clamped in a neutral or unloaded position.

12 Suspension arm (Sedan) – removal and installation

1 Jack-up the rear of the vehicle and support it firmly on suitable stands. Chock the front wheels.
2 Remove the appropriate roadwheel.
3 Refer to Chapter 8, and remove the propeller shaft.
4 Disconnect the hydraulic brake pipe from the rear wheel cylinder and the flexible hose. Remove the pipe from the suspension arm. Before dismantling the pipe, place a sheet of polythene over the mouth of the rear reservoir of the master cylinder. This will create a vacuum in the hydraulic system and prevent excessive fluid loss. Be sure to plug the open end of the disconnected hose.
5 Disconnect the parking brake cable from the operating lever on the backplate. Remove the cable assembly from the suspension arm.
6 Refer to Chapter 8, Section 3, and remove the rear axle stub, wheel bearings and oil seal.
7 Disconnect the shock absorber at its lower mounting.
8 Remove the suspension arm by loosening the two pivot pin nuts, removing them, and driving out the pivot pins with a suitable drift.
9 Renewing the pivot pin bushes will require the use of a bush extractor. The old bushes are drawn out and new ones are installed using the same tool.
10 Installation of the suspension arm is a reversal of removal but ensure that all nuts and bolts are tightened to the specified torque. The weight of the vehicle must be on the rear wheels before the suspension arm pivot pins are tightened.
11 Ensure that the rear hydraulic line is bled and that all relevant adjustments are carried out before driving the vehicle.

13 Shock absorber (Station Wagon) – removal and installation

1 Drive the vehicle over an inspection pit, or alternatively, raise the rear end and support it on suitable stands.
2 Working beneath the vehicle, loosen and remove the nut that secures the shock absorber to the leaf spring seat stud.
3 Loosen and remove the two bolts securing the upper mounting bracket to the bodyframe. The shock absorber can now be detached from its lower stud mounting and removed, together with the upper mounting bracket, from the vehicle.
4 With the shock absorber removed, visually check it for any evidence of oil leakage or cracks. Hold the assembly vertically, and fully extend and depress the piston rod several times. Unless there is a definite resistance with smooth operation in both directions, the unit must be renewed. Check also the rubber bushes for damage or wear and, if necessary, renew them.
5 Installation of the shock absorber is a reversal of removal but ensure that the upper mounting bolts are tightened to the specified torque. Tighten the lower mounting nut to the specified torque with the weight of the vehicle on the rear wheels.

14 Rear spring (Station Wagon) – removal and installation

1 Jack-up the rear of the vehicle and firmly support it under the bodyframe with suitable stands.
2 Remove the appropriate roadwheel.
3 Loosen and remove the nut securing the shock absorber to its lower mounting stud. Detach the shock absorber from the mounting stud.
4 Loosen and remove the four nuts which secure the two U-bolts in position. Carefully tap the ends of the U-bolts upwards, and detach the lower spring seat and the seating pad.
5 At this stage, place a suitable jack beneath the axle case and lift the axle until its weight is off the spring.
6 Working at the back end of the leaf spring, undo and remove the two nuts retaining the shackle plate in position. Using a suitable drift, and taking great care not to damage the shackle pin threads, drive out the shackle pin assembly.
7 Working at the front end of the leaf spring, undo and remove the nut retaining the shackle pin. Using a suitable drift, drive out the shackle pin. The leaf spring can now be removed.
8 Using a wire brush, clean all rust and dirt from the spring leaves and check for any fractures or cracks in the leaves. Examine the shackles and pins, U-bolts and spring seat for wear, cracks, distortion

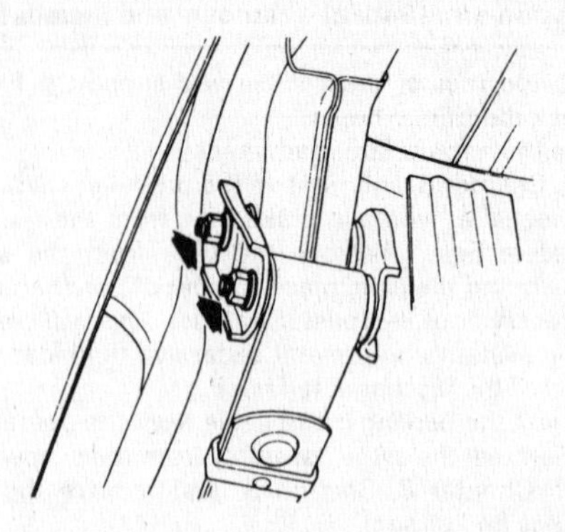

Fig. 11.27 The rear shackle retaining nuts arrowed (Sec 14)

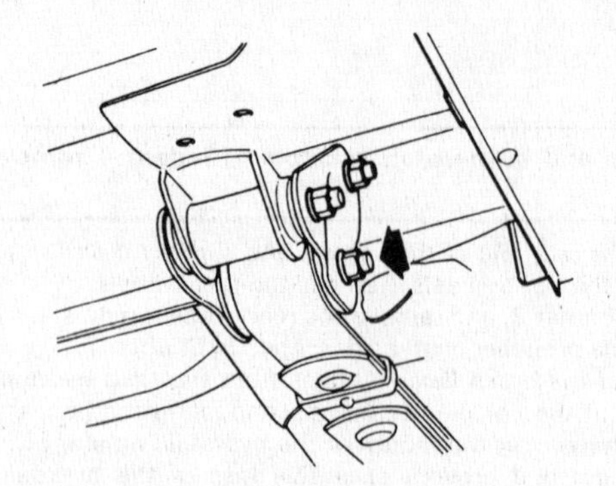

Fig. 11.28 The front shackle pin retaining nut arrowed (Sec 14)

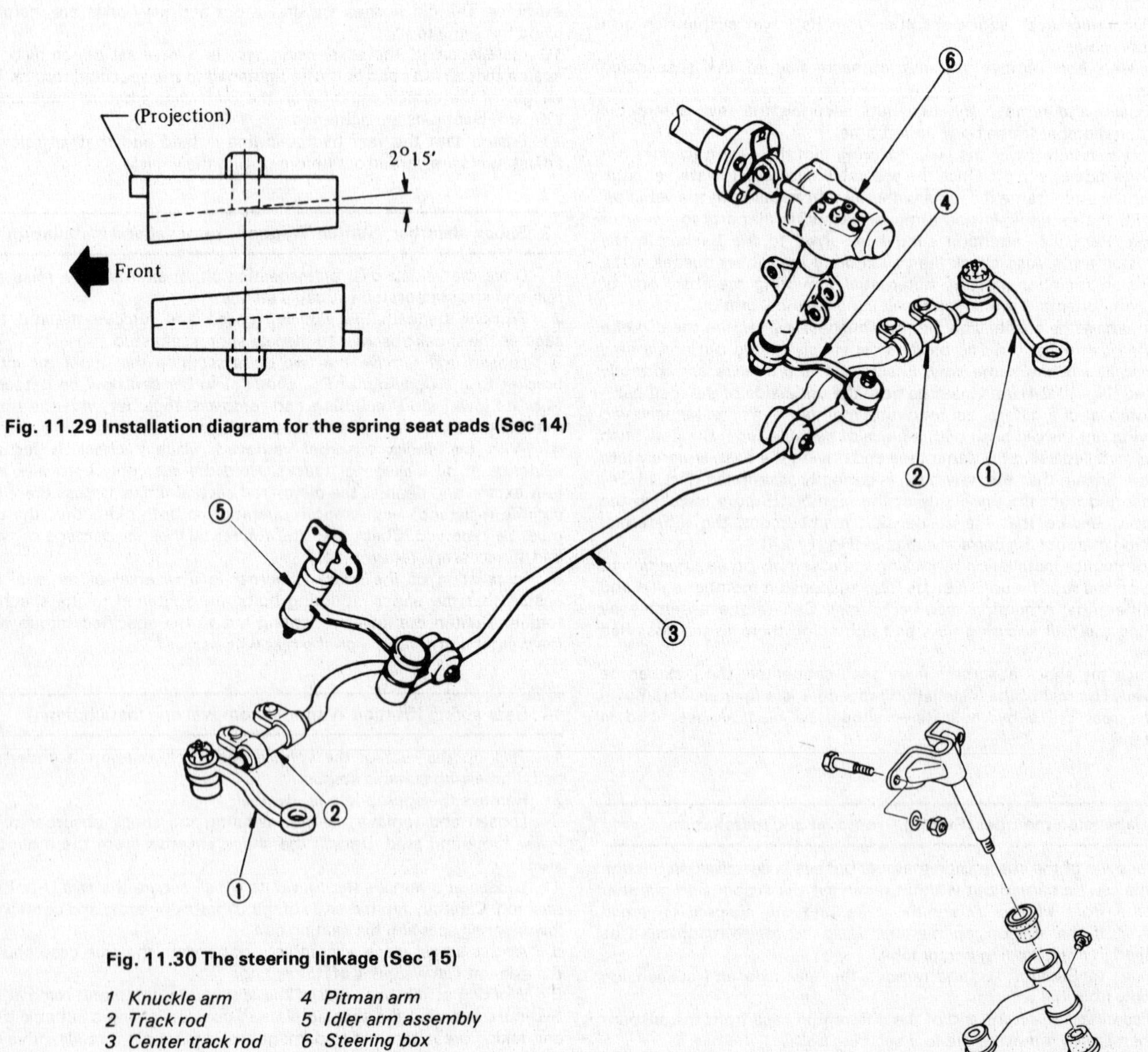

Fig. 11.29 Installation diagram for the spring seat pads (Sec 14)

Fig. 11.30 The steering linkage (Sec 15)

1 Knuckle arm
2 Track rod
3 Center track rod
4 Pitman arm
5 Idler arm assembly
6 Steering box

Fig. 11.31 The idler arm assembly (Sec 15)

or damaged threads. If faulty parts are found, replace them with new ones. Inspect all rubber components for wear, damage or separation, renewing any parts that are faulty.
9 Installation of the rear leaf spring is a reversal of removal but the following points must be observed:

(a) *Prior to installation, coat the shackle pins and bushes with a soap and water solution.*
b) *The spring seat pads are tapered and must be installed as shown in Fig. 11.29.*
(c) *All nuts and bolts must be tightened to the specified torque, but the weight of the vehicle must be on the rear wheels before tightening the shackles and the shock absorber lower mounting nut.*

15 Steering linkage – inspection, dismantling and reassembly

1 Wear in the steering gear and linkage is indicated when there is considerable movement in the steering wheel without corresponding movement at the roadwheels. Wear is also indicated when the vehicle tends to wander off the line one is trying to steer. There are three main steering groups to examine in such circumstances. These are the wheel bearings, the linkage joints and bushes, and the steering box itself.
2 First jack-up the front of the car and support it on stands under the sideframe members so that both front wheels are clear of the ground.
3 Grip the top and bottom of the wheel and try to rock it. It will not take any great effort to be able to feel any play in the wheel bearing. If this play is very noticeable it would be as well to adjust it straight away as it could confuse further examinations. It is also possible that during this check, play may be discovered in the lower suspension balljoint (at the foot of the suspension strut). If this is the case the balljoint will need renewal.
4 Next grip each side of the wheel and try rocking it laterally. Steady pressure will, of course, turn the steering but an alternated back and forward pressure will reveal any loose joint. If some play is felt it would be easier to get assistance from someone, so that while one person rocks the wheel from side to side, the other can look at the joints and bushes on the track-rods and connections. Excluding the steering box itself there are eight places where the play may occur. The two outer balljoints on the two outer track-rods are the most likely, followed by the two inner joints on the same rods, where they join to the center rod. Any play in these means renewal of the balljoint. Next are the two swivel bushes, one at each end of the center track rod. Finally check the steering box Pitman arm balljoint and the one on the idler arm which supports the center track-rod on the side opposite the steering box. This unit is bolted to the side of the frame member and any play calls for renewal of the bushes.
5 To check the steering box, first make sure that the bolts holding the steering box to the side-frame member are tight. Then get another person to help examine the mechanism. One should look at, or get hold of, the Pitman arm at the bottom of the steering box while the other turns the steering wheel a little way from side-to-side. The amount of lost motion between the steering wheel and the Pitman arm indicates the degree of wear somewhere in the steering box mechanism. This check should be carried out with the wheels first of all in the straight-ahead position and then at nearly full lock on each side. If the play only occurs noticeably in the straight-ahead position then the wear is most probably in the worm and/or nut. If it occurs at all positions of the steering, then the wear is probably in the sector shaft bearing. Oil leaks from the unit are another indication of such wear. In either case the steering box will need removal for closer examination and repair.
6 The balljoints on the two outer track-rods and the swivel bushes on the center track-rod are all installed into their respective locations by means of a taper pin in a tapered hole, and secured by a self-locking or castellated nut. In the case of the four balljoints (two on each of the outer track-rods), they are also screwed onto the rod and held by rod clamps.
7 To remove a taper pin first remove the self-locking nut. On occasion the taper pins have been known to simply pull out. More often they are well and truly wedged in position and a clamp or slotted steel wedges may be driven between the ball unit and the arm to which it is attached. Another method is to place the head of a hammer (or other solid metal article) on one side of the hole in the arm into which the pin is fitted. Then hit it smartly with a hammer on the opposite side. This has the effect of squeezing the taper out and usually works, provided one can get a good swing at it.
8 When the taper pin is free, grip the shank of the joint, loosen the rod clamp and screw the joint off the rod. It is important to count the number of turns needed to unscrew the balljoint so that, when reassembling, the same number of turns are used to ensure the correct tracking.
9 If the center track-rod bushes require that the track-rod be renewed then it will be necessary to detach the inner balljoints of the outer track-rods from it. The two swivel joints can then be removed from the Pitman arm and idler arm respectively and the unit removed.
10 Any play in the idler assembly can be remedied by renewal of the rubber bushes.
11 When any part of the steering linkage is renewed it is advisable to have the alignment of the steering checked at a garage with the proper equipment.

16 Steering wheel – removal and installation

1 Disconnect the battery ground cable
2 Working at the rear side of the steering wheel, loosen and remove the screws retaining the horn pad in position.
3 Lift the horn pad from the steering wheel and disconnect the horn pad electrical supply lead.
4 Undo and remove the central nut retaining the steering wheel hub to the column shaft.
5 Using a suitable scriber, mark a line across the center of the splined shaft and the steering wheel hub, to ensure the correct relationship on installation.
6 A special wheel extractor will have to be manufactured to withdraw the steering wheel. This can be carried out using three bolts and a suitable block of metal. Most home mechanics will have a block of metal with a bolt screwed through the center of it. All that is needed then is two bolts that will screw into the holes in the steering wheel hub. These two bolts should then be located into slots cut into the metal block. It is then quite simple to withdraw the steering wheel from the column. **Note:** *On no account strike the end of the steering column with a hammer as this will damage the collapsible shaft within the assembly.*
7 Installation of the steering wheel is a reversal of removal; ensure that the steering wheel is tightened to the specified torque.
8 After installation, check that the horn works and that the steering wheel operation is smooth.

17 Steering column – removal and installation

1 Loosen and remove the bolt securing the wormshaft and flexible coupling to the steering box.
2 Refer to Section 16, and remove the steering wheel.
3 Loosen and remove the screws retaining the two halves of the column shell together. Lift away the shells.
4 Refer to Chapter 10, and remove the turn signal switch assembly

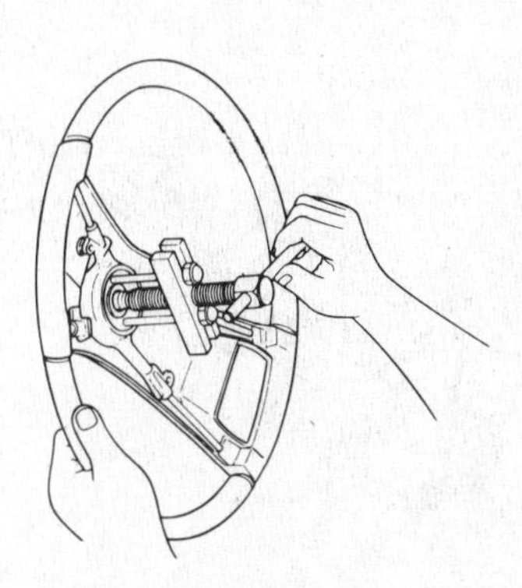

Fig. 11.32 Using an extractor to withdraw the steering wheel (Sec 16)

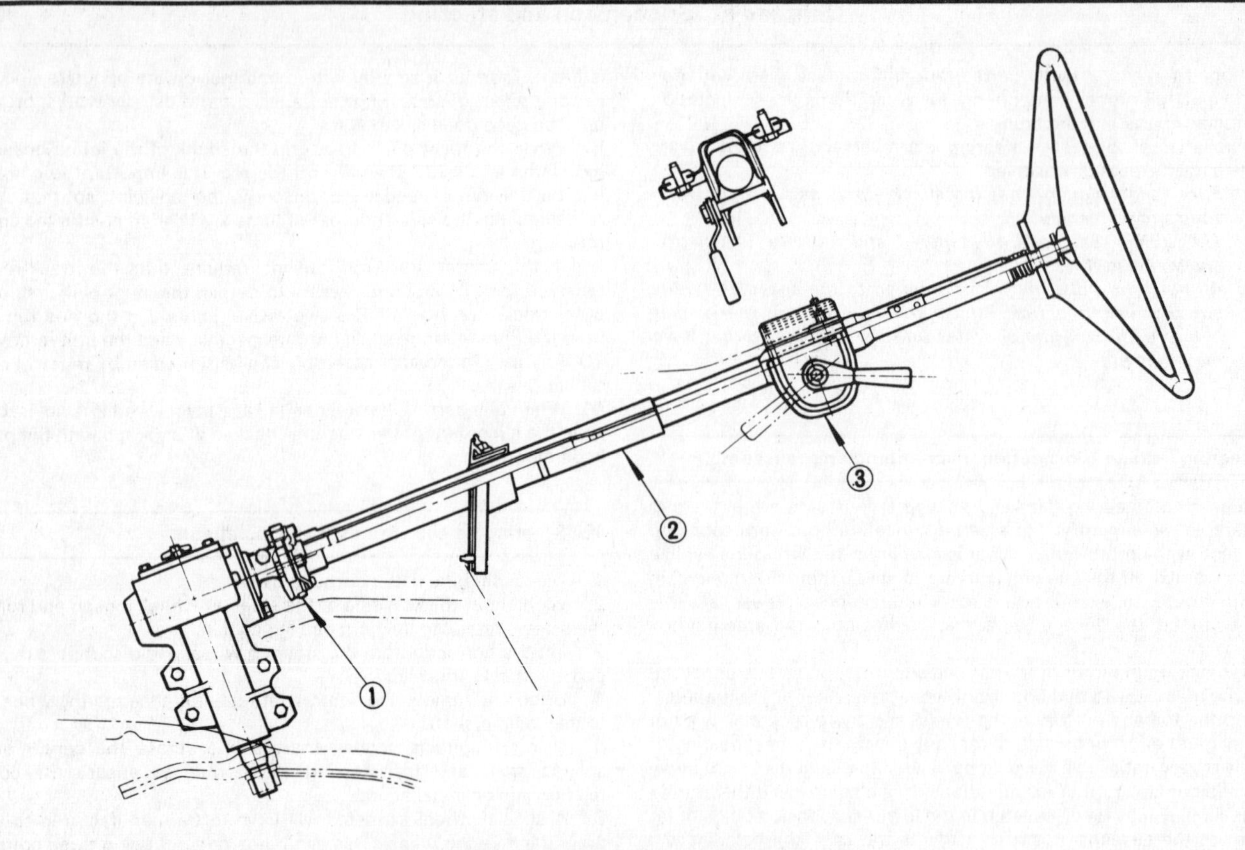

Fig. 11.33 The steering column assembly (Sec 17)

1 Flexible coupling *2 Column assembly*
3 Tilt mechanism

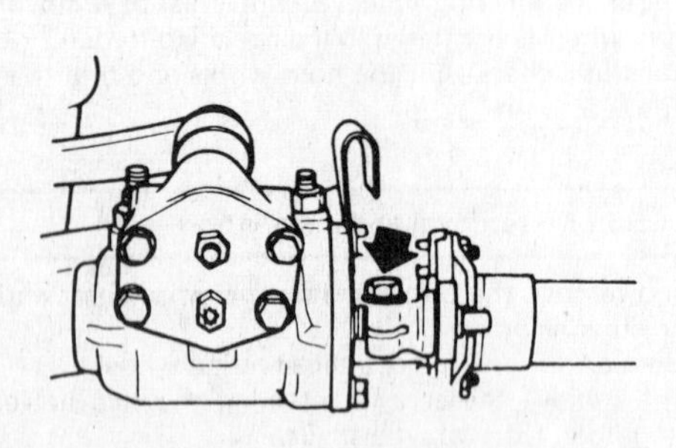

Fig. 11.34 The flexible coupling bolt arrowed (Sec 17)

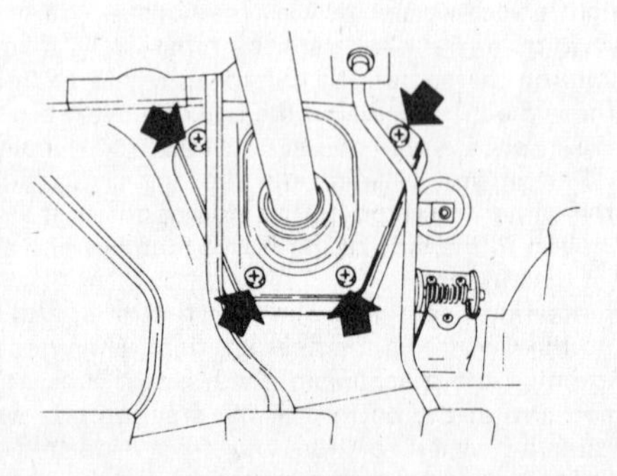

Fig. 11.35 The column flange bracket screws arrowed (Sec 17)

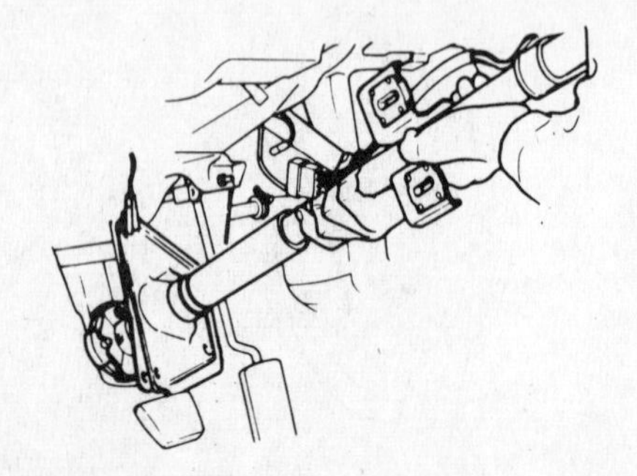

Fig. 11.36 Withdrawing the steering column assembly (Sec 17)

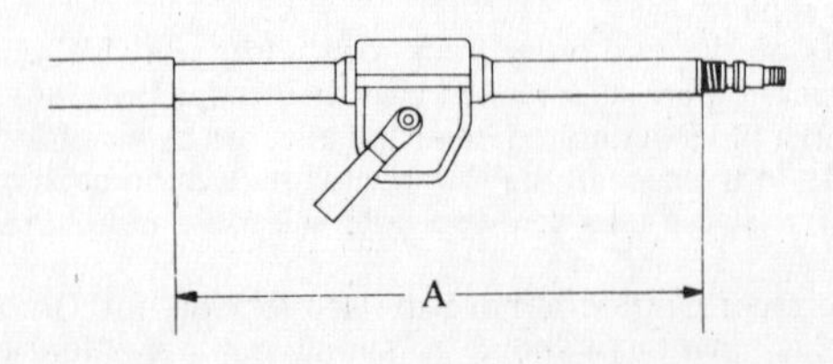

Fig. 11.37 Measuring dimension A (Sec 18)

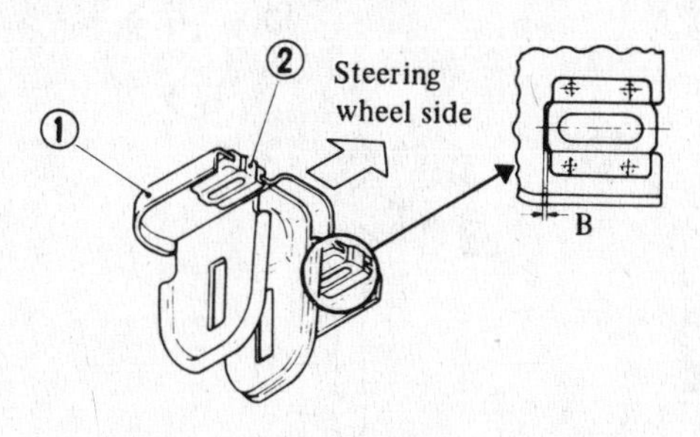

Fig. 11.38 Measuring dimension B (Sec 18)

1 Column clamp 2 Block

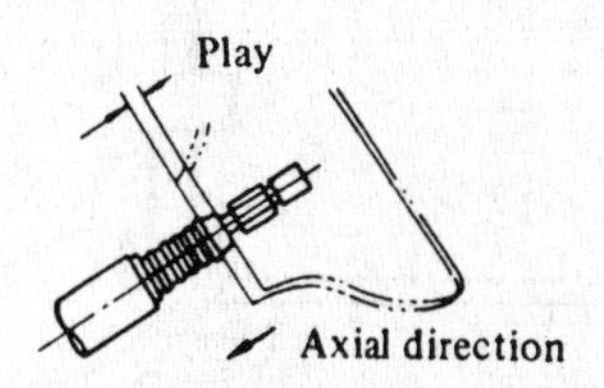

Fig. 11.39 Testing for axial play in the steering column (Sec 18)

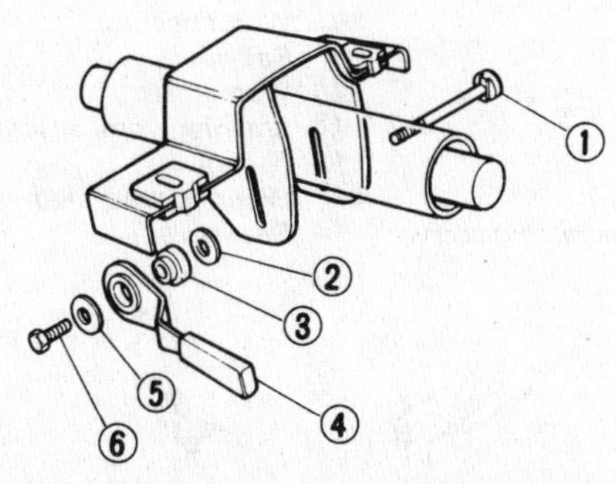

Fig. 11.40 Exploded view of the tilt mechanism (Sec 19)

1 Adjusting bolt
2 Washer
3 Adjusting lever nut
4 Locking lever
5 Washer
6 Bolt (left-hand thread)

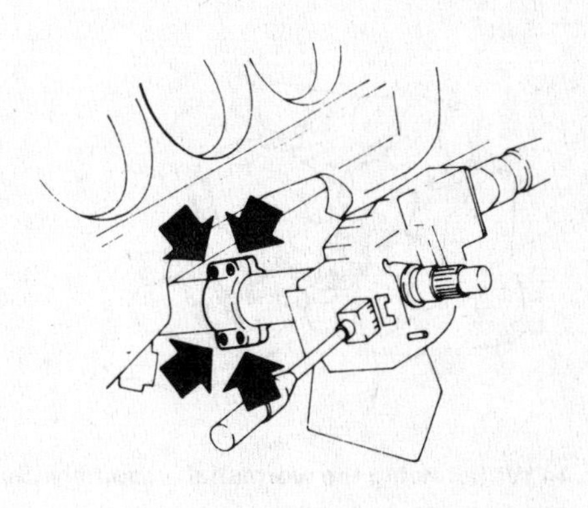

Fig. 11.41 The steering lock self-shear screws arrowed (Sec 20)

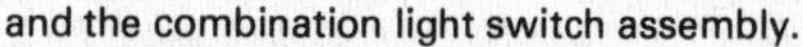

and the combination light switch assembly.

5 Working beneath the instrument panel, loosen and remove the four screws retaining the steering column flange to the engine compartment firewall.

6 Support the upper part of the steering column, then loosen and remove the bolts securing the column clamp/tilt mechanism to the pedal mounting bracket. The steering column assembly can then be withdrawn into the interior of the vehicle.

7 For servicing the steering column assembly, refer to Section 18.

8 Commence installation by ensuring that the front roadwheels are in the straight-ahead position.

9 Carefully slide the assembly into position, align the bolt hole through the wormshaft and flexible coupling, install the bolt but do not tighten at this stage.

10 To support the upper end of the assembly, locate the column clamp/tilt mechanism in position but do not tighten the bolts at this stage.

11 Align the holes in the column flange bracket with their counterparts in the firewall. Assemble the four retaining screws and tighten them to the specified torque.

12 Once the column flange bracket is secured in position, proceed to tighten the column clamp/tilt mechanism bolts followed by the flexible coupling bolt to the specified torque.

13 The remainder of the installation procedure is a reversal of removal.

18 Steering column – servicing

1 With the steering column removed from the vehicle as described in Section 17, proceed to check the assembly in the following manner.

2 If the steering wheel will not turn smoothly, and it is known that the steering box, linkage and suspension are in order, check the column bearings for damage or rough running. If greasing the bearings has no effect upon the rough running, the entire column assembly should be renewed.

3 If the vehicle has been involved in a light collision the following checks should be made:

Measure the distance A in Fig. 11.37, which should be 16.27 in (413.5 mm). If the dimension is less than stated, renew the column assembly complete.

Examine the block inside the column clamp bracket (Fig. 11.38); dimension B should be zero, and if any clearance exists here it indicates that the column has been crushed and should be renewed complete.

As a final check, grip the splined end of the steering column and try to move it in an axial direction. There should be no movement.

19 Tilt mechanism – removal and installation

1 The tilt mechanism is designed to allow the steering wheel to be moved up or down as required; the maximum movement at the steering wheel is 1.18 in (30.0 mm).

2 If, for any reason, the tilt mechanism has to be removed, this will involve removal of the steering column assembly, as described in Section 17.

3 The clamp-bolt assembly, however, can be dismantled with the mechanism in position. To do this, proceed as described in the following paragraphs.

4 Loosen and remove the bolt that retains the locking lever in position. Lift away the washer and the locking lever. Note that the locking lever retaining bolt has a left-hand thread.

5 Unscrew and remove the locking lever nut and remove the washer. The adjusting bolt can now be pushed out.

6 Reassembly and installation of the tilt mechanism is a reversal of removal.

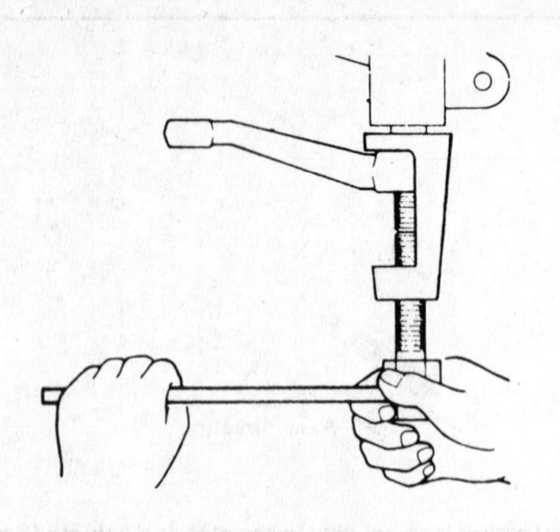

Fig. 11.42 Removing the Pitman arm from the sector shaft (Sec 21)

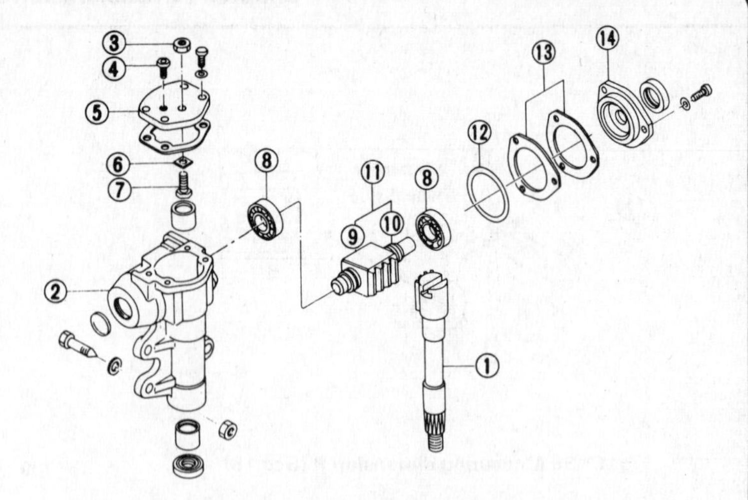

Fig. 11.43 Exploded view of the steering gear assembly (Sec 22)

1	*Sector shaft*	8	*Worm bearing*
2	*Housing*	9	*Ball nut*
3	*Locknut*	10	*Wormshaft*
4	*Filler plug*	11	*Steering worm assembly*
5	*Top cover*	12	*O-ring*
6	*Adjusting shim*	13	*Worm bearing shim*
7	*Sector shaft adjusting screw*	14	*Rear cover*

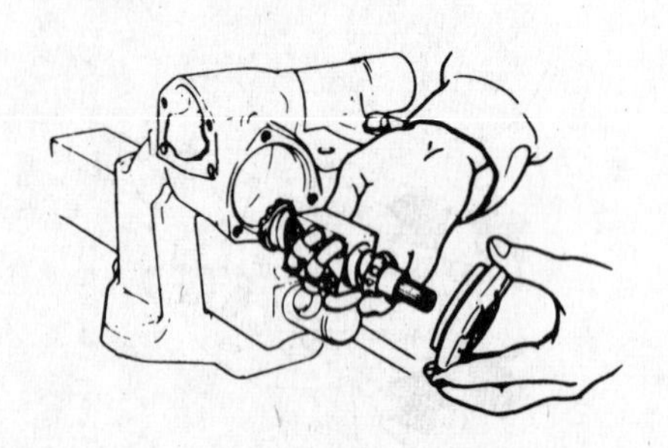

Fig. 11.44 Withdrawing the wormshaft assembly (Sec 22)

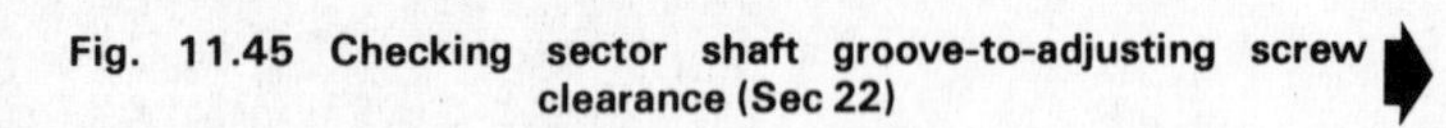

Fig. 11.45 Checking sector shaft groove-to-adjusting screw clearance (Sec 22)

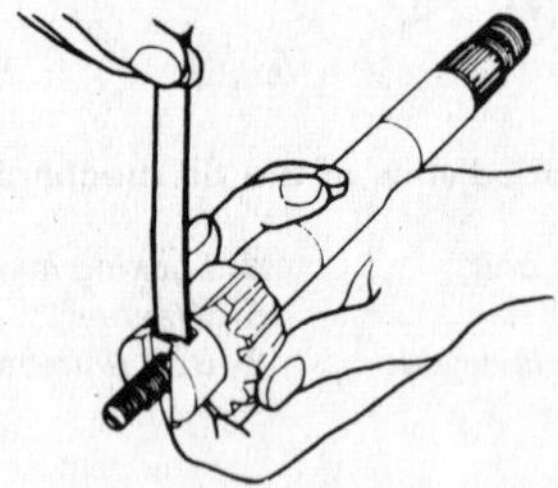

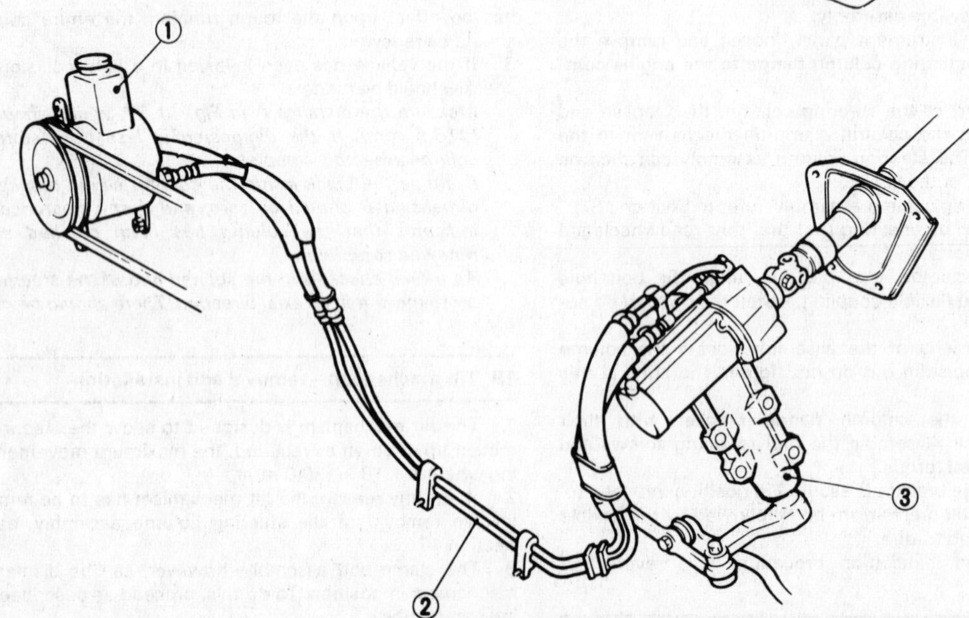

Fig. 11.46 The component parts of the power steering system (Sec 23)

1 Oil pump assembly *2 Hydraulic hoses*
3 Power steering gear assembly

20 Steering lock – removal and installation

1 In order to make the steering lock mechanism tamper-proof, self-shear type screws are used for installation.
2 To remove the shear screws will involve very careful drilling of the screws. Start by center punching a dot in the center of the screw. Select a drill that is about the diameter of the root of the thread. Very carefully, ensuring that the drill is square to the screw, drill out the center of the screw. This operation should be repeated at the other three screws.
3 Once the shear screws have been drilled out the lock mechanism can be removed.
4 To install the steering lock mechanism new shear type screws must first be purchased.
5 Align the steering lock hole in the jacket tube with the mating portion on the steering lock.
6 Install the screws just tight enough to testing the locking mechanism alignment and, if all is well, firmly tighten the screws so that their heads shear.

21 Steering gear – removal and installation

1 Set the wheels in the straight-ahead position, then loosen and remove the bolt securing the wormshaft to the flexible coupling at the steering box.
2 Working at the steering box, loosen and remove the nut securing the Pitman arm to the end of the sector shaft; mark the relationship of the Pitman arm to the sector shaft, then extract the Pitman arm.
3 Unscrew and remove the four bolts which secure the steering gear housing to the bodyframe sidemember. Withdraw the steering gear from the engine compartment.
4 Installation is a reversal of removal but align the Pitman arm to the sector shaft before installing the Pitman arm.
5 Tighten all bolts to the specified torque.
6 If the steering gear has been dismantled, refill the steering box with the correct grade and quantity of oil.

22 Steering gear – dismantling and reassembly

1 Remove the filler plug, invert the unit and drain out the oil.
2 Secure the steering box in a vise then loosen the locknut from the adjusting screw on the top cover and unscrew the adjuster two or three times.
3 Unscrew the top cover bolts, and remove the cover and the gasket.
4 To release the cover from the adjuster screw, turn the slot in the end of the screw clockwise. Withdraw the sector shaft.
5 Remove the rear cover (three bolts) followed by the bearing adjusting shim(s) and the worm assembly. Do not remove the nut from the worm, or allow the nut to run from end to end of the worm, or the ball guides will be damaged.
6 Inspect all components for wear or damage, and renew as necessary. If the sector shaft needle bearings are worn, do not remove them but renew the steering gear housing complete.
7 Always renew both oil seals at time of major overhaul.
8 Clean all components and lubricate them prior to reassembly.
9 Insert the worm assembly into the steering box.
10 Install the rear cover with O-ring and bearing shim(s), (thicker shims to steering box side). If new components have been used, the worm bearing pre-load must be checked and adjusted. To do this, attach a spring balance to a piece of cord wrapped around the splines of the worm pinion, and check the starting torque which should be 4 to 7 lbf in (4.0 to 8.0 kgf cm). Add or remove shims as necessary from the following shim thicknesses which are available:

0.0300 in (0.762 mm)
0.0100 in (0.254 mm)
0.0050 in (0.127 mm)
0.0020 in (0.050 mm)

11 Insert the adjusting screw into the T-shaped groove in the end of the sector shaft. Using a feeler blade, check the play between the bottom of the groove and the lower face of the adjusting screw. Use shims if necessary to provide an endplay of between 0.0004 and 0.0012 in (0.01 and 0.03 mm). Shims are available in the following thicknesses:

0.0620 to 0.0630 in (1.575 to 1.600 mm)
0.0610 to 0.0620 in (1.550 to 1.575 mm)
0.0600 to 0.0610 in (1.525 to 1.550 mm)
0.0591 to 0.0600 in (1.500 to 1.525 mm)
0.0581 to 0.0591 in (1.475 to 1.500 mm)
0.0571 to 0.0581 in (1.450 to 1.475 mm)

12 Turn the wormshaft by hand until the nut is in the center of its travel, then install the sector shaft complete with adjusting screw. Make sure that the center tooth of the sector shaft engages with the center groove of the nut. Take great care not to cut or damage the lips of the oil seals during these operations.
13 Install the top cover using sealant between the gasket faces. Turn the adjusting screw in a counter-clockwise direction (in order to install the cover to the steering box mating faces) by means of the screwdriver slot.
14 With the top cover bolts tightened to the correct torque, turn the adjusting screw clockwise with finger pressure only and then tighten the locknut temporarily.
15 Temporarily engage the Pitman arm onto the sector shaft splines and move the sector shaft over its complete arc of travel in both directions, to check for smooth operation.
16 Move the Pitman arm to the steering straight-ahead position (in alignment with the worm pinion) and after releasing the top cover adjusting screw locknut, turn the adjusting screw until any sector shaft endfloat just disappears. Turn the adjusting screw a further 1/6th of a turn and fully tighten the locknut.

23 Power steering – maintenance and adjustment

1 Normal maintenance of the power steering system consists mainly of periodically checking the fluid level in the reservoir, keeping the pump drivebelt tension correct and visually checking the hoses for any evidence of fluid leakage. It will also be necessary, after a system component has been removed, to bleed the system as described in Section 26.
2 If the operational characteristics of the system appear to be suspect, and the maintenance and adjustment mentioned in this Section is in order, the vehicle should be taken along to a Datsun dealer, who will have the necessary equipment to check the pressure in the system and the operational torque needed to turn the steering wheel. These two operations are considered beyond the scope of the home mechanic, in view of the high working pressure of the system and the special tools required.
3 If the checks mentioned in paragraph 2 prove that either the oil pump assembly or the steering gear assembly are at fault, a new pump or steering gear assembly will have to be purchased as it is not possible to overhaul them.
4 With regard to the oil pump drivebelt tension, this is adjusted by loosening the idler pulley locknut and turning the adjustment bolt as necessary. The correct tension exists when the belt deflects between

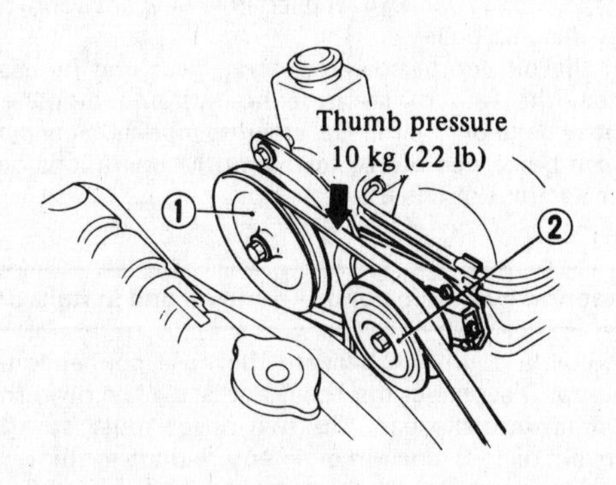

Fig. 11.47 The oil pump pulley (1) and the idler pulley (2) (Sec 23)

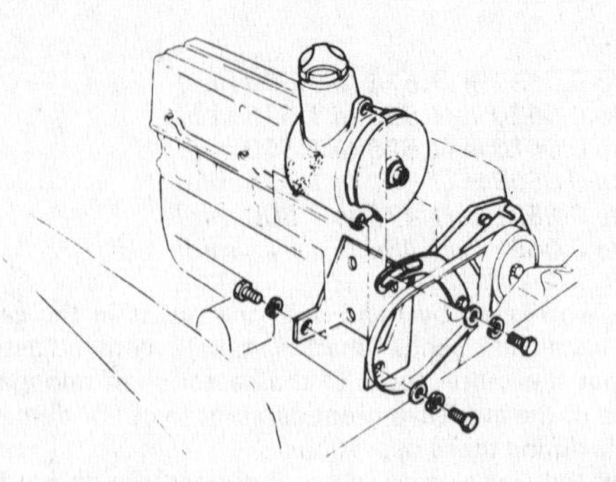

Fig. 11.48 Removing the power steering hydraulic pump (Sec 24)

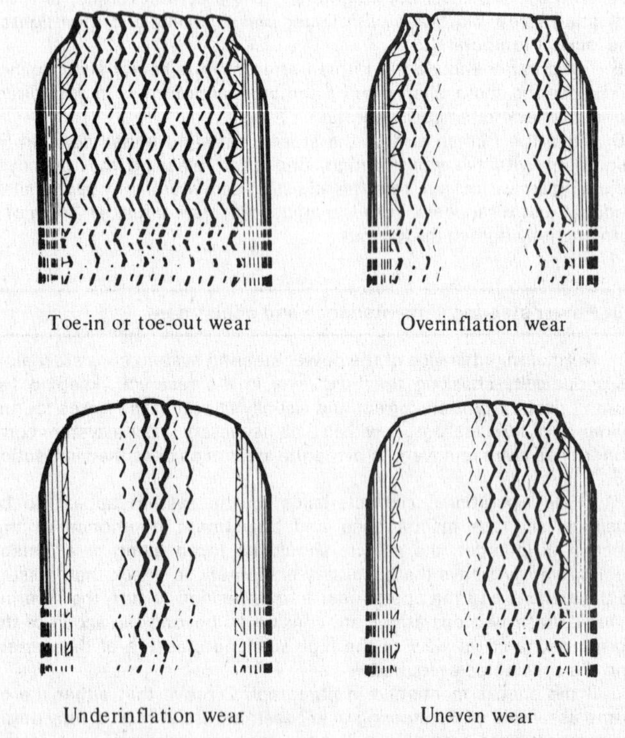

Fig. 11.50 Typical tire wear patterns (Sec 28)

0.31 and 0.47 in (8.0 and 12.0 mm), with a thumb pressure of approximately 22 lbf (10.0 kgf) applied midway between the idler pulley and the oil pump pulley.

5 Excluding the oil pump, power steering gear and its associated pressure hoses, the rest of the steering system components are identical to those used on the manual steering models. Servicing these components can be carried out by following the operations described in the relevant Sections of this Chapter.

24 Power steering hydraulic pump – removal and installation

1 Place a suitable drain pan beneath the hose connections at the back of the pump. Disconnect the hoses and allow the oil in the pump assembly to drain into the pan. The two hoses must be effectively blanked to prevent oil loss, and dirt or air entering the system.

2 Loosen and remove the adjustment locking bolt.

3 Working beneath the pump, slacken the two mounting bolts sufficiently to allow the pump to be pushed in towards the engine block. In this position the drivebelt can be removed from the pump assembly.

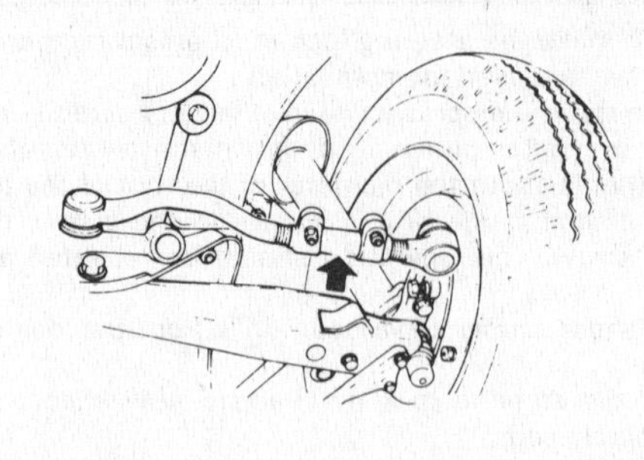

Fig. 11.49 The track-rod adjusting tube clamp arrowed (Sec 27)

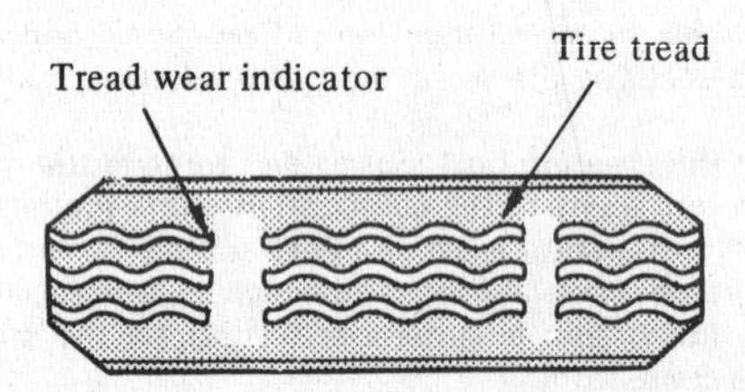

Fig. 11.51 The tread wear indicators (Sec 28)

4 Remove the two mounting bolts and lift away the pump.

5 If necessary, the drive pulley can be removed by unscrewing the retaining nut, and using a standard puller to withdraw the pulley from the shaft.

6 Installation of the oil pump assembly is a reversal of removal but before tightening the mounting bolts ensure that the correct drivebelt tension exists, as described in Section 23.

7 Once installation is complete, refill the pump reservoir with the correct grade and quantity of oil. If necessary, bleed the system as described in Section 26.

25 Power steering gear assembly – removal and installation

1 To aid reassembly, place the front wheels in the straight-ahead position. Loosen and remove the bolt securing the universal joint to the worm shaft at the steering box.

2 Place a suitable drain pan beneath the steering box hose connections, disconnect the hoses and allow the oil to drain from the hoses and the steering box into the pan. Plug the open ends of the hoses and the open ports at the steering box.

3 Unscrew the nut and remove the washer securing the Pitman arm to the sector shaft.

4 Scribe a line across the sector shaft and the Pitman arm to ensure correct installation.

5 Withdraw the Pitman arm from the sector shaft (see Fig. 11.42, Section 21).

6 Loosen and remove the bolts securing the steering box to the body sideframe and remove the unit from the vehicle.

7 Installation of the steering box assembly is a reversal of removal but observe the alignment marks on the sector shaft and Pitman arm. Screw in the universal joint-to-wormshaft bolt finger-tight. Tighten the steering box-to-sideframe bolts to the specified torque, then tighten the universal joint-to-wormshaft bolt to 24 to 28 lbf ft (3.3 to 3.9 kgf m).

8 When installation is complete, refill the oil pump reservoir with new oil of the correct grade and quantity. Bleed the air from the system, as described in Section 26.

26 Hydraulic system – bleeding

Whenever a hose in the power steering hydraulic system has been disconnected, it is quite probable, no matter how much care was taken to prevent air entering the system, that the system will need bleeding. To do this, proceed as described in the following paragraphs.

1 First, ensure that the reservoir level is correct; if necessary, add fluid to bring the level to the mark on the dipstick. If new fluid is added, it should be allowed to remain undisturbed in the reservoir for at least two minutes.

2 Raise the front end of the vehicle until the front wheels are just clear of the ground.

3 Quickly turn the steering wheel all the way to the right lock and then the left lock. Do not allow the lock stoppers to be struck with a bang. Try to gauge the end of the lock and only lightly touch the lock stoppers. This operation should be repeated several times.

4 Now check the reservoir fluid level again, as detailed in paragraph 1.

5 Start the engine and allow it to idle.

6 Repeat the operations described in paragraph 3. With the steering wheel in a full lock position, quickly open the bleed nipple at the oil pump and allow a little fluid to escape. This operation should be repeated several times on both locks, until the fluid escaping from the bleed nipple is free from air bubbles. The bleed nipple should only be opened for a few seconds and then shut again.

7 If the air bleeding is insufficient, the oil reservoir will be extremely foamy and the pump will be noisy. In this case, allow the foam in the reservoir to disperse, recheck the level again and repeat the entire bleeding process.

8 If it becomes obvious, after several attempts, that the system cannot be satisfactorily bled, there is quite probably a leak in the system. Visually check the hoses and their connections for leaks. If no leaks are evident, the problem could be in the steering box itself and the only solution is to have the entire system checked by a Datsun dealer.

27 Front wheel alignment and steering angles

1 Accurate front wheel alignment is essential for good steering and slow tire wear. Before considering the steering angle, check that the tires are correctly inflated, that the front wheels are not buckled, the hub bearings are not worn or incorrectly adjusted and that the steering linkage is in good order, without slackness or wear at the joints.

2 Wheel alignment consists of four factors:

Camber, is the right angle at which the front wheels are set from the vertical when viewed from the front of the car. Positive camber is the amount (in degrees) that the wheels are tilted outwards at the top from the vertical.

Castor is the right angle between the steering axis and a vertical line when viewed from each side of the car. Positive castor is when the steering axis is inclined rearward.

Steering axis inclination is the angle, when viewed from the front of the car, between the vertical and an imaginary line drawn between the upper and lower suspension leg pivots.

Toe-in is the amount by which the distance between the front inside edges of the road wheels (measured at hub height) is less than the diametrically opposite distance measured between the rear inside edges of the front road wheels.

3 Due to the need for special gap gauges and correct weighting of the car suspension it is not within the scope of the home mechanic to check steering angles. Indeed, the steering angles (except toe-in) are all set in production and cannot be altered. They should be checked by a service station, however, after any part of the steering or the front end of the car has been damaged in an accident. Front wheel tracking (toe-in) checks are best carried out with modern setting equipment but a reasonably accurate alternative and adjustment procedure may be carried out as described in the following paragraphs.

4 Place the car on level ground with the wheels in the straight-ahead position.

5 Obtain or make a toe-in gauge. One may be easily made from tubing, cranked to clear the oil pan and bellhousing, having an adjustable nut and setscrew at one end.

6 Using the gauge, measure the distance between the two inner wheel rims at hub height at the rear of the wheels.

7 Rotate the wheels (by pushing the car backwards or forwards) through 180° (half a turn) and again using the gauge, measure the distance of hub height between the two inner wheel rims at the front of the wheels. This measurement should be between 0 and 0·08 in (0 and 2·0 mm) less than that previously taken at the rear of the wheel and represents the correct toe-in.

8 Where the toe-in is found to be incorrect, slacken the adjusting tube clamp nuts on each outer track-rod and rotate the tube clamp an equal amount until the correct toe-in is obtained. Tighten the clamp nuts, ensuring that the balljoints are held in the center of their arc of travel during tightening.

28 Roadwheels and tires

1 Whenever the roadwheels are removed it is a good idea to clean the insides of the wheels to remove accumulations of mud and in the case of the front ones, disc pad dust.

2 Check the condition of the wheel for rust and repaint if necessary.

3 Examine the wheel stud holes. If these are tending to become elongated or the dished recesses in which the nuts seat have worn or become overcompressed, then the wheel will have to be renewed.

4 With a roadwheel removed, pick out any embedded flints from the tread and check for splits in the sidewalls or damage to the tire carcass generally.

5 Where the depth of tread pattern reveals the tread wear indicator, the tire must be renewed.

6 Rotation of the roadwheels to even-out wear is a worthwhile idea if the wheels have been balanced off the car. Include the spare wheel in the rotational pattern.

7 If the wheels have been balanced on the car then they cannot be moved round the car as the balance of wheel, tire and hub will be upset.

8 It is recommended that wheels are re-balanced halfway through the life of the tires to compensate for the loss of tread rubber due to wear.

9 Finally, always keep thc tires (including the spare) inflated to the recommended pressures and always replace the dust caps on the tire valve. Tire pressures are best checked first thing in the morning when the tires are cold.

29 Fault diagnosis – suspension and steering

Symptom	Reason/s
Steering feels vague, car wanders and floats at speed	Tire pressures uneven Shock absorbers worn Spring broken Steering gear balljoints badly worn Suspension geometry incorrect Steering mechanism play excessive Front suspension and rear axle pick-up points out of alignment
Stiff and heavy steering	Tire pressures too low

No grease in swivel joints
No grease in steering and suspension balljoints
Front wheel toe-in incorrect
Suspension geometry incorrect
Steering gear incorrectly adjusted too tightly
Steering column badly misaligned

Wheel wobble and vibration	Wheels nuts loose
	Front wheels and tires out of balance
	Steering balljoints badly worn
	Hub bearings badly worn
	Steering gear play excessive
	Front springs weak or broken

Chapter 12 Bodywork and fittings

Refer to Chapter 13 for specifications and information applicable to 1980 through 1984 models.

Contents

Specifications

Dimensions	**Sedan**	**Station Wagon**
Overall length	183.5 in (4660 mm)	185.6 in (4715 mm)
Overall width	64.8 in (1645 mm)	64.8 in (1645 mm)
Overall height	54.9 in (1395 mm)	56.1 in (1425 mm)
Wheelbase	104.3 in (2650 mm)	104.3 in (2650 mm)
Gross vehicle weight rating (GVWR):		
All models	3550 lb (1610 kg)	3680 lb (1669 kg)
Gross axle weight rating (GAWR):		
Front, all models	1795 lb (814 kg)	1770 lb (803 kg)
Rear, all models	1755 lb (796 kg)	1910 lb (866 kg)

1 General description

The body and underframe, on both the Sedan and the Station Wagon, are of unitary, all-welded steel construction. At the front end, the fenders, hood, cowl top grille, radiator grille, front apron and bumper are all detachable. At the rear, a tailgate is fitted to the Station Wagon, and a trunk lid to the Sedan. Both the tailgate and the trunk lid are opened with the assistance of torsion bar springs.

The front and rear bumpers are of the impact-absorbing type, being mounted on gas-filled telescopic shock absorbers.

A heating and ventilating system is fitted as standard equipment, and an air conditioning system can be specified as optional equipment.

2 Maintenance – bodywork and underframe

1 The condition of your car's bodywork is of considerable importance as it is on this that the secondhand value will mainly depend. It is much more difficult to repair neglected bodywork than to renew mechanical assemblies. The hidden portions of the body, such as the fender arches, the underframe and the engine compartment are equally important, although obviously not requiring such frequent attention as the immediately visible paintwork.

2 Once a year or every 12 000 miles (20 000 km) it is a sound scheme to visit your local dealer and have the underside of the body steam-cleaned. All traces of dirt and oil will be removed and the underside can then be inspected carefully for rust, damaged hydraulic pipes, frayed electrical wiring and similar maladies. The front suspension should be greased on completion of this job.

3 At the same time, clean the engine and the engine compartment either using a steam-cleaner or a water-soluble cleaner.

4 The fender arches should be given particular attention as undersealing can easily come away here, and stones and dirt thrown up from the roadwheels can soon cause the paint to chip and flake, and so allow rust to set in. If rust is found, clean down to the bare metal and apply an anti-rust paint.

5 The bodywork should be washed once a week or when dirty. Thoroughly wet the car to soften the dirt and then wash down with a soft sponge and plenty of clean water. If the surplus dirt is not washed off very gently, in time it will wear down paint.

6 Spots of tar or bitumen coating thrown up from the road surfaces are best removed with a cloth soaked in gasoline.

7 Once every six months, give the bodywork and chromium trim a thoroughly good wax polish. If a chromium cleaner is used to remove rust on any of the car's plated parts remember that the cleaner also removes part of the chromium, so use it sparingly.

3 Maintenance – upholstery and carpets

1 Mats and carpets should be brushed or vacuum-cleaned regularly to keep them free of grit. If they are badly stained, remove them from the car for scrubbing or sponging and make quite sure they are dry before they are put back.
2 Seats and interior trim panels can be kept clean by a wipe over with a damp cloth. If they do become stained (which can be more apparent on light colored upholstery) use a little liquid detergent and a soft brush to scour the grime out of the grain of the material. Do not forget to keep headlining clean in the same way as the upholstery.
3 When using liquid cleaners inside the car do not over-wet the surfaces being cleaned. Excessive dampness could get into the seams and padded interior causing stains, offensive odours or even rot. If the inside of the car gets wet accidentally, it is worthwhile taking some trouble to dry it out properly particularly where carpets are involved. *Do not leave electric heaters inside the car for this purpose.*

4 Bodywork repairs – minor damage

See photo sequences on pages 174 and 175

Repair of minor scratches in the car's bodywork

If the scratch is very superficial, and does not penetrate to the metal of the bodywork, repair is very simple. Lightly rub the area of the scratch with a paintwork renovator, or a very fine cutting paste, to remove loose paint from the scratch and to clear the surrounding bodywork of wax polish. Rinse the area with clean water.

Apply touch-up paint to the scratch using a thin paint brush; continue to apply thin layers of paint until the surface of the paint in the scratch is level with the surrounding paintwork. Allow the new paint at least two weeks to harden, then blend it into the surrounding paintwork by rubbing the paintwork, in the scratch area, with a paintwork renovator or a very fine cutting paste. Finally, apply wax polish.

An alternative to painting over the scratch is to use a paint transfer. Use the same preparation for the affected area, then simply pick a patch of a suitable size to cover the scratch completely. Hold the patch against the scratch and burnish its backing paper; the patch will adhere to the paintwork, freeing itself from the backing paper at the same time. Polish the affected area to blend the patch into the surrounding paintwork.

Where the scratch has penetrated right through to the metal of the bodywork, causing the metal to rust, a different repair technique is required. Remove any loose rust from the bottom of the scratch with a pocket knife, then apply rust inhibiting paint to prevent the formation of rust in the future. Using a rubber or nylon applicator fill the scratch with bodystopper paste. If required, this paste can be mixed with cellulose thinners to provide a very thin paste which is ideal for filling narrow scratches. Before the stopper paste in the scratch hardens, wrap a piece of smooth cotton rag around the top of a finger. Dip the finger in cellulose thinners and then quickly sweep it across the surface of the stopper paste in the scratch; this will ensure that the surface of the stopper paste is slightly hollowed. The scratch can now be painted over as described earlier in this Section.

Repair of dents in the car's bodywork

When deep denting of the car's bodywork has taken place, the first task is to pull the dent out, until the affected bodywork almost attains its original shape. There is little point in trying to restore the original shape completely, as the metal in the damaged area will have stretched on impact and cannot be reshaped fully to its original contour. It is better to bring the level of the dent up to a point which is about $\frac{1}{8}$ inch below the level of the surrounding bodywork. In cases where the dent is very shallow anyway, it is not worth trying to pull it out at all.

If the underside of the dent is accessible, it can be hammered out gently from behind, using a mallet with a wooden or plastic head. Whilst doing this, hold a suitable block of wood firmly against the impact from the mallet blows and thus prevent a large area of the bodywork from being 'belled-out'.

Should the dent be in a section of the bodywork which has double skin or some other factor making it inaccessible from behind, a different technique is called for. Drill several small holes through the metal inside the dent area – particularly in the deeper sections. Then screw long self-tapping screws into the holes just sufficiently for them to gain a good 'key' in the metal. Now the dent can be pulled out by pulling on the protruding heads of the screws with a pair of pliers.

The next stage of the repair is the removal of the paint from the damaged area, and from an inch or so of the surrounding sound bodywork. This is accomplished most easily by using a wire brush or abrasive pad on a power drill, although it can be done just as effectively by hand using sheets of abrasive paper. To complete the preparation for filling, score the surface of the bare metal with a screwdriver or the tang of a file, or alternatively, drill small holes in the affected area. This will provide a really good 'key' for the filler paste.

To complete the repair see the Section on filling and respraying.

Repair of rust holes or gashes in the car's bodywork

Remove all paint from the affected area and from an inch or so of the surrounding sound bodywork, using an abrasive pad or a wire brush on a power drill. If these are not available a few sheets of abrasive paper will do the job just as effectively. With the paint removed you will be able to judge the severity of the corrosion and therefore decide whether to renew the whole panel (if this is possible) or to repair the affected area. Replacement body panels are not as expensive as most people think and it is often quicker and more satisfactory to fit a new panel than to attempt to repair large areas of corrosion.

Remove all fittings from the affected area except those which will act as a guide to the original shape of the damaged bodywork (eg headlamp shells etc). Then, using tin snips or a hacksaw blade, remove all loose metal and any other metal badly affected by corrosion. Hammer the edges of the hole inwards in order to create a slight depression for the filler paste.

Wire brush the affected area to remove the powdery rust from the surface of the remaining metal. Paint the affected area with rust inhibiting paint; if the back of the rusted area is accessible treat this also.

Before filling can take place it will be necessary to block the hole in some way. This can be achieved by the use of one of the following materials: *Zinc gauze, Aluminium tape or Polyurethane foam.*

Zinc gauze is probably the best material to use for a large hole. Cut a piece to the approximate size and shape of the hole to be filled, then position it in the hole so that its edges are below the level of the surrounding bodywork. It can be retained in position by several blobs of filler paste around its periphery.

Aluminium tape should be used for small or very narrow holes. Pull a piece off the roll and trim it to the approximate size and shape required, then pull off the backing paper (if used) and stick the tape over the hole; it can be overlapped if the thickness of one piece is insufficient. Burnish down the edges of the tape with the handle of a screwdriver or similar, to ensure that the tape is securely attached to the metal underneath.

Polyurethane foam is best used where the hole is situated in a section of bodywork of complex shape, backed by a small box section (eg. where the sill panel meets the rear fender arch – most cars). The usual mixing procedure for this foam is to put equal amounts of fluid from eachof the two cans provided in the kit, into one container. Stir until the mixture begins to thicken, then quickly pour this mixture into the hole, and hold a piece of cardboard over the larger apertures. Almost immediately the polyurethane will begin to expand, gushing out of any small holes left unblocked. When the foam hardens it can be cut back to just below the level of the surrounding bodywork with a hacksaw blade.

Having blocked off the hole the affected area must now be filled and sprayed – see Section on bodywork filling and respraying

Bodywork repairs – filling and respraying

Before using this Section, see the Sections on dent, deep scratch, rust hole and gash repairs.

Many types of bodyfiller are available, but generally speaking those proprietary kits which contain a tin of filler paste and a tube of resin hardener are best for this type of repair. A wide, flexible plastic or nylon applicator will be found invaluable for imparting a smooth and well contoured finish to the surface of the filler.

Mix up a little filler on a clean piece of card or board – use the hardener sparingly (follow the maker's instructions on the pack) otherwise the filler will set very rapidly.

Using the applicator, apply the filler paste to the prepared area; draw the applicator across the surface of the filler to achieve the correct contour and to level the filler surface. As soon as a contour that approximates the correct one is achieved, stop working the paste – if you carry on too long the paste will become sticky and begin to 'pick-up' on the applicator.

Continue to add thin layers of filler paste at twenty-minute intervals until the level of the filler is just proud of the surrounding bodywork.

Once the filler has hardened, excess can be removed using a Surform plane or Dreadnought file. From then on, progressively finer grades of abrasive paper should be used, starting with a 40 grade 'wet and dry' paper. Always wrap the abrasive paper around a flat rubber, cork, or wooden block – otherwise the surface of the filler will not be completely flat. During the smoothing of the filler surface the 'wet-and-dry' paper should be periodically rinsed in water – this will ensure that a very smooth finish is imparted to the filler at the final stage.

At this stage the repair area should be surrounded by a ring of bare metal, which in turn should be encircled by the finely 'feathered' edge of the good paintwork. Rinse the repair area with clean water, until all of the dust produced by the rubbing-down operation is gone.

Spray the whole repair area with a light coat of primer; this will show up any imperfections in the surface of the filler. Repair these imperfections with fresh filler paste or bodystopper, and once more smooth the surface with abrasive paper. If bodystopper is used, it can be mixed with cellulose thinners to form a really thin paste which is ideal for filling small holes. Repeat this spray-and-repair procedure until you are satisfied that the surface of the filler and the feathered edge of the paintwork are perfect. Clean the repair area with clean water and allow to dry fully.

The repair area is now ready for final spraying. Paint spraying must be carried out in a warm, dry, windless and dust-free atmosphere. This condition can be created artificially if you have access to a large indoor working area, but if you are forced to work in the open, you will have to pick your day very carefully. If you are working indoors, dousing the floor in the work area with water will 'lay' the dust which would otherwise be in the atmosphere. If the repair area is confined to one body panel, mask off the surrounding panels; this will help to minimise the effects of a slight mis-match in paint colours. Bodywork fittings (eg. chrome strips, door handles etc) will also need to be removed or masked off. Use genuine masking tape and several thicknesses of newspaper for the masking operations.

Before commencing to spray, agitate the aerosol can thoroughly, then spray a test area (an old tin, or similar) until the technique is mastered. Cover the repair area with a thick coat of primer; the thickness should be built up using several thin layers of paint rather than one thick one. Using 400 grade 'wet-and-dry' paper, rub down the surface of the primer until it is really smooth. While doing this, the work area should be thoroughly doused with water, and the 'wet-and-dry' paper periodically rinsed in water. Allow to dry before spraying on more paint.

Spray on the top coat, again building up the thickness by using several thin layers of paint. Start spraying in the centre of the repair area and then using a circular motion, work outwards until the whole repair area and about 2 inches of the surrounding original paintwork is covered. Remove all masking material 10 to 15 minutes after spraying on the final coat of paint. Allow the new paint at least two weeks to harden fully; then, using a paintwork renovator or a very fine cutting paste, blend the edges of the new paint into the existing paintwork. Finally, apply wax polish.

5 Bodywork repairs – major damage

Where serious damage has occurred or large areas need renewal due to neglect, it means certainly that completely new sections or panels will need welding in and this is best left to professionals. If the damage is due to impact it will also be necessary to completely check the alignment of the bodyshell structure. Due to the principle of construction the strength and shape of the whole can be affected by damage to a part. In such instances the services of a Datsun agent with specialist checking jigs are essential. If a body is left misaligned it is first of all dangerous as the car will not handle properly and secondly uneven stresses will be imposed on the steering, engine and transmission, causing abnormal wear or complete failure. Tire wear may also be excessive.

6 Maintenance – hinges and locks

1 Oil the hinges of the hood, trunk and doors with a drop or two of light oil periodically. A good time is after the car has been washed.
2 Oil the hood release catch pivot pin and the safety catch pivot pin periodically.
3 Do not over-lubricate door latches and strikers. Normally a little oil on the rotary cam spindle alone is sufficient.

7 Doors – rattles and their rectification

1 Check first that the door is not loose at the hinges and that the latch is holding the door firmly in position. Check also that the door aligns with the aperture in the body.
2 If the hinges are loose or the door is out of alignment it will be necessary to reset the hinge positions, as described in Section 19.
3 If the latch is holding the door properly it should hold the door tightly when fully latched and the door should align with the body. If it is out of alignment it needs adjustment as described in Section 22. If loose, some part of the lock mechanism must be worn out and requires renewal.
4 Other rattles from the door would be caused by wear or looseness in the window glass regulator, the glass channels and sill strips, or the

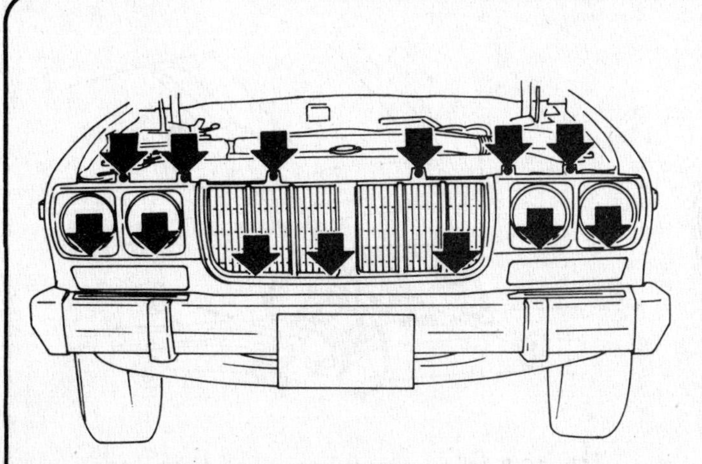

Fig. 12.1 The front grille and headlamp finisher attachment points (Sec 8)

Fig. 12.2 The front apron attachment points (Sec 9)

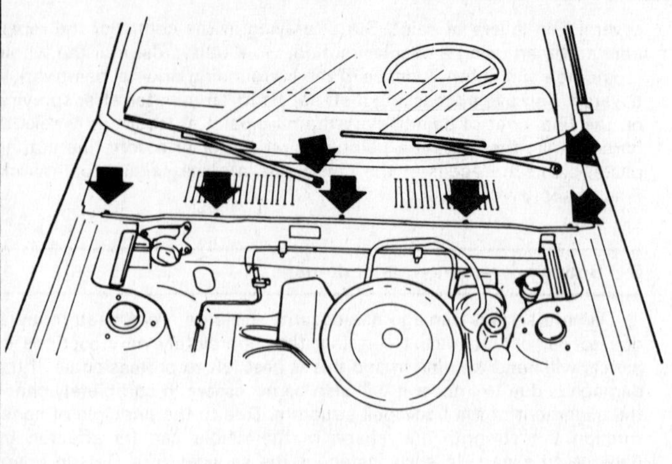

Fig. 12.3 The cowl top grille attachment points (Sec 10)

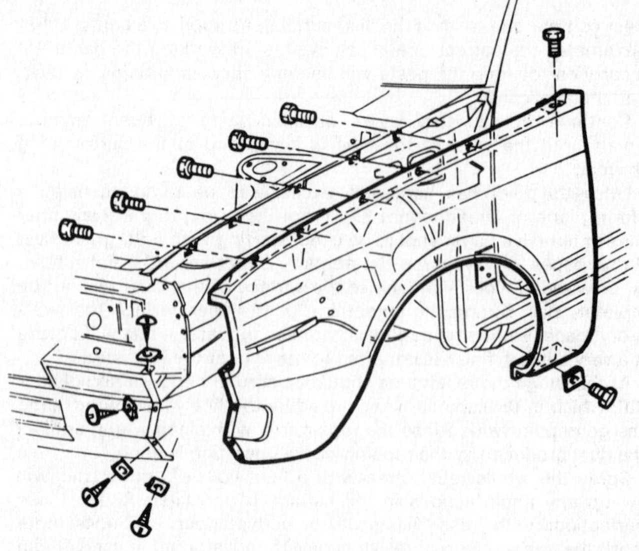

Fig. 12.4 The front fender attachment points (Sec 11)

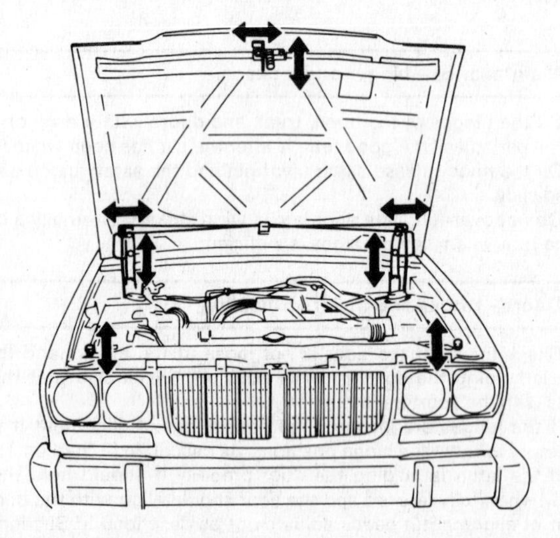

Fig. 12.5 The hood adjustment directions arrowed (Sec 12)

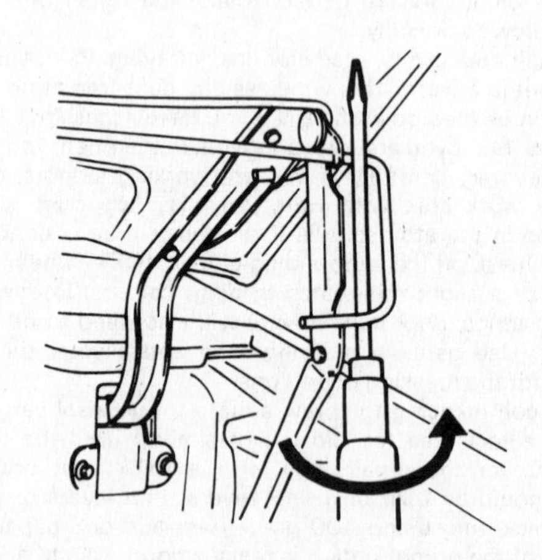

Fig. 12.6 Removing the hood torsion bar (Sec 12)

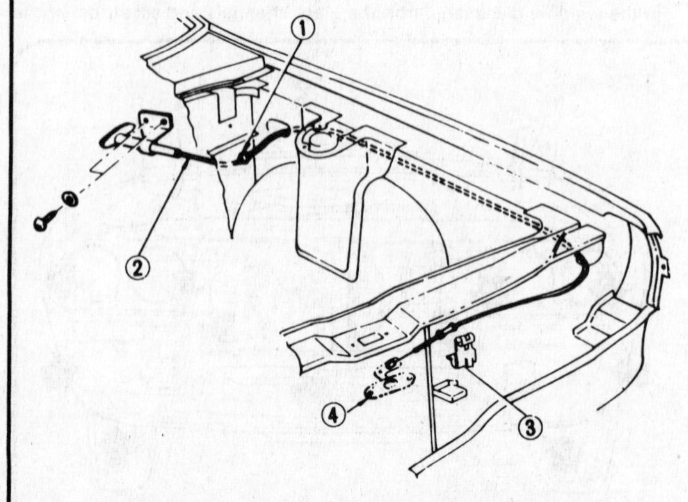

Fig. 12.7 The hood lock and control cable (Sec 13)

1 Grommet
2 Hood lock control cable
3 Cable clamp
4 Hood lock

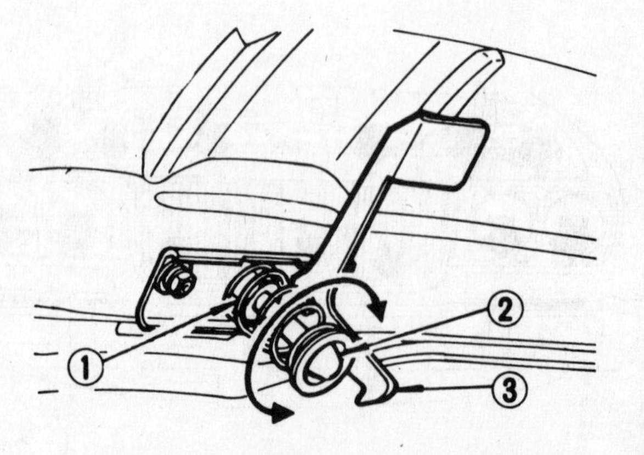

Fig. 12.8 Adjusting the hood lock plunger (Sec 13)

1 Locknut
2 Lock plunger
3 Safety catch

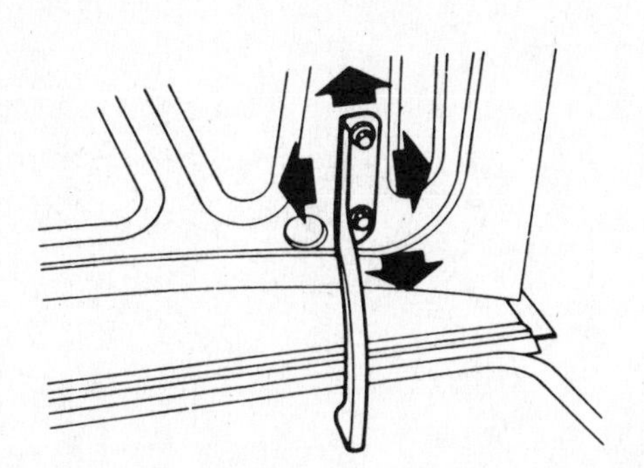

Fig. 12.9 Adjustment directions of the trunk hinge (Sec 14)

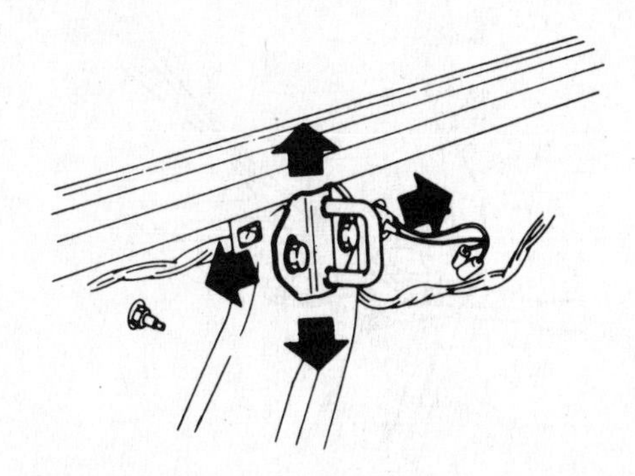

Fig. 12.10 Adjustment directions of the trunk lock striker (Sec 14)

door buttons and interior latch release mechanism.

8 Radiator grille and headlamp finishers – removal and installation

1 Open the engine compartment hood.
2 Loosen and remove the thirteen screws that retain the grille and headlamp finishers in position. The grille can now be lifted away from the front of the car.
3 Installation of the radiator grille and headlamp finishers is a reversal of removal.

9 Front apron – removal and installation

1 Refer to Section 8, and remove the radiator grille and headlamp finishers.
2 Refer to Section 25, and remove the front bumper.
3 Remove the headlamps and front parking combination units as described in the relevant Sections of Chapter 10.
4 Eleven screws attach the front apron to the car and removal of these will enable the apron to be removed.
5 Installation of the front apron is a reversal of removal.

10 Cowl top grille – removal and installation

1 To remove the cowl top grille, first refer to Chapter 10, Section 34, and remove the wiper arms.
2 Remove the five screws that secure the grille in position, then lift the grille from the car. **Note**: *Great care must be taken, when removing the grille, to avoid scratching the car's paintwork.*
3 Installation of the cowl top grille is a reversal of removal.

11 Front fender – removal and installation

1 Disconnect the battery ground cable.
2 Remove the cowl top grille, as described in Section 10.
3 Remove the front bumper (Section 25).
4 Remove the radiator grille and headlamp finishers (Section 8).
5 Loosen and remove the five bolts securing the fender to the hood ledge.
6 Loosen and remove the four screws securing the fender to the front apron.
7 Loosen and remove the two bolts that secure the rear of the fender to the sill inner panel (at the bottom of the fender), and the top of the fender at the cowl top grille. The fender can now be removed from the vehicle, but great care must be taken to ensure that no damage occurs to the paintwork of the fender (if undamaged) and the surrounding areas.
8 Installation of a front fender is a reversal of removal, but take great care not to damage the paintwork of the fender and surrounding areas. When removing and installing the right-hand fender, take care to ensure that the windshield wiper tube is not damaged.

12 Hood – removal, installation and adjustment

These operations call for the help of an assistant if the bodywork is not to be damaged

1 Raise the hood and cover the tops of the fenders with suitable protection to prevent damage.
2 Mark round the hinge plates which are bolted to the hood lid.
3 Support the hood, and unscrew and remove the bolts which secure the hinge plates to the hood.
4 Very carefully lift the hood away.
5 If necessary, the hood counterbalance torsion bar can be removed after disengaging the ends from the hinges using a screwdriver to lever them out.
6 Installation of the hood is a reversal of removal, but the hood must be aligned within its aperture so that there is an even gap all round when it is closed. This is achieved by sliding the hinges on their bolts (bolts screwed only finger-tight) in the appropriate direction.
7 Closure height of the front opening edge can be adjusted by releasing the locknuts on the bump stops and screwing them in or out.
8 The hood lock male component should be adjusted to align correctly with the female component, and in conjunction with the bump stops it should be adjusted to give positive hood closure without any rattle. This is carried out by altering the length of the dovetail bolt once the locknut has been loosened. Make sure that any adjustment has not affected the positive engagement of the safety catch.

13 Hood lock and control cable – removal and installation

1 Refer to Section 8 and remove the radiator grille and headlamp finishers.
2 Undo and remove the two bolts and spring washers securing the lock to the front panel.
3 Disconnect the cable from the lock arm.
4 Undo and remove the two bolts and washers securing the lock to the dashboard side trim.
5 Release the outer cable from the support clips and pull the cable into the passenger compartment. Take care as it passes through the bulkhead grommet.
6 The lock plunger and safety catch may be removed from the underside of the hood by undoing and removing the two securing bolts, spring and plain washers.
7 Installation of the lock is the reverse sequence to removal. It may be necessary to adjust the lock. Before tightening the lock securing bolts line it up with the plunger. Do not lock the hood at this stage as it could be difficult to open again! Tighten the lock securing bolts.
8 To adjust the plunger, slacken the locknut and using a screwdriver in the end of the plunger turn in the required direction. Lock by tightening the locknut.
9 Lubricate the lock and plunger with a little grease or engine oil.

14 Trunk lid (Sedan) – removal and installation

1 Open the trunk lid and, using a soft pencil, mark the outline of the

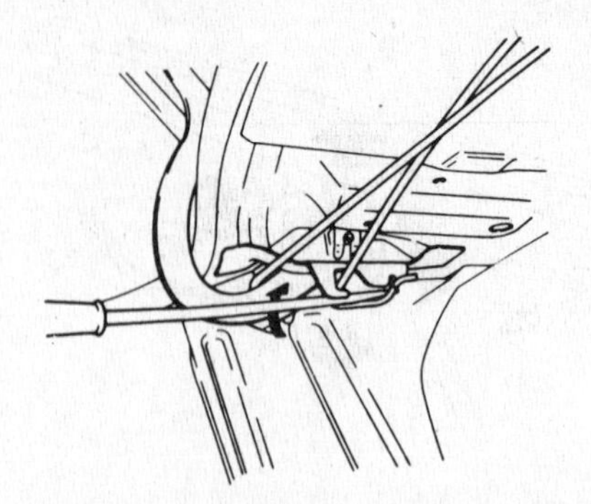

Fig. 12.11 Removing the trunk lid torsion bar (Sec 15)

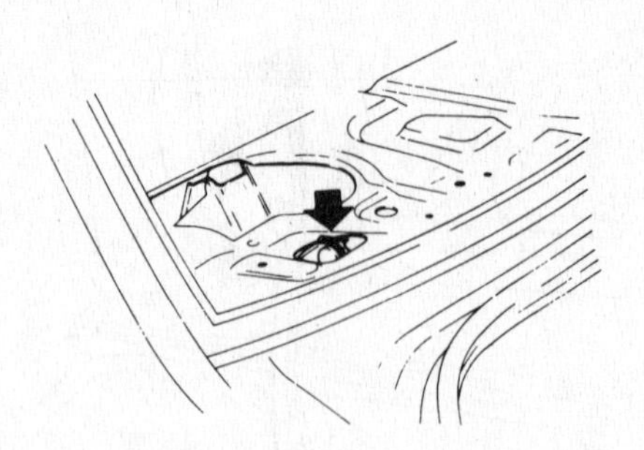

Fig. 12.12 Location of a hinge clamp at the rear parcel shelf (Sec 16)

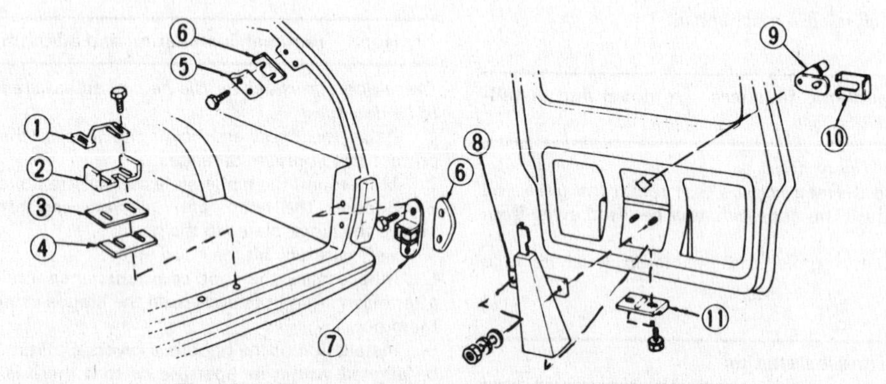

Fig. 12.13 The tailgate lock and striker assembly (Sec 17)

1 Striker
2 Bracket
3 Friction plate
4 Shim
5 Bump rubber
6 Shim
7 Down position stopper
8 Tailgate lock
9 Lock cylinder
10 Clip
11 Tailgate striker catch

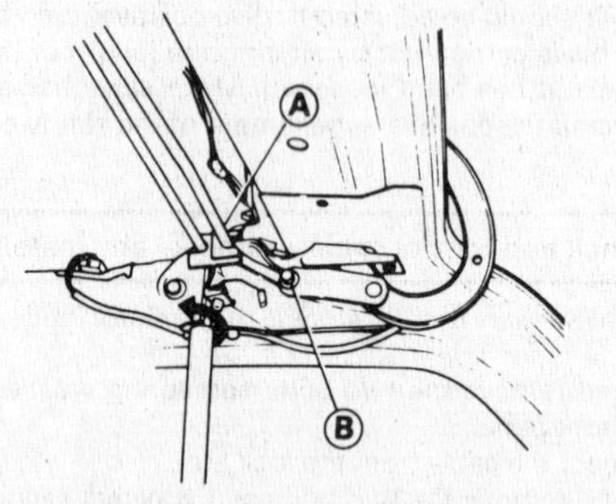

Fig. 12.14 Removing a torsion bar (Sec 18)

A and B are grease points

Fig. 12.15 Adjusting the position of a front door hinge (Sec 19)

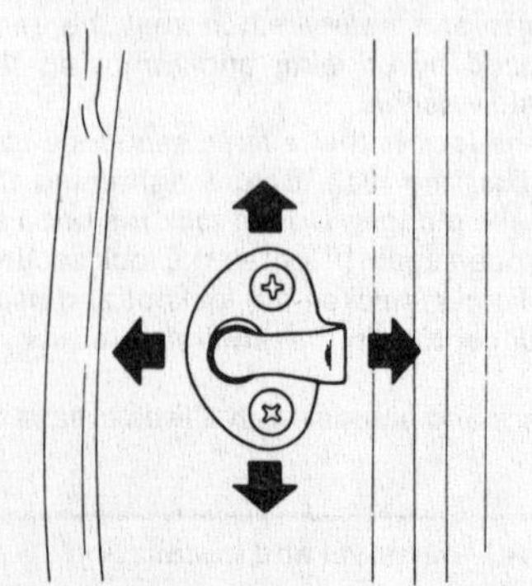

Fig. 12.16 The door lock striker adjustment directions (Sec 19)

hinges on the lid to act as a datum for installing.

2 With the assistance of a second person hold the trunk lid in the open position and then remove the two bolts, spring and plain washers to each hinge.

3 Lift away the trunk lid.

4 Refitting the trunk lid is the reverse sequence to removal. If necessary, adjust the position of the hinges relative to the lid until the lid is centralised in the aperture.

5 To obtain a watertight fit between the trunk lid and weatherstrip, move the striker up-and-down or side-to-side as necessary.

15 Trunk lid torsion bar (Sedan) – removal and installation

1 Open the trunk lid and support it with a suitable piece of wood.

2 Using a screwdriver, lever the torsion bar from the torsion bar

bracket at each hinge location. Care should be taken during this operation as the spring action of the torsion bar, when released from the brackets, will be very strong.
3 Installation is a reversal of removal. To obtain the correct tension on the torsion bar, a series of notches is provided on the bracket assembly.

16 Trunk lid hinge (Sedan) – removal and installation

1 Remove the torsion bars as described in Section 15.
2 Remove the trunk lid as described in Section 14.
3 Remove the rear parcel shelf trim.
4 Pull out the hinge clamps from both hinges and remove them from the car.
5 Installation is a reversal of removal, but ensure that correct alignment of the trunk lid is obtained, as described in Section 14.

17 Tailgate (Station Wagon) – removal and installation

1 Open the tailgate and support it in the open position with a piece of wood. The assistance of a second person is required to hold the tailgate as it is detached from the hinges.
2 Using a soft pencil, mark the outline of the hinge on the tailgate to act as a datum for installing.
3 Undo and remove the bolts and washers securing each hinge to the tailgate and carefully lift the tailgate away.
4 Installing the tailgate is the reverse sequence to removal,
5 Should it be necessary to adjust the position of the tailgate in the aperture, it may be moved up-or-down and side-to-side at the tailgate-to-hinge securing bolts. The fore-and-aft movement adjustment is obtained by slackening the bolts securing the tailgate hinge to the body.

18 Tailgate torsion bar (Station Wagon) – removal and installation

1 Open the tailgate and support it in the open position with a piece of wood.
2 Undo and remove the fixing that secures the tailgate hinge cover to body lift away the cover.
3 Undo and remove the screws fixing the headlining rear end to the tailgate aperture rail panel. Detach the headlining.
4 Using a suitable pry bar, detach the left-hand torsion bar from the bracket.
5 Remove the right-hand torsion bar from the bracket in a similar manner to the left-hand torsion bar.
6 Installation of the assembly is the reverse sequence to removal, but if, when opening or closing the tailgate, squeaking is heard, apply a smear of grease to items A and B in Fig. 12.14.

19 Front and rear door – removal and installation

Front door

1 Fully open the door and, using suitable wood-blocks, support it both front and rear. *Note: Place rags or newspapers between the blocks and the door to protect the door panel lower edge from damage.*
2 Loosen and remove the bolts securing the door to the hinges. Lift away the door.
3 Installation of the front door is a reversal of removal but, if necessary, adjust it as described later in this Section, to obtain a satisfactory fit.

Rear door

4 The procedure for removal of a rear door in identical to that described for a front door in paragraph 1 through 3.

Adjustment

5 Correct door alignment may be obtained by adjusting the position of the door hinge and/or the lock striker.
6 The door hinge or the striker may be moved up-and-down, as well as fore-and-aft, by slackening the securing bolts and moving the lock striker in the required direction. It will be found necessary to use a cranked wrench when adjusting the hinges.

20 Door trim and interior handles – removal and installation

1 Open the door and, using a metal hook, withdraw the window regulator handle retaining spring clip.
2 Undo the self-tapping sheet metal screw securing the lock handle finisher. Lift away the finisher.
3 Undo and remove the two screws securing the armrest to the door inner panel. Lift away the armrest.
4 The door trim inner panel may now be detached from the door inner panel. Using a knife or hacksaw blade (with the teeth ground down) inserted between the door trim panel and door inner panel, carefully ease each clip from its hole in the door inner panel.
5 When all clips are free lift away the door trim panel.
6 Inspect the dust and splash shields to ensure that they are correctly installed and also not damaged.
7 Inspect the trim panel retaining clips and inserts for excessive corrosion or damage. Obtain new as necessary.
8 Installation of the door trim panel is the reverse sequence to removal. With the door glass up the regulator handle should be pointing upwards 60° to the front.

21 Door window glass and regulator – removal and installation

Front

1 Refer to Section 20, and remove the door interior handles and trim panel.
2 Carefully pry out the outside moulding.
3 Loosen the front and rear sash attaching screws.
4 Slacken the nuts securing the glass guide to the glass side guide channel, working through the opening of the door panel.
5 Very carefully lift the glass upwards and out of the door.
6 Loosen and remove the bolts securing the window regulator baseplate and guide channel to the door panel.
7 Manipulate the regulator assembly out through the opening of the door panel.
8 Installation of the front window glass and regulator is a reverse of removal. Be sure to adjust the fore-and-aft movement of the glass in the sashes; firmly tighten the sash securing screws when satisfactory adjustment is obtained.

Rear

9 Lower the door glass.
10 Unscrew and remove the door lock knob and grommet.
11 Refer to Section 20, and remove the door interior handles and trim panel.
12 Carefully pry the outside moulding from its location.
13 Peel back the rubber weatherstrip at the top edge of the door, and loosen and remove the two sash attaching screws.
14 Loosen and remove the two sash securing screws at the door panel.
15 Loosen and remove the bolts attaching the glass backplate to the guide channel.
16 Carefully lift the glass upwards and out of the door.
17 Loosen and remove the screws attaching the guide channel and regulator base to the door panel.
18 Manipulate the regulator assembly through the lower aperture of the door panel.
19 Installation of the regulator and glass is a reversal of removal.

22 Door lock and controls – removal and installation

1 Refer to Section 21, and remove the appropriate window glass.
2 Undo and remove the screws that secure the interior door handle assembly to the inner panel.
3 Disconnect the rods between the interior door handle and the door lock. The interior door handle and rods can now be lifted away.
4 Loosen and remove the bellcrank attaching screw.
5 Disconnect the rod connecting the key cylinder to the door lock.

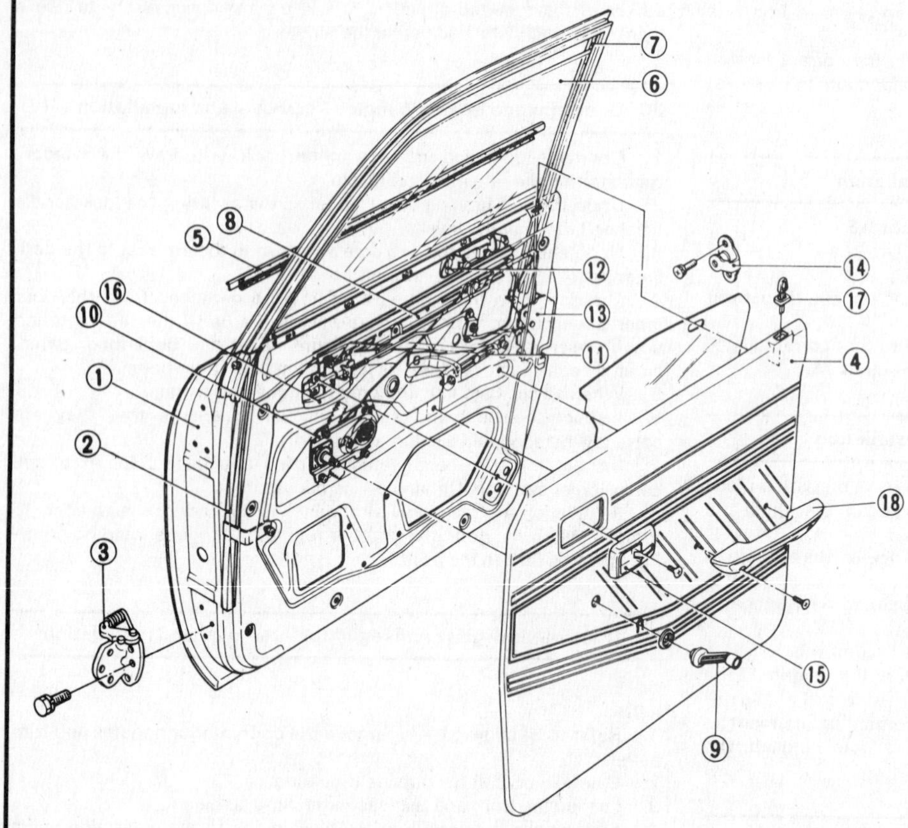

Fig. 12.17 The trim panel removed from the front door (Sec 20)

1 Door
2 Front glass guide
3 Door hinge
4 Trim panel
5 Outside moulding
6 Glass
7 Weatherstrip
8 Glass side guide channel
9 Regulator handle
10 Regulator
11 Panel side guide channel
12 Outside door handle
13 Door lock
14 Door lock striker
15 Interior door handle finisher
16 Interior door handle
17 Door lock knob
18 Armrest

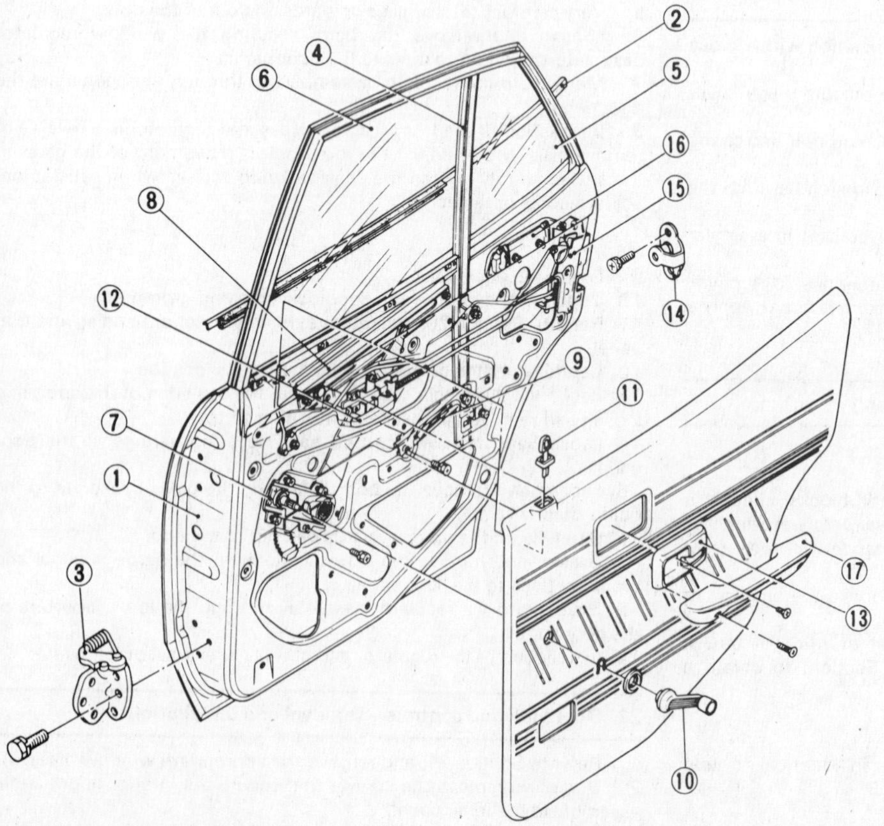

Fig. 12.18 The trim panel removed from the rear door (Sec 20)

1 Door
2 Outside moulding
3 Door hinge
4 Rear sash
5 Rear partition glass
6 Rear door glass
7 Regulator
8 Glass side guide channel
9 Panel side guide channel
10 Regulator handle
11 Door lock knob
12 Interior handle
13 Interior handle finisher
14 Door lock striker
15 Door lock
16 Outside handle
17 Armrest

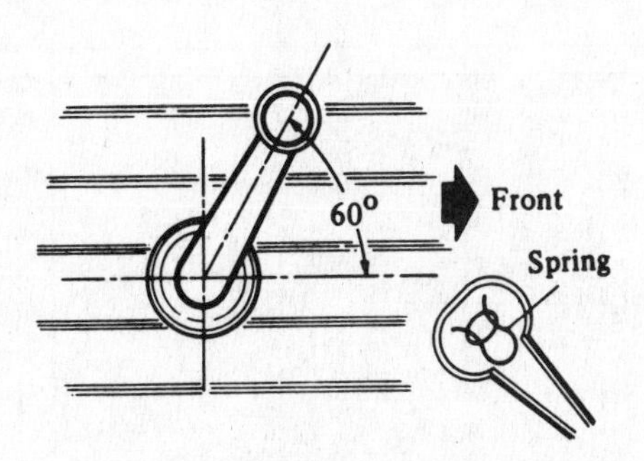

Fig. 12.19 Correct installation of the window regulator arm (Sec 20)

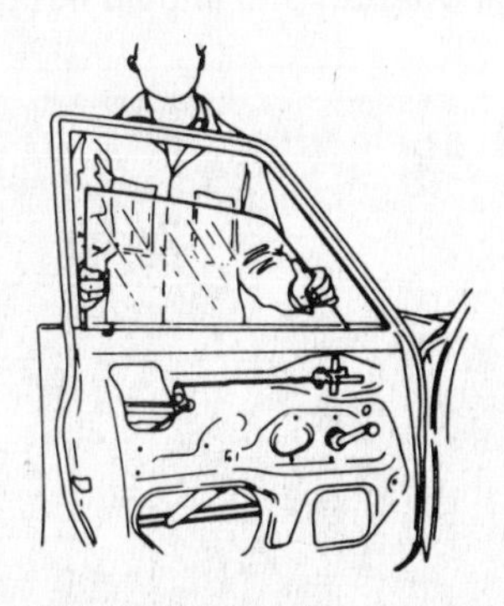

Fig. 12.20 Removing the front door glass (Sec 21)

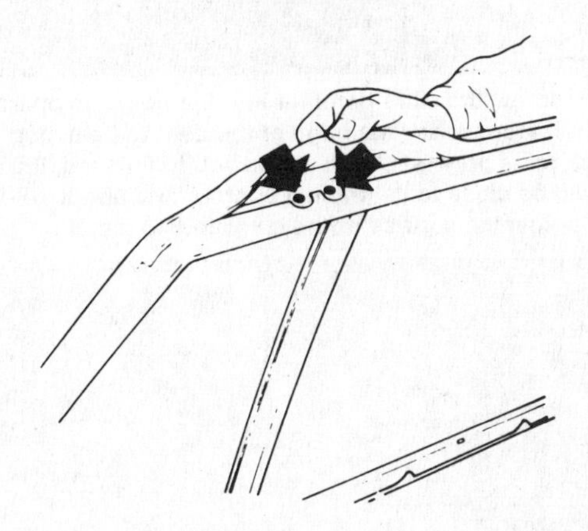

Fig. 12.21 The upper sash attaching screws (Sec 21)

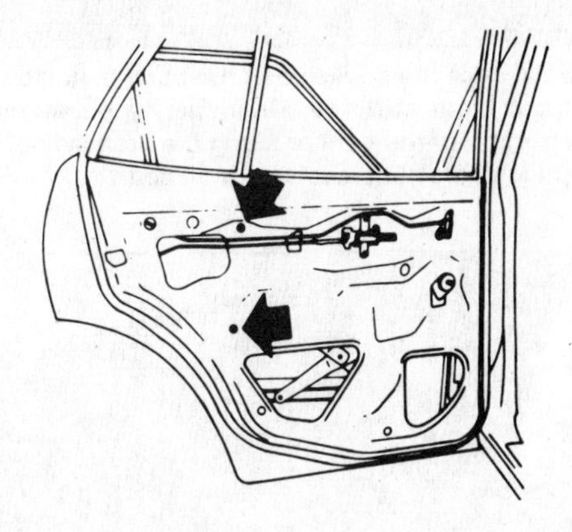

Fig. 12.22 The lower sash attaching screws (Sec 21)

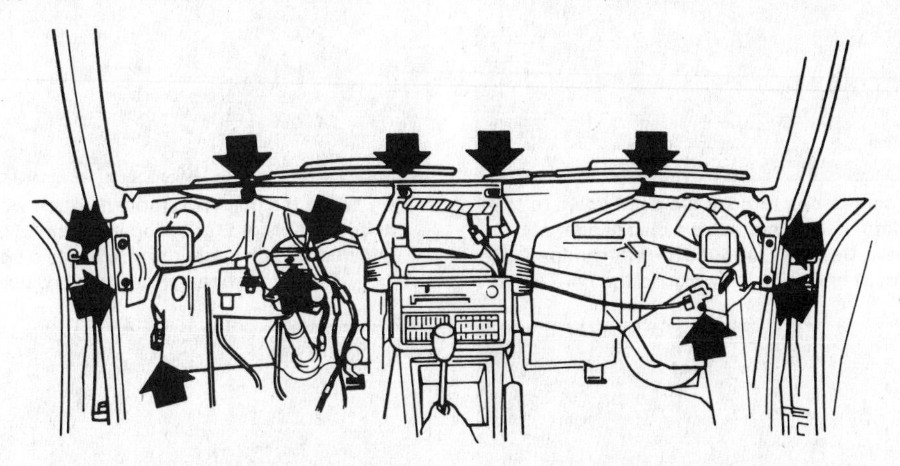

Fig. 12.23 The instrument panel attachment points (Sec 24)

6 Disconnect the rod connecting the outside handle to the door lock at the outside handle lever.
7 Unscrew and remove the bolts securing the lock assembly to the door. Remove the lock assembly.
8 If necessary, undo and remove the nuts securing the outside handle in position and remove it from the door.
9 Installation of the door lock and controls is the reverse sequence to removal. Lubricate all moving parts with a little engine oil.
10 Should adjustment be necessary, this must be carried out before the trim panel is installed. The correct clearance between the door lock lever and adjustment nut must be less than 0·039 in (1·0 mm).
11 To adjust the interior door handle free-play, move the interior door handle assembly fore-and-aft in the elongated holes until there is a free-play of 0·039 in (1·0 mm).
12 The door lock striker can be moved both vertically and horizontally to align with the door lock catch.

23 Windshield glass – removal and installation

If you are unfortunate enough to have a windshield shatter, fitting a replacement is one of the few jobs that the average owner is advised to leave to a body repair or windshield specialist. The reason for this is that the glass is not located in the body in the normal manner, but uses a special sealer and clips.

24 Instrument panel – removal and installation

1 Disconnect the battery ground cable.

These photos illustrate a method of repairing simple dents. They are intended to supplement *Body repair - minor damage* in this Chapter and should not be used as the sole instructions for body repair on these vehicles.

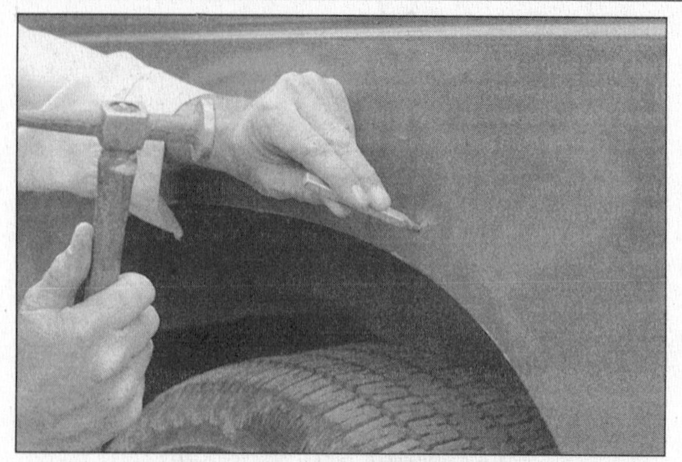

1 If you can't access the backside of the body panel to hammer out the dent, pull it out with a slide-hammer-type dent puller. In the deepest portion of the dent or along the crease line, drill or punch hole(s) at least one inch apart . . .

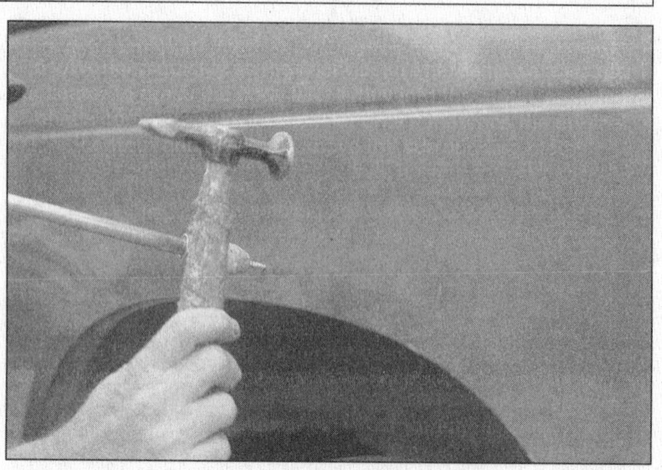

2 . . . then screw the slide-hammer into the hole and operate it. Tap with a hammer near the edge of the dent to help 'pop' the metal back to its original shape. When you're finished, the dent area should be close to its original contour and about 1/8-inch below the surface of the surrounding metal

3 Using coarse-grit sandpaper, remove the paint down to the bare metal. Hand sanding works fine, but the disc sander shown here makes the job faster. Use finer (about 320-grit) sandpaper to feather-edge the paint at least one inch around the dent area

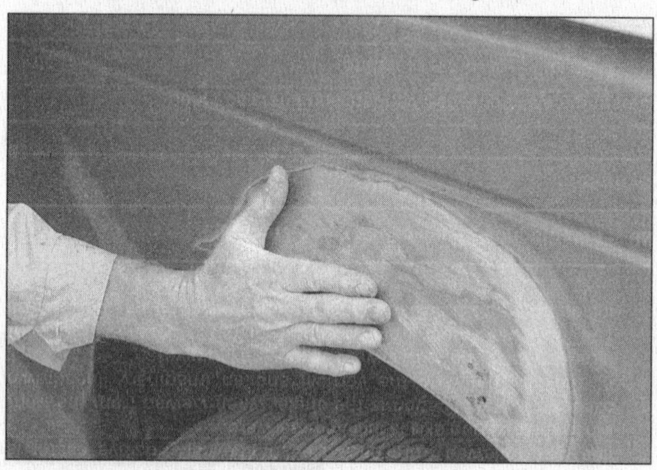

4 When the paint is removed, touch will probably be more helpful than sight for telling if the metal is straight. Hammer down the high spots or raise the low spots as necessary. Clean the repair area with wax/silicone remover

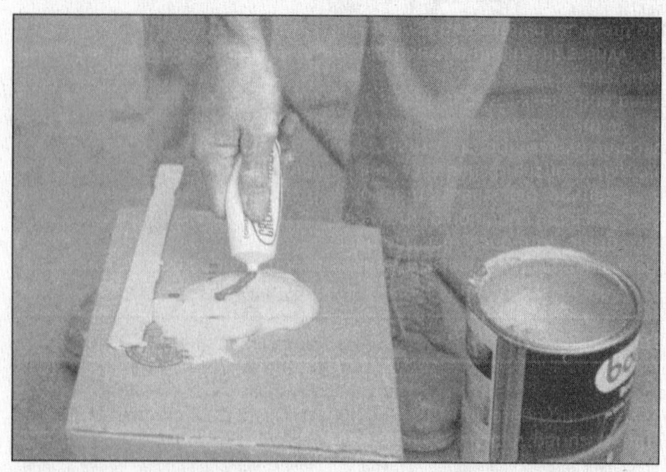

5 Following label instructions, mix up a batch of plastic filler and hardener. The ratio of filler to hardener is critical, and, if you mix it incorrectly, it will either not cure properly or cure too quickly (you won't have time to file and sand it into shape)

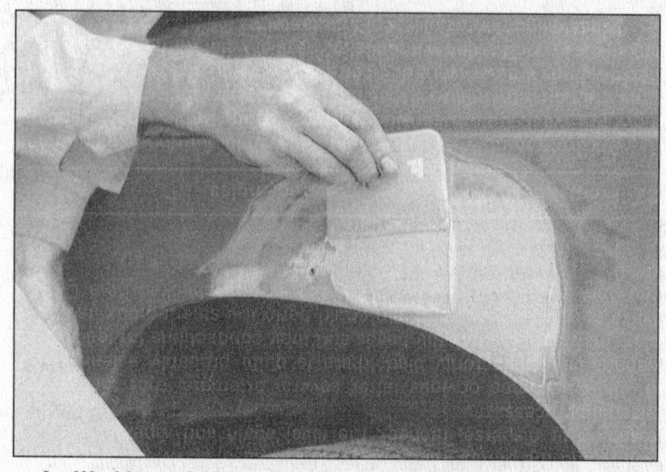

6 Working quickly so the filler doesn't harden, use a plastic applicator to press the body filler firmly into the metal, assuring it bonds completely. Work the filler until it matches the original contour and is slightly above the surrounding metal

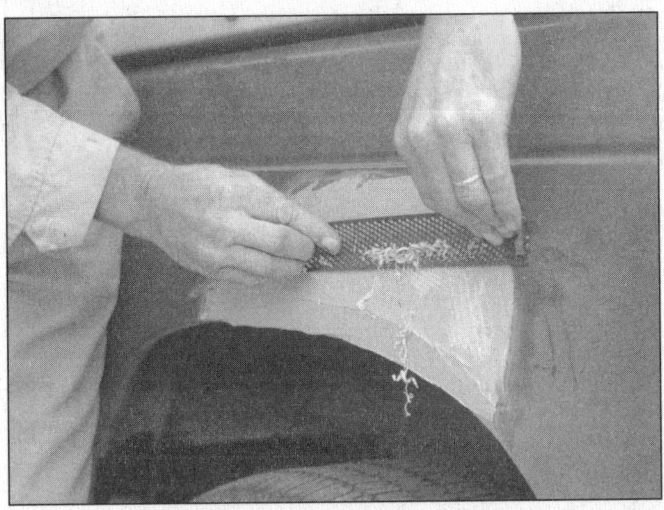

7 Let the filler harden until you can just dent it with your fingernail. Use a body file or Surform tool (shown here) to rough-shape the filler

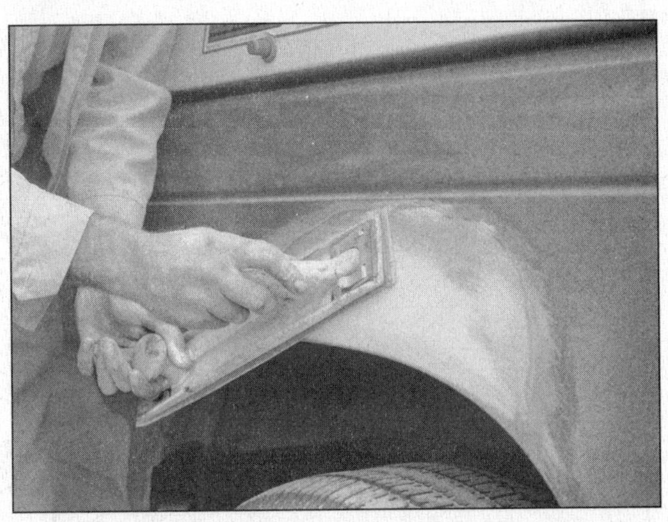

8 Use coarse-grit sandpaper and a sanding board or block to work the filler down until it's smooth and even. Work down to finer grits of sandpaper - always using a board or block - ending up with 360 or 400 grit

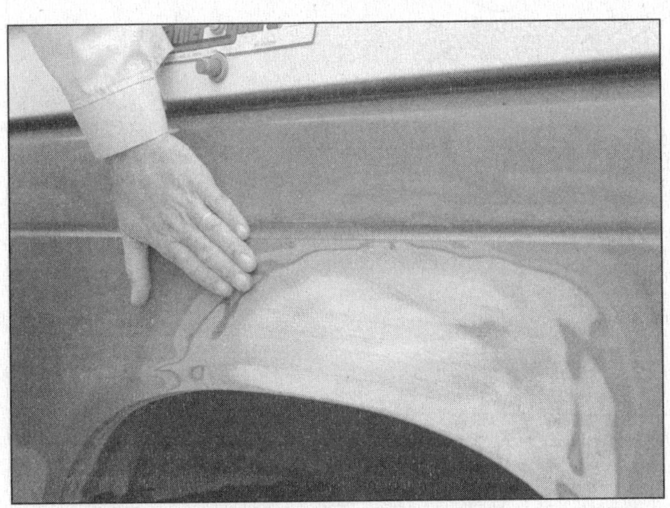

9 You shouldn't be able to feel any ridge at the transition from the filler to the bare metal or from the bare metal to the old paint. As soon as the repair is flat and uniform, remove the dust and mask off the adjacent panels or trim pieces

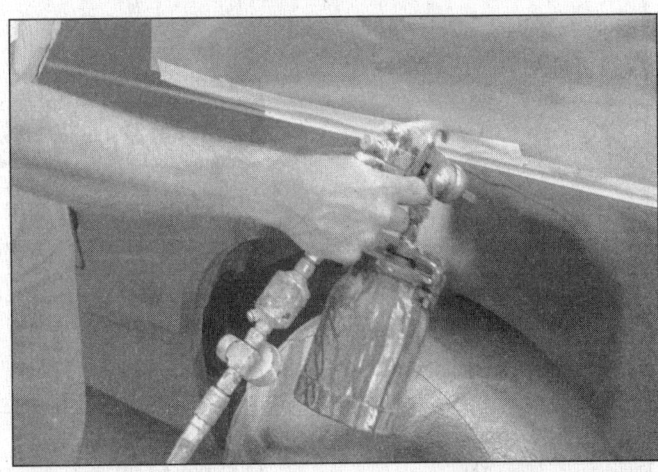

10 Apply several layers of primer to the area. Don't spray the primer on too heavy, so it sags or runs, and make sure each coat is dry before you spray on the next one. A professional-type spray gun is being used here, but aerosol spray primer is available inexpensively from auto parts stores

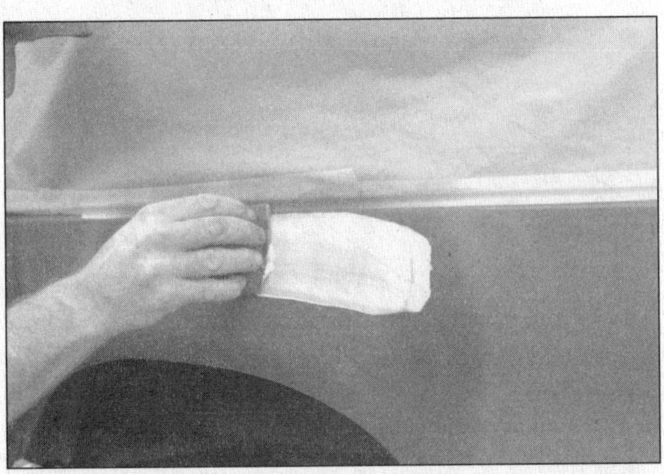

11 The primer will help reveal imperfections or scratches. Fill these with glazing compound. Follow the label instructions and sand it with 360 or 400-grit sandpaper until it's smooth. Repeat the glazing, sanding and respraying until the primer reveals a perfectly smooth surface

12 Finish sand the primer with very fine sandpaper (400 or 600-grit) to remove the primer overspray. Clean the area with water and allow it to dry. Use a tack rag to remove any dust, then apply the finish coat. Don't attempt to rub out or wax the repair area until the paint has dried completely (at least two weeks)

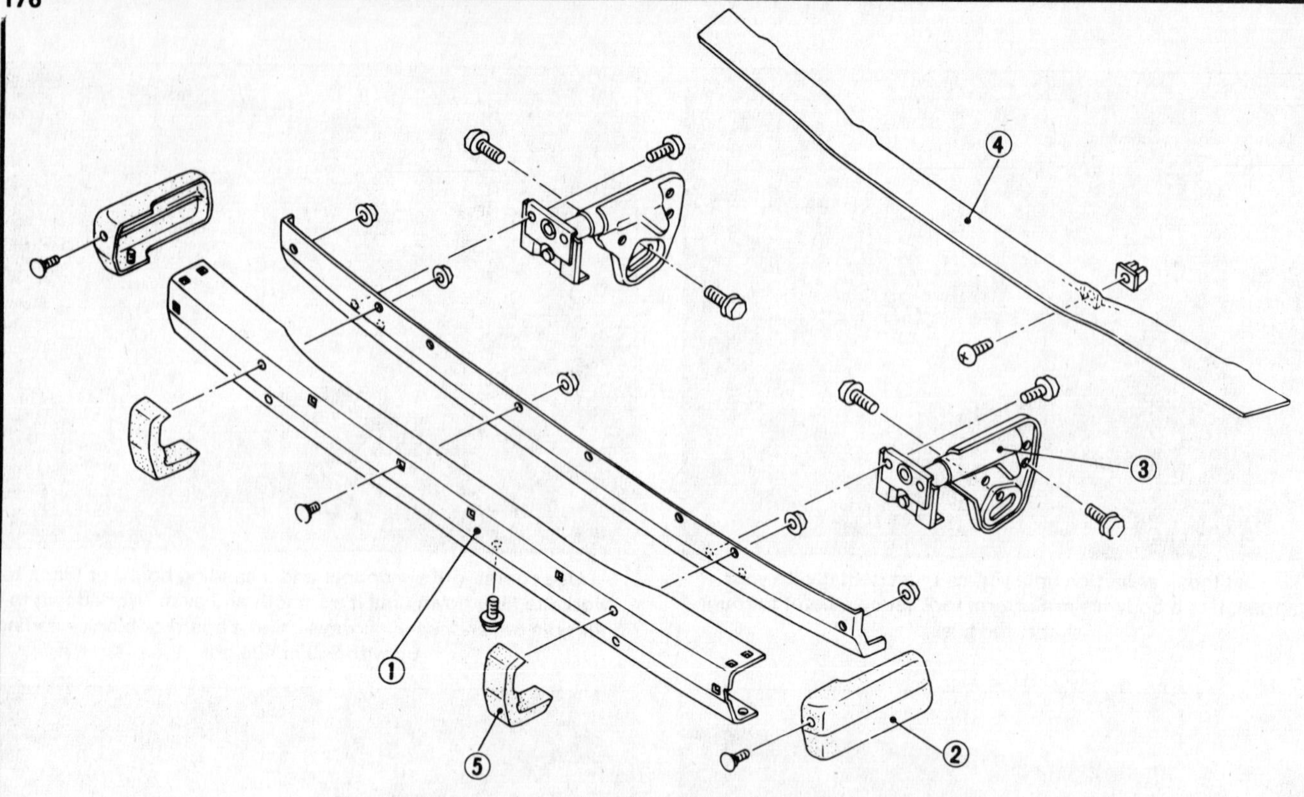

Fig. 12.24 An exploded view of the front bumper assembly (Sec 25)

1 Bumper
2 Side piece
3 Shock absorber
4 Finisher panel (attached to body)
5 Overrider

251.5 (9.90)
72.5 to 79.5
(2.854 to 3.130)

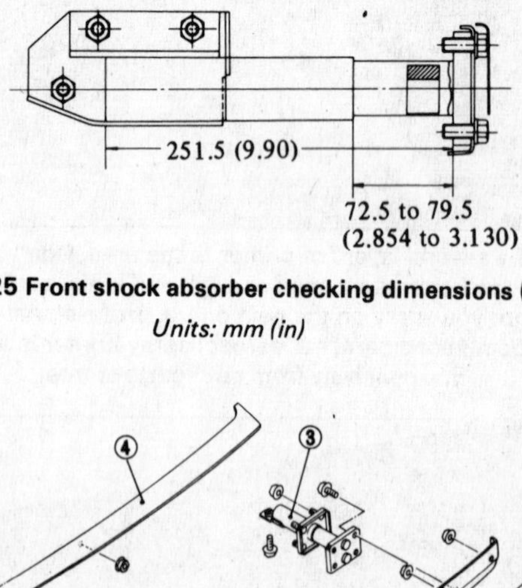

Fig. 12.25 Front shock absorber checking dimensions (Sec 25)

Units: mm (in)

Fig. 12.26 Checking front bumper height H (Sec 25)

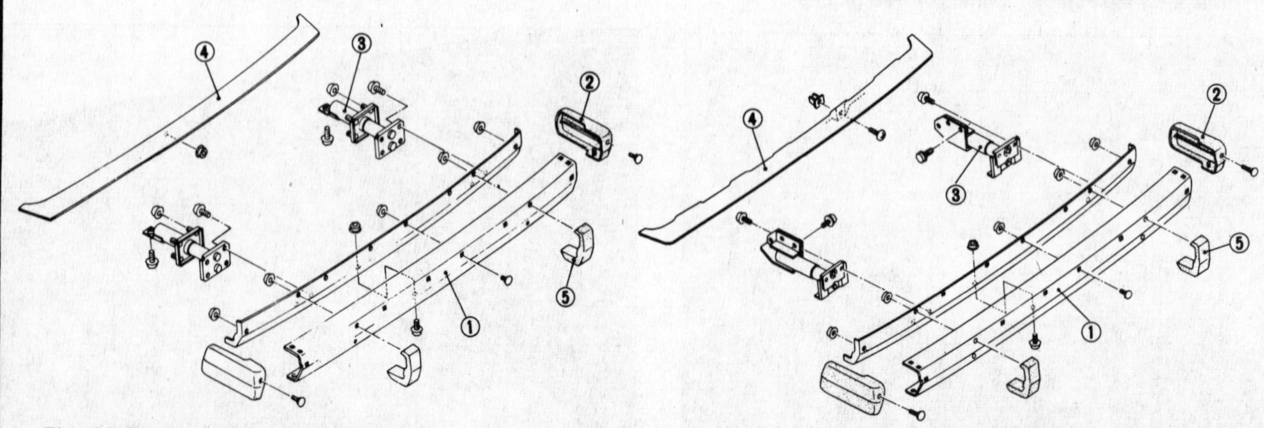

Fig. 12.27 An exploded view of the rear bumper on the Sedan (Sec 26)

1 Bumper
2 Side piece
3 Shock absorber
4 Finisher panel (attached to body)
5 Overrider

Fig. 12.28 An exploded view of the rear bumper on the Station Wagon (Sec 26)

1 Bumper
2 Side piece
3 Shock absorber
4 Finisher panel (attached to body)
5 Overrider

251.5 (9.90)
72.5 to 79.5
(2.854 to 3.130)
(Sedan)
251.5 (9.90)
72.5 to 79.5
(2.854 to 3.130)
(Station Wagon)
Unit: mm (in)

Fig. 12.29 Rear shock absorber checking dimensions (Sec 26)

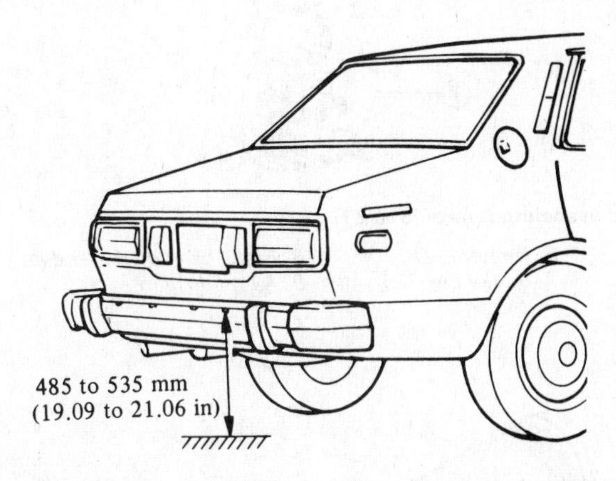

Fig. 12.30 Rear bumper height setting dimension (Sedan) (Sec 26)

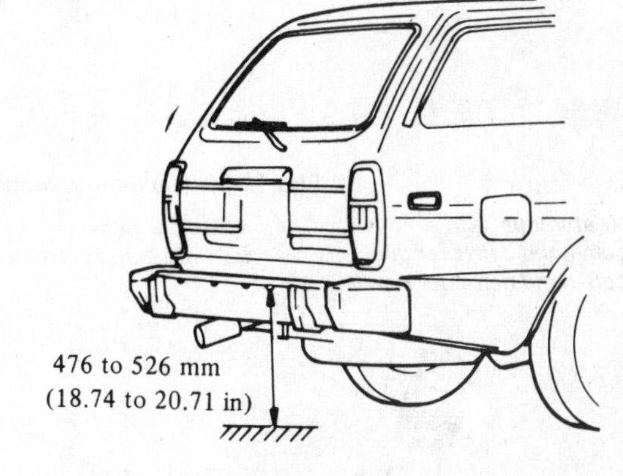

Fig. 12.31 Rear bumper height setting dimension (Station Wagon) (Sec 26)

2 Loosen and remove the bolts securing the instrument panel to the upper dash panel.
3 Refer to Chapter 11, and remove the steering wheel and shell covers.
4 Loosen and remove the bolts securing the steering column to the pedal pivot bracket.
5 Loosen and remove the bolt securing the instrument panel to the pedal pivot bracket.
6 Loosen and remove the bolts attaching the package tray to the instrument panel.
7 Reach behind the instrument panel and disconnect the speedometer drive cable from the speedometer.
8 Remove the manual choke wire knob.
9 Loosen and remove the bolts attaching the instrument panel to the dash side brackets.
10 Disconnect the harness connectors at both ends of the instrument panel.
11 Lift the instrument panel from the vehicle.
12 Installation of the instrument panel is a reversal of removal, but ensure that any electrical connections are correctly made before reconnecting the battery.

25 Front bumper – removal and installation

1 Loosen and remove the nuts attaching the front bumper to the shock absorbers. Carefully lift away the bumper.
2 The shock absorbers can be removed by loosening and removing the bolts attaching them to the vehicle side members.
3 The only check that can be made on the shock absorbers, is to measure their free length (see Fig. 12.25).
4 Installing the shock absorbers and the front bumper is a reversal of removal. It will be noted that the shock absorber mounting bolt holes are elongated to allow up-and-down movement. Measure the distance from the ground to the top of the bumper at either end of the bumper. The correct height should be between 18·60 and 20·57 in (472·5 and 522·5 mm) for the Sedan, and between 18·68 and 20·65 in (474·5 and 524·5 mm) for the Station Wagon.
5 When a satisfactory height has been obtained, firmly tighten the mounting bolts.

Note: *The shock absorbers are pressurized and no attempt should be made to dismantle them. Never puncture them or apply heat to them.*

26 Rear bumper – removal and installation

1 The procedures for removing and installing the rear bumper, both on the Sedan and Station Wagon, are identical to the procedures detailed for the front bumper, described in Section 25.
2 When installation of the rear bumper is complete, the height should be adjusted (see Fig. 12.30 and Fig. 12.31).

27 Heating and ventilation system – description

The heater/ventilator unit is located at the base of the instrument panel, forward of the center console. When the vehicle is in motion, air enters the cowl top grille just below the windshield, and passes through the heater matrix which is heated by coolant from the engine cooling system.

When the car is stationary, a booster fan can be switched on to suck air into the unit. Demister hoses are installed behind the instru-

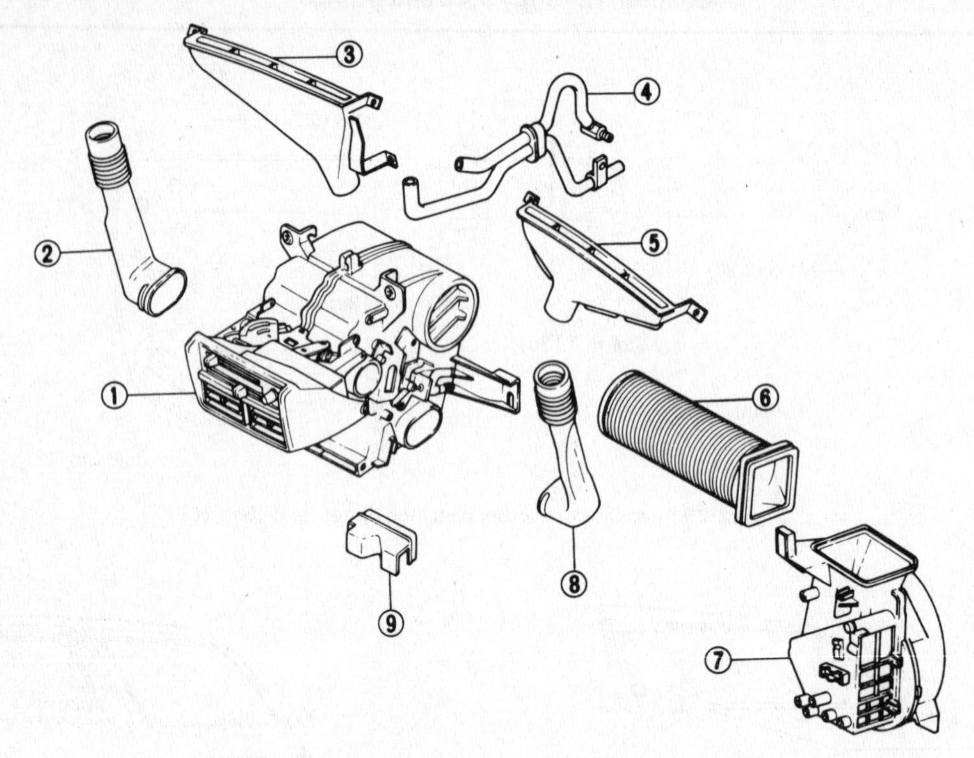

Fig. 12.32 The heater/ventilator unit and associated parts (Sec 27)

1 Heater unit
2 Left-hand defroster duct
3 Left-hand defroster nozzle
4 Heater hose
5 Right-hand defroster nozzle
6 Heater duct
7 Intake box
8 Right-hand defroster duct
9 Air guide plate

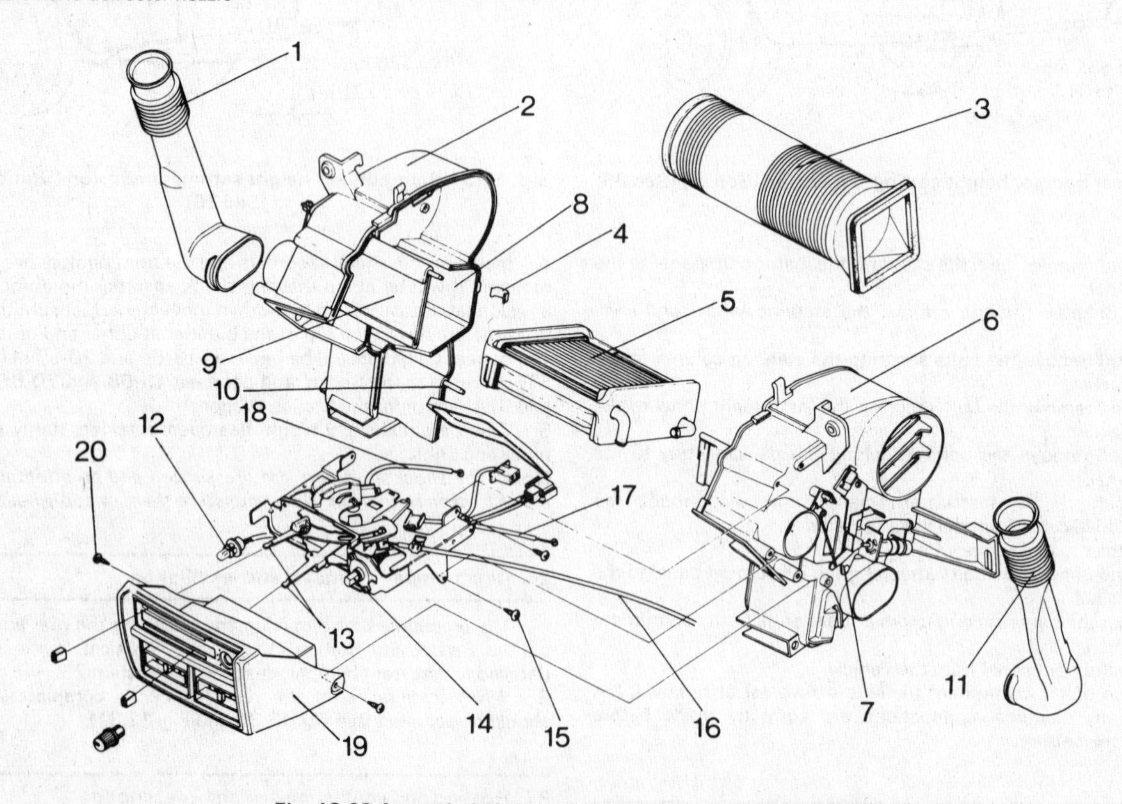

Fig. 12.33 An exploded view of the heater assembly (Sec 29)

1 Defroster nozzle
2 Heater case
3 Heater duct
4 Heater case clip
5 Heater matrix
6 Heater case
7 Flap controls
8 Flap shaft
9 Mixing flap
10 Floor flap shaft
11 Defroster nozzle
12 Panel illumination bulb
13 Control levers
14 Boost fan switch
15 Control lever assembly screw
16 Air intake box control cable
17 Water valve control cable
18 Mixing flap control cable
19 Control panel and ventilation grille assembly
20 Control panel screw

ment panel leading to nozzles which keep the windshield free from mist or frost. Control levers are attached to the panel just above the center console and provide any desired combination of air direction and temperature.

28 Heater unit – removal and installation

1 Disconnect the battery ground cable.
2 Refer to Chapter 2, and drain the cooling system.
3 Remove the center console box and the console box bracket, by loosening and removing the securing screws.
4 Remove the front floor mat.
5 Loosen and remove the heater duct attaching screws. Lift out the heater ducts from each side of the unit.
6 Loosen the hose clamps and disconnect the heater inlet and outlet hoses. A suitable drain pan should be placed beneath these hoses to catch any coolant left in them.
7 Disconnect and remove the defroster hoses from each side of the heater unit.
8 Disconnect the intake flap control cable from the air intake box.
9 Disconnect the electrical harness connector.
10 Loosen and remove the heater unit retaining bolts. The heater unit can now be removed from the vehicle.
11 Installation of the heater unit is a reversal of removal.

29 Heater unit – dismantling and reassembly

1 With the heater unit removed from the vehicle as described in Section 28, proceed to dismantle the unit as described in the following paragraphs.
2 Pull off the two heater control lever knobs and the boost fan control knob.
3 Loosen and remove the two screws securing the control panel and center ventilator assembly to the heater case. Lift the control panel and center ventilator assembly from the heater unit.
4 Undo and remove the four screws securing the control lever assembly to the heater unit.
5 Before the control lever assembly can be removed the three control cables will have to be disconnected. One of the cables is attached to the control flap at the air intake box and the other two are connected to the flap controls at the heater unit.
6 Unscrew and remove the screws that secure the flap shafts in position.
7 Remove the clips that secure the two halves of the heater unit case together. Separate the two halves.
8 The heater matrix can now be lifted out.
9 If the reason for dismantling the heater unit is a leaking matrix, it is best to obtain an exchange matrix. If the matrix is firmly blocked, try flushing water through it from a hose. If this fails, the matrix should be renewed.
10 Reassembly of the heater unit is a reversal of dismantling, but the control lever cables should be adjusted, if necessary, as described in Section 30.

30 Heater control assembly – removal, installation and adjustment

1 Disconnect the battery ground cable.
2 Remove the console box and console box bracket screws, and lift away the center console.
3 Pull off the control lever knobs and the boost fan control knob.
4 Working at each side of the control panel and ventilator grille assembly, loosen and remove the two attaching screws. Remove the control panel and ventilator grille assembly.
5 Disconnect the three control cables, one from the air intake box flap shaft linkage, and two from the heater unit floor flap shaft and the mixing flap shaft linkages.
6 Undo and remove the four screws securing the control assembly to the heater unit. Remove the control assembly.
7 Installation of the heater control assembly is a reversal of removal.
8 With the control assembly installed, the cable controls should be connected in the following manner:

(a) Set the TEMP lever on the control lever assembly to the HOT position. Set the mixing flap at the uppermost position, then firmly secure the cable to the linkage. As a double check, when the TEMP lever is in the COLD position, the water valve must be fully closed.
(b) Set the AIR lever of the control assembly to the RECIRC position. Set the floor flap to the shut position, then firmly secure the cable to the linkage.
(c) Now set the AIR lever to the heat position. Set the ventilator door to the shut position, then firmly secure the control cable to the linkage with the securing screw.

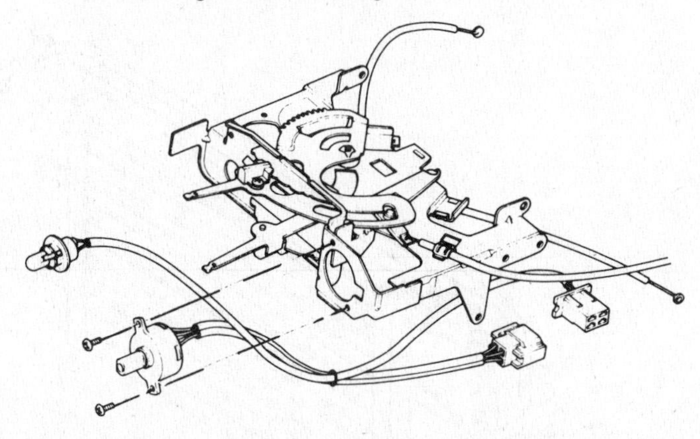

Fig. 12.34 The heater control assembly (Sec 30)

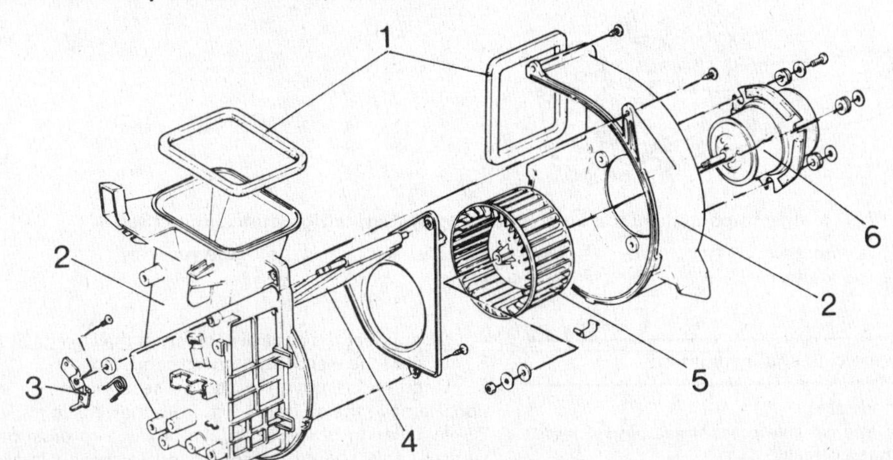

Fig. 12.35 An exploded view of the air intake box (Sec 31)

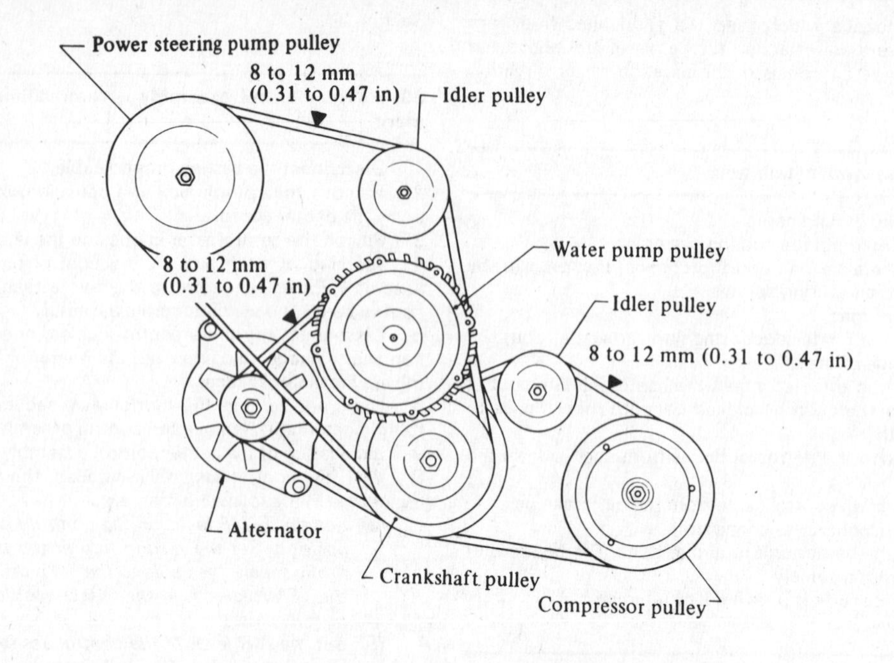

Fig. 12.36 Drivebelt tension adjusting diagram (Sec 32)

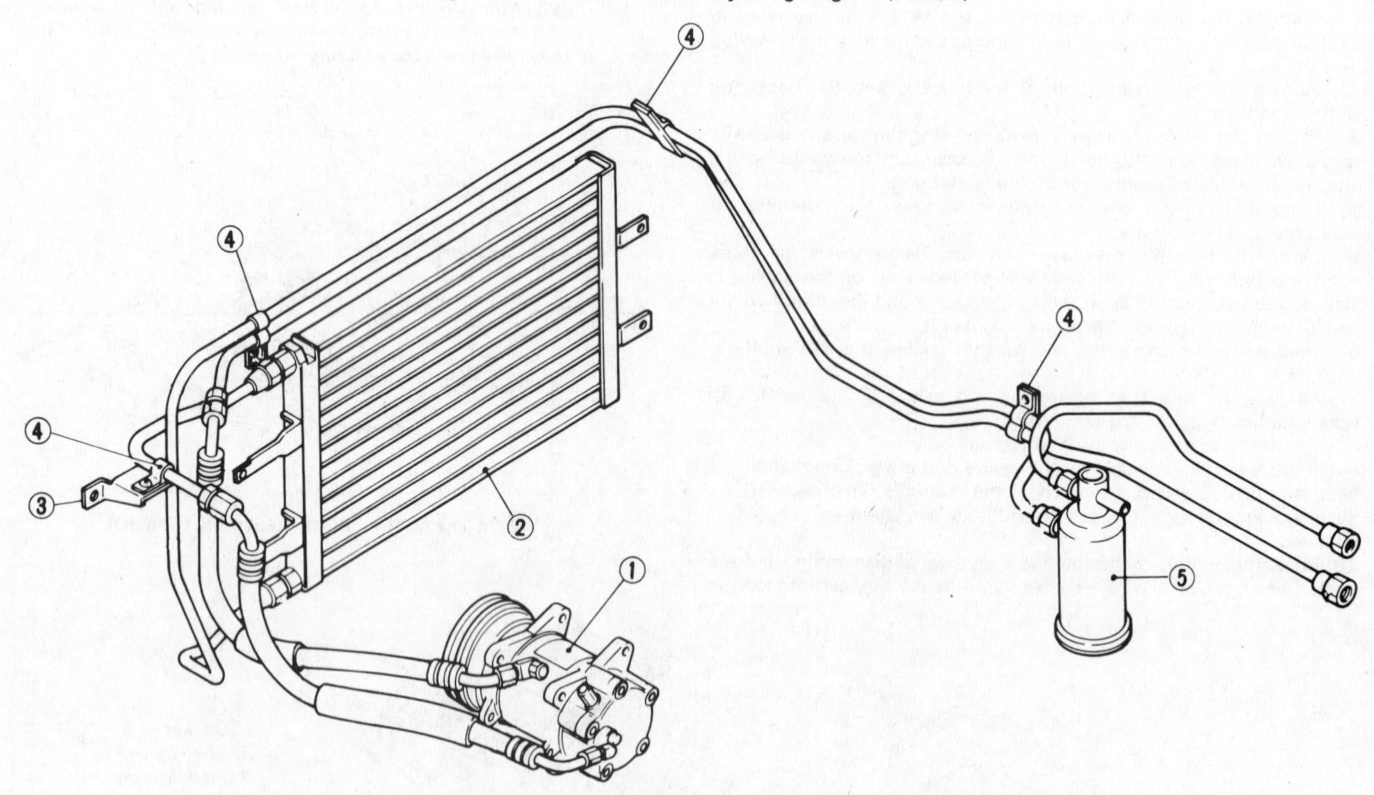

Fig. 12.37 Main components of the air conditioning system refrigeration circuit (Sec 32)

1 Compressor
2 Condensor
3 Clamp bracket
4 Tube clamp
5 Receiver dryer

31 Air intake box – removal, servicing and installation

1 Disconnect the battery ground cable.

2 Remove the center console and console bracket securing screws. Remove the center console from the vehicle.

3 Disconnect and remove the heater duct between the heater unit and the air intake box.

4 Disconnect the control cable from the linkage at the air intake box.

5 Unplug the electrical harness connector.

6 Undo and remove the screws securing the air intake box in position. Remove the air intake box from the vehicle.

7 Installation of the air intake box is a reversal of removal, but the control cable should be connected to the air intake box linkage, as described in Section 30.

8 To dismantle the intake box and remove the fan motor, proceed as

described in the following paragraphs.
9 Undo and remove the screws securing the two halves of the intake box together. Separate the two halves.
10 Undo the nut at the end of the fan motor shaft, remove the washers and draw off the fan from the motor shaft.
11 Now loosen and remove the two nuts and one bolt securing the motor to the case. Lift out the motor.
12 If the motor operation is sluggish, test it by connecting it directly to a fully charged battery. If unsatisfactory operation is still experienced the motor should be taken to an automobile electrician for inspection.
13 To install the motor and assemble the air intake box is a reversal of removal.

32 Air conditioning system – description and maintenance

1 An air conditioning system is optionally available on the Datsun 810 models. The system incorporates a heater and refrigeration unit with blower assembly.
2 The refrigeration section comprises an evaporator mounted to the rear and below the instrument panel, a belt-driven compressor installed in the engine compartment and a condenser mounted just in front of the radiator.
3 Due to the nature of the refrigeration gases employed in the system, no servicing can be undertaken by the home mechanic other than a few maintenance tasks which are described later in this Section. Where any part of the refrigeration circuit must be disconnected to facilitate other repair work, then the system must be discharged (and later re-charged) by professional refrigeration engineers having the necessary equipment.
4 Maintenance tasks which can be carried out safely include checking of the compressor driving belt tension. There should be a total deflection of between 0·31 and 0·47 in (8·0 and 12·0 mm) of the belt at a point midway between the compressor pulley and the idler pulley. The tension is adjusted by moving the position of the idler pulley.
5 Examine all the hose connections of the system. Any trace of oil at these points indicates leakage of refrigerant and the joint connections should be tightened.
6 Where the car is seldom used, the engine must be run at approximately 1500 rev/min at least once a month to keep the compressor and the system generally in good working order.
7 If it is suspected that the amount of refrigerant in the system is incorrect, start the engine and hold it at a steady speed of 1500 rev/min. Set the AIR lever in the A/C position and switch on the blower to maximum speed. Check the sight glass after an interval of about five minutes. The sight glass is located on the receiver drier. If a continuous stream of bubbles or mist is observed, then there is very little refrigerant left in the system. Where some bubbles are seen at intervals of 1 or 2 seconds then there is insufficient refrigerant in the system. The system is correctly charged when conditions within the sight glass are almost transparent with a few bubbles appearing if the engine speed is raised or lowered. If the system required recharging, this must be carried out professionally.

33 Air conditioner/heater – component removal and installation

1 Although it is not possible for the home mechanic to disconnect any part of the air conditioner refrigeration circuit, the following components can be removed for renewal or repair.

Blower motor

2 Disconnect the battery ground cable.
3 Remove the two clips (by turning counter-clockwise) and three screws that retain the package tray in position. Remove the package tray.
4 Refer to Chapter 10, and remove the cluster lid.
5 Pull off the vacuum hoses from the air intake flap actuators. Disconnect the harness plug.
6 Loosen and remove the three screws securing the blower housing assembly in position. Lift out the assembly.
7 Undo and remove the screws and detach the clips that retain the two halves of the assembly together. Separate the two halves of the housing.
8 Undo and remove the nut, lift off the washers and draw the fan off the motor shaft.
9 Undo and remove two nuts and one bolt, then remove the blower motor from the housing.
10 Reassembly and installation of the blower motor is a reversal of removal.

Heater unit

11 The heater unit used on air conditioned vehicles is identical to the unit used in non-air conditioned vehicles. Removal of this unit is described in Section 28, and dismantling is described in Section 29. The only difference is two vacuum hoses which must be disconnected from the vacuum switch on the control lever assembly.

Air conditioner control assembly

12 The air conditioner control assembly is identical to the heater control assembly used on non-air conditioned vehicles, so reference should be made to Section 30. When removing the control assembly,

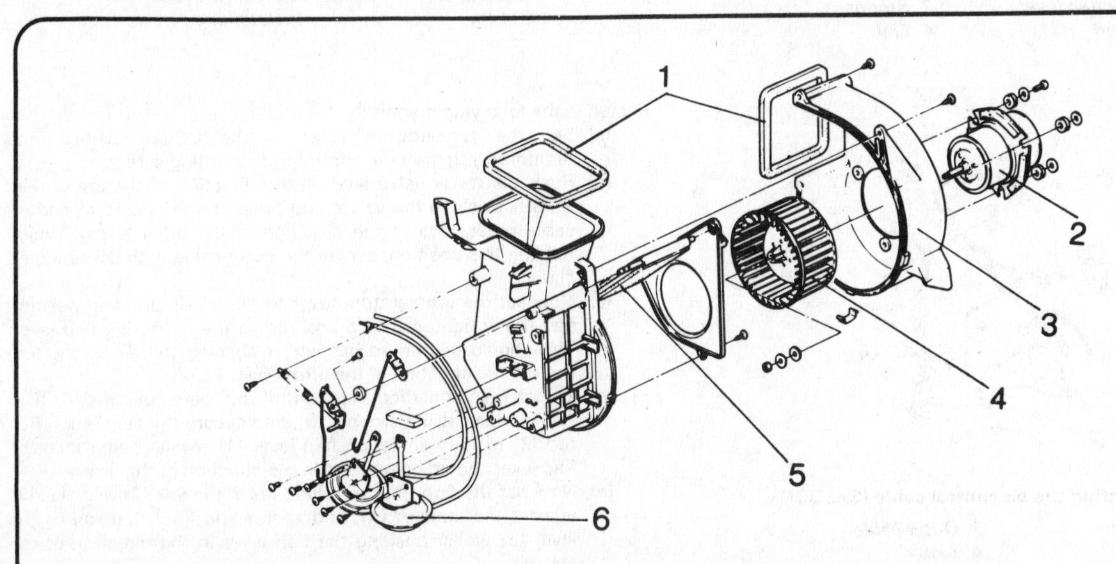

Fig. 12.38 An exploded view of the blower motor housing (Sec 33)

1 Gasket
2 Motor
3 Housing
4 Fan
5 Flap
6 Vacuum actuator capsules

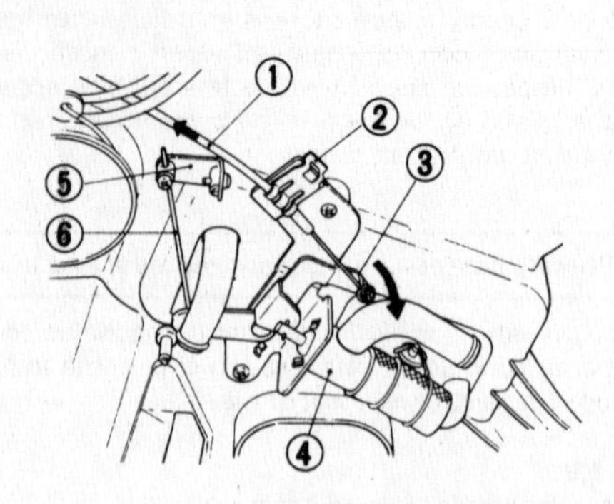

Fig. 12.39 Setting the temperature control cable (Sec 33)

1 Outer cable
2 Clip
3 Water valve lever
4 Water valve
5 Air mixing flap lever
6 Control rod

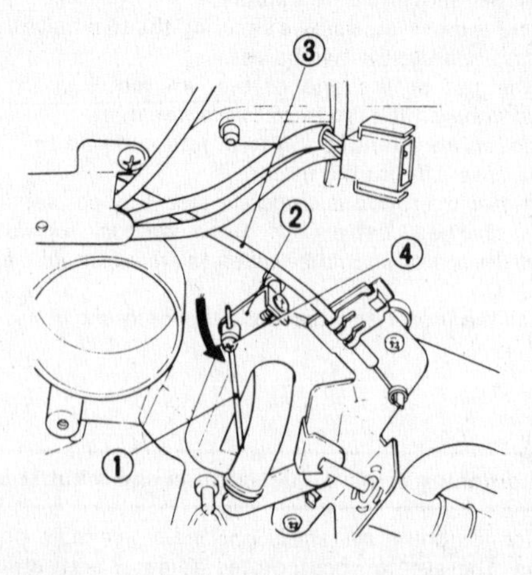

Fig. 12.40 Setting the air mixing flap control rod (Sec 33)

1 Control rod
2 Air mixing flap lever
3 Temperature control cable
4 Securing screw

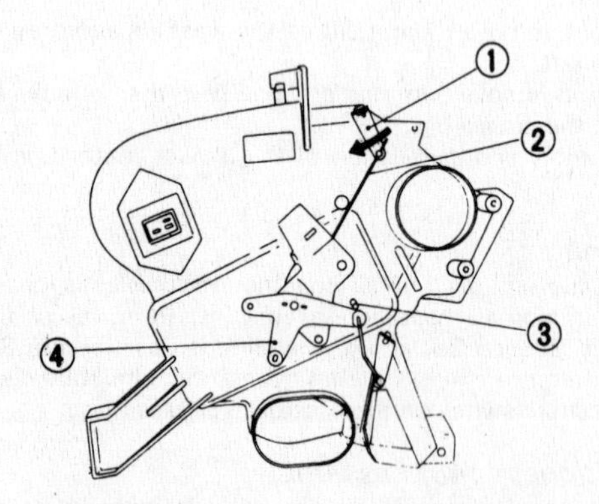

Fig. 12.41 Setting the ventilator flap control rod (Sec 33)

1 Ventilator flap lever
2 Control rod
3 Stopper
4 Link

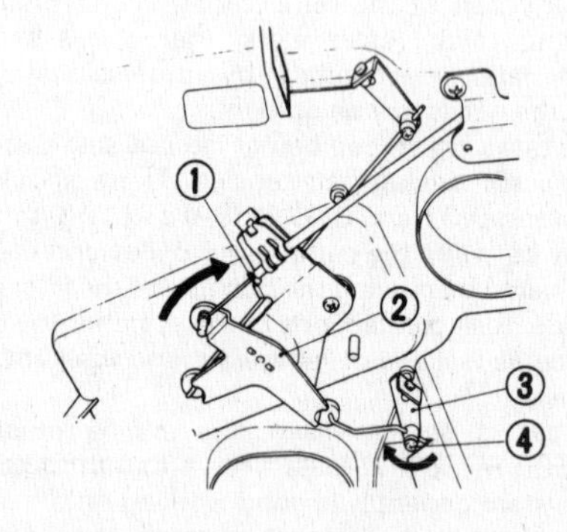

Fig. 12.42 Setting the floor flap control rod (Sec 33)

1 Stopper
2 Link
3 Floor flap lever
4 Retaining screw

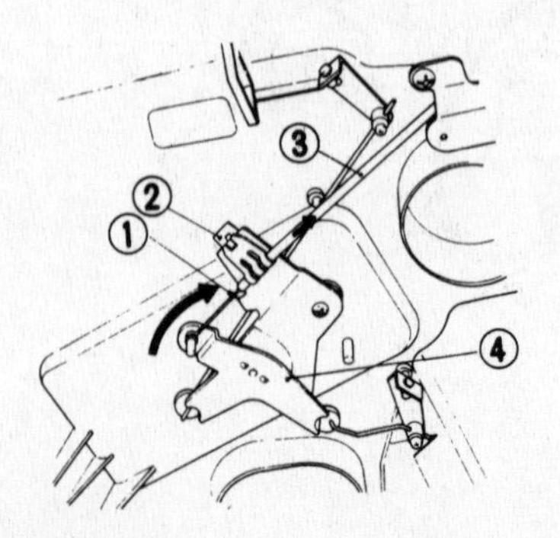

Fig. 12.43 Setting the air control cable (Sec 33)

1 Stopper
2 Clip
3 Outer cable
4 Link

the only difference that will be found is a vacuum switch and two vacuum hoses connected to it. These hoses must be disconnected.

13 When installing the control assembly, the cables must be connected in the following manner:

(a) Set the temperature lever in the COLD position and temporarily tighten the control rod mounting screw.
(b) Push the water valve lever in the direction of the arrow (Fig. 12.39) to close the valve, and press the temperature control cable outer case in the direction of the other arrow. Whilst holding this position, secure the outer cable with the retaining clip.
(c) Now set the temperature lever to the maximum heat position and firmly tighten the control rod to the air mixing flap lever. This should be carried out whilst pushing the air mixing flap lever in the direction of the arrow (Fig. 12.40).
(d) To set the ventilation flap control rod, press the link (4) (Fig. 12.41) against the stopper (3), and secure the ventilation flap rod (2) to the ventilation flap lever (1); whilst doing this, the flap lever should be pressed in the direction of the arrow.
(e) Now set the floor flap rod. First press the link (2) (Fig. 12.42) against the stopper (1), and secure the floor flap rod to the lever (3) whilst pushing the flap lever in the direction of the arrow.
(f) To set the air control cable, select the control lever in the A/C position. Press the link (4) (Fig. 12.43) against the stopper (1) and secure the air control cable with the clip (2), whilst pushing the cable's outer case in the direction of the arrow.

Chapter 13 Supplement: Revisions and information on later North American models

Contents

1 Introduction

This supplement contains specifications and service procedure changes that apply to 810 and Maxima models produced for sale in North America during the 1980 through 1984 model years. Also included is information related to previous models that was not available at the time of the original publication of this manual.

Where no differences (or very minor differences) exist between the later models and previous models, no information in given. In such instances, the original material included in Chapters 1 through 12 should be used.

2 Specifications

Note: *The specifications listed here include only those items which differ from those listed in Chapters 1 through 12. For information not specifically listed here, refer to the appropriate Chapter.*

Engine

Crankpin diameter	1.7701 to 1.7706 in (44.961 to 44.974 mm)
Piston diameter	3.2663 to 3.2665 in (82.965 to 82.970 mm)

Torque wrench settings

	lbf ft	kgf m
Cylinder head bolts (1983 and 1984)		
Stage 1	22	3.0
Stage 2	58	8.0
Stage 3	58 to 65	8.0 to 9.0
Connecting rod big end nut (1983 and 1984)	22 to 27	3.0 to 3.8
Camshaft locating bolts	4.3 to 7.2	0.6 to 1.0

Emissions

BCDD Operating pressure specified range	22.44 in Hg to 29.92 in Hg (570 mm Hg to 760 mm Hg)

Ignition system

Distributor type	
1980	
Manual and automatic transmission (California models)	D6K8-01
Automatic transmission (non-California models)	D6K9-07
Manual transmission (non-California models)	K6K9-10
1981	
USA	D6K9-22
Canada	D6K80-04
1982	D6K81-02
1983 and 1984	D6K82-02
Air gap	0.012 to 0.020 in (0.3 to 0.5 mm)
Carbon point length	
1980 and 1981	0.47 in (12 mm)
1982 through 1984	0.39 in (10 mm)
Ignition timing (BTDC) at idle speed	
1980 and 1981	
Manual transmission	10° at 700 rpm
Automatic transmission (selector in Drive)	10° at 650 rpm
1981 through 1984	
Manual transmission	8° at 700 rpm
Automatic transmission (selector in Drive)	8° at 650 rpm
Centrifugal advance	
1980	0° at 600 rpm or 8.5° at 1250 rpm
1981 through 1984	0° at 650 rpm or 9° at 1350 rpm
Ignition coil type	
1980	CIT-43
1981 through 1984	CIT-30
Primary resistance at 20°C (68°F)	0.84 to 1.02 ohms
Secondary resistance at 20°C (68°F)	8.2 to 12.4 ohms
Spark plugs	
Type	BP6ES-11 or BPR6ES-11
Gap	0.039 to 0.043 in (1.0 to 1.1 mm)

Clutch

Clutch pedal height	
1980	6.91 in (175 mm)
1981 and 1982	7.17 to 7.32 in (182 to 186 mm)
1983 and 1984	
With damper	6.89 to 7.28 in (175 to 185 mm)
Without damper	6.69 to 7.09 in (170 to 180 mm)
Clutch master cylinder bore-to-piston clearance	Less than 0.0059 in (0.15 mm)

Torque wrench settings

	lbf ft	kgf m
Clutch damper-to-bracket bolt	2.2 to 4.3	0.3 to 0.6
Clutch tube flare nut	11 to 13	1.5 to 1.8
Clutch operating cylinder bleeder screw	5.1 to 6.5	0.7 to 0.9
Cylinder-to-housing	22 to 30	3.1 to 4.1
Damper cover-to-bracket	2.2 to 4.3	0.3 to 0.6
Pedal stopper locknut	5.8 to 8.7	0.8 to 1.2

Manual transmission

Type	FS5W71B, five forward speeds and reverse

Synchromesh	All forward speeds	
Gear ratios		
1st	3.321:1	
2nd	2.044:1	
3rd	1.408:1	
4th	1.000:1	
5th	0.752:1	
Reverse	3.382:1	
Oil capacity	4-1/4 US pints (2.0 liters)	
Baulk ring/gear clearance	**in**	**(mm)**
Standard	0.0472 to 0.0630	(1.20 to 1.60)
Wear limit	0.031	(0.08)
Snap-ring sizes		
Main drive gear bearing	0.0681	(1.73)
	0.0709	(1.80)
	0.0736	(1.87)
	0.0764	(1.94)
	0.0792	(2.01)
	0.0819	(2.08)
Mainshaft front	0.055	(1.4)
	0.059	(1.5)
	0.63	(1.6)
Mainshaft rear end bearing	0.043	(1.1)
	0.047	(1.2)
	0.051	(1.3)
	0.055	(1.4)
Counter drive gear	0.055	(1.4)
	0.059	(1.5)
	0.063	(1.6)
Gear end play	**in**	**(mm)**
1st main gear	0.0106 to 0.0134	(0.27 to 0.34)
2nd main gear	0.0047 to 0.0075	(0.12 to 0.19)
3rd main gear	0.0051 to 0.0146	(0.13 to 0.37)
5th main gear	0.0039 to 0.0067	(0.10 to 0.17)
Reverse idler gear	0.0020 to 0.0197	(0.05 to 0.50)

Torque wrench settings

	lbf ft	kgf m
Transmission-to-engine	32 to 43	3.1 to 4.1
Mainshaft locknut	101 to 123	14 to 17
Counter gear locknut	72 to 94	10 to 13
Rear extension-to-case	12 to 15	1.6 to 2.1
Front cover-to-case	12 to 15	1.6 to 2.1
Filler plug	18 to 25	2.5 to 3.5
Ball pin	14 to 25	2.0 to 3.5
Striking lever locknut	6.5 to 8.7	0.9 to 1.2
Check ball plug	14 to 18	2.0 to 2.5

Automatic transmission

Type	L4N71B, three forward speeds plus overdrive and reverse
Stall speed	1800 to 2100 rpm
Fluid type	DEXRON
Capacity	
1980	8.0 US quarts (5.5 liters)
1981 and 1982	6-1/2 US quarts (6.1 liters)
1983 and 1984	7-3/8 US quarts (7.01 liters)
Gear ratios	
1st	2.458:1
2nd	1.458:1
3rd	1.000:1
Overdrive	0.686:1
Reverse	2.182:1

Torque wrench settings

	lbf ft	kgf m
Driveplate-to-crankshaft bolts	101 to 116	14 to 16
Driveplate-to-torque converter	29 to 36	4 to 5
Converter housing-to-engine	29 to 36	4 to 5

Propeller shaft

Torque wrench settings

	lbf ft	kgf m
Propeller shaft-to-differential	17 to 24	2.4 to 3.3
Center bearing locknut	181 to 212	25 to 30
Center bearing support-to-bracket nut	6.5 to 8.0	0.9 to 1.1
Center bearing bracket-to-body	26 to 35	3.6 to 4.8

Rear axle

Torque wrench settings

	lbf ft	kgf m
Drive pinion nut		
Sedan	65 to 72	9 to 10
Station wagon	94 to 217	13 to 30
Ring gear bolt		
Sedan	65 to 72	9 to 10
Station wagon	58 to 72	8 to 10
Companion flange	17 to 24	2.4 to 3.3
Oil drain and filler plugs		
Sedan	29 to 43	4 to 6
Station wagon	43 to 72	6 to 10
Side bearing cap bolt	36 to 43	5 to 6
Differential carrier	18 to 25	2.5 to 3.5
Side retainer bolt	6.5 to 8.7	0.9 to 1.2
Rear cover bolt	29 to 36	4 to 5
Rear cover nut	43 to 58	6 to 8
Axleshaft-to-side gear locknut	23 to 31	3.2 to 4.3

Braking system

System type

1980 through 1982	Front disc, rear drum
1983 and 1984	Four wheel disc

Disc brake

Front	
Minimum permissible pad thickness	0.079 in (2 mm)
Maximum permissible disc runout	0.0059 in (0.15 mm)
Minimum permissible disc thickness	0.630 in (16 mm)
Rear	
Minimum permissible pad thickness	0.079 in (2 mm)
Maximum permissible disc runout	0.0059 in (0.15 mm)
Minimum permissible disc thickness	0.339 in (8.6 mm)

Master cylinder

Allowable clearance between cylinder and piston	0.0059 in (0.15 mm)

Brake pedal

Free height	
1981 and 1982	6.38 to 6.77 in (162 to 172 mm)
1983 and 1984	6.46 to 6.69 in (164 to 170 mm)
Free play	0.039 to 0.197 in (1 to 5 mm)

Torque wrench settings

	lbf ft	kgf m
Master cylinder/booster		
Master cylinder-to-booster	5.8 to 8.0	0.8 to 1.1
Master cylinder secondary piston stopper bolt		
Nabco	1.1 to 2.2	0.15 to 0.30
Tokico	1.4 to 2.5	0.20 to 0.35
Check valve plug	33 to 40	4.5 to 5.5
Booster-to-dash panel	5.8 to 8.0	0.8 to 1.1
Front disc brake		
Baffle plate	2.3 to 3.2	0.32 to 0.44
Torque member fixing bolt	53 to 72	7.3 to 9.9
Torque member-to-cylinder body	12 to 15	1.6 to 2.1
Front disc fixing bolt	36 to 51	5.0 to 7.0
Rear disc brake		
Baffle plate	2.3 to 3.2	0.32 to 0.44
Caliper pin bolt	16 to 23	2.2 to 3.2
Caliper fixing bolt	28 to 38	3.9 to 5.3
Rear drum brake		
Backing plate	16 to 20	2.2 to 2.7
Wheel cylinder nut	4.3 to 5.8	0.6 to 0.8
Parking brake		
Control lever-to-body	6.5 to 8.7	0.9 to 1.2
Adjuster locknut	2.2 to 2.9	0.3 to 0.4

Electrical system

Bulbs

Headlamp (Sealed beam)	
Inner	4651
Outer (High/Low)	4652
Front turn signal lamp	1156
Front side marker (clearance) lamp	67
Rear side marker lamp	158

Rear combination lamp	
Stop/tail	1157
Turn signal	1156
Back-up	1156
License plate lamp	67
Spot lamp	67
Luggage compartment lamp	158
Combination meter lamps	
Illumination and brake warning	158
Warning	158
Turn signal and beam indicator	158
Radio illumination	158
Heater/air conditioner control illumination	158
Clock illumination	158
Glove box	158
Selector lever illumination	158

Suspension and steering

Coil spring free length

Front	
GL	13.37 in (339.5 mm)
Deluxe	12.78 in (324.5 mm)
Rear	
Sedan	15.22 in (386.5 mm)
Station wagon	12.28 in (312 mm)

Rear wheel alignment

Toe-in	-0.217 to 0.177 in (-5.5 to 4.5 mm)
Camber	1°15′ to 2°45′

Steering

Number of turns of steering wheel (lock-to-lock)	
Manual steering	4.0
Power steering	3.1
Steering angles (vehicle unladen)	
Camber	-20′ to 1°10′
Caster	2°55′ to 4°25′
Kingpin inclination	11°25′ to 12°55′
Toe-in	-0.04 to 0.04 in (-1 to 1 mm)

Torque wrench settings

Steering	**lbf ft**	**kgf m**
Steering wheel nut	27 to 38	3.8 to 5.2
Jacket tube-to-dash	5.8 to 8.0	0.8 to 1.1
Steering column mounting bracket	9 to 13	1.3 to 1.8
Lower joint-to-rubber coupling	29 to 36	4.0 to 5.0
Lower joint-to-pinion gear	24 to 28	3.3 to 3.9
Steering gear side-rod-to knuckle arm	40 to 72	5.5 to 10
Side-rod locknut	58 to 72	8 to 10
Gear housing clamp bolt	25 to 33	3.5 to 4.5
Boot clamp securing bolt	0.1 to 0.4	0.002 to 0.05
Power steering pump-to-bracket	20 to 27	2.7 to 3.7
Power steering pump hose	29 to 36	4.0 to 5.0
Front suspension		
Balljoint-to-knuckle arm	71 to 88	9.8 to 12.2
Knuckle arm-to-strut	53 to 72	7.3 to 9.9
Strut-to-hood ledge	18 to 25	2.5 to 3.5
Piston rod locknut	43 to 54	6.0 to 7.5
Tension rod bushing nut	33 to 40	4.5 to 5.5
Tension rod-to-lower arm	33 to 40	4.5 to 5.5
Stabilizer bar bracket	20 to 27	2.7 to 3.7
Suspension crossmember-to-body	37 to 50	5.1 to 6.9
Crossmember-to-transverse link	58 to 80	8.0 to 11.0
Transverse link-to-balljoint	29 to 40	4.0 to 5.5
Knuckle arm-to-side-rod	40 to 72	5.5 to 10.0
Tension rod bracket	20 to 27	2.7 to 3.7
Wheel bearing nut	18 to 22	2.5 to 3.0
Rear suspension		
Sedan		
Shock absorber lower nut	43 to 58	6 to 8
Shock absorber upper nut	19 to 29	2.6 to 4.0
Suspension member mounting bracket locknut	58 to 72	8 to 10
Suspension member mounting bracket-to-body bolt	23 to 31	3.2 to 4.3
Differential mounting bracket locknut	43 to 65	6 to 9
Differential mounting bracket	61 to 83	8.5 to 1.5
Differential gear carrier fitting nut	43 to 58	6 to 8

Torque wrench settings (continued)

Stabilizer bar fixing nut	12 to 15	1.6 to 2.1
Stabilizer bar mounting nut	12 to 15	1.6 to 2.1
Suspension arm pin nut	58 to 72	8 to 10
Station wagon (1980 through 1982)		
Shock absorber lower nut	6.5 to 8.7	0.9 to 1.2
Shock absorber upper nut	6.5 to 8.7	0.9 to 1.2
U-bolt nut	43 to 47	6.0 to 6.5
Spring mount, front pin and shackle nut	43 to 47	6.0 to 6.5
Station wagon (1983 and 1984)		
Shock absorber lower nut	26 to 35	3.6 to 4.8
Shock absorber upper nut	6.5 to 8.7	0.9 to 1.2
Upper link	58 to 72	8 to 10
Lower link	58 to 72	8 to 10
Link damper	22 to 29	3.0 to 4.0

3 Engine

Releasing fuel line pressure

1 Remove the relay bracket and start the engine.
2 Unplug the fuel pump relay harness connector.
3 After the engine stalls, crank it over two or three more times.
4 Turn the ignition switch to the Off position and plug in the fuel pump relay harness connector.

4 Fuel, exhaust and emissions control systems

Fuel injection system

1 The checking and component replacement procedures for later models are the same as described in Chapter 3, except as noted below.

Releasing fuel system pressure

2 To release the built-up fuel pressure, follow the procedure described in Section 3 of this Chapter.

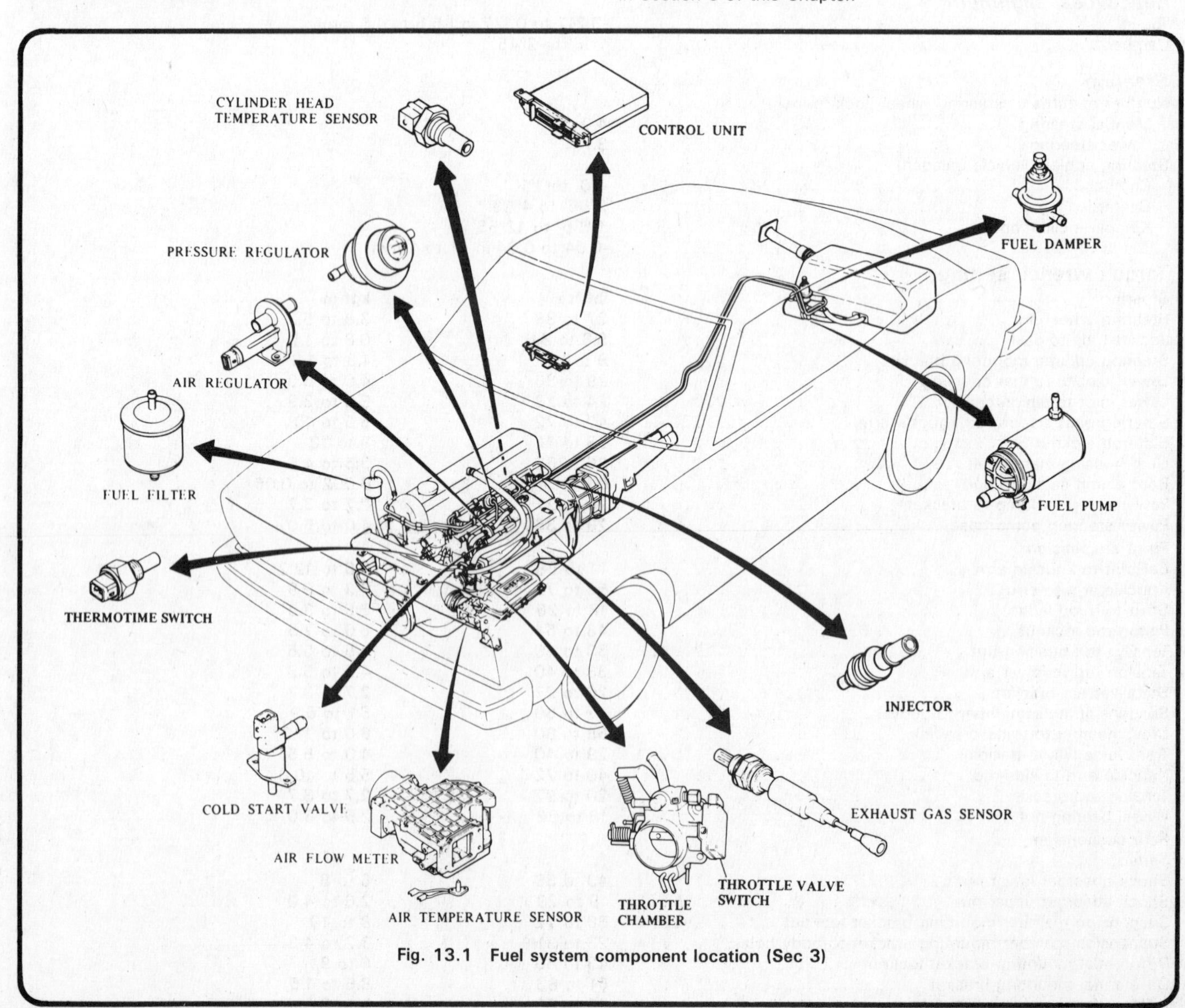

Fig. 13.1 Fuel system component location (Sec 3)

Preparation for testing

3 On later models, the fuel injection system 35-pin connector is checked for resistance and correct voltage with a multimeter (to diagnose faults).

4 Turn the ignition off and disconnect the battery negative cable and the lead wire from the "S" terminal of the starter motor.

5 Disconnect the cold start harness connector.

6 Make sure the air flow meter flap can be pushed manually from the air cleaner side.

7 Unplug the exhaust gas sensor connector (if equipped) and ground it with a jumper wire.

8 Carefully unplug the 35-pin connector, taking care not to bend the pins.

Control unit check (1981 through 1984)

9 Ground the ohmmeter negative lead and touch the positive lead to pin number 15, followed by pins 19, 20 and 22 to check for continuity. If there is no continuity, there is a short circuit in the wiring harness.

Airflow meter potentiometer check

10 Measure the resistance between terminals 33 and 34 as shown in the illustration. The resistance should be between 100 and 400 ohms.

11 Measure the resistance between terminals 34 and 35 as shown in the illustration. The resistance should be between 200 and 500 ohms.

Airflow meter flap check

12 Check the resistance between terminals 32 and 34 while moving the airflow meter flap. If the reading is zero (0), the airflow meter is

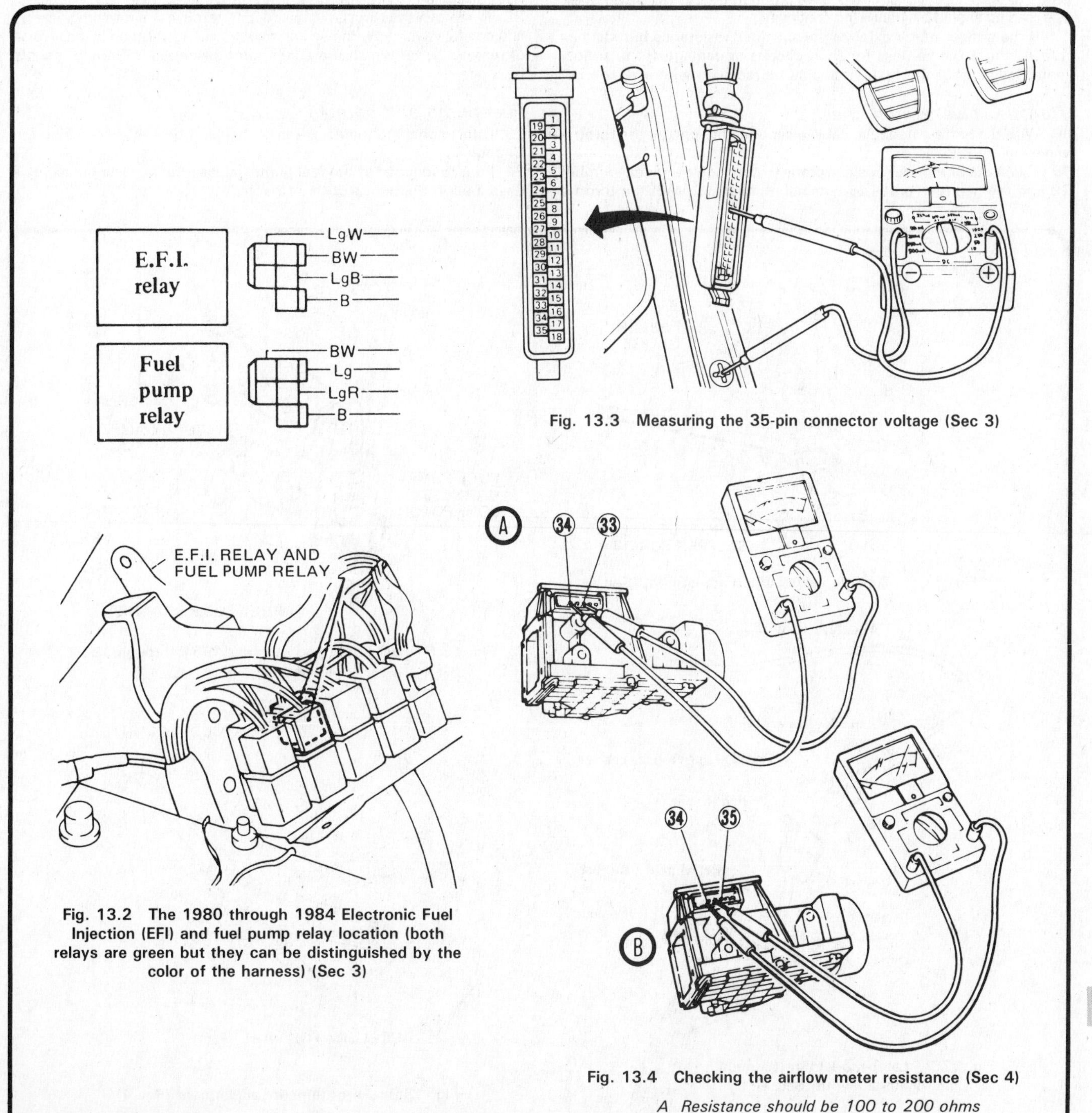

Fig. 13.2 The 1980 through 1984 Electronic Fuel Injection (EFI) and fuel pump relay location (both relays are green but they can be distinguished by the color of the harness) (Sec 3)

Fig. 13.3 Measuring the 35-pin connector voltage (Sec 3)

Fig. 13.4 Checking the airflow meter resistance (Sec 4)

A Resistance should be 100 to 200 ohms
B Resistance should be 200 to 400 ohms

operating properly.

Airflow meter insulation resistance check

13 Ground the negative lead of the ohmmeter to the airflow meter body and check for resistance by touching the positive lead to any of the following terminals: 32, 33, 34 or 35. The resistance should be zero (0).

Airflow meter air temperature sensor check

14 With the intake air temperature above 68°F (20°C), touch the ohmmeter positive probe to terminal 25 and the negative probe to number 34. The meter should read below 2900 ohms.

15 If the temperature is below 68°F (20°C), the meter reading should be 2100 ohms or above.

16 To check the insulation resistance, touch the ohmmeter positive probe to terminal 25 and the negative probe to the airflow meter body. The reading should be infinite (no continuity).

17 If the airflow meter exhibits the specified resistance but still has a fault, check the harness for short circuits or damage. If the airflow meter fails any of the tests, it should be replaced with a new one.

Throttle valve switch check

18 With the battery negative cable disconnected, unplug the throttle valve switch connector.

19 Touch the ohmmeter positive probe to connector terminal number 29 and the negative probe to terminal number 30. With the throttle depressed there should be no continuity and with it released there should be continuity. To adjust the switch to attain proper operation, turn the throttle valve stopper screw until a clearance of 0.012 inch (0.3 mm) is obtained between the screw and valve shaft lever as shown in the illustration. Repeat the test, adjusting as necessary.

Full throttle switch check

20 On 1980 through 1984 models, the full throttle contacts can be checked. Contact the number 24 terminal with the ohmmeter positive probe and the number 30 terminal with the negative probe. With the throttle released there should be no continuity and with it depressed, continuity should exist.

21 If the switch fails either test and the adjustment in Step 19 does not correct the fault, the unit must be replaced with a new one.

22 To remove the throttle valve switch, remove the retaining screws and carefully withdraw the switch toward you. Installation is the reverse of removal, after which the adjustment described in Step 19 should be made.

Fuel pump and damper

23 Later models require the same checking procedure described in Chapter 3.

24 Prior to removal of the fuel pump, release the fuel line pressure as described in Section 3 of this Chapter.

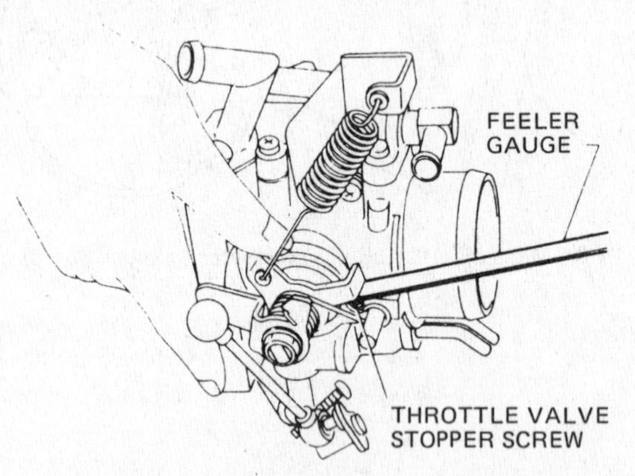

Fig. 13.5 Throttle valve switch adjustment (Sec 4)

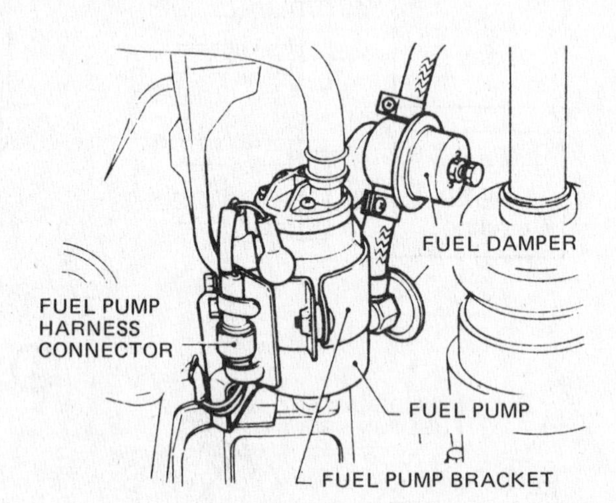

Fig. 13.6 Fuel pump and damper (1981 through 1984 models) (Sec 4)

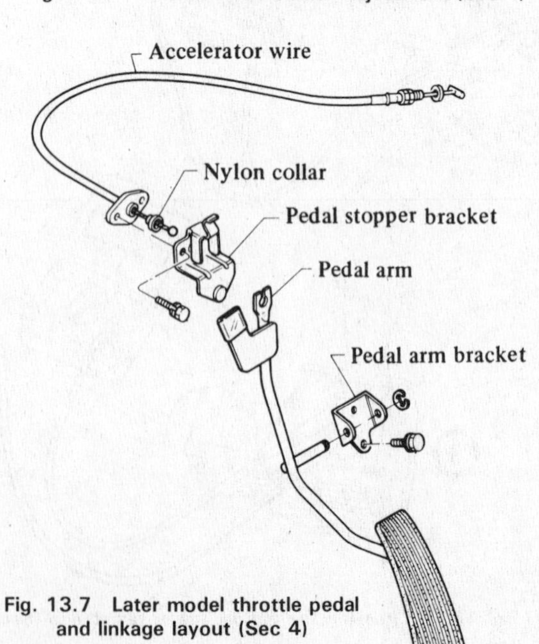

Fig. 13.7 Later model throttle pedal and linkage layout (Sec 4)

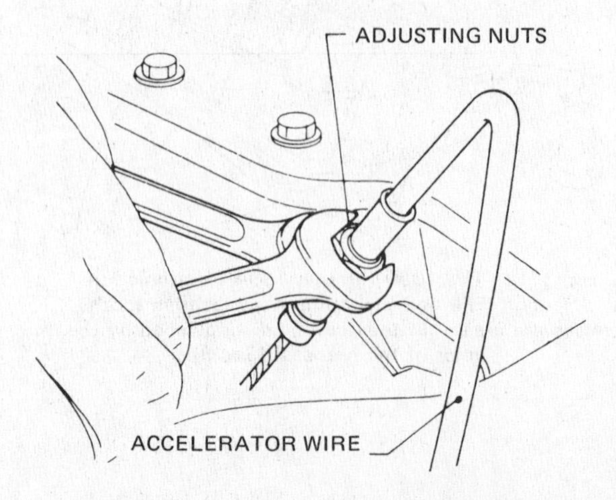

Fig. 13.8 Throttle pedal adjustment (Sec 4)

25 The removal procedure for later models is unchanged except that on 1981 through 1984 models the fuel pump and damper are removed and installed as a unit.

Accelerator linkage — 1981 through 1984 models

Adjustment

26 On these models, pedal height adjustment is not necessary because it is determined by a stopper.

27 To adjust the pedal free play, use two wrenches to turn the adjustment nuts (as shown) until the free play at the center of the pedal pad is between 0.04 and 0.08 inch (1 to 2 mm).

Accelerator wire removal and installation

28 Loosen the adjusting nuts and disconnect the accelerator wire in the engine compartment.

29 Remove the instrument panel lower cover for access and then push the nylon collar toward the end of the wire. Disconnect the wire from the pedal arm.

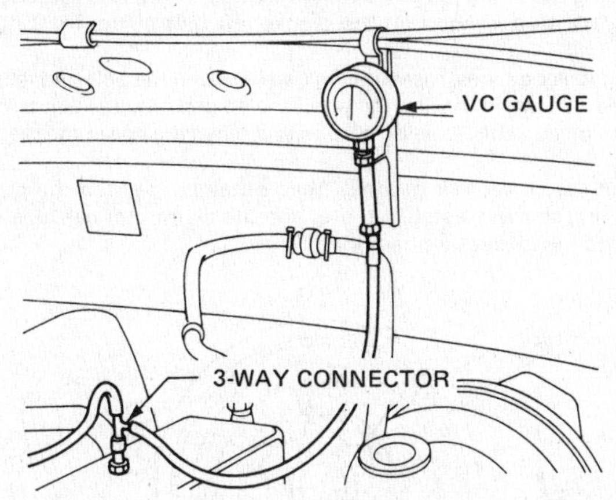

Fig. 13.9 Vacuum gauge connected for BCDD adjustment (Sec 4)

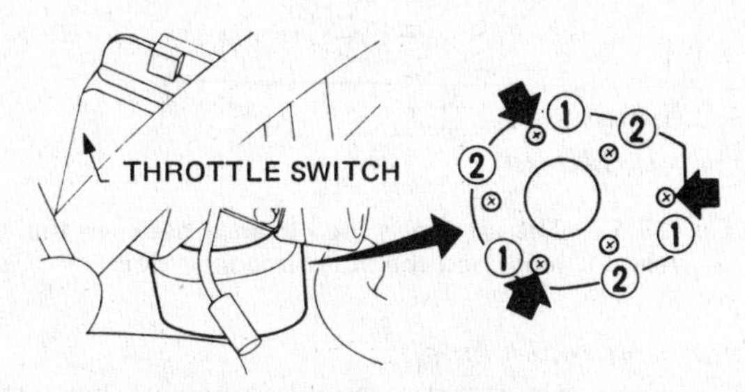

Fig. 13.10 BCDD removal (Sec 4)

1 Screws retaining the BCDD
2 Screws which should **not** *be removed*

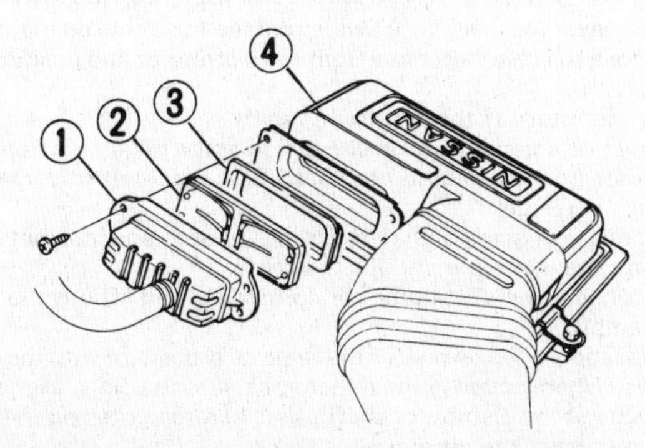

Fig. 13.11 AIS Filter and valve components (Sec 4)

1 Air induction valve case
2 Air induction valve
3 Valve filter
4 Air cleaner assembly

30 Remove the pedal stopper bracket and extract the wire into the engine compartment.

31 Installation is the reverse of removal.

Boost controlled deceleration device (BCDD) adjustment

32 Connect a vacuum hose as shown in the illustration.

33 With the transmission in Neutral and the engine at normal operating temperature, run the engine to approximately 3000 to 3500 rpm and quickly close the throttle.

34 The manifold vacuum reading should abruptly rise to approximately -23.62 in Hg (-600 mm Hg) or above and then gradually decrease and level off. This level is called the operating pressure. The vacuum will then drop from the operating pressure to idling pressure.

35 The operating pressure should be within the specified range. If it is not, turn the BCDD adjusting screw counterclockwise to lower it and clockwise to raise it.

36 Repeat the operation until the operating pressure is within the specified range.

37 If the BCDD cannot be adjusted to achieve the proper operating pressure, it should be replaced with a new one.

38 Replacement of the BCDD is a simple matter of removing the mounting screws, referring to the illustrations.

Mixture ratio feedback system

39 The mixture ratio feedback system coordinates fuel injection and catalytic converter systems to maintain the proper emissions level. An exhaust gas sensor located in the exhaust manifold monitors the gas composition and sends an electrical signal to the electronic fuel injection control unit to control the mixture. A three-way catalytic converter is activated to change the hydrocarbons, carbon monoxide and oxides of nitrogen into carbon dioxide, water and nitrogen.

40 The oxygen sensor should be checked every 30000 miles (48000 km). After the first 30000 miles, the sensor warning light on the instrument panel will automatically come on when the ignition switch is turned on. The wiring should be disconnected from the light after checking the sensor after 30000 miles has elapsed.

41 Inspect the wire and connector at the sensor to make sure they are not cracked, damaged or broken.

42 The control unit located under the front passenger seat has an inspection lamp which indicates if the sensor is operating properly.

43 With the engine at normal operating temperature, run it at approximately 2000 rpm for about two minutes with no load.

44 The inspection lamp on the control unit should flash on and off more than five times during a ten-second period. If it does not, have the sensor checked by your dealer or a properly equipped shop.

Air induction system (AIS)

Description

45 The purpose of the AIS system is to reduce hydrocarbons in the exhaust by drawing fresh air directly into the exhaust manifold. The fresh, oxygen-rich air helps burn the remaining hydrocarbons before they are expelled as exhaust.

46 The components of this simple system include an air filter and a one-way reed valve, both of which are located in the air cleaner housing, and a metal EAI tube with a rubber hose connection running to the manifold.

47 The opening and closing of the exhaust valves creates vacuum pulses which draw in fresh air through the filter and valve assembly. The reed valve only admits air to be drawn into the manifold and prevents exhaust back-pressure from forcing the gases back out.

Checking

48 The simplicity of the system makes it a very reliable one which seldom causes problems. Periodic checks should be made, however, of the condition of the components to be sure there are no leaks or cracks in the system.

49 Two simple functional tests can be performed to ensure that the system is operating properly. Disconnect the rubber hose from the metal EAI tube (the engine must be completely cooled down first). For the first test, attempt to suck air through the rubber hose and then attempt to blow air through the hose. The reed valve should allow you to suck air in through the hose but prevent you from blowing it back out. If you are able to blow air through the hose, the valve is defective and should be replaced with a new one. If you are not able to suck air in, check for a clogged air filter or hose. If this is not the case, replace the valve.

50 For the second test, start the engine and allow it to idle. With the engine idling, hold your hand over the open end of the metal EAI tube. There should be a steady stream of air being sucked into it. Have an assistant depress the throttle. As the engine gains speed, the suction should increase. If this does not happen, there are either leaks or a blockage in the tube.

Valve and filter replacement

51 After removing the valve case-to-air cleaner retaining screws, the valve and filter can be removed and replaced. Installation is the reverse of removal.

EAI tube removal and installation

52 Loosen the pipe-to-exhaust manifold securing nut, remove the screws securing the bracket and rubber hose clamp and remove the pipe. Installation is the reverse of removal.

Positive crankcase ventilation (PCV) system

Description

53 The positive crankcase ventilation, or PCV as it is more commonly called, reduces hydrocarbon emissions by circulating fresh air through the crankcase. This air combines with blow-by gases, or gases blown past the piston rings during compression. The combination is then sucked into the intake manifold to be reburned by the engine.

54 This process is achieved by using one air pipe running from the duct on the rocker arm cover, a one-way PCV valve located on the left underside of the intake manifold and a second air pipe running from the crankcase to the PCV valve.

55 During partial throttle operation of the engine, the vacuum created in the intake manifold is great enough to suck the gases from the crankcase through the PCV valve and into the manifold. The PCV valve allows the gases to enter the manifold but will not allow them to pass in the other direction.

56 The ventilating air is drawn into the rocker arm cover from the air duct and then into the crankcase.

57 Under full-throttle operation, the vacuum in the intake manifold is not great enough to suck in the gases. Under this condition, the blow-by gases flow backward into the rocker arm cover, through the air tube and into the air duct where they are carried into the intake manifold in the normal air intake flow.

Checking

58 The PCV system can be checked quickly and easily for proper operation. This system should be checked regularly as carbon and gunk deposited by the blow-by gases will eventually clog the PCV valve and/or system hoses. When the flow of the PCV system is reduced or stopped, common symptoms are rough idling or a reduced engine speed at idle.

59 To check for proper vacuum in the system, disconnect the rubber air hose where it exits the air duct.

60 With the engine idling, place your thumb lightly over the end of the hose. You should feel a slight vacuum. The suction may be heard as your thumb is released. This will indicate that air is being drawn all the way through the system. If a vacuum is felt, the system is functioning properly.

61 If there is very little vacuum or none at all at the end of the hose, the system is clogged and must be inspected further.

62 With the engine still idling, disconnect the vent tube from the PCV valve. Now place your finger over the end of the valve and feel for suction. You should feel a relatively strong vacuum at this point. This indicates that the valve is good.

63 If no vacuum is felt at the PCV valve, remove the valve from the intake manifold, shake it and listen for a clicking sound. If the valve does not click freely, replace it.

64 If a strong vacuum is felt at the PCV valve (yet there is still no vacuum during the test described in paragraph 64), then one of the system's vent tubes is probably clogged. Both should be removed and blown out with compressed air.

65 If, after cleaning the vent tubes, there is still no suction at the air pipe, there is a blockage in an internal passage. Correction requires disassembly of the engine.

66 When purchasing a new PCV valve, make sure it is the proper one. An incorrect PCV valve may pull too much or too little vacuum, possibly causing damage to the engine.

PCV valve replacement

67 The PCV valve is located on the underside of the intake manifold. It is connected to a short rubber hose which leads into the crankcase.

68 Loosen the clamp that secures the hose to the PCV valve and disconnect the hose from the valve.

69 Unscrew the valve from the intake manifold.

70 Compare the old valve with the new one to make sure they are the same.

71 Screw the new valve into the manifold and connect the hose to the valve.

5 Ignition system

Distributor air gap adjustment

1 Disengage the two spring clips and remove the cap from the distributor.

2 Remove the rotor from the end of the distributor shaft.

3 Position one of the raised segments of the reluctor directly opposite the pole piece protruding from the pickup coil. This can be done easily by removing the spark plugs (to relieve compression) and rotating the crankshaft with a wrench on the crankshaft bolt at the front of the engine.

4 Using feeler gauges, measure the gap between the pole and reluctor segment as shown in the illustration. If the air gap requires adjustment, loosen the pickup coil retaining screws and move the coil in the required direction.

5 When the correct air gap has been obtained, tighten the pickup coil retaining screws; install the rotor and the distributor cap, ensuring that the cap is correctly positioned.

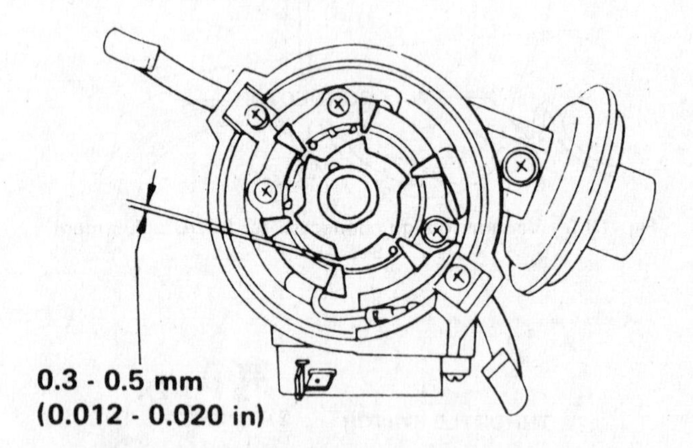

Fig. 13.12 The air gap is the distance between the reluctor teeth and the stator magnet (Sec 5)

IC ignition unit replacement

6 The IC ignition unit is located on the outside of the distributor body and is removed by simply disconnecting the wiring connector and removing the two mounting screws.

7 Installation is the reverse of removal.

Stator and magnet (pickup assembly) replacement

8 Disconnect the ignition coil wire from the top of the distributor cap.

9 Remove the distributor cap from the distributor and position it out of the way.

10 Pull the rotor off the distributor shaft.

11 Using two screwdrivers on either side of the reluctor, carefully pry the reluctor from the distributor shaft. Be careful not to damage the teeth on the reluctor.

12 Remove the screws that retain the stator and magnet to the distributor breaker plate and lift them off.

13 Disconnect the distributor wiring harness and lift out the pickup coil assembly.

14 Installation is the reverse of the removal procedure with the following notes. When installing the reluctor, be sure the pin is aligned with the flat side of the distributor shaft. Also, before tightening the stator, check and adjust the air gap.

Distributor removal

15 Remove the ignition coil wire that leads from the distributor to the coil.

16 Remove the distributor cap from the distributor and position it out of the way.
17 Mark the rotor in relationship to the distributor housing.
18 Disconnect the wires leading to the IC ignition unit.
19 Remove the vacuum line from the vacuum advance canister.
20 Remove the two distributor mounting screws and lift out the distributor. **Note:** *If possible, do not rotate the engine while the distributor is out of place.*

Distributor installation

If the engine was not rotated after removal

21 Position the rotor in the location it was in when the distributor was removed.
22 Lower the distributor down into the engine. To mesh the groove in the bottom of the distributor shaft with the drive spindle (which turns the distributor shaft), it may be necessary to turn the rotor slightly.
23 With the base of the distributor all the way down against the engine block and the mounting screw holes lined up, the rotor should be pointing to the mark made on the distributor housing during removal. If the rotor is not in alignment with the mark, repeat the previous steps.
24 Install the two distributor mounting screws and tighten them securely.
25 Connect the vacuum line to the vacuum advance canister.
26 Reconnect the wiring connector to the IC unit.
27 Reinstall the distributor cap and connect the ignition coil wire.
28 Check the ignition timing as described in Chapter 1.

If the engine was rotated after removal

29 Set the number 1 piston at TDC. Refer to Chapter 1, if necessary.
30 Temporarily install the distributor cap on the distributor. Note where the number 1 spark plug terminal inside the cap is located in relation to the distributor body and make a mark on the outside of the body at this point.
31 Remove the distributor cap again. Rotate the rotor until it is aligned with the mark. In this position it should be firing the number one

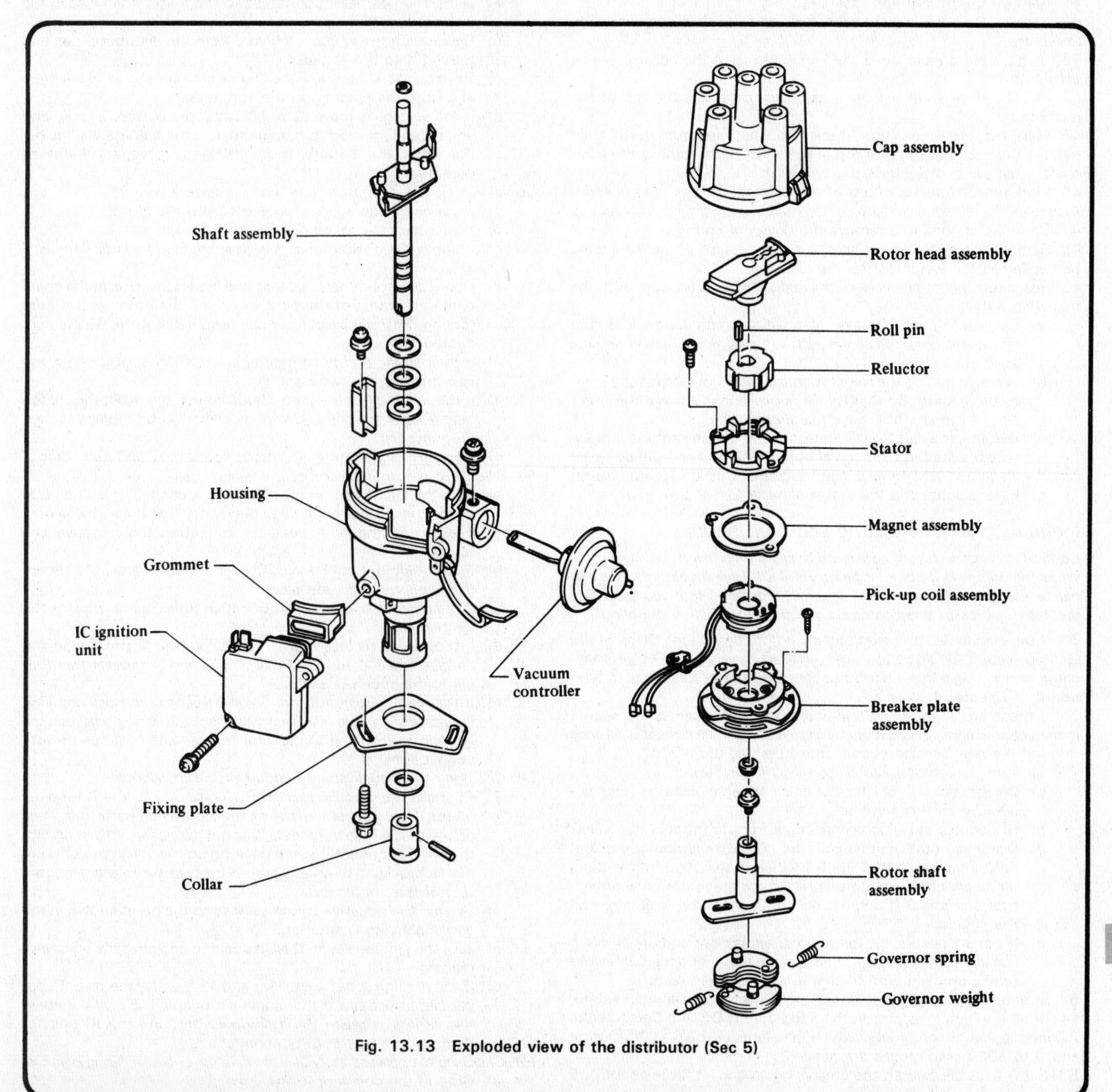

Fig. 13.13 Exploded view of the distributor (Sec 5)

13

spark plug.
32 Proceed with installation as detailed in Steps 22 through 28. Disregard the rotor mark references in Step 23.

Distributor dismantling and reassembly

33 Remove the distributor.
34 Remove the distributor cap.
35 Pull off the rotor.
36 Disconnect the wiring connector leading to the IC unit. Remove the unit's two mounting screws and lift it off.
37 Remove the stator attaching screws and lift out the stator and magnet.
38 Remove the screws that retain the vacuum canister and lift it off.
39 Using two screwdrivers placed on either side of the reluctor, carefully pry it off the distributor shaft. Be careful not to damage the reluctor teeth.
40 Drive the roll pin out of the reluctor.
41 Remove the pickup coil assembly.
42 Remove the breaker plate set screws and lift out the breaker plate assembly.
43 Using a pin punch, drive the knock pin from the collar and pull off the collar.
44 Pull the rotor shaft assembly and driveshaft out the top of the distributor.
45 Mark the relative position of the rotor shaft and driveshaft. Then remove the packing from the top of the rotor shaft, remove the rotor shaft setscrew and separate the two shafts.
46 Mark the relationship of one of the governor springs to its bracket. Also mark the relationship of one of the governor weights to its pivot pin.
47 Carefully unhook and remove the governor springs.
48 Remove the governor weights. A small amount of grease should be applied to the weights after removal.
49 Reassembly is the reverse of the disassembly procedure with the following notes:

a) Be sure to correctly align all positioning marks made during disassembly so that all the parts are assembled in their original position.
b) When installing the reluctor on the shaft, drive the roll pin into the reluctor so that its slit is positioned toward the outer end of the shaft. Always use a new roll pin.
c) Before installing the IC unit onto the distributor body, make sure the mating surfaces of both the IC unit and the body are clean and free from dirt or moisture. This is very important.
d) Before tightening the stator plate, adjust the air gap.

Ignition system — testing and fault finding

Caution: *Never touch your bare hand to any high tension cables or parts while the engine is running or being cranked. Even insulated parts can cause a shock if they are moist. It is recommended that you wear dry, insulated gloves or wrap the part in a dry cloth before handling.*

50 If the engine will turn over but will not start, the first check of the ignition system should be to visually inspect the condition of the spark plugs, spark plug wires, distributor cap and rotor as described in Section 2 of Chapter 4.
51 If these are all in good condition, and the plug wires are secure in their connections, the next check should be to see if current is flowing through the high tension circuit, thus sparking the plugs.

a) Turn the ignition switch to the Off position.
b) Disconnect the EFI fusible link connector located near the positive battery terminal.
c) Disconnect the wiring connector leading to the cold start valve.
d) Disconnect the ignition coil wire from the distributor cap and hold it approximatly 3/16 to 1/4 inch (4 to 5 mm) from a clean metal area of the engine. Have an assistant crank the engine over and check if a spark occurs between the coil wire and the engine.
e) If a spark occurs, the ignition system is okay, and the problem lies in another system. If no spark occurs, or occurs intermittently, proceed with further ignition system tests.

52 In order to accurately diagnose problems in the Datsun system, a voltmeter which measures in the 0 to 20 volts DC and 0 to 10 volts AC ranges, and an ohmmeter which measures in the 0 to 1000 ohm and 0 to 5000 ohm ranges are needed.
53 If it is possible to start the engine, do so and let it run about 5 to 15 minutes with the hood closed to bring all the components to normal operating temperature.
54 *Checking the battery voltage with no load.*

a) With the ignition key in the Off position, connect a voltmeter so its positive lead is on the positive battery terminal and its negative lead is on the negative battery terminal.
b) Note the reading on the voltmeter. If the reading is between 11.5 volts and 12.5 volts, the battery is okay and you should proceed to paragraph 55.
c) If the reading is below 11.5 volts, the battery is discharged. It should be brought to a full charge either by running the engine or by using a battery charger. If the car has been used on a regular basis and there is no obvious cause for the battery to be discharged (such as leaving the lights on), then the condition of the battery, charging system and starting system should be checked out.

55 *Checking the battery voltage while the engine is cranking.*

a) Leave the voltmeter connected to the battery cable as in the last paragraph.
b) Disconnect the ignition coil wire from the distributor cap and ground it to the engine.
c) Have an assistant crank the engine over for about 15 seconds and note the reading on the voltmeter.
d) If the voltage is more than 9.6 volts, the battery is okay and you should proceed to paragraph 7. If the voltage was below 9.6 volts the battery is insufficiently charged. Refer to paragraph 54.

56 *Checking the distributor cap and secondary winding.*

a) Disconnect the spark plug wires from the plugs.
b) Disconnect the ignition coil wire from the coil.
c) Remove the distributor cap, with spark plug and coil wires still attached.
d) Connect an ohmmeter so that one lead is inserted in the spark plug end of one of the plug wires and the other lead is contacting the inner distributor cap terminal that the wire is connected to.
e) If the reading on the ohmmeter is less than 30000 ohms, the cap terminal and wire are okay.
f) If the reading is more than 30000 ohms, the resistance is too high. Check the cap and wire individually and replace the appropriate part.
g) Repeat this test on each of the spark plug and coil wires.

57 *Checking the ignition coil secondary circuit.*

a) With the ignition key in the off position, connect the ohmmeter so that one lead is contacting the center high tension wire connector and the other lead is contacting the negative coil terminal.
b) If the reading is between 8200 and 12 400 ohms, the secondary coil windings are okay.
c) If the ohmmeter reading is not within these specs, replace the ignition coil.
d) If the reading is less than 1 volt below the batttery cranking voltage (measured in paragraph 55) and is greater than 8.6 volts, the circuit is okay.
e) If the reading is more than 1 volt below the battery cranking voltage and/or below 8.6 volts, inspect the wiring between the ignition switch and the IC unit for damage or loose or dirty connections.

58 *Checking the power supply circuit at the distributor.*

a) Connect the positive lead of the voltmeter to the B terminal on the IC ignition unit in the distributor (when performing this or any of the following tests, it is not necessary to disconnect the wiring connector when hooking up the voltmeter or ohmmeter, providing they have probes that can be inserted into the rear of the connector).
b) Ground the negative voltmeter lead to the point shown in the accompanying illustration.
c) Turn the ignition key to the On position and note the voltmeter reading.
d) If the reading is between 11.5 and 12.5 volts, the power supply circuit is okay. If the reading is below 11.5 volts, inspect the wiring between the ignition switch and the IC unit for damage or loose or dirty connections.

59 *Checking the power supply circuit while the engine is being cranked.*

a) Ground the coil wire to the engine.

b) Connect the voltmeter as in paragraph 58.
c) Have an assistant crank the engine over for about 15 seconds and note the reading on the voltmeter.
d) If the reading is less than 1 volt below the battery cranking voltage (measured in paragraph 55) and is greater than 8.6 volts, the circuit is okay.
e) If the reading is more than 1 volt below the battery cranking voltage and/or below 8.6 volts, inspect the wiring between the ignition switch and the IC unit for damage or loose or dirty connections.

60 *Checking the ignition primary circuit.*

a) Connect the voltmeter so the positive lead is connected to the C terminal on the IC unit and the negative lead contacts the point shown in the illustration. Turn the ignition key to the On position and note the voltmeter reading. If between 11.5 and 12.5 volts is shown, the circuit is okay. If less than 11.5 volts is shown, proceed to paragraph 61.

61 *Checking the ignition coil primary circuit.*

a) With the ignition key in the Off position and the ignition coil wire removed from the coil, connect an ohmmeter (set in the 1x range) between the positive and negative terminals on the ignition coil.
b) If the ohmmeter reading is between 0.84 and 1.02 ohms the coil is okay and the ignition switch and the wiring running between the ignition switch, the coil and the IC unit should be checked for damage or loose or dirty connections.
c) If the ohmmeter reading is not within these specs, the ignition coil should be replaced.

62 *Checking the IC unit ground circuit.*

a) Remove the ignition coil wire from the distributor cap and ground it to the engine.
b) Connect a voltmeter so its negative lead is connected to the negative battery terminal, and the positive lead is connected to the point shown in the illustration.

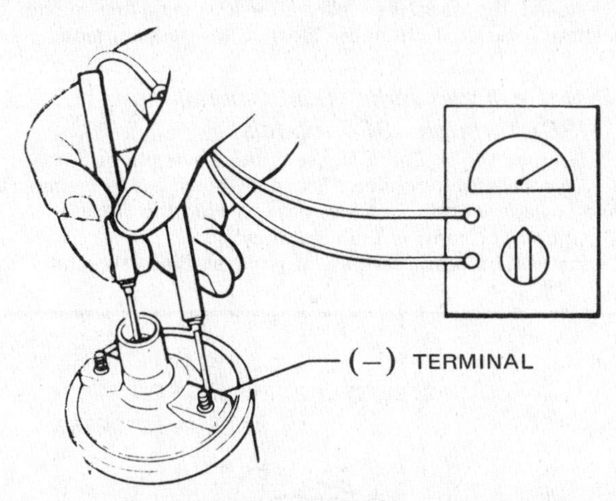

Fig. 13.14 Proper ohmmeter connections for checking the secondary circuit of the ignition coil (Sec 5)

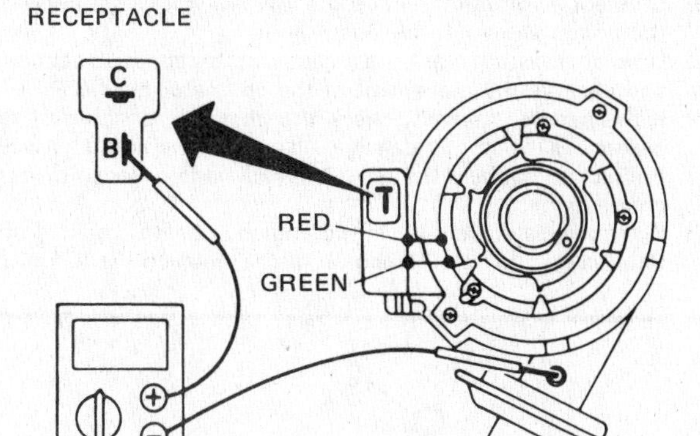

Fig. 13.15 Proper voltmeter connections for checking the power supply circuit (Sec 5)

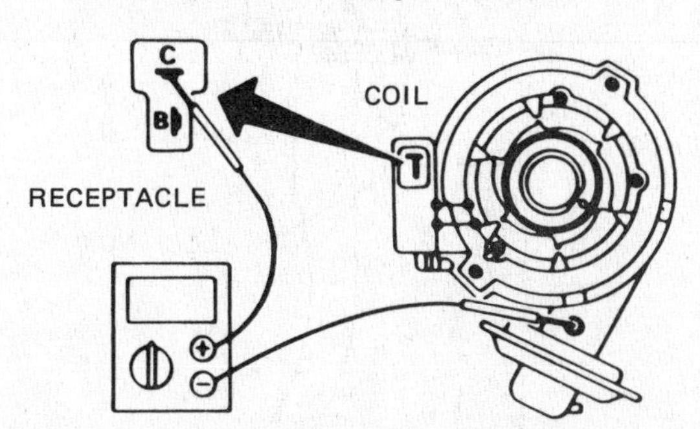

Fig. 13.16 Proper voltmeter connections for checking the ignition primary circuit (Sec 5)

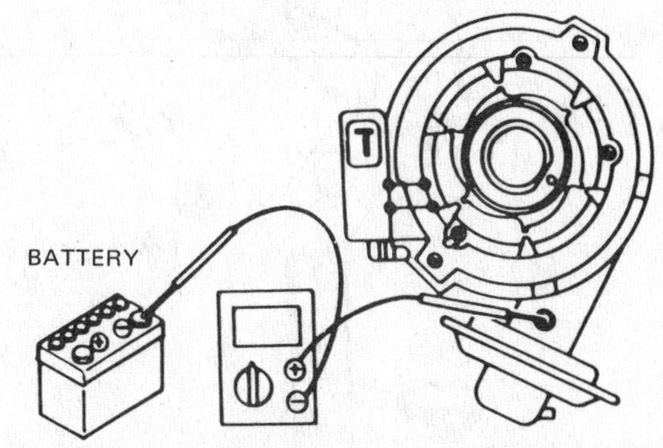

Fig. 13.17 Proper voltmeter connections for checking the IC unit ground circuit (Sec 5)

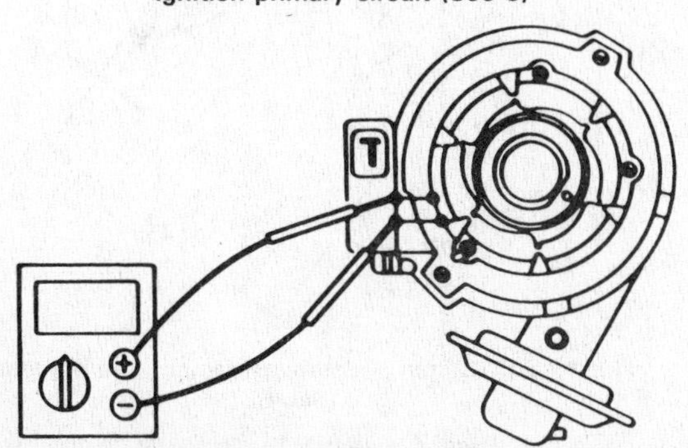

Fig. 13.18 Proper voltmeter connections for checking the pickup coil resistance (Sec 5)

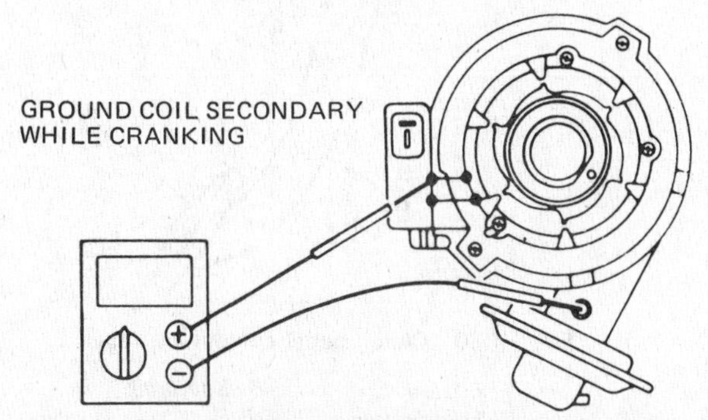

Fig. 13.19 Proper voltmeter connections for checking the pickup coil output (Sec 5)

c) Have an assistant crank the engine over for about 15 seconds, and note the voltmeter reading.
d) If more than 0.5 volts is shown, check the grounding of the distributor, the wiring between the battery and the chassis ground and the negative battery connection.
e) If the reading shown 0.5 volts or less further checks are still needed. Proceed to pararaph 63.

63 *Checking the pickup coil resistance.*
a) For this check the engine should be at normal operating temperature.
b) With the ignition switch in the Off position, connect an ohmmeter (set in the 100x range) as shown in the illustration and note the reading.
c) If approximately 400 ohms is shown, the pickup coil resistance is okay.
d) If the reading is substantially above or below 400 ohms, inspect the pickup coil and its wiring for damage or loose or dirty connections.

64 *Checking the pickup coil output*
a) The engine should still be at normal operating temperature.
b) Connect a voltmeter (set on the low AC volt scale) between the points shown in the illustration.
c) Have an assistant crank the engine over for about 15 seconds, and observe the movement of the voltmeter needle.
d) If the needle is steady, check the physical condition of the pickup coil reluctor for damage. Also check the wiring between the pickup coil and the IC unit for damage or loose or dirty connections.
e) If the needle wavers while the engine is being cranked, and there is still no spark being produced, replace the IC unit.

6 Clutch

Clutch pedal removal, installation and adjustment (1981 through 1984 models)

1 Remove the lower instrument panel cover.
2 Pry off the snap-ring and take out the clevis pin.
3 Remove the fulcrum pin, followed by the clutch pedal.
4 Inspect the pedal bushing and nut, clevis pin, snap-ring, pedal, pedal pad and stopper and return spring for wear, damage or fatigue. Replace any damaged or worn components. Because the bushing is a press fit, the entire pedal assembly will have to be replaced if it is worn.
5 Installation is the reverse of removal, taking care to apply multipurpose grease to all sliding surfaces and to install the clevis pin on the left of the clutch pedal.
6 Referring to the accompanying illustration, adjust the pedal to the specified height by loosening the pedal stopper locknut and turning the nut. Tighten the locknut after adjustment.
7 Adjust the pedal free play after loosening the master cylinder pushrod locknut. Tighten the locknut after adjustment.

Master cylinder removal and installation (1981 through 1984 models)

8 Remove the snap-ring and pull the clevis pin out.
9 Disconnect the line with a flare nut wrench and be prepared to catch fluid spillage with a container or rags. Plug the line.
10 Unbolt and remove the master cylinder.
11 Installation is the reversal of removal. Bleed the clutch system.

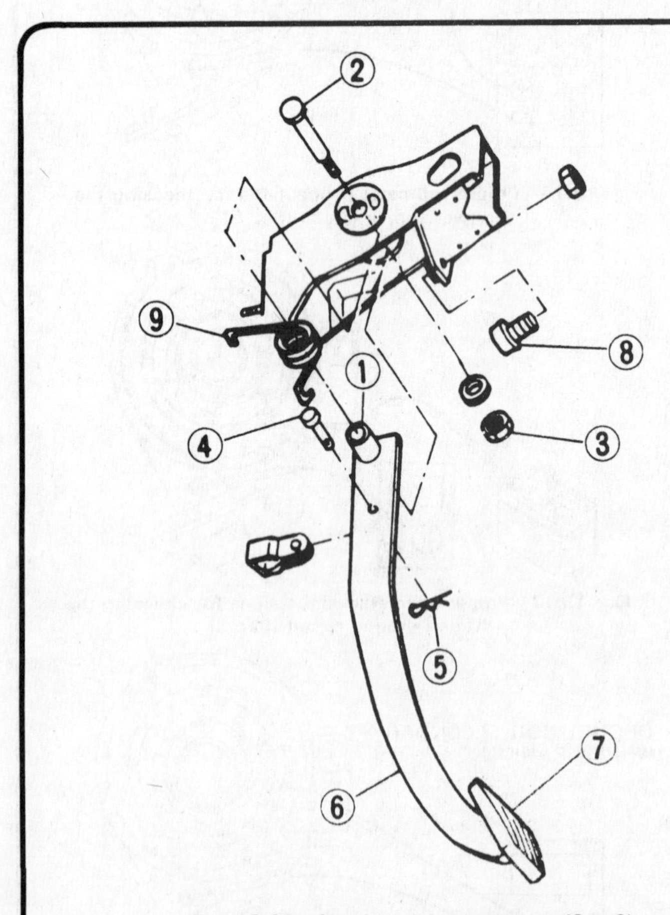

Fig. 13.20 Clutch pedal components (Sec 6)

1 Bushing
2 Fulcrum pin
3 Nut
4 Clevis pin
5 Snap-ring
6 Pedal
7 Pedal pad

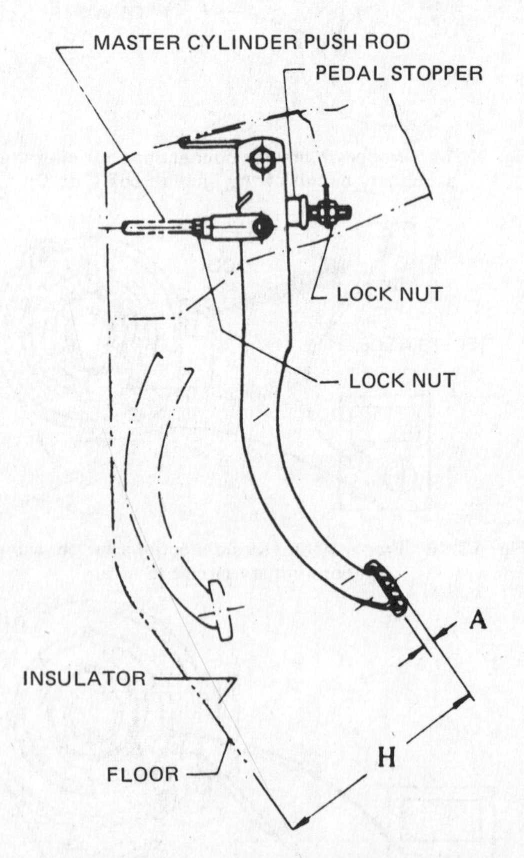

Fig. 13.21 Clutch pedal measurement and adjustment (Sec 6)

A Pedal free play *B Pedal height*

Clutch damper removal and installation (1981 through 1984 models)

12 Disconnect and plug the clutch line at the clutch damper with a flare nut wrench.
13 Remove the damper from the bracket.
14 Installation is the reverse of removal, taking care to bleed the clutch hydraulic system and adjust the pedal height and free play.

7 Manual and automatic transmission

5-speed transmission overhaul

Initial disassembly

1 With the transmission removed, thoroughly clean the external surfaces.

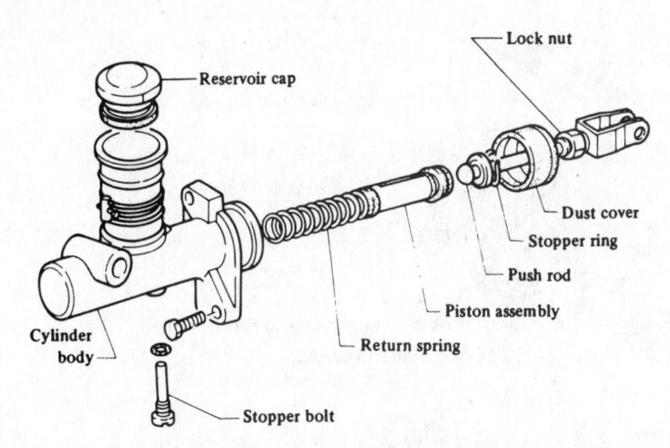

Fig. 13.22 Clutch master cylinder component layout (Sec 6)

2 Remove the rubber dust boot from the withdrawal lever opening in the clutch housing.
3 Remove the release bearing and hub together with the withdrawal lever.
4 Remove the reverse lamp switch, neutral switch, top gear switch, and/or overdrive gear switch, as equipped.
5 Remove the screw that retains the speedometer gear to the rear extension housing and lift out the gear.
6 Pry off the E-ring from the stopper guide pin and drive out the pin.
7 Remove the return spring plug and lift out the return spring and plunger.
8 Remove the two screws that retain the reverse check cover to the housing and lift out the reverse check sleeve.

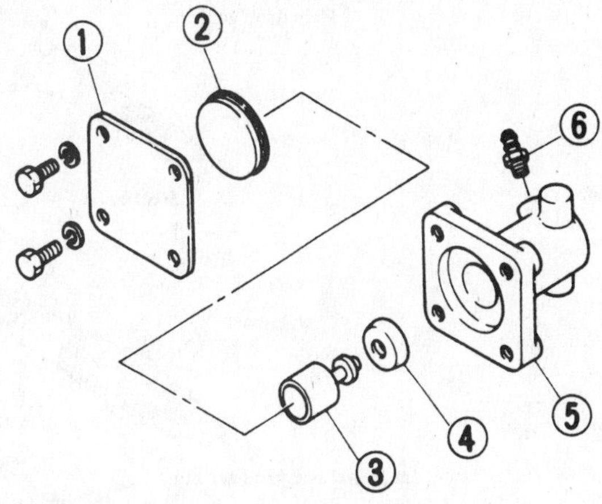

Fig. 13.23 Clutch damper components (Sec 6)

1 Damper cover
2 Damper rubber
3 Piston
4 Piston cup
5 Cylinder
6 Bleeder screw

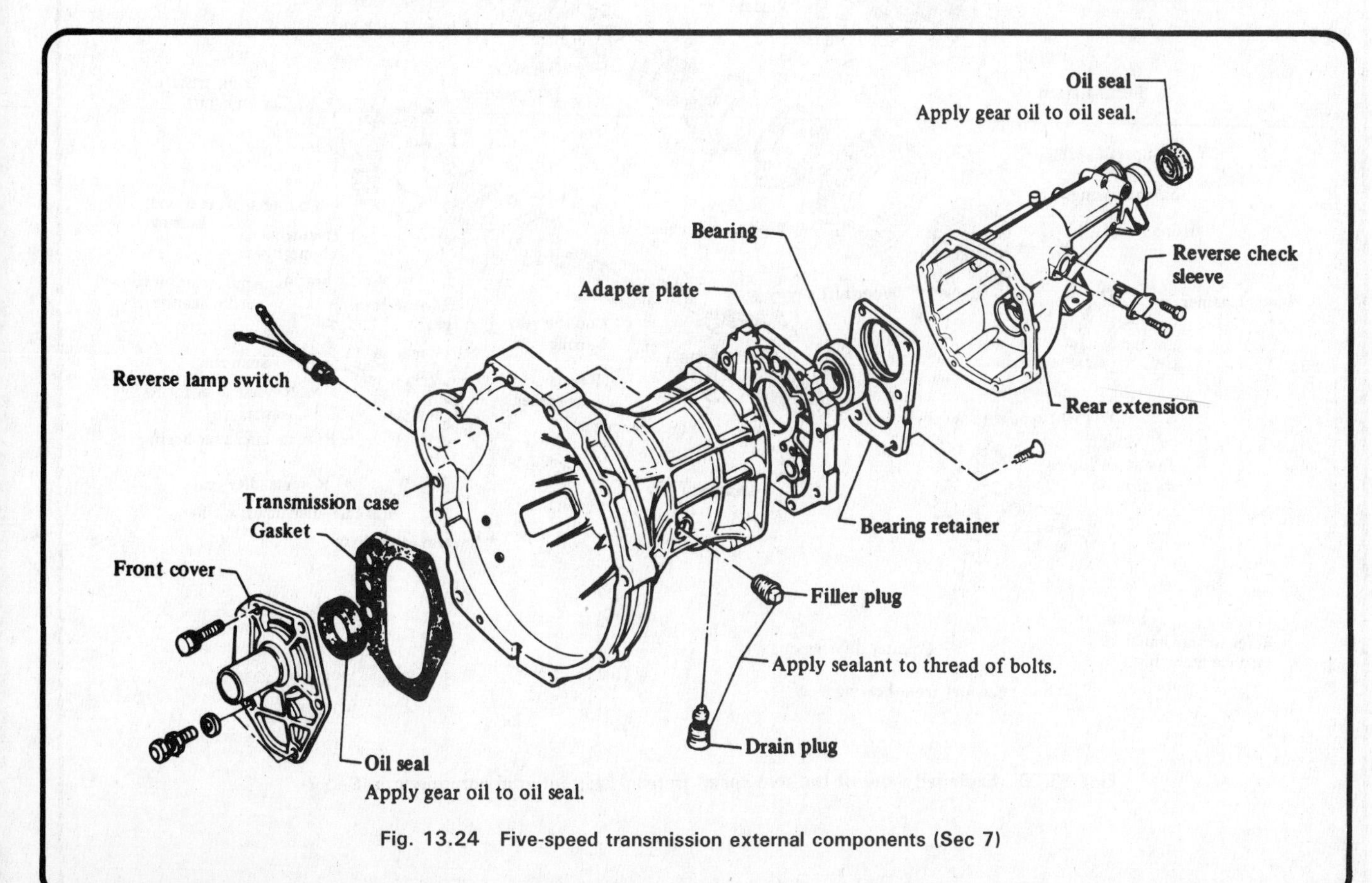

Fig. 13.24 Five-speed transmission external components (Sec 7)

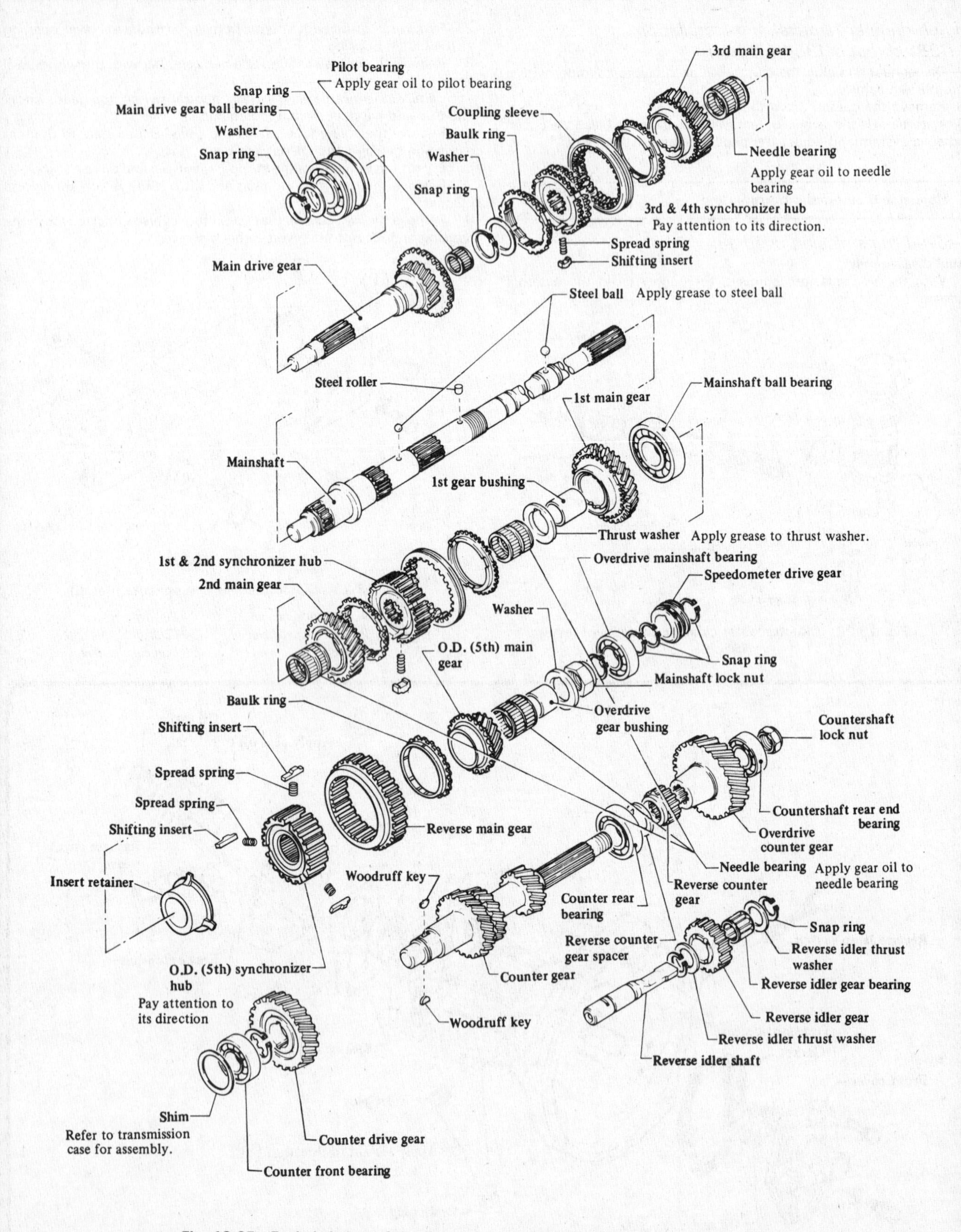

Fig. 13.25 Exploded view of the five-speed transmission internal components (Sec 7)

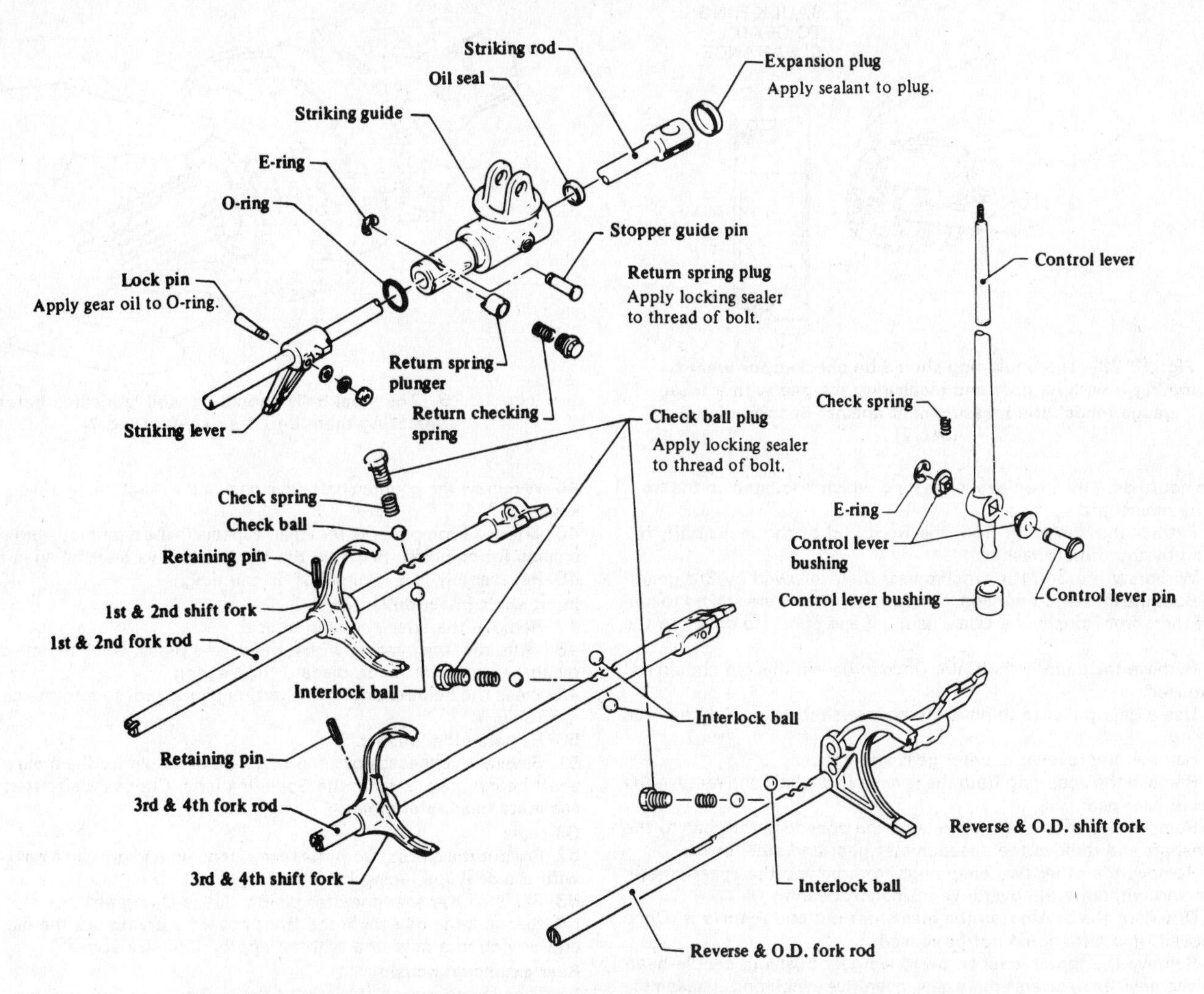

Fig. 13.26 Exploded view of the five-speed transmission shift rod components (Sec 7)

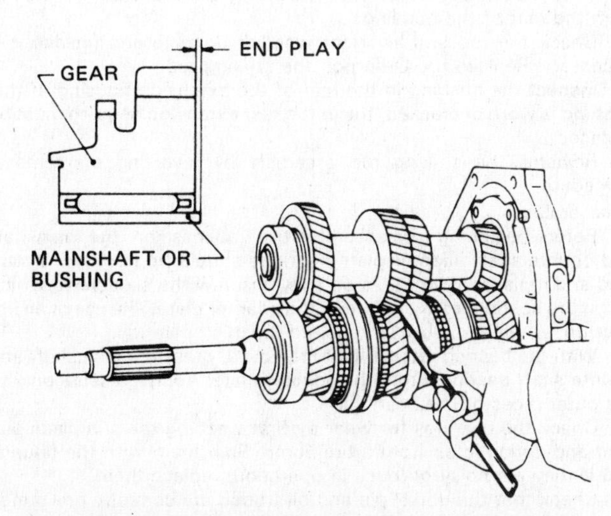

Fig. 13.27 Gear end play should be measured both before and after disassembling the transmission to check for wear and proper installation (Sec 7)

9 Unscrew the securing bolts, and drive the rear extension housing from the main transmission casing, using a soft mallet.
10 Unscrew and remove the front cover retaining bolts. Remove the front cover and extract the countershaft bearing shim and the input shaft bearing snap-ring.
11 Drive off the one-piece bellhousing/transmission casing from the adapter plate.
12 Make up a suitable plate and bolt it to the transmission adapter plate and then secure the support plate in a vise.
13 Drive out the securing pins from each of the shift forks, using a suitable thin drift.
14 Unscrew and remove the three detent ballplugs.
15 Withdraw the selector rods from the adapter plate.
16 Catch the shift forks and extract the balls and springs as the selector rods are withdrawn. The four smaller balls are the interlock balls.
17 At this point, inspect the gears and shafts for any wear, chipping or cracking.
18 Also, use a feeler gauge between each mainshaft gear to determine the amount of gear and end play that exists. Compare the results to the Specifications.
19 If the gear end play is not within specifications, or if the gears or shafts show signs or wear or damage, the gear assemblies should be disassembled and the defective parts replaced.
20 Lock the gears and draw the front bearing from the countershaft with a puller.
21 Extract the now exposed snap-ring from the countershaft.
22 Withdraw the countershaft gear together with the input shaft. Take

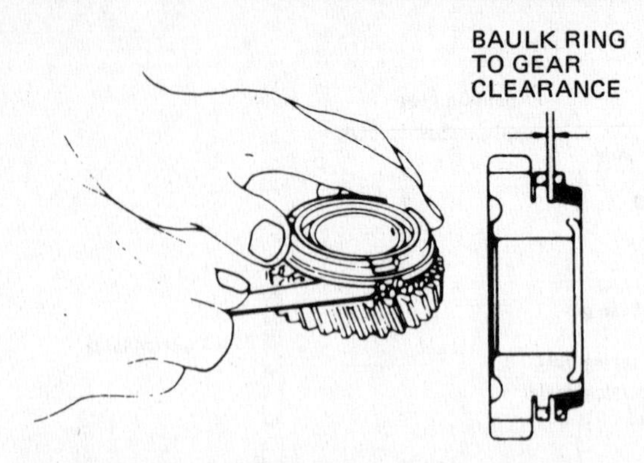

Fig. 13.28 The baulk ring should be checked for wear by mating it with its gear and measuring the gap with a feeler gauge (check the measurement against Specifications) (Sec 7)

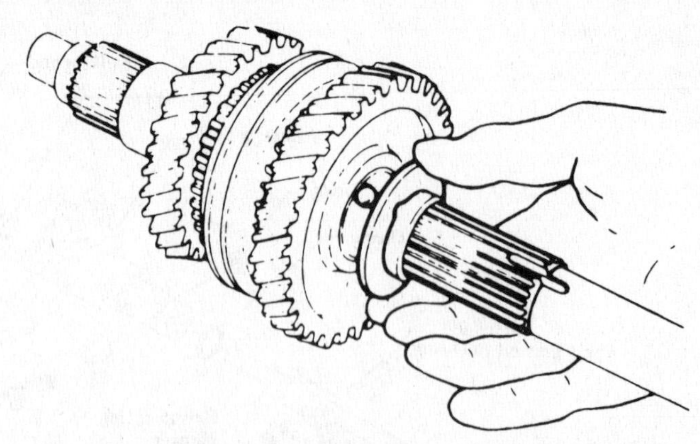

Fig. 13.29 The steel balls should be well lubricated before installing them into their shafts (Sec 7)

care not to drop the needle roller bearing which is located on the front of the mainshaft.
23 Extract the snap-ring from the front end of the mainshaft, followed by the thrust washer.
24 Withdraw the 3rd/4th synchronizer unit, followed by 3rd gear.
25 Both the mainshaft nut and the countershaft nut are staked to prevent them from loosening. Use a hammer and punch to drive out the staking.
26 Remove the countershaft nut. Once removed, this nut should not be reused.
27 Use a gear puller to remove the countershaft overdrive gear and bearing.
28 Remove the reverse counter gear and spacer.
29 Remove the snap-ring from the reverse idler shaft and remove the reverse idler gear.
30 Remove the snap-ring that retains the speedometer gear to the mainshaft and remove the speedometer gear and steel ball.
31 Remove the other two snap-rings from behind the speedometer gear and withdraw the overdrive mainshaft bearing.
32 Drive out the staking on the mainshaft nut and remove it. Once removed, this nut should not be reused.
33 Remove the thrust washer, overdrive gear bushing, needle bearing, overdrive gear, reverse main gear, overdrive synchronizer assembly and insert retainer.
34 Drive the mainshaft and countershaft assemblies simultaneously from the adapter plate, using a soft-faced hammer.

Mainshaft

35 Carefully examine the gears and shaft splines for chipping of the teeth or wear and then dismantle the gear train into its component parts, replacing any worn or damaged items.
36 Examine the shaft itself for scoring or grooving, also the splines for twist, taper or general wear.
37 Examine the synchromesh units for cracks or wear or general looseness in the assembly and replace them if evident, particularly if there has been a history of noisy gearchange or where the synchromesh can be easily 'beaten'.
38 Press the baulk ring tight against the synchromesh cone and measure the gap between the two components. If it is less than specified, replace the components (refer to Specifications and accompanying figure).
39 When reassembling the synchromesh unit ensure that the ends of the snap-ring on opposite sides of the units do not engage in the same slot.
40 Begin assembly of the mainshaft by installing the 2nd gear needle bearing, 2nd gear, the baulk ring followed by the 1st/2nd synchromesh unit, noting carefully the direction of installing the latter.
41 Now install the 1st gear baulk ring, needle bearing, steel ball, thrust washer, bushing and 1st gear. Be sure the steel ball is well greased when installed.

Countershaft

42 The countershaft front bearing was removed at the time of dismantling the transmission into major units.
43 The countershaft rear bearing was left in position in the adapter plate.
44 Withdraw the countershaft drive gear and extract the two Woodruff keys.
45 Check all components for wear, especially the gear teeth and shaft splines for chipping. Reinstall the Woodruff keys and the snap-ring.
46 Reassembly is a reversal of dismantling.

Input shaft (main drive gear)

47 Remove the snap-ring and spacer.
48 Withdraw the bearing with a puller or a press. Once removed (by means of its outer race), discard the bearing.
49 Press the bearing onto the shaft, applying pressure to the center race only.
50 Reinstall the washer.
51 Several thicknesses of snap-rings are available for the main input shaft bearing, as listed in the Specifications. Choose a size that will eliminate bearing end play.

Oil seals

52 Pry out the oil seal from the rear extension and drive in a new one, with the seal lips facing in.
53 Reinstall the speedometer pinion sleeve O-ring seal.
54 Reinstall the oil seal in the front cover by prying out the old one and driving in a new one with an appropriate size socket.

Rear extension housing

55 Loosen the nut on the end of the striking rod lock pin until it is half off the threads.
56 Using the nut as a guide, drive the lock pin from the striking rod with a punch.
57 Slide the striking lever from the striking rod and withdraw the rod from the rear of the housing.
58 Check the rod and lever for wear or damage and replace it if necessary. Replace the O-ring on the striking rod.
59 Inspect the bushing in the rear of the extension housing. If this bushing is worn or cracked, the entire rear extension housing must be replaced.
60 Reinstall the striking rod assembly by reversing the removal procedure.

Reassembly

61 Before beginning to reassemble the transmission, the mainshaft and countershaft adapter plate bearings should be removed, examined and replaced if worn. To do this, unscrew the six screws which retain the bearing retainer plate to the adapter plate. The use of an impact driver will probably be required for this operation.
62 With the bearing retainer plate removed, press the mainshaft and countershaft bearings from the adapter plate. Apply pressure only to the outer races of the bearings.
63 Check the bearings for wear by first washing them in clean solvent and drying in air from a tire pump. Spin them with the fingers, and if they are noisy or loose in operation, replace them.
64 Check that the dowel pin and oil trough are correctly positioned on the adapter plate.
65 Tap the mainshaft bearing lightly and squarely into position in the adapter plate.
66 Drive the reverse idler shaft into the adapter plate so that 2/3rds of its length is projecting rearwards. Ensure that the cutout in the shaft is positioned to receive the edge of the bearing retainer plate.

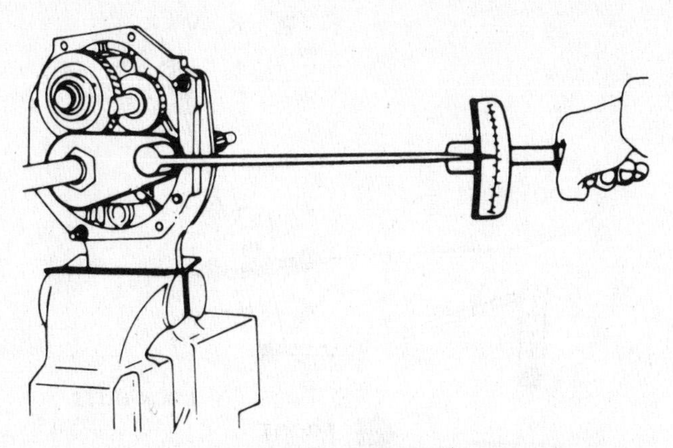

Fig. 13.30 The mainshaft locknut should be tightened using a torque wrench and adapter as shown (Sec 7)

67 Install the bearing retainer plate and tighten the screws to the specified torque.
68 Stake each screw in two places to prevent them from loosening.
69 Tap the countershaft rear bearing into position in the adapter plate.
70 Press the mainshaft assembly into position in the bearing in the adapter plate. Support the rear of the bearing center track during this operation.
71 Press the countershaft assembly into position in the bearing in the adapter plate. Again support the rear of the bearing center race during this operation.
72 Install the needle bearing, 3rd gear, baulk ring and the 3rd/4th synchromesh unit to the front of the mainshaft.
73 Install the thrust washer, and then choose a snap-ring from the sizes listed in the Specifications that will minimize end play.
74 Insert the needle pilot bearing into its recess at the end of the input shaft.
75 Mesh the countershaft drive gear with the 4th gear on the input shaft. Push the drive gear and input shaft onto the countershaft and mainshaft simultaneously, but a piece of tubing will be needed to drive the countershaft gear into position while supporting the rear end of the countershaft.
76 Select a countershaft drive gear snap-ring from the sizes listed in the Specifications, so that the gear end play will be minimized.
77 Using an appropriate size socket, drive the front bearing onto the countershaft.
78 Install the reverse counter gear spacer onto the rear of the countershaft.
79 Install the snap-ring, thrust washer, needle bearing, reverse idler gear, reverse idler thrust washer and rear snap-ring onto the reverse idler shaft.
80 Onto the rear side of the mainshaft, install the insert retainer, synchronizer assembly, reverse gear, overdrive gear bushing, needle bearing and baulk ring.
81 Install the reverse counter gear on the countershaft.
82 Mesh the overdrive gear with the overdrive counter gear and install them on their respective shafts with the overdrive gear on the mainshaft and the overdrive counter gear on the countershaft.
83 Apply grease to the steel ball and install it and the thrust washer onto the rear of the mainshaft.
84 Install a new locknut onto the rear of the mainshaft and torque it to specs. **Note:** *In order to accurately tighten the nut to its torque specifications, a wrench adapter should be used as shown in the accompanying figure. Used with the adapter, the torque reading on the wrench will not be accurate and should be converted to the correct torque by referring to the chart shown.*
85 Install the countershaft rear end bearing onto the countershaft.
86 Install the countershaft locknut and torque it to specs.
87 Use a hammer and punch to stake both the mainshaft and countershaft locknuts so they engage the grooves in their respective shafts.
88 Once again, measure the gear end play as described in Step 18.
89 Fit a snap-ring onto the mainshaft and then install the overdrive mainshaft bearing.
90 Choose a snap-ring from the sizes listed in the Specifications to eliminate end play of the mainshaft rear bearing.

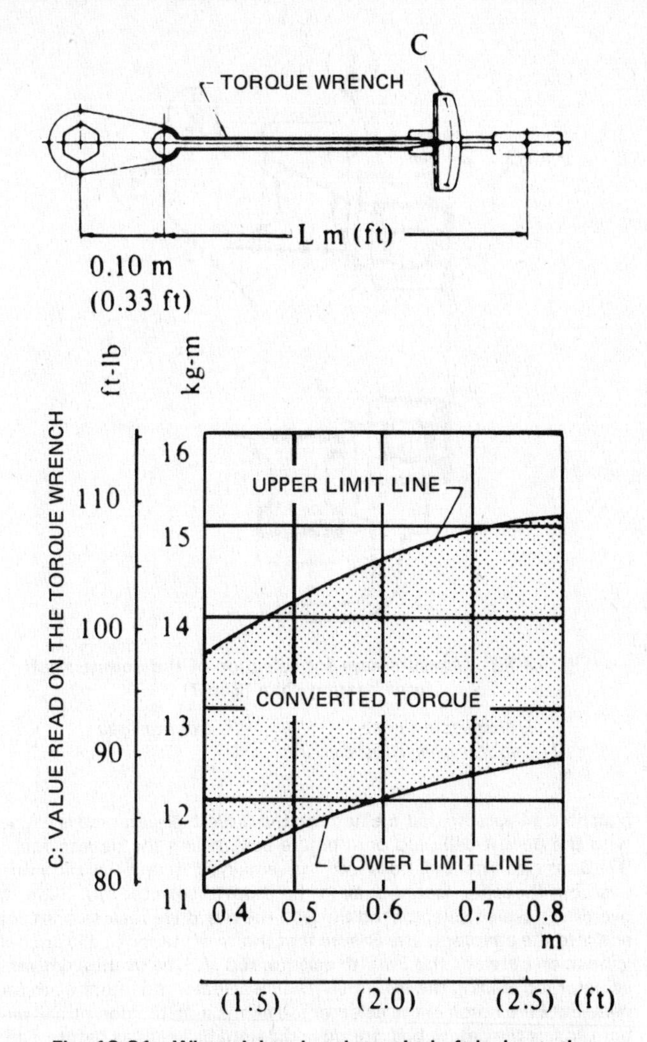

Fig. 13.31 When tightening the mainshaft locknut, the true torque can be found by matching the torque reading on the wrench with the chart (Sec 7)

91 Install the next snap-ring, then grease the steel ball and install the ball and the speedometer drive gear onto the mainshaft. Finally, install the last snap-ring.
92 Locate the 1st/2nd shift fork onto the 1st/2nd synchronizer unit, (the long end of the shift fork must be toward the countershaft). Now locate the 3rd/4th shift fork onto the 3rd/4th synchronizer unit, (the long end of the shift fork must be the opposite side to the 1st/2nd shift fork).
93 Locate the overdrive reverse shift fork onto the overdrive synchronizer so that the upper rod hole is in line with the 3rd/4th shift fork.
94 Slide the 1st/2nd selector rod through the adapter plate and into the 1st/2nd shift fork; align the hole in the rod with the hole in the fork and drive in a new retaining pin.
95 Align the notch in the 1st/2nd selector rod with the detent (check) ball bore, then install the detent (check) ball, spring and screw in the detent ball plug. Apply a little thread sealant to the detent ball plug.
96 Now invert the adapter plate assembly (hold the 3rd/4th and OD/reverse shift forks in position) so that the check ball plug assembled at Step 95, is lowermost. Drop two interlock balls into the 3rd/4th detent ball plug hole and, using a suitable thin probe, push them up against the 1st/2nd selector rod (if the adapter plate is correctly positioned, the interlock balls will drop into position). Slide the 3rd/4th selector rod through the upper hole of the OD/reverse shift fork and the adapter plate, ensuring that the interlock balls are held between this selector rod and the 1st/2nd selector rod, and into the 3rd/4th shift fork. Align the holes in the shift fork and selector rod, and drive in a new retaining pin. Now install a detent ball, spring and detent ball plug (with thread sealant applied) to the 3rd/4th detent ball

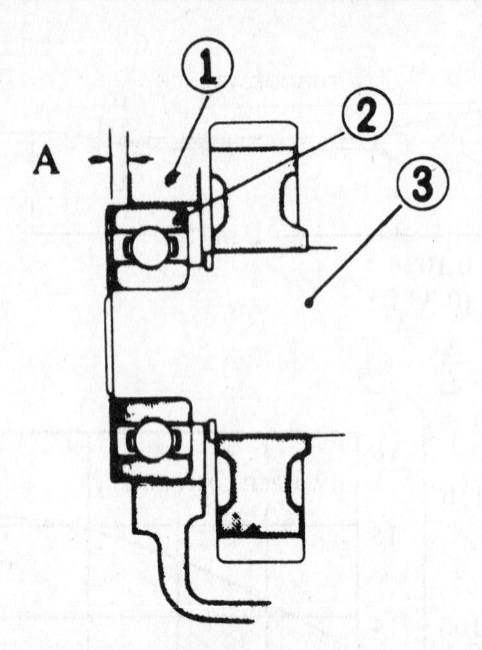

Fig. 13.32 Measurement for selection of the countershaft front bearing shim (Sec 7)

1 Transmission case
2 Counter gear front bearing
3 Counter gear

plug bore. Make sure that the notch in the 3rd/4th selector rod is aligned with the detent ball plug bore before assembling the detent ball.

97 Drop two interlock balls into the remaining detent ball plug bore, ensuring that they locate against the 3rd/4th selector rod. Slide the overdrive/reverse selector rod through the overdrive reverse shift fork and into the adapter plate. Ensure that the two interlock balls are held in position between the 3rd/4th selector rod and the overdrive/reverse selector rod, sliding the overdrive/reverse selector rod into the adapter plate until the notch in the selector rod aligns with the detent ball plug bore. Insert the detent ball, spring and detent ball plug as before. Drive in a new retaining pin to retain the overdrive/reverse shift fork to the overdrive/reverse selector rod.

98 Finally, tighten the three detent ball plugs to the specified torque.

99 Thoroughly oil the entire assembly and check to see that the selector rods operate correctly and smoothly.

100 Clean the mating faces of the adapter plate and the transmission casing and apply gasket sealant to both surfaces.

101 Tap the transmission casing into position on the adapter plate using a soft-faced hammer. Take particular care that it engages correctly with the input shaft bearing and countershaft front bearing.

102 Fit the outer snap-ring to the input shaft bearing.

103 Clean the mating faces of the adapter plate and rear extension housing and apply gasket sealant.

104 Arrange the shift forks in their neutral mode and then lower the rear extension housing onto the adapter plate so that the striking lever engages correctly with the selector rods.

105 Fit the bolts which secure the sections of the transmission together and tighten them to the specified torque.

106 Measure the amount by which the countershaft front bearing protrudes from the transmission casing front face. Use feeler blades for this and then select the appropriate shims after reference to the following table:

Measurement	Shim (thickness)
0.1150/0.1185 in (2.92/3.01 mm)	0.0236 in (0.6 mm)
0.1189/0.1124 in (3.02/3.11 mm)	0.0197 in (0.5 mm)
0.1228/0.1264 in (3.12/3.21 mm)	0.0157 in (0.4 mm)
0.1268/0.1303 in (3.22/3.31 mm)	0.0118 in (0.3 mm)
0.1207/0.1343 in (3.32/3.41 mm)	0.0079 in (0.2 mm)
0.1346/0.1382 in (3.42/3.51 mm)	0.0039 in (0.1 mm)

107 Stick the shim in position using a dab of thick grease, then fit the front cover to the transmission casing (within the clutch bellhousing)

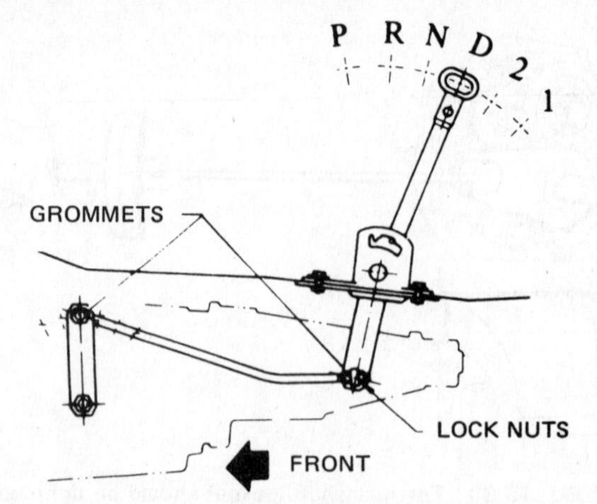

Fig. 13.33 Location of the selector rod locknut (Sec 7)

complete with a new gasket and taking care not to damage the oil seal as it passes over the input shaft splines.

108 Tighten the securing bolts to the specified torque, making sure that the bolt threads are coated with gasket sealant to prevent oil seepage.

109 Complete the reassembly by reversing the steps described in Steps 1 through 8 of this Section.

Automatic transmission

General description

110 The automatic transmission used on 1980 through 1982 models is virtually the same as on earlier models. On 1983 and 1984 models, a three-speed transmission incorporating an overdrive and utilizing a lock-up torque converter is used.

Selector linkage adjustment

111 To check the manual shift linkage adjustment, move the shifter through the entire range of gears. You should be able to feel the detents in each gear position. If these detents are not felt or if the pointer is not properly aligned with the correct gear, the shift linkage should be adjusted in the following manner.

112 With the engine off, place the shift lever in the Drive position.

113 Working underneath the car, loosen the locknuts shown in the accompanying illustration.

114 Move the shift lever so that it is correctly aligned with the D position.

115 Move the selector lever on the transmission so that it is also correctly aligned with the Drive position.

116 Tighten the locknuts and recheck the levers. There should be no tendency for the selector rod to push or pull one rod against the other.

117 Run the shifter through the entire range of gear positions again. If there are still problems with alignment, the grommets connecting the selector rod with the levers may be worn or damaged and should be replaced with new ones.

Kickdown switch and downshift solenoid – checking

118 The kickdown switch, coupled with the downshift solenoid, causes the transmission to downshift when the accelerator pedal is fully depressed. This is to provide extra power when passing. If the transmission is not downshifting upon full throttle, the system should be inspected.

119 With the engine off but the ignition on, depress the accelerator pedal all the way and listen for a click just before the pedal bottoms.

120 If no click is heard, locate the kickdown switch at the upper post of the accelerator pedal. Loosen the locknut and, with the pedal still depressed, extend the switch until it makes contact with the post and clicks. The switch should click only as the pedal bottoms. If it clicks too soon, it will cause the transmission to downshift on part throttle.

121 Tighten the locknut and recheck the adjustment.

122 If the kickdown switch adjustment is correct but the transmission still will not downshift, check to see if current is reaching the switch and, with a continuity tester, check that the switch is passing current through it.

123 If the switch checks out okay but the problem still persists, have the downshift solenoid tested and replaced if necessary.

Starter inhibitor and back-up lamp switch – checking and adjustment

124 The inhibitor switch performs two functions. It provides current

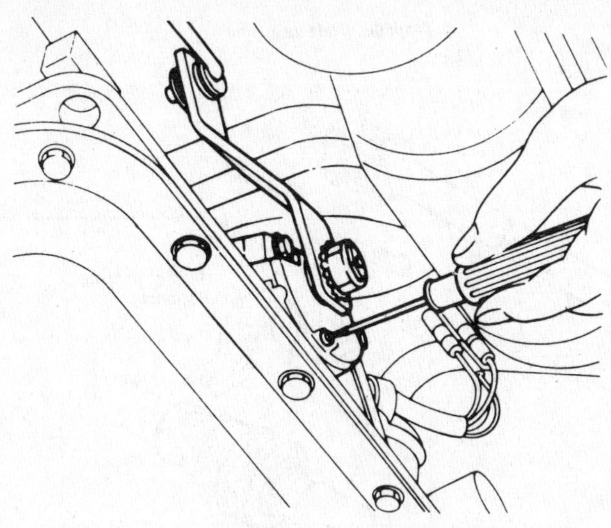

Fig. 13.34 The inhibitor switch is adjusted by removing the screw and aligning the screw hole with the rotor hole underneath it (Sec 7)

to the back-up lights when the transmission is in the Reverse position. It also prevents the car from being started in any gear position except Park or Neutral. If the back-up lights fail to operate, or if the car will not start when the shifter is in the middle of the Park or Neutral positions, the inhibitor switch should be checked and, if necessary, adjusted.
125 Locate the inhibitor switch on the right side of the transmission and connect a continuity tester to the black and yellow wires.
126 With the engine off, have an assistant run the gear shifter through the entire range of gear positions. The tester should show current passing through the switch only when the shifter is in the Park or Neutral positions.
127 Now connect the continuity tester to the red and black wires. With this arrangement, the tester should show current passing through the switch only when the shifter is in the Reverse position.
128 If the continuity tests did not give the results described above, the switch should be adjusted as follows:

a) Place the selector lever on the transmission in the Neutral (vertical) position.
b) Remove the screw shown in the illustration.
c) Loosen the inhibitor switch attaching bolts.
d) Using a thin rod or piece of wire, align the screw hole with the hole in the rotor behind the switch by moving the switch. Holding this alignment, retighten the inhibitor switch attaching bolts.
e) Remove the alignment rod or wire and install the screw.

Vacuum diaphragm rod – adjustment

129 The vacuum diaphragm and the length of its rod affect the shift patterns of the transmission. If the transmission is not shifting at precisely the right points, a different vacuum diaphragm rod may have to be installed.
130 Disconnect the vacuum hose from the vacuum diaphragm on the left side of the transmission.
131 Remove the vacuum diaphram.
132 Be sure the vacuum throttle valve is pushed into the valve body as far as possible and measure the distance L in the accompanying figure with a depth gauge.
133 Once this measurement is taken, use the chart below to determine the correct vacuum diaphragm rod length.

Measured depth L in (mm)	Rod length in (mm)
Under 1.0059 (25.55)	1.42 (29.0)
1.0098 to 1.0256 (25.65 to 26.05)	1.161 (29.5)
1.0295 to 1.0453 (26.15 to 26.55)	1.181 (30.0)
1.0492 to 1.0650 (26.65 to 27.05)	1.201 (30.5)
Over 1.0689 (27.15)	1.110 (31.0)

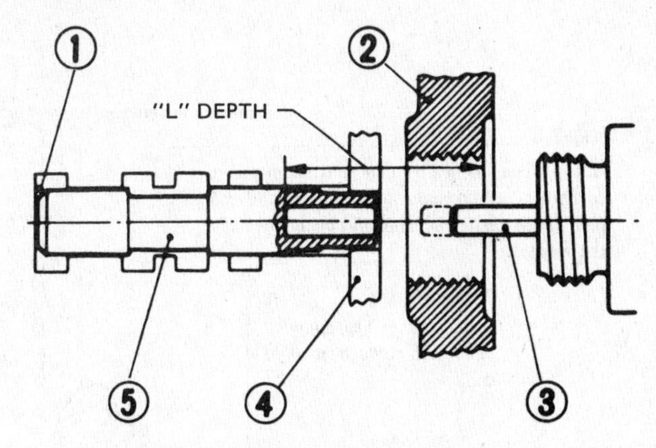

Fig. 13.35 Measurement of the vacuum diaphragm rod (Sec 7)

1 Note sealed valve body
2 Transmission case wall
3 Diaphragm rod
4 Valve body side plate
5 Vacuum throttle valve

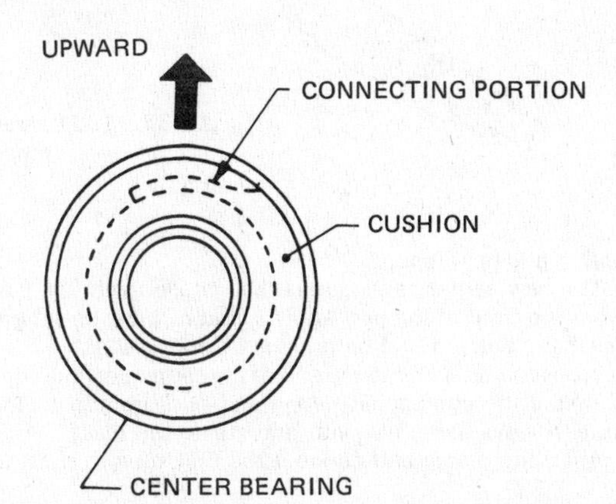

Fig. 13.36 Correct installation of the propeller shaft center bearing on the bracket (Sec 8)

8 Propeller shaft

Propeller shaft – removal and installation (1981 through 1984 models)

1 Raise the vehicle and support it securely on jackstands.
2 Scribe or paint matching marks on the universal joint and differential carrier flanges.
3 Remove the nuts and bolts and separate the propeller shaft from the differential.
4 Remove the center bearing bracket and bolts.
5 Withdraw the propeller shaft from the transmission, taking care to plug the extension housing to prevent fluid leakage.
6 Remove the propeller shaft from the vehicle.
7 Installation is the reverse of removal, using the alignment marks to position the shaft exactly as it was before removal. The center bearing must be installed on the bracket with the contract surface of the cushion facing up.

Center bearing – dismantling and reassembly (1981 through 1984 models)

8 Scribe or paint match marks on the flanges and separate the second tube from the first tube.
9 Mark the relative positions of the flange and shaft so they can be installed in the same position.
10 Remove the locking nut with a socket and holding tool as shown in Chapter 7.
11 Remove the companion flange with a puller.
12 It will be necessary to have the center bearing pressed out by a

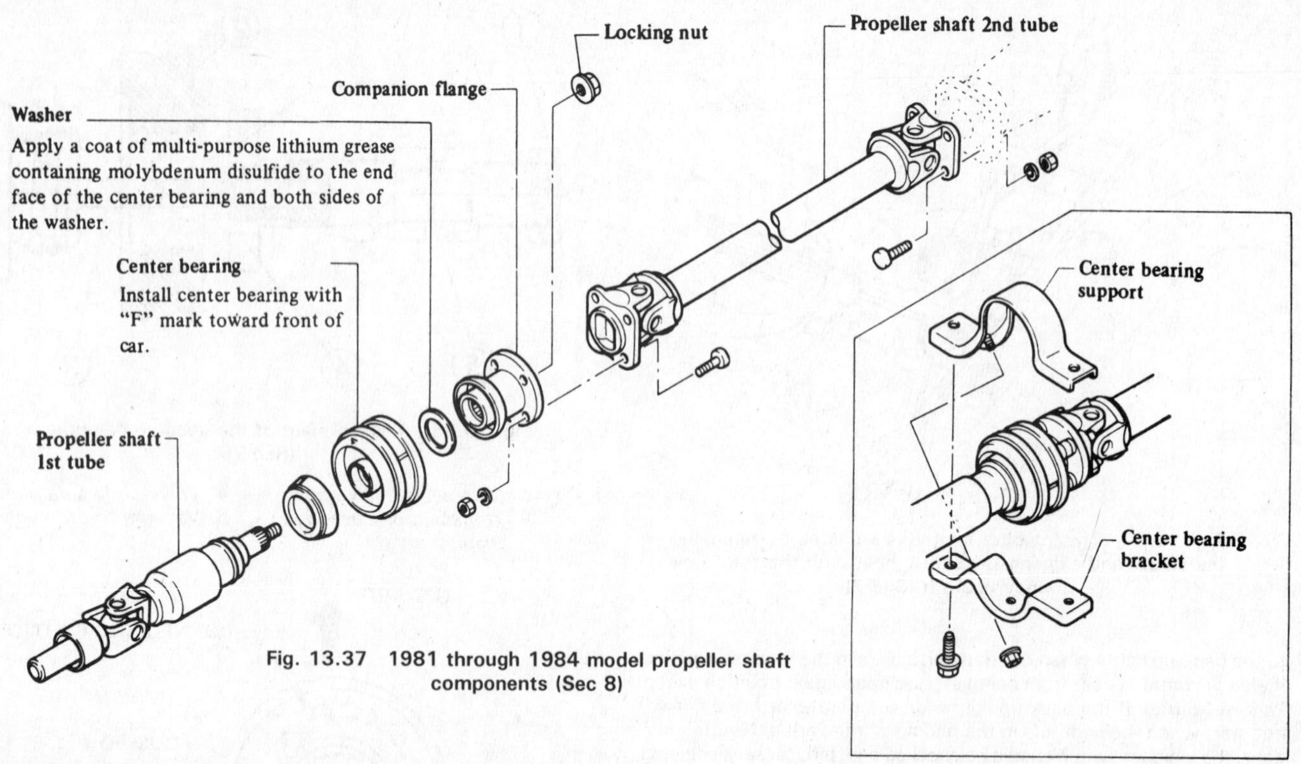

Fig. 13.37 1981 through 1984 model propeller shaft components (Sec 8)

properly equipped shop.

13 The new center bearing must be installed with the F mark facing toward the front of the car. Apply a coat of moly-based grease to the face of the bearing and both sides of the washer.

14 Insert the washer into the end of the center bearing, align the mark, and install the companion flange and nut. Tighten to the specified torque. Always use a new nut and stake it in place.

15 Align the marks and connect the first tube to the second tube.

9 Rear axle

General description

Sedan models

The rear axle driveshaft, stub, bearing and oil seal assemblies are new on 1983 and 1984 models. The driveshaft can be removed but special tools are required for overhaul of the driveshaft, stub, bearing and oil seal assemblies, so it should be left to your dealer or a properly equipped shop.

Station wagon models

The axle design is the same as on previous models, though the suspension is coil spring instead of the leaf spring type on 1983 and 1984 models (See the Suspension and Steering Section of this Chapter).

Driveshaft (1983 and 1984 models)

Checking

1 Inspect the length of each driveshaft for damage, cracks or corrosion. Check the boots for cracks, abrasion, leaking lubricant or loose bands. Clean any oil, grease or dirt from the boots using soap and water.

Removal

2 Block the front wheels, raise the rear of the vehicle and support it securely on jackstands.

3 Remove the bolts retaining the driveshaft to the companion flange on the wheel side.

4 Extract the driveshaft from the differential carrier by carefully prying with a bar to dislodge the circlip from the differential side gear.

5 Lower the driveshaft carefully from the vehicle, taking care not to damage the oil seals.

Installation

6 Installation is the reverse of removal, making sure to fully insert the driveshaft into the differential until the circlip snaps in place.

7 Lower the vehicle.

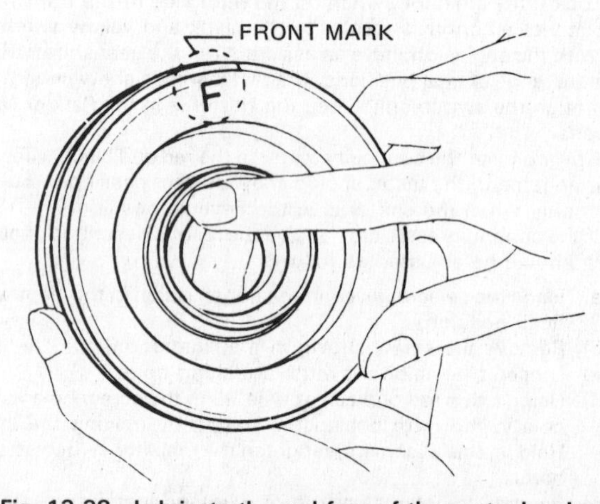

Fig. 13.38 Lubricate the end face of the center bearing prior to installation (Sec 8)

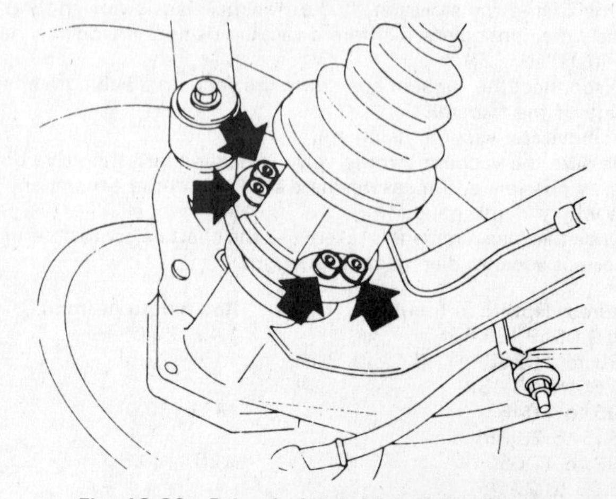

Fig. 13.39 Driveshaft retaining bolts (Sec 9)

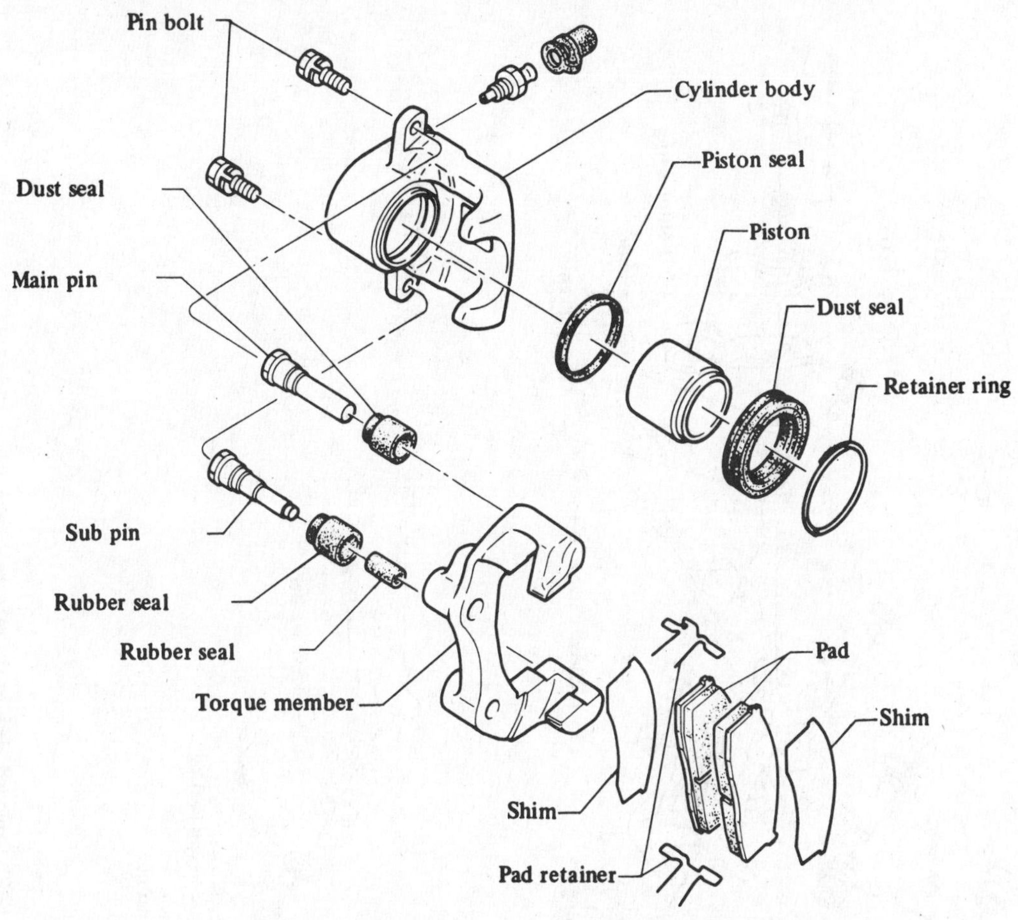

Fig. 13.40 Exploded view of the front disc brake (Sec 10)

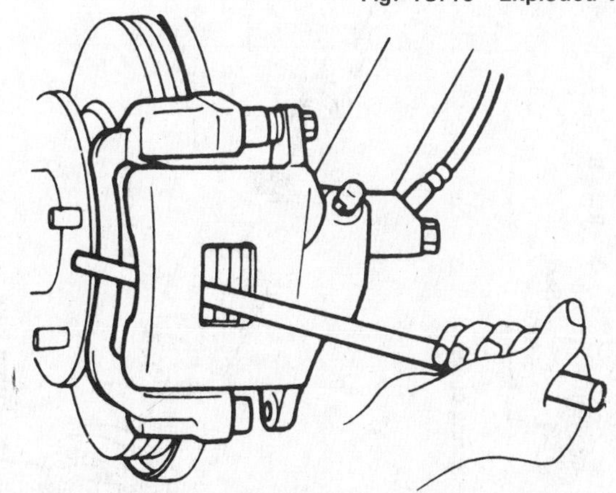

Fig. 13.41 Use a centering tool to pull the caliper out (Sec 10)

10 Braking system

General description

1981 through 1984 models use a front disc brake and master cylinder of a different design than that described in Chapter 9. Sedan models for 1981 through 1984 feature disc brakes at the rear.

Front disc brake pads — 1981 through 1984 models

Inspection

1 Raise the front of the vehicle, support it securely on jackstands and remove the front wheel.

2 Remove the lower sub pin and rotate the cylinder body up.

3 Measure the friction material remaining on the brake pad and compare it to the Specifications. Replace the pads as axle sets (both wheels on the same axle) if they are worn beyond the limits.

4 Inspect the caliper for rust, cracks, damage and torn or leaking seals. Replace the caliper with a new or rebuilt unit if it is damaged or leaking.

Removal

5 Lift out the brake pad retainer, the inner and outer shims and the pad. Clean the ends of the piston and the area surrounding the pin bolts. **Caution:** *Do not breathe the dust from the brake pads because it may contain asbestos, which is dangerous to your health.*

Installation

6 Install the inner pad, pull the cylinder body out and install the outer pad, shim and retainer.

7 Rotate the cylinder body into position, install the sub pin and tighten it to the specified torque.

8 Install the wheels and lower the vehicle.

Rear disc brake pads — 1981 through 1984 models

Removal

9 Raise the rear of the vehicle and support it securely on jackstands.

10 Remove the wheel and tire

11 Remove the pin bolts.

12 Remove the pad springs.

13 Lift out the brake pads and shim.

14 If the pads are glazed, damaged, fouled with oil or grease or worn beyond their specified limit (see Specifications in this Chapter), they should be replaced. **Note:** *Always replace all four pads on the axle (two in each brake assembly) at the same time, and do not mix different pad materials.*

15 Prior to installation, use brake fluid to clean the end of the piston and the area around the pin bolts. Be careful not to get fluid on the rotor.

16 Using needle-nose pliers as shown in the illustration, rotate the

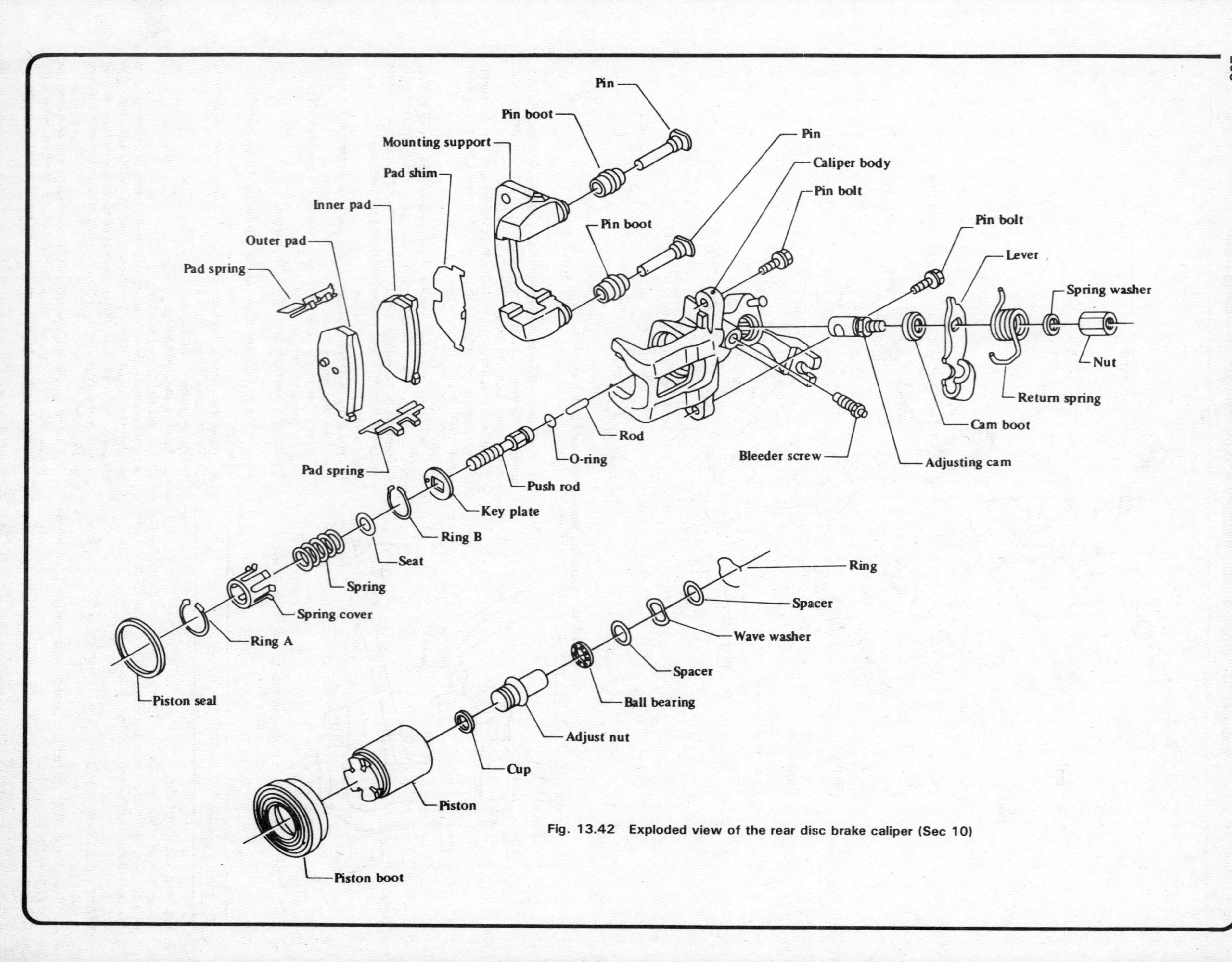

Fig. 13.42 Exploded view of the rear disc brake caliper (Sec 10)

piston clockwise to retract it into the cylinder body.
17 Apply a small amount of silicone-based grease to the contact area between the pads and the mounting support.

Installation

18 Install the new pads, shim and pad springs into position.
19 Reinstall the cylinder body on the mounting support and torque the pin bolts to 20 ft-lb (2.7 m-kg).
20 Depress the brake pedal several times to settle the pads into their proper positions.
21 Reinstall the wheel and tire.
22 Lower the vehicle.

Front disc brake caliper — 1981 through 1984 models

Removal

23 Disconnect and plug the brake hose.
24 Unbolt and remove the caliper.

Overhaul

25 Remove the main and sub pins and separate the cylinder body from the torque member.
26 Carefully pry the dust cover from the piston bore. Use a wooden or plastic tool to prevent scratching the bore.
27 The piston can be removed from the bore by slowly applying compressed air to the fluid port until the piston is forced out. Alternatively, strike the cylinder body sharply against a block of wood to dislodge the piston. With either method, pad the cylinder body with cloths to prevent damage to the piston or caliper.
28 Check the interior of the cylinder body for rust, wear, contamination and scoring, replacing it with a new unit if necessary. Minor imperfections can be removed by polishing with fine emery cloth. Inspect the torque member for wear, cracks, damage or rust. Replace it with a new one if necessary. Check the piston for scoring, rust, contamination, wear or damage. A badly damaged piston must be replaced with a new one as the surface is plated and polishing with emery cloth will make it completely unserviceable. Check the main and sub pins and rubber bushings for wear, cracks and damage, replacing any unserviceable components.
29 Lubricate the piston seal and grooves with clean brake fluid and install the seal.
30 Lubricate the piston, bore and dust cover with clean brake fluid. Install the dust seal on the piston and then insert the dust seal into the groove on the cylinder body. Install the piston, referring to the accompanying illustration.
31 Lubricate the main and sub pins with high-temperature disc brake grease. Connect the cylinder body and torque member and install the pins, tightening to the specified torque.

Installation

32 Installation is the reverse of removal. Bleed the brakes as described in Chapter 9.

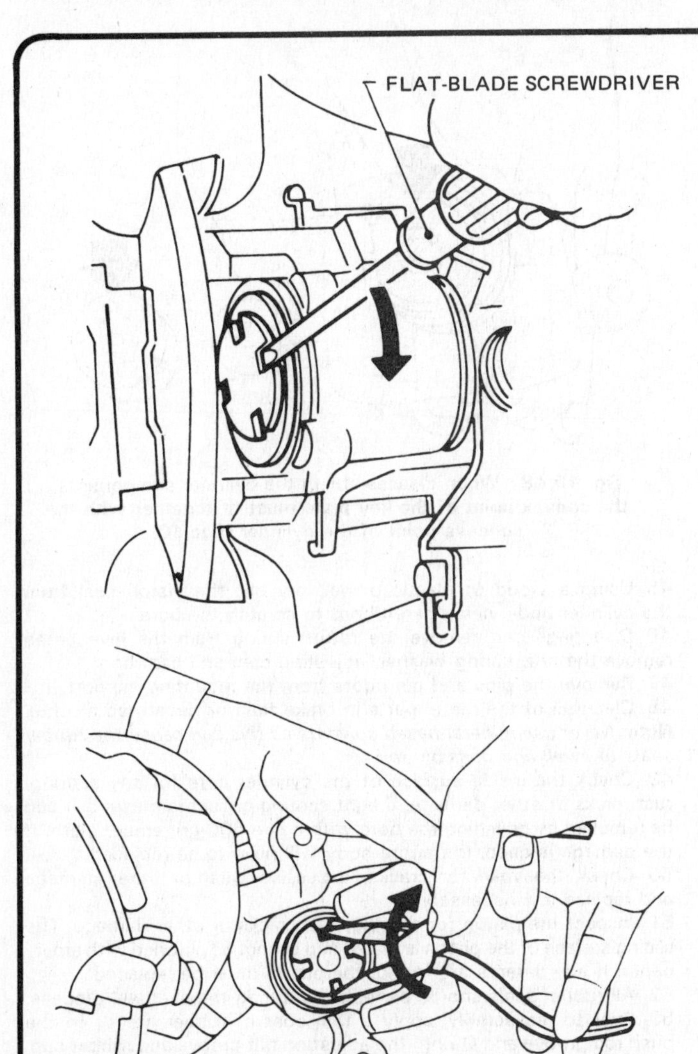

Fig. 13.43 Needle-nose pliers or a screwdriver can be used to retract or remove the piston by rotating it (Sec 10)

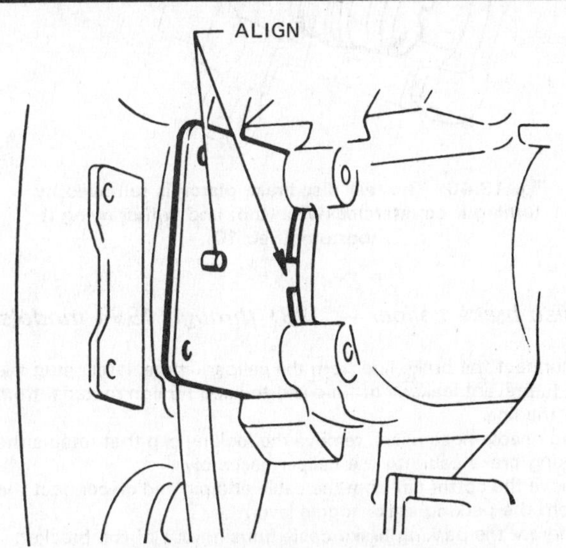

Fig. 13.44 When installing the inner rear disc pad, be sure the locating pin engages with the outer piston notch (Sec 10)

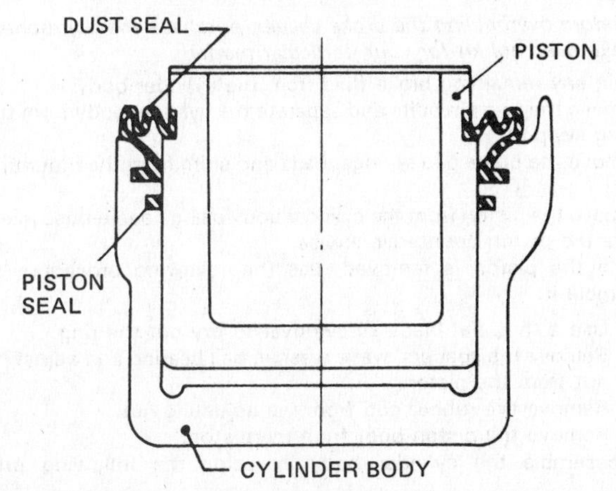

Fig. 13.45 Proper installation of the front disc brake piston (Sec 10)

13

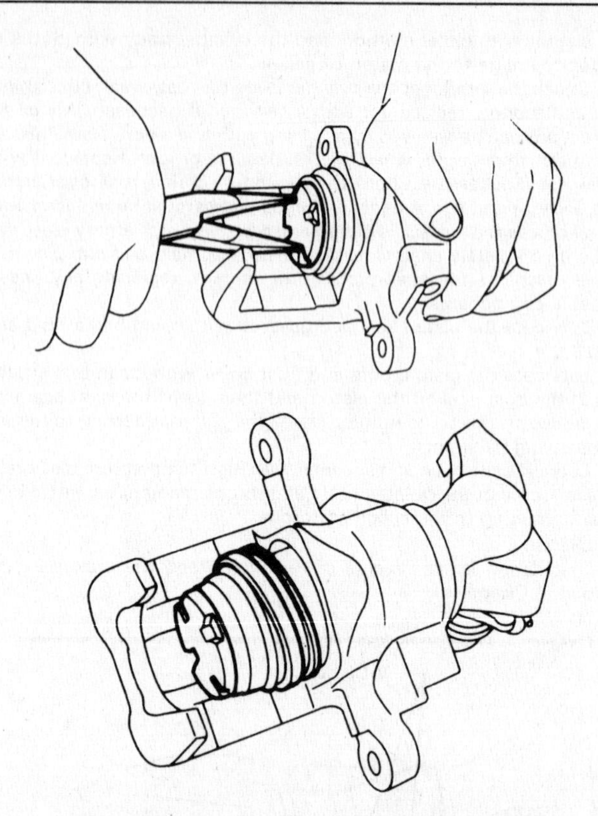

Fig. 13.46 The rear disc brake piston is removed by turning it counterclockwise (top) and withdrawing it (bottom) (Sec 10)

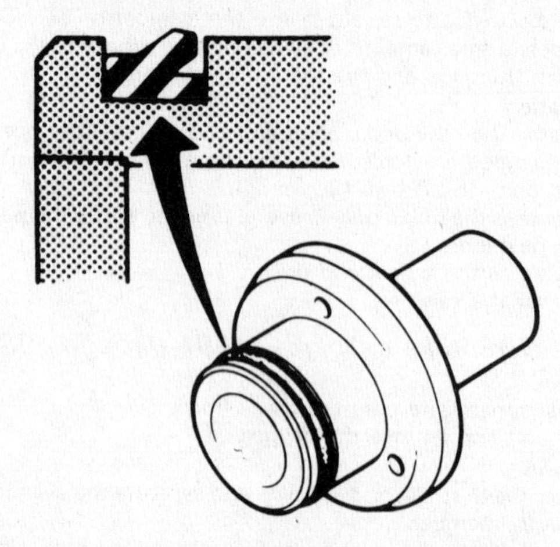

Fig. 13.47 The rubber cup should be installed on the adjusting nut so the seal lip faces in the direction shown (Sec 10)

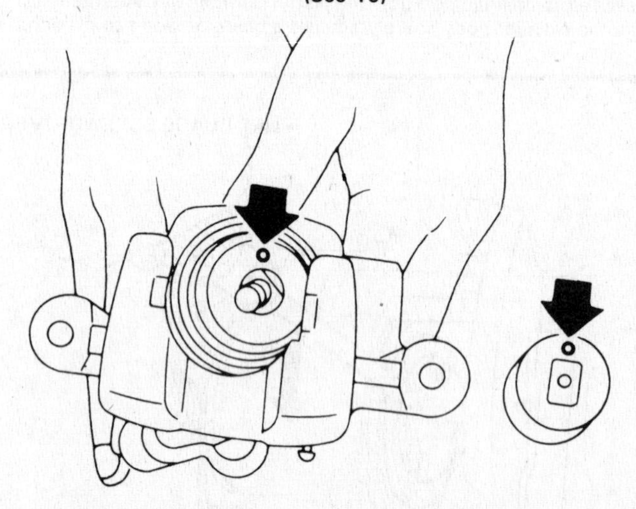

Fig. 13.48 When disassembling the cylinder components, the convex point of the key plate must fit together with the concave point of the cylinder (Sec 10)

Rear disc brake caliper — 1981 through 1984 models

Removal

33 Disconnect the brake line from the caliper. Immediately plug the opening to prevent leakage of fluid and to keep foreign material from entering the line.
34 Using needle-nose pliers, remove the locking clip that retains the rear parking brake cable to the caliper assembly.
35 Remove the cotter pin from the cable end pin and disconnect the cable from the parking brake toggle lever.
36 Withdraw the parking brake cable housing out of the bracket.
37 Remove the two caliper mounting bolts located on the rear of the caliper assembly.
38 Remove the caliper assembly.

Overhaul

Note: *Before overhauling the brake caliper, purchase the appropriate rear brake overhaul kit for your particular model.*

39 Drain any remaining brake fluid from the cylinder body.
40 Remove the two pin bolts and separate the cylinder body from the mounting support.
41 Remove the brake pad springs, pads and shim from the mounting support.
42 Remove the piston from the cylinder body using needle-nose pliers to rotate the piston counterclockwise.
43 Once the piston is removed, use the following procedure to disassemble it:

a) Use a thin, flat-blade screwdriver to pry out the ring.
b) Remove the spacers, wave washer, ball bearing and adjusting nut from the piston.
c) Remove the rubber cup from the adjusting nut.
d) Remove the piston boot from the piston.

44 Disassemble the cylinder body by using the following procedure:

a) Using pliers, pry out ring A and then remove the spring cover, spring and seat.
b) Pry out ring B and remove the key plate, pushrod and rod.
c) Remove the O-ring from the pushrod.

45 Using a wood or plastic dowel, pry out the piston seal from the cylinder body, being careful not to scratch the bore.
46 Disengage and remove the return spring from the lever, then remove the nut, spring washer, adjusting cam and cam boot.
47 Remove the pins and pin boots from the mounting support.
48 Clean all of the metal parts in brake fluid or denatured alcohol. **Note:** *Never use mineral-based solvents as this can cause the rubber seals to swell and possibly fail.*
49 Check the inside surface of the cylinder bore for any scoring, rust, nicks or other damage. If light scoring or rust is present, it can be removed by polishing the bore with a fine 600-grit emery cloth. If the damage is deep, the entire body will have to be replaced.
50 Check the yoke for cracks, excessive wear or other damage and replace it if necessary.
51 Inspect the piston for scoring, rust, nicks or other damage. The sliding surface of the piston is plated and cannot be polished with emery paper. If any defects are found, the piston must be replaced.
52 All rubber seals should be replaced during the overhaul process.
53 Prior to reassembly, apply a thin coat of rubber grease to the push rod groove and O-ring, the adjusting nut groove and rubber cup, the piston seal, the inside of the piston boot and the sliding surfaces of the piston and pins.
54 Install the rubber cup on the adjusting nut so the lip faces in the direction shown in the illustration.
55 When assembling the cylinder body components, fit the push rod into the square hole of the key plate and fit the convex point of the key plate so it engages with the concave point of the cylinder.

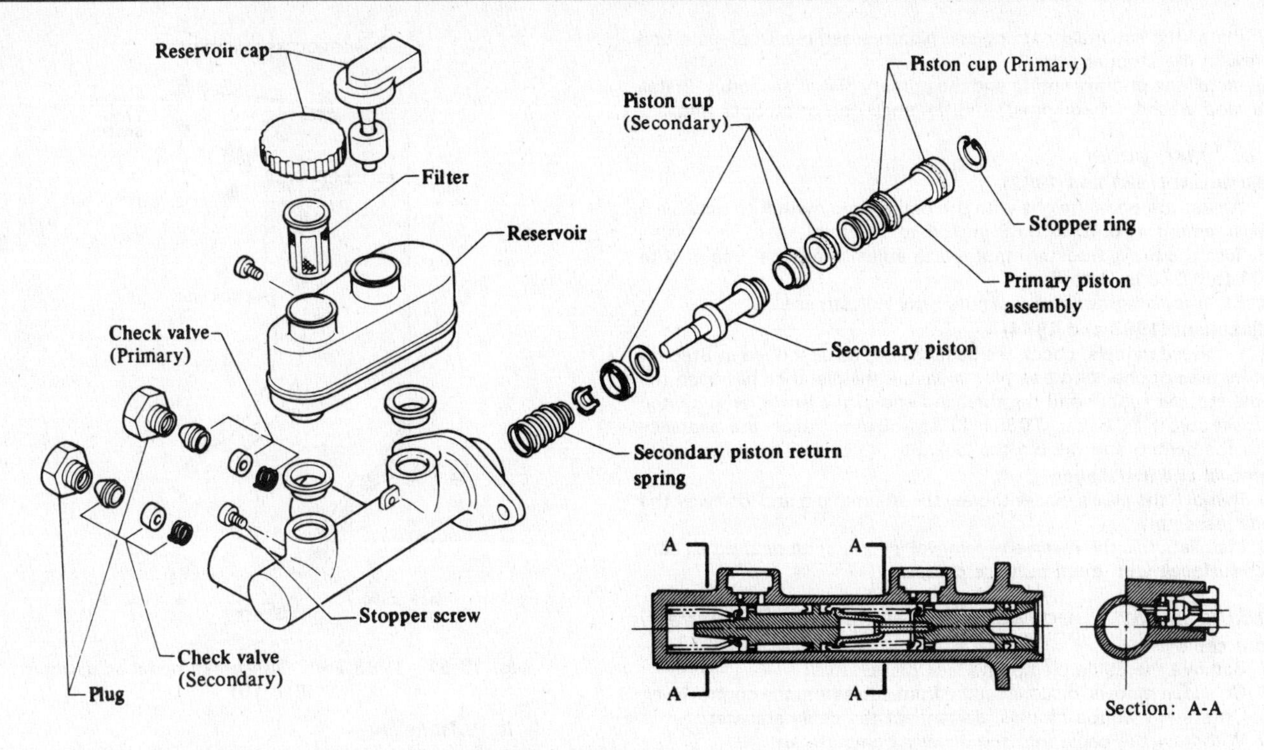

Fig. 13.49 Exploded view of the master cylinder used on 1981 through 1984 models (Sec 10)

Installation

56 Installation is the reverse of removal, but be sure to bleed the brake system. After installation, depress the brake pedal several times to automatically adjust the clearance between the pads and rotor.

Rear brake disc – removal and installation

57 Remove the brake caliper.

58 Remove the disc by sliding it off the hub and wheel studs. If the disc will not separate easily from the hub, apply penetrating oil where the hub and disc meet and tap on the disc with a soft-faced hammer.

59 Installation is the reverse of removal.

Master cylinder

Removal and installation

60 Disconnect the negative battery cable.

61 Place newspapers or rags under the master cylinder to catch any leaking brake fluid. **Note:** *Be sure not to let brake fluid touch your skin or the painted surfaces of the car.*

62 Disconnect the electrical lead(s) going to the reservoir caps.

63 Loosen the nuts securing the two brake lines to the master cylinder.

64 Remove the two nuts that secure the master cylinder to the vacuum servo unit.

65 Carefully lift the cylinder off its mounting studs, remove the brake lines from the cylinder, and immediately place your fingers over the holes to prevent leakage of fluid. Lift the cylinder out of the engine compartment.

66 Plug the fluid lines to prevent further leakage of fluid. If the master cylinder needs to be disassembled, refer to paragraph 69.

67 Installation is the reverse of the removal procedure.

68 Bleed the entire brake system as described in Chapter 9.

Overhaul

69 Obtain a master cylinder rebuild kit. **Note:** *Some models use one of two different makes of master cylinder, either a Nabco or Tokico. Since there is no interchangeability of parts between the two models, be sure you get the appropriate rebuild kit for your car.*

70 Clean away external dirt and then remove the reservoir caps and filters and empty the fluid.

71 From the end of the master cylinder, pry out the snap-ring or stopper ring. Remove the stop washer (if equipped), the primary piston and the spring.

72 Insert a rod to depress the secondary piston and then unscrew the stopper screw from the master cylinder. Release the rod and withdraw the secondary piston.

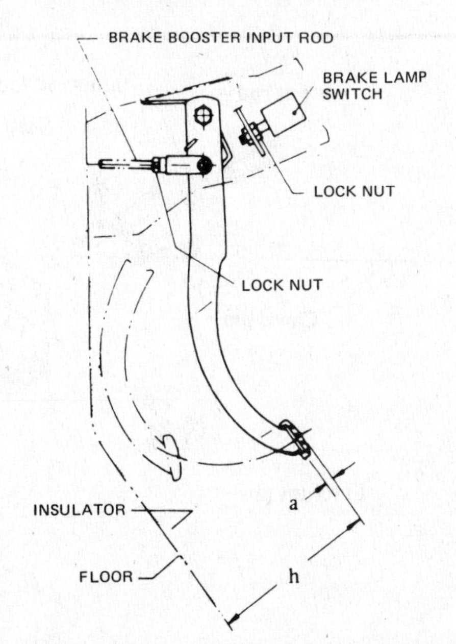

Fig. 13.50 1981 and 1982 brake pedal adjustment (Sec 10)

a Free play *h Pedal height*

73 The check valve assemblies can be removed by unscrewing the check valve plugs.

74 At this stage, inspect the surfaces of the pistons and cylinder bores for scoring or 'bright' areas. If these are evident, replace the complete master cylinder.

75 Wash all components in clean brake fluid or denatured alcohol – nothing else.

76 Do not detach the reservoirs unless absolutely necessary.

77 Commence reassembly by manipulating the new seals into position by using the fingers only. Be sure the seal lips are facing in the proper directions.

78 Dip all internal components in clean hydraulic fluid before reassembly.

79 Install the secondary spring and piston assembly. Depress it and screw in the stopper screw.
80 Install the primary spring and the primary piston assembly. Install the stop washer (if equipped) and the snap-ring or stopper ring.

Foot-brake pedal

Adjustment (1981 and 1982)

81 Adjust the pedal height with the brake lamp switch to achieve a measurement of 6.46 to 6.69 in (164 to 170 mm).
82 Turn the brake booster input rod to adjust the pedal free play to 0.04 to 0.020 in (1 to 5 mm).
83 Be sure to tighten the locknuts after adjustment.

Adjustment (1983 and 1984)

84 On these models, check the pedal height as described in Step 81 but instead of checking free play, measure the distance between the pedal stopper rubber and the threaded end of the brake lamp switch to make sure it is 0.012 to 0.039 in (0.3 to 1.0 mm). Adjust the clearance with the switch and tighten the locknut.

Removal and installation

85 Remove the clevis pin, withdraw the fulcrum pin and lift away the pedal assembly.
86 Installation is the reverse of removal after first lubricating all contact surfaces with multi-purpose grease.

Parking brake — renewal (1981 through 1984 models)

Front cable

87 Remove the cable clamp and lock plate.
88 On sedan models, disconnect the front cable from the control lever.
89 On station wagon models, disconnect the cable adjuster.
90 Withdraw the cable into the driver's compartment.
91 Installation is the reverse of removal, however all sliding surfaces

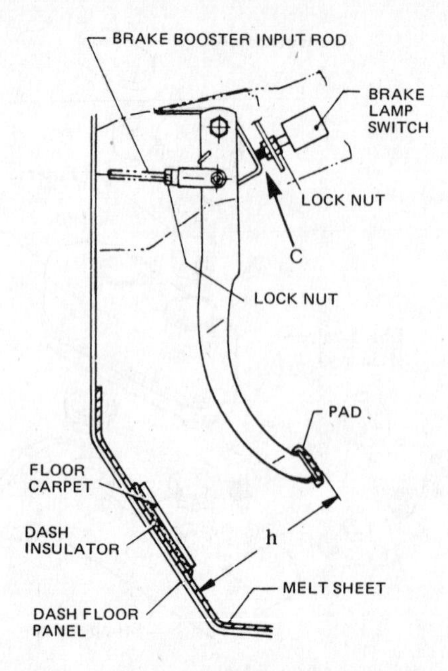

Fig. 13.51 1983 and 1984 brake pedal adjustment (Sec 10)

h Pedal height
c Pedal stopper-to-brake lamp switch end clearance

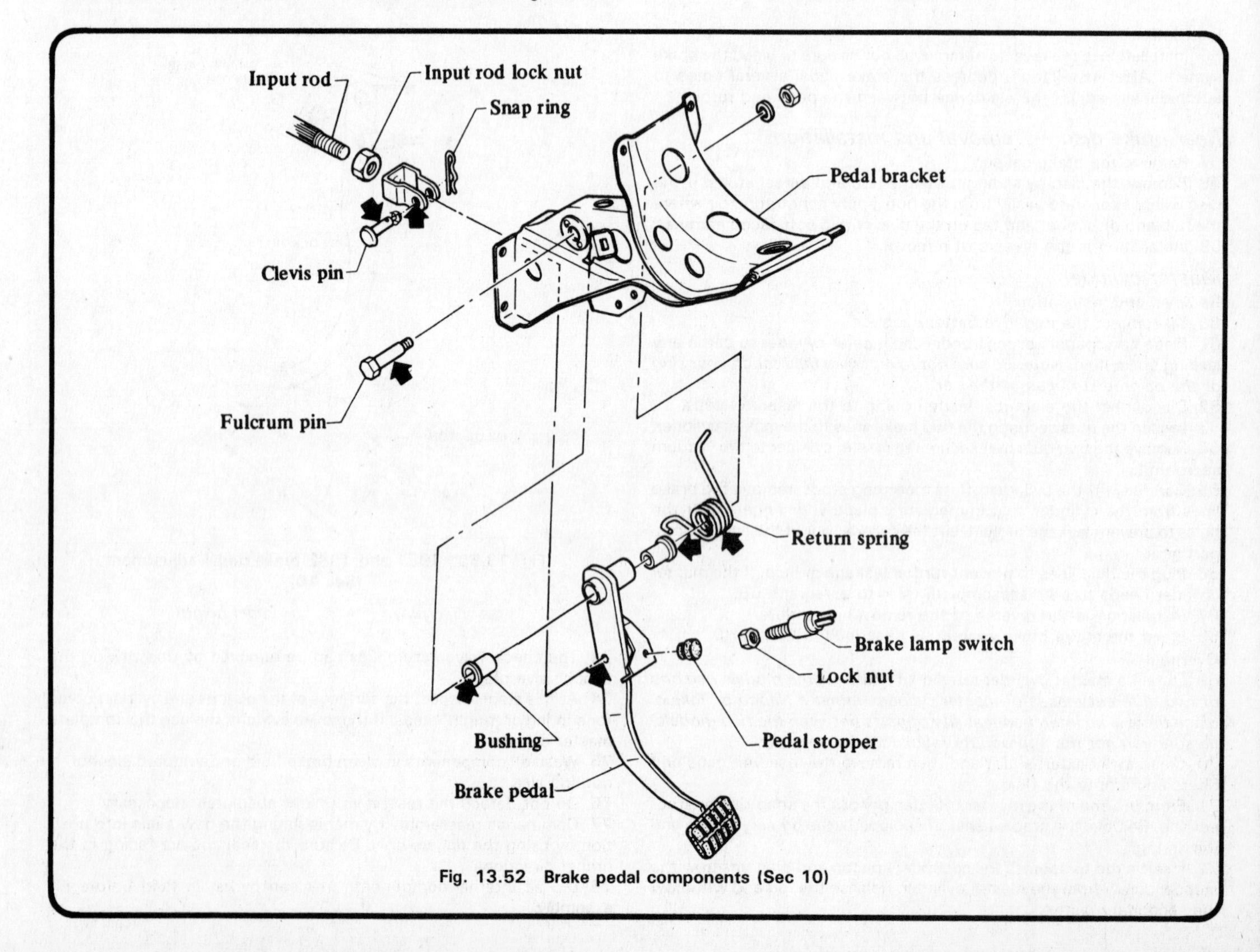

Fig. 13.52 Brake pedal components (Sec 10)

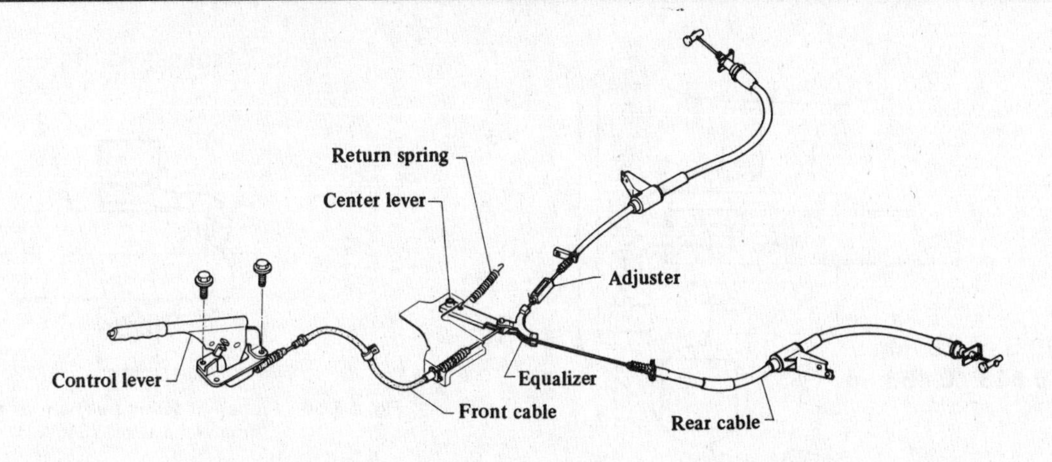

Fig. 13.53 Typical 1981 through 1984 parking brake cable layout (Sec 10)

should first be lubricated with multi-purpose grease. Adjust the parking brake lever to the specified number of clicks by loosening the adjuster locknut and turning the adjuster.

Rear cable (Sedan)

92 Disconnect the adjuster, remove the lock plate and bracket-fixing bolt followed by the rear cable-fixing bolts.

93 Remove the lock plate and disconnect the rear cable from the lever by removing the cotter pin at the rear wheel.

94 Installation is the reverse of removal. Lubricate all sliding surfaces with multi-purpose grease prior to installation and adjust the lever travel.

Rear cable (Station wagon)

95 Disconnect the cable adjuster and remove the lock plate at the axle housing.

96 Remove the rear cable fixing bracket from the axle housing.

97 Remove the return spring and clevis pin from the brake backing plate and remove the cable.

98 Installation and assembly is the reverse of removal. Lubricate the cable and mechanism sliding surfaces with multi-purpose grease prior to installation and adjust the cable tension with the adjuster after installation.

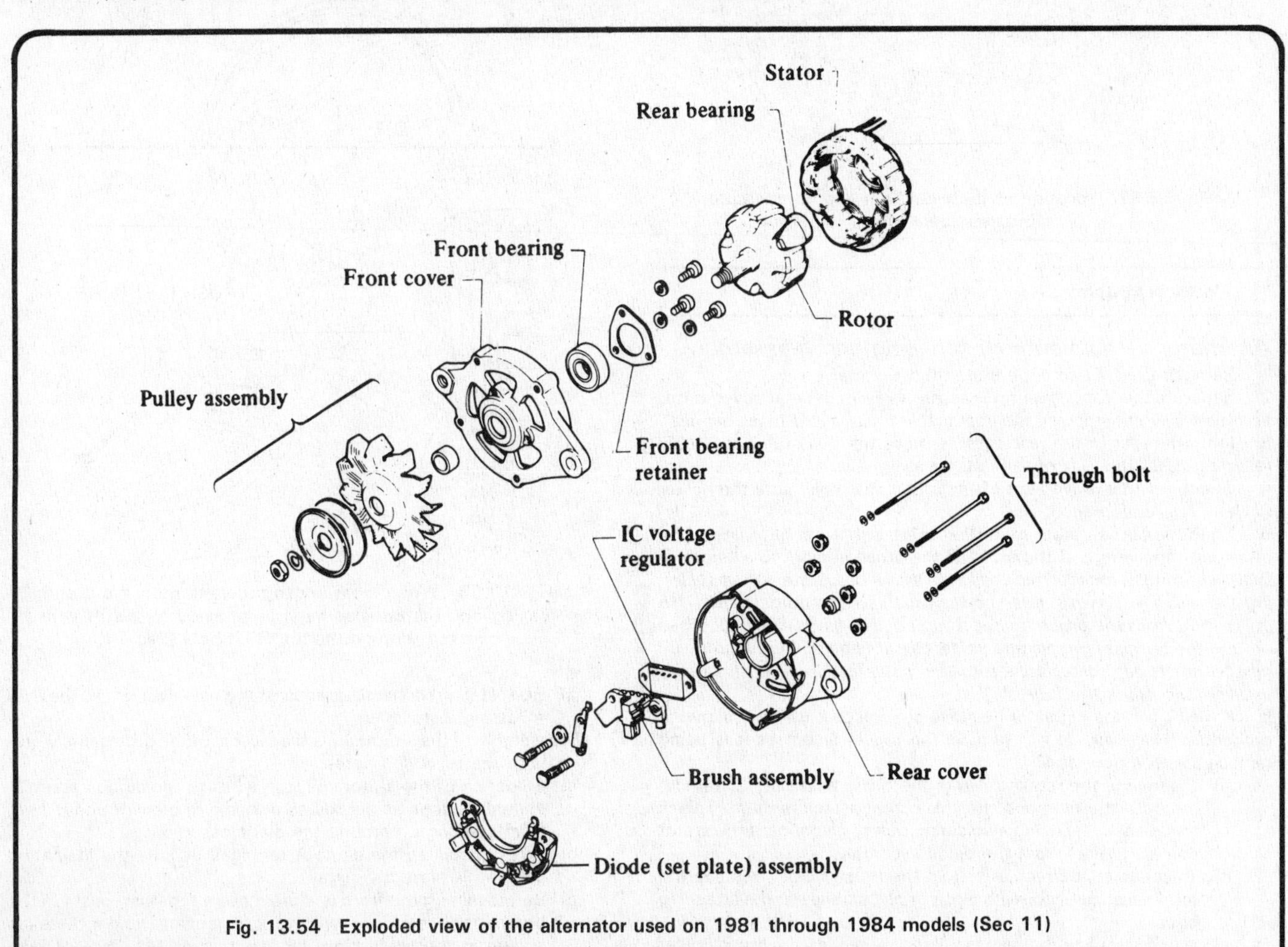

Fig. 13.54 Exploded view of the alternator used on 1981 through 1984 models (Sec 11)

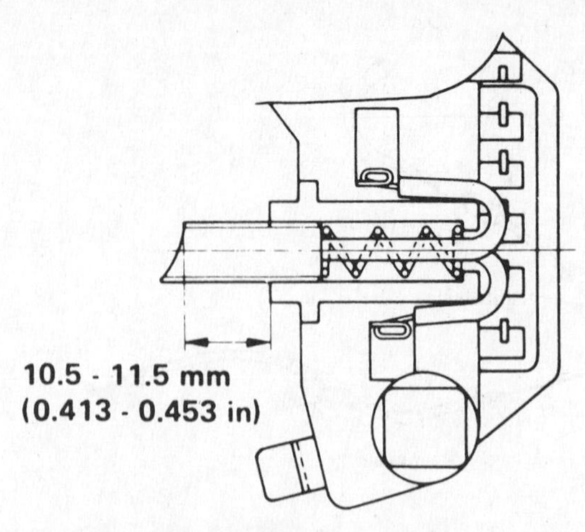

Fig. 13.55 When installing the new alternator brush, position it as shown before soldering (Sec 11)

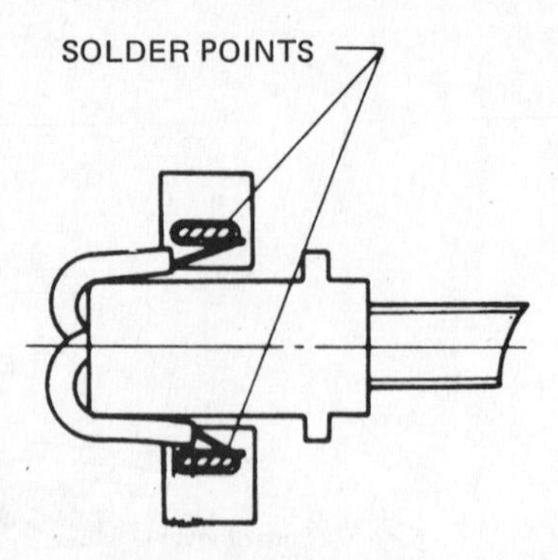

Fig. 13.56 Proper soldering position for the alternator brush lead wires (Sec 11)

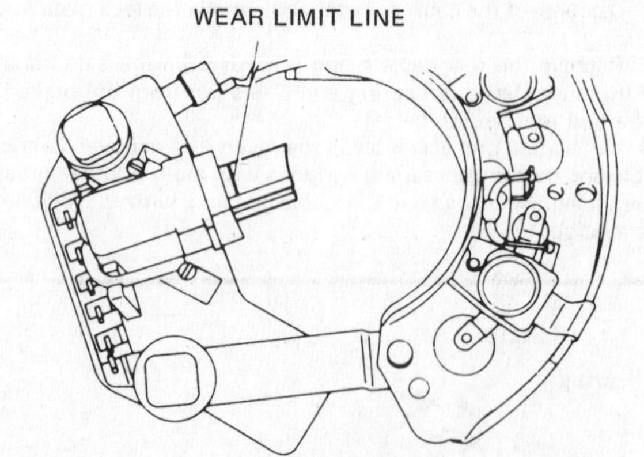

Fig. 13.57 Location of the wear limit line on alternator brushes (Sec 11)

11 Electrical system

Alternator — dismantling, servicing and reassembly

1 Remove the four through-bolts from the rear cover.
2 Separate the front cover/rotor assembly from the rear cover/stator assembly by lightly tapping the front bracket with a soft-faced hammer.
3 From the rear of the rear cover, remove the five stator assembly retaining nuts, then lift out the stator assembly.
4 Check for free movement of the brush and make sure the holder is clean and undamaged.
5 Check for brush wear by noting the brush wear limit line or by measuring the length of the brush. If the brush is worn to a length of 0.28 in (7 mm), it must be replaced with a new one. **Note:** *When soldering the brush lead wires, position the brush so it extends about 7/16 in (11 mm) from the brush holder. Then coil the lead wire 1-1/2 times around the terminal groove and solder the outside of the terminal. Be careful not to let solder adhere to the insulating tube as this could weaken and crack the tube.*
6 If the IC voltage regulator needs to be replaced, use the following procedure. However, do not remove the regulator unless it is being replaced with a new one.

a) Disengage the regulator from the diode assembly by removing both the rivet and the solder that attaches them. This is made easier by using a soldering gun to disconnect the stator coil lead wires from the diode assembly.
b) To separate the regulator from the brush holder, remove the terminal solder and, with a pair of pliers, take out the attaching bolts.
c) When installing the new regulator, place it on the brush holder and press-fit the bolts into place by either using a hand press or by carefully tapping it in.

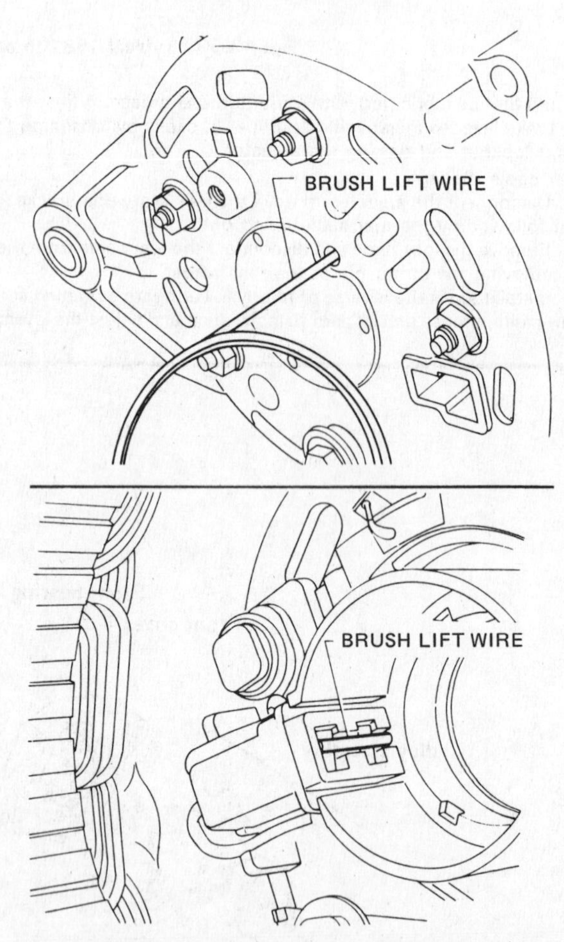

Fig. 13.58 When reassembling the alternator front and rear covers, the brushes must be retained by a stiff wire inserted through the brush lift hole (Sec 11)

d) Re-solder all connections and install a new rivet. Stake the rivet following installation.

7 Reassembly of the alternator is the reverse of the disassembly procedure, with the following notes:

a) Soldering of the stator coil lead wires to the diode assembly should be done as quickly as possible to prevent undue heat from building up around the diode assembly.
b) When installing the diode A terminal, be sure the insulating bushing is correctly installed.
c) Before joining the front and rear covers together, push up the brush in the rear cover with your fingers and hold it there (as shown in the illustration) by inserting a stiff piece of wire through the brush lift hole from the outside. After the front and rear covers have been joined, the wire can be removed by

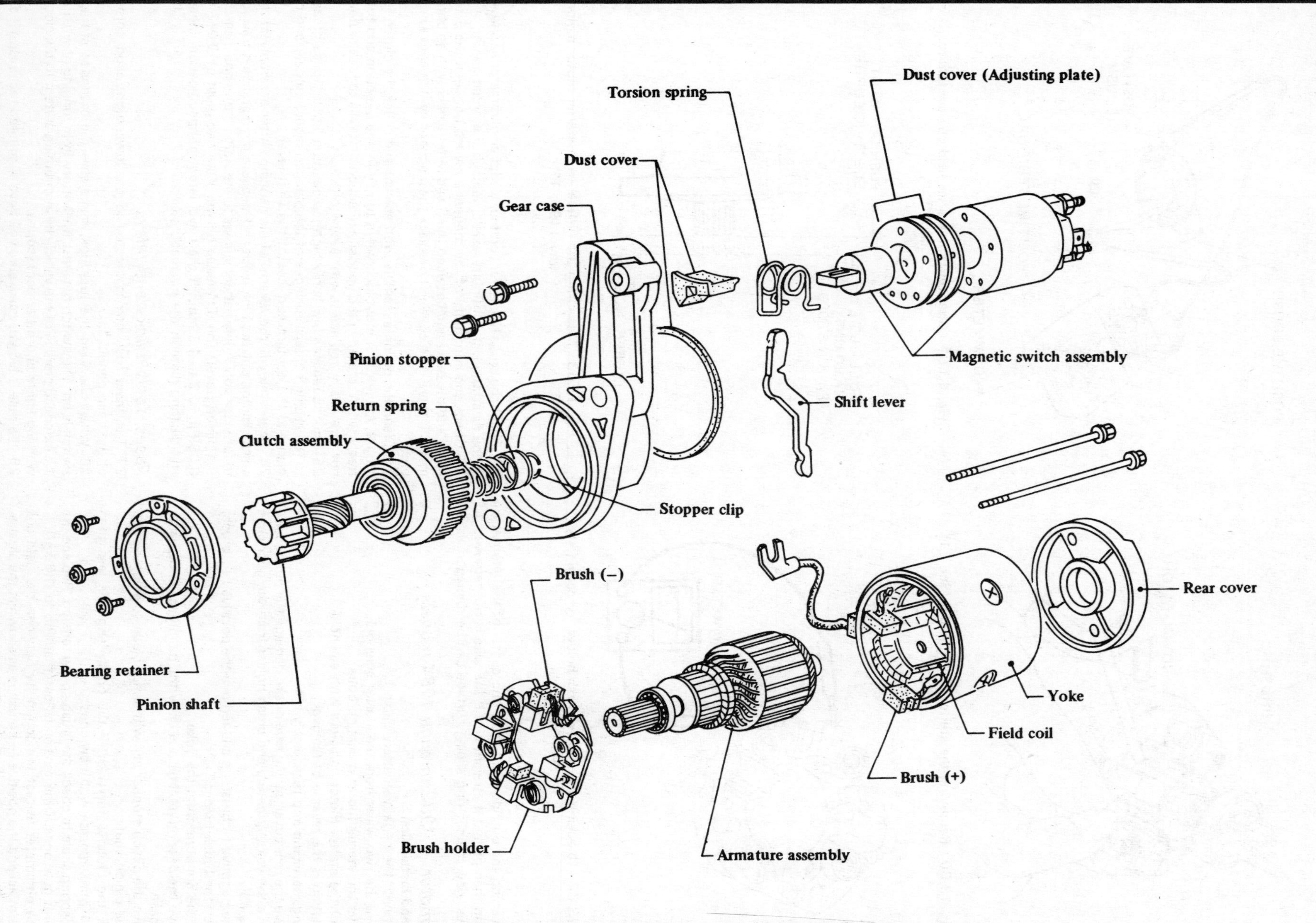

Fig. 13.59 Exploded view of the 1982 through 1984 starter motor (Sec 11)

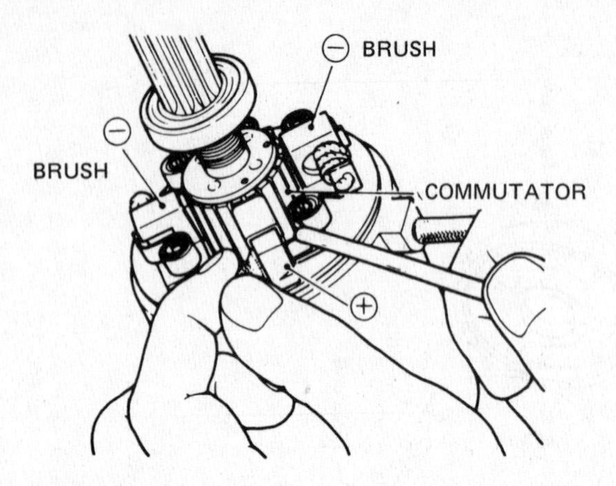

Fig. 13.60 Removing the starter brush holder (Sec 11)

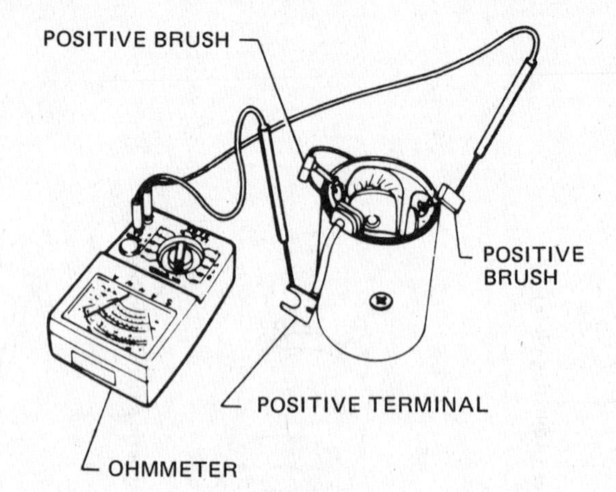

Fig. 13.61 Checking the starter field coil continuity (Sec 11)

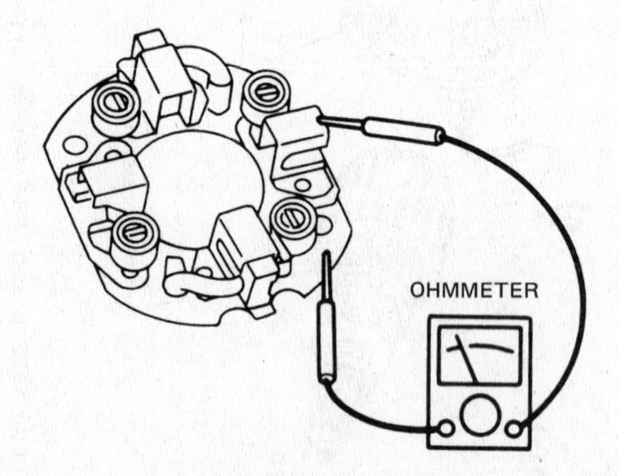

Fig. 13.62 Checking the starter brush holder for continuity (Sec 11)

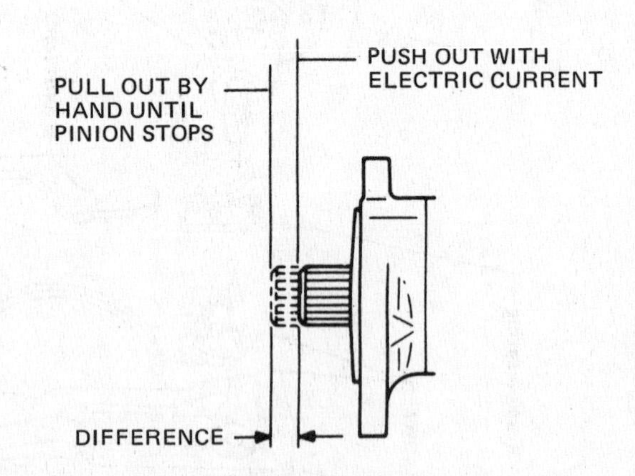

Fig. 13.63 Measuring the difference in starter motor pinion travel (Sec 11)

pushing the outside end toward the center of the alternator and then pulling it straight out. If the wire is not removed in this way, the slip-ring sliding surface can be damaged.

Starter motor – 1982 through 1984 models

Removal and installation

8 Disconnect the negative battery cable.
9 Remove the two wires from the starter solenoid.
10 Remove the starter motor mounting bolts.
11 Lift out the starter motor complete with solenoid.
12 Installation is the reverse of removal.

Dismantling, servicing and reassembly

13 Remove the rear cover, taking care not to damage the dust cover.
14 Remove the yoke, armature and brush holder from the gear case as an assembly.
15 Lift the negative (-) brush up and remove the positive (+) brush, followed by the brush holder.
16 Draw the armature from the yoke.
17 Remove the bearing retainer and draw the pinion assembly from the gear case.
18 Remove the pinion stopper clip with a flat-blade screwdriver and remove the pinion shaft.
19 Check the brushes for wear. If the brush length is less than 0.43 in (11 mm), replace with new ones.
20 Use a spring scale to check the brush spring tension. Replace the springs with new ones if the tension is less than 3.5 lb (1.6 kgf).
21 Use an ohmmeter to test the continuity between the field coil positive terminal and the positive (+) brushes as shown in the illustration. If there is no continuity, the field coil must be replaced by your dealer or a properly equipped shop.
22 Check the brush holder for continuity with an ohmmeter as shown in the illustration. If continuity exists, replace the brush holder.
23 Check the continuity of the magnetic switch between the S terminal and the body and between the S and M terminals with an ohmmeter. If there is no continuity, replace the magnetic switch with a new one.
24 Clean the pinion assembly with solvent and check the clutch for proper operation. It must lock when turned in one direction and turn freely when rotated in the opposite direction. If it fails to lock or does not turn easily, replace it with a new one.
25 Prior to reassembly, apply high-temperature, lithium-base grease to the sliding contact surfaces of the plunger, reduction gear and pinion and the contact surfaces of the shift lever.
26 Reassembly is the reverse of dismantling. After assembly of the starter motor, pull the pinion out and measure the distance it traveled to full extension. Apply battery voltage to the starter motor and measure the pinion travel. The difference between the two must be 0.012 to 0.059 in (0.3 to 1.5 mm). If the travel is out of specification, install an adjusting plate in the magnetic switch.

Fuses, fusible links and relays

27 The main fuse box is located in the lower right-hand corner of the dash panel.
28 When checking the fuse box, make sure the extended storage switch is on. The switch prevents the battery from running down during extended storage of the vehicle by interrupting current to the clock, warning lights and other accessories.
29 If a fuse is blown, make sure you have eliminated the reason it did so before installing a new fuse.
30 In addition to fuses, the electrical circuits incorporate fusible links

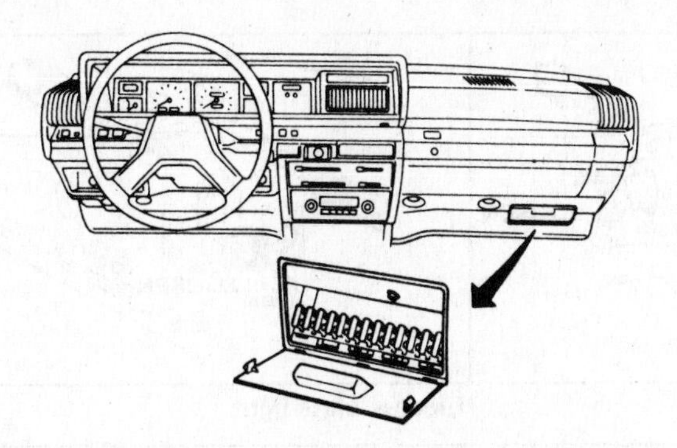
Fig. 13.64 Fusebox location (1981 through 1984 models) (Sec 11)

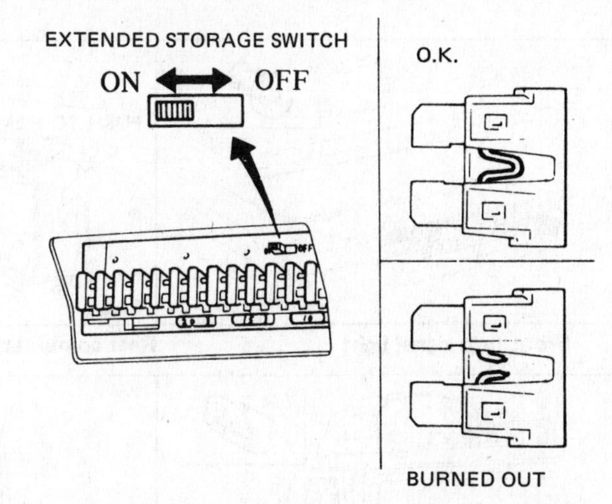

Fig. 13.65 Extended storage switch and reading of fuses (Sec 11)

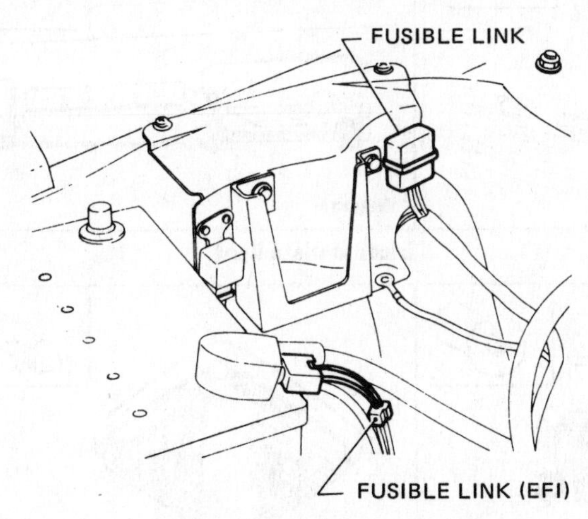

Fig. 13.66 Fusible link locations (Sec 11)

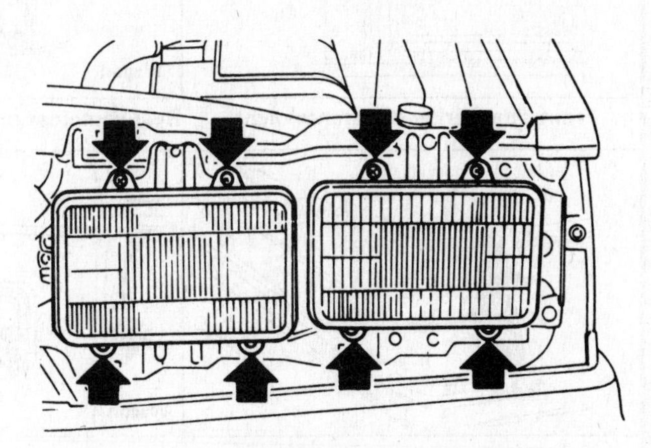
Fig. 13.67 Headlamp retaining screws (1981 through 1984 models)

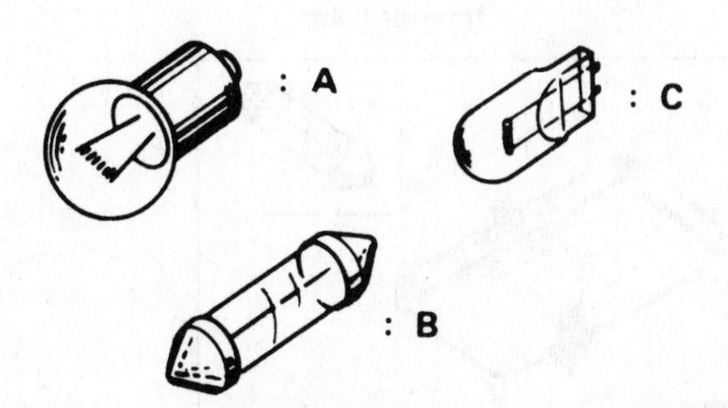

Fig. 13.68 The three types of bulbs used on 1981 through 1984 models (Sec 11)

for overload protection. The links are used in circuits which are not ordinarily fused, such as the ignition and Electronic Fuel Injection (EFI) circuits. The fusible links are located in the engine compartment and are removed by unplugging the connectors. If the link is melted, the entire fusible link harness should be replaced, but only after checking and correcting the electrical fault that caused it. Never wrap a fusible link in plastic tape and make sure it does not come in contact with other wiring harness or plastic or rubber components.
31 The various electrical relays are grouped together in the engine compartment or under the dash. If a faulty relay is suspected, it can be removed and tested by a dealer or other qualified shop. Defective relays must be replaced as a unit.

Headlamps – removal and installation (1981 through 1984 models)

32 Disconnect the negative battery cable.
33 Remove the headlamp finisher screws and pry the finisher off to remove it.
34 Loosen the retaining ring screws and remove the headlamp from the mounting ring. Unplug the electrical connector and remove the headlamp.
35 Installation is the reverse of removal, taking care to install the headlamp with the TOP mark in the proper location.

Exterior and interior lamps

36 Lamps other than the headlamps are of the three types shown in the illustration. After removing the lens or cover, these bulbs are removed as follows:

Type A Press or turn the bulb counterclockwise
Type B Pull the bulb from the holder clips
Type C Pull the bulb out of its socket

37 Refer to the accompanying illustrations for the details of each bulb replacement.

Combination switch – removal and installation (1981 through 1984 models)

38 Disconnect the battery ground cable.
39 Remove the steering wheel.
40 Remove the steering column cover and disconnect the combination switch wires.
41 Loosen the retaining screw and lift the combination switch from the steering column.
42 When installing the switch, align the protrusion on the switch body with the hole on the steering column.

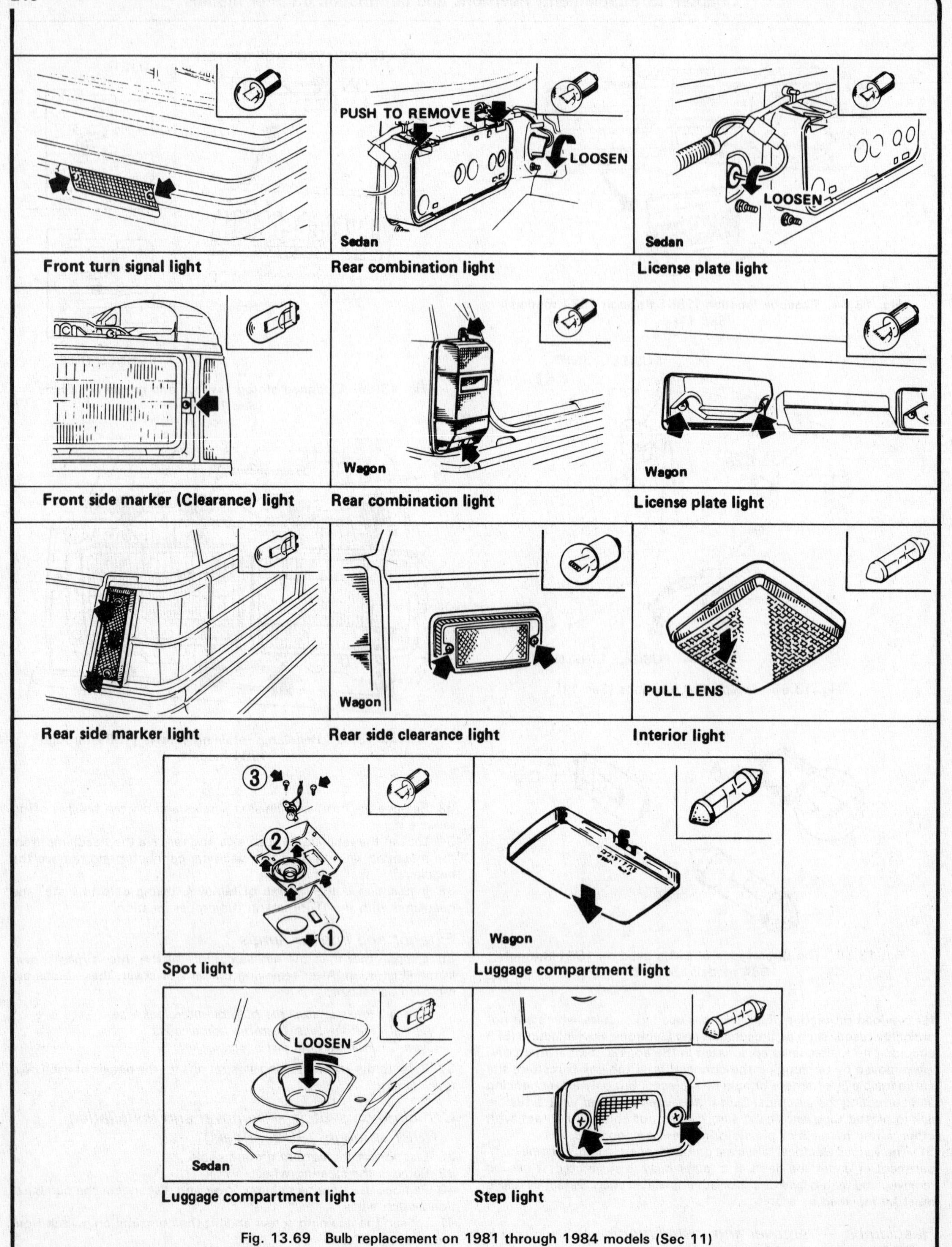

Fig. 13.69 Bulb replacement on 1981 through 1984 models (Sec 11)

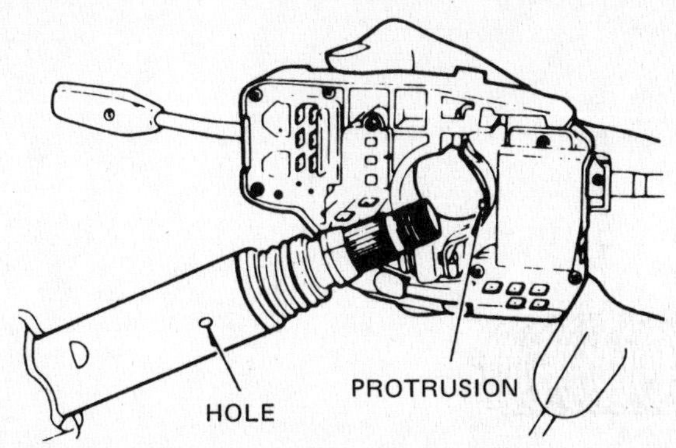

Fig. 13.70 Align the protrusion on the combination switch when installing (Sec 11)

Instrument panel and cluster lid — removal and installation

43 Disconnect the battery ground cable.
44 Remove the instrument lower cover, remove the fuse block and disconnect the wiring connectors from the junction block.
45 Remove the steering wheel, shell cover and combination switch.
46 Loosen the tilt column adjusting lever and lower the column out of the way.
47 Disconnect the speedometer cable, harness connectors and antenna cable.
48 Remove the choke control knob.
49 Referring to the illustration, remove the cluster lid A and instrument pad A.
50 Remove the center console.
51 Remove the heater control panel, followed by the heater control unit screws and radio.
52 Remove the defroster grille, followed by the heater nozzle on the driver's side.
53 Remove the instrument panel retaining bolts and withdraw the panel from the vehicle.
54 Installation is the reverse of removal.

Wiring diagrams

Wiring diagrams for 1980 through 1984 models are included at the end of this Supplement.

12 Suspension and steering

General description

The suspension on later models is virtually the same as on earlier models, with detail changes. Beginning in 1981, rack and pinion steering with optional power assist is used on all models.

Station wagon models in 1983 and 1984 were equipped with coil spring rear suspension instead of the previous leaf spring design. Also, later independent rear suspensions feature a stabilizer bar.

Rear axle and suspension assembly (Sedan) — removal and installation

1 This procedure is the same as described in Chapter 11 except that the differential and suspension mounting brackets are different. Remove the bolts and nuts securing these components in the order shown in the illustration.

Shock absorber (Station Wagon) — removal and installation

2 The procedure is the same as for earlier leaf spring models except that the shock absorber nut is tightened to the specified torque before the vehicle weight is lowered onto it.

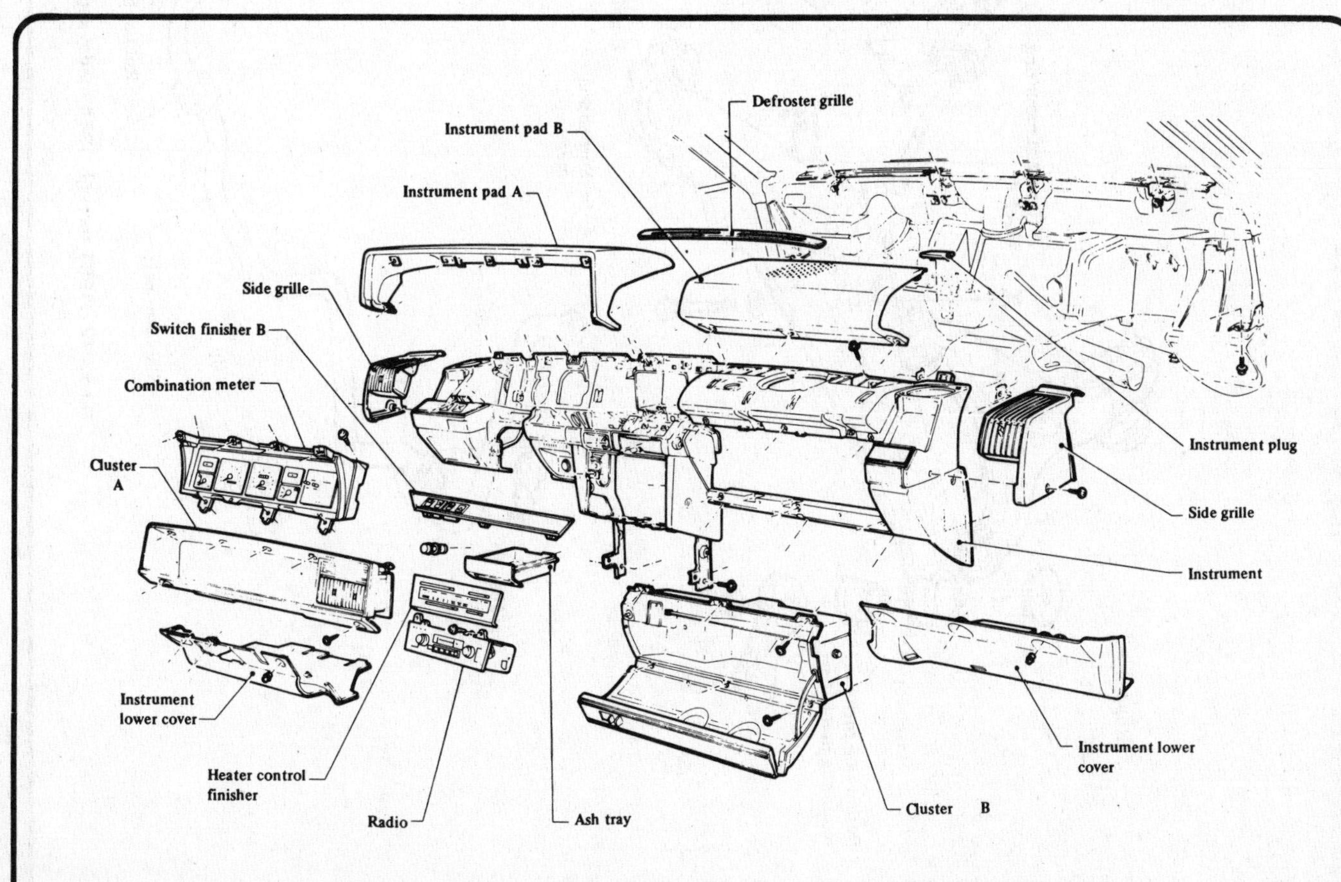

Fig. 13.71 Instrument panel components (1981 through 1984 models) (Sec 11)

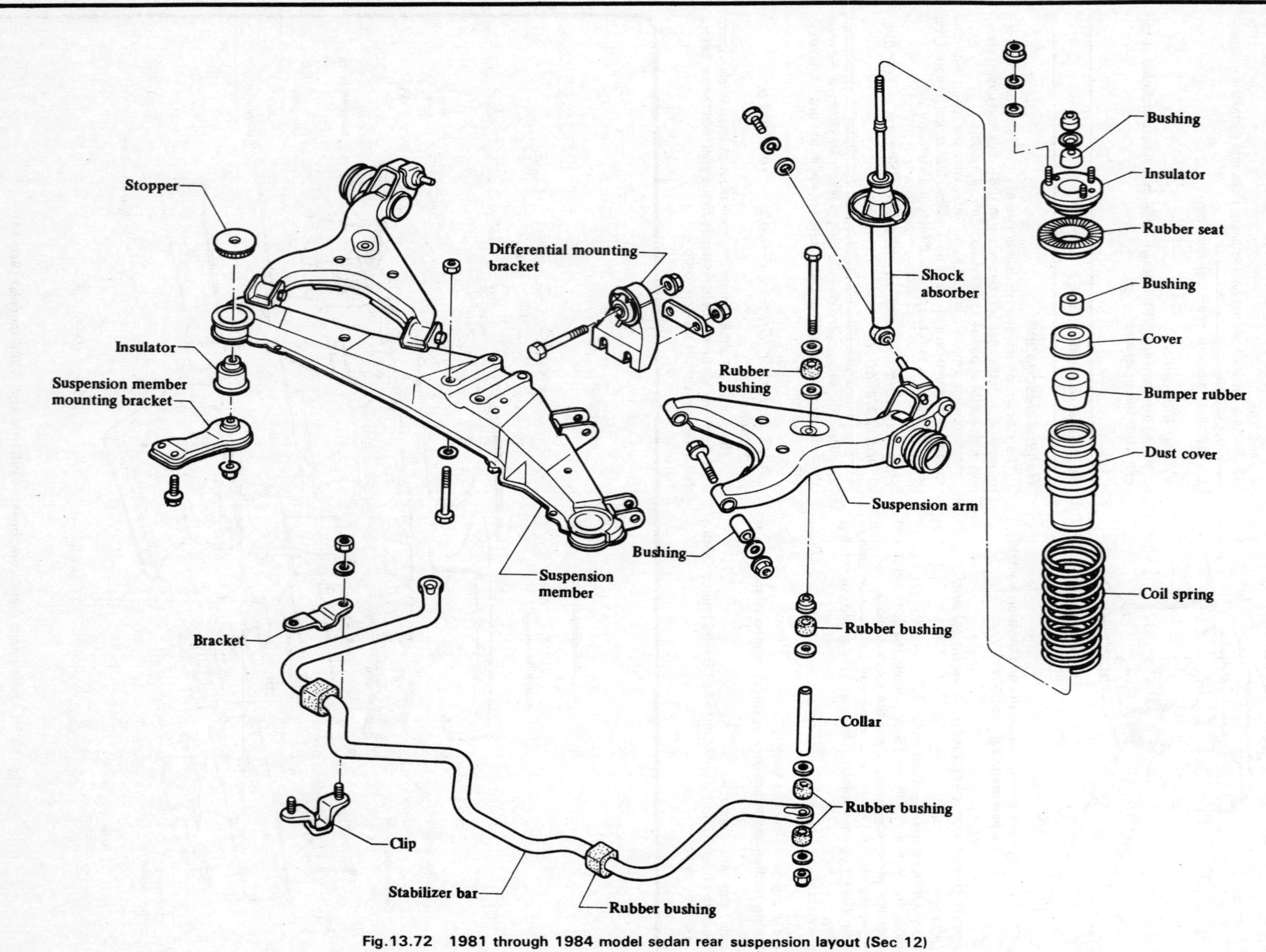

Fig.13.72 1981 through 1984 model sedan rear suspension layout (Sec 12)

Rear spring (Station Wagon) — removal and installation

3 Block the front wheels, raise the rear of the vehicle, support it securely on jackstands and remove the wheels.
4 Support the axle under the center section with a jack.
5 Disconnect the lower end of the shock absorber.
6 Slowly lower the jack and remove the spring and seat.

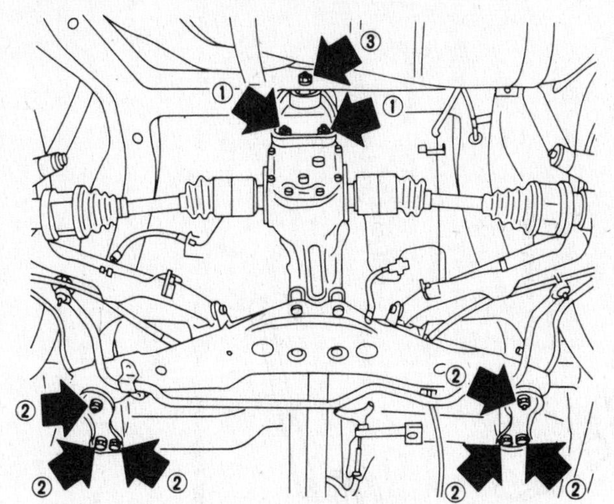

Fig. 13.73 Sedan rear suspension and differential mounting fastener removal sequence (Sec 12)

7 Installation is the reverse of removal, making sure the flat side of the coil spring is facing down, with the upper side matched to the vehicle side spring seat.

Upper link (Station Wagon) — removal and installation

8 Remove the securing nuts and bolts and remove the link.
9 Install one end of the link on the chassis, make sure the link is level and install the nut and bolt, tightening securely.
10 Install the other end of the link to the axle.
11 Tighten the nuts and bolts to 58 to 72 lbf ft (8.0 to 10.0 kgf m).

Lower link (Station Wagon) — removal and installation

12 Remove the spring, disconnect the link damper (right-hand link) and remove the securing bolts and nuts from either end of the link. Remove the link from the vehicle.
13 Installation is the reverse of removal. Install the nuts and bolts but do not tighten them to the specified torque of 58 to 72 lbf ft (8.0 to 10.0 kgf m) until the vehicle weight is lowered onto the suspension.

Link damper (Station Wagon) — inspection and replacement

14 Check the shock absorber-like link damper for leaking fluid, damage and cracked or deformed rubber bushings.
15 Remove the retaining nuts and bolts, noting the direction in which it is installed, and remove the damper.
16 Install the new damper in the original direction and tighten the nuts and bolts to 22 to 29 lbf ft (3 to 5 kgf m).

Rear axle (Station Wagon) — removal and installation

17 With the rear of the vehicle supported securely on jackstands, remove the rear wheels and support the axle housing under the center

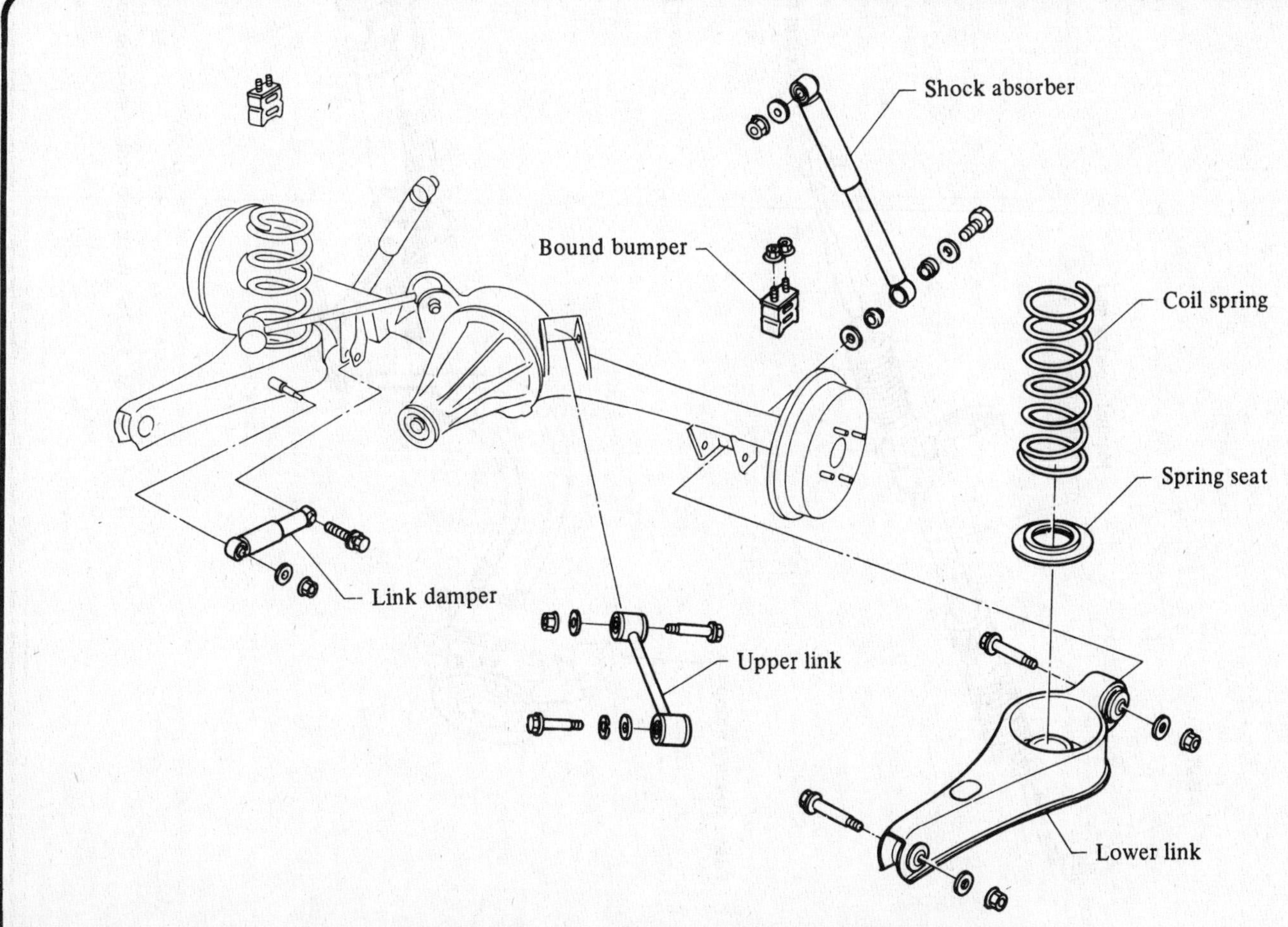

Fig. 13.74 1983 and 1984 station wagon coil spring rear suspension layout (Sec 12)

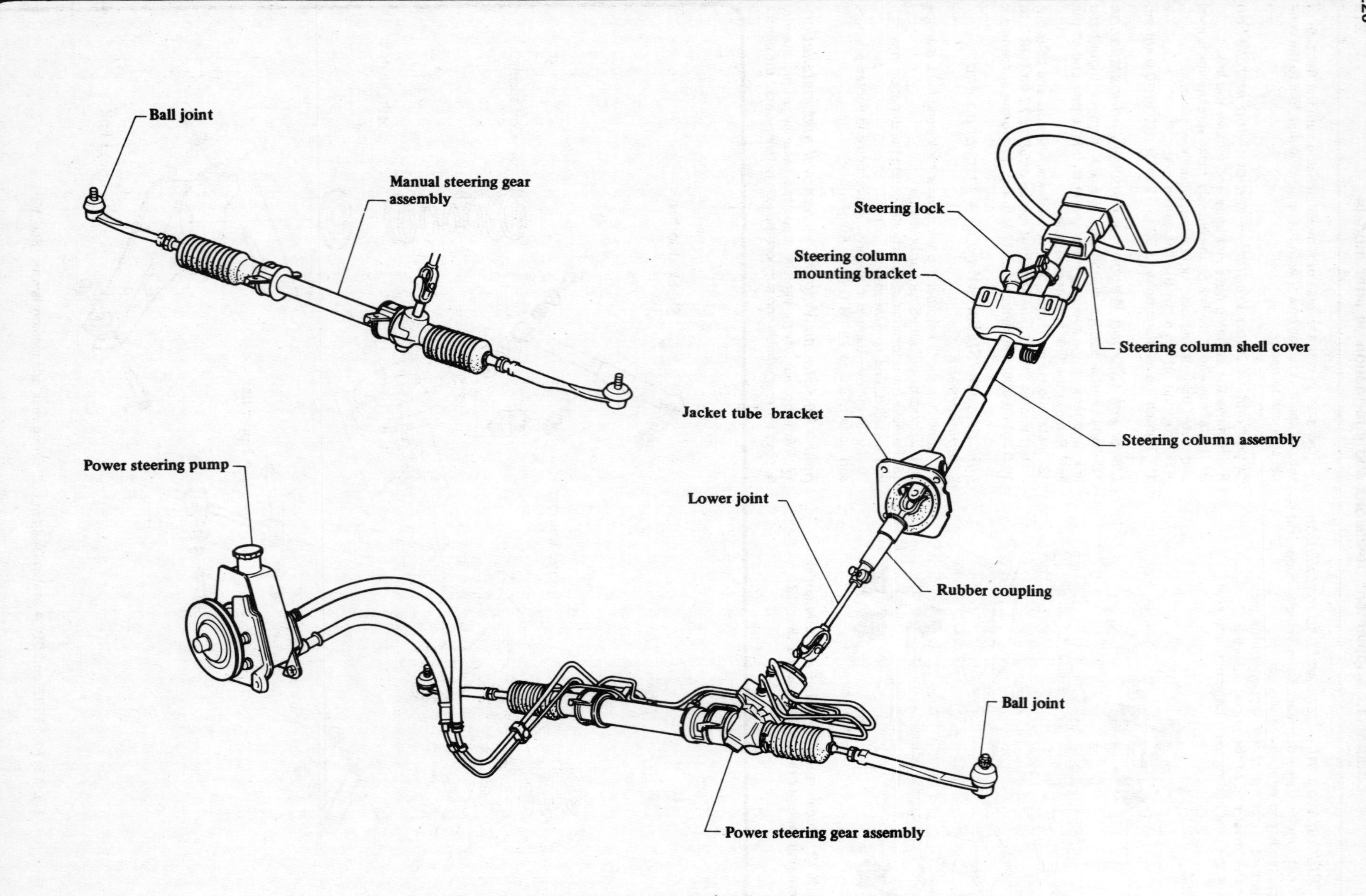

Fig. 13.75 Rack and pinion steering component layout (Sec 12)

section with a jack.

18 Disconnect the axle attaching points in the sequence shown in the accompanying illustration.

19 Lower the axle from the vehicle with the jack.

20 Installation is the reverse of removal.

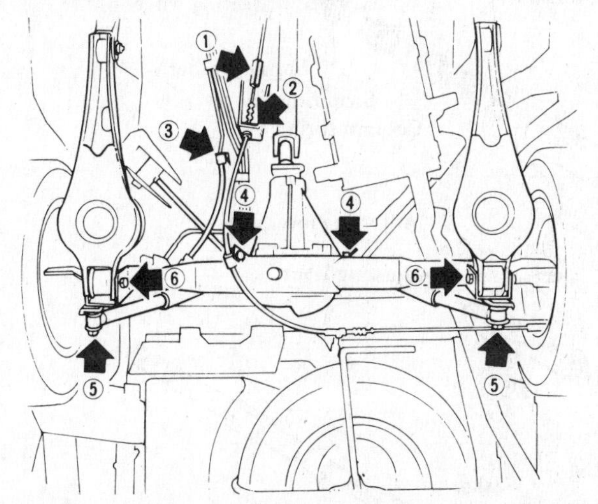

Fig. 13.76 Station wagon rear axle mounting bolt removal sequence (1983 and 1984) (Sec 12)

Rack and pinion steering and linkage – inspection

21 Check for play at the outer ends of the tie-rods, or excessive play within the rack and pinion itself. Also check for grease leakage from the rack and pinion boots.

Steering column – removal and installation (1981 through 1984 models)

22 Disconnect the negative battery cable.

23 Remove the lower joint-to-rubber coupling bolt.

24 Remove the steering wheel as described in Chapter 11.

25 Remove the steering column shell covers, followed by the com bination switch.

26 Remove the heater duct for access and then remove the steering column tube bracket and dust cover from the dash panel.

27 Remove the steering column mounting bracket and withdraw the column assembly into the passenger compartment. Remove the lower joint from the steering gear pinion.

28 Installation is the reverse of removal with attention paid to the following points:

a) The wheels should be set in a straight ahead position.
b) Insert the lower joint into the pinion with the punch mark aligned with the projection on the spacer and facing the right side as shown in the illustration.
c) Install the steering column assembly from the passenger compartment so the assembly fits into the lower joint with the bolt cutout as shown in the illustration.
d) Tighten all bolts securely during installation. After installation, tighten the bolts to the specified torque.

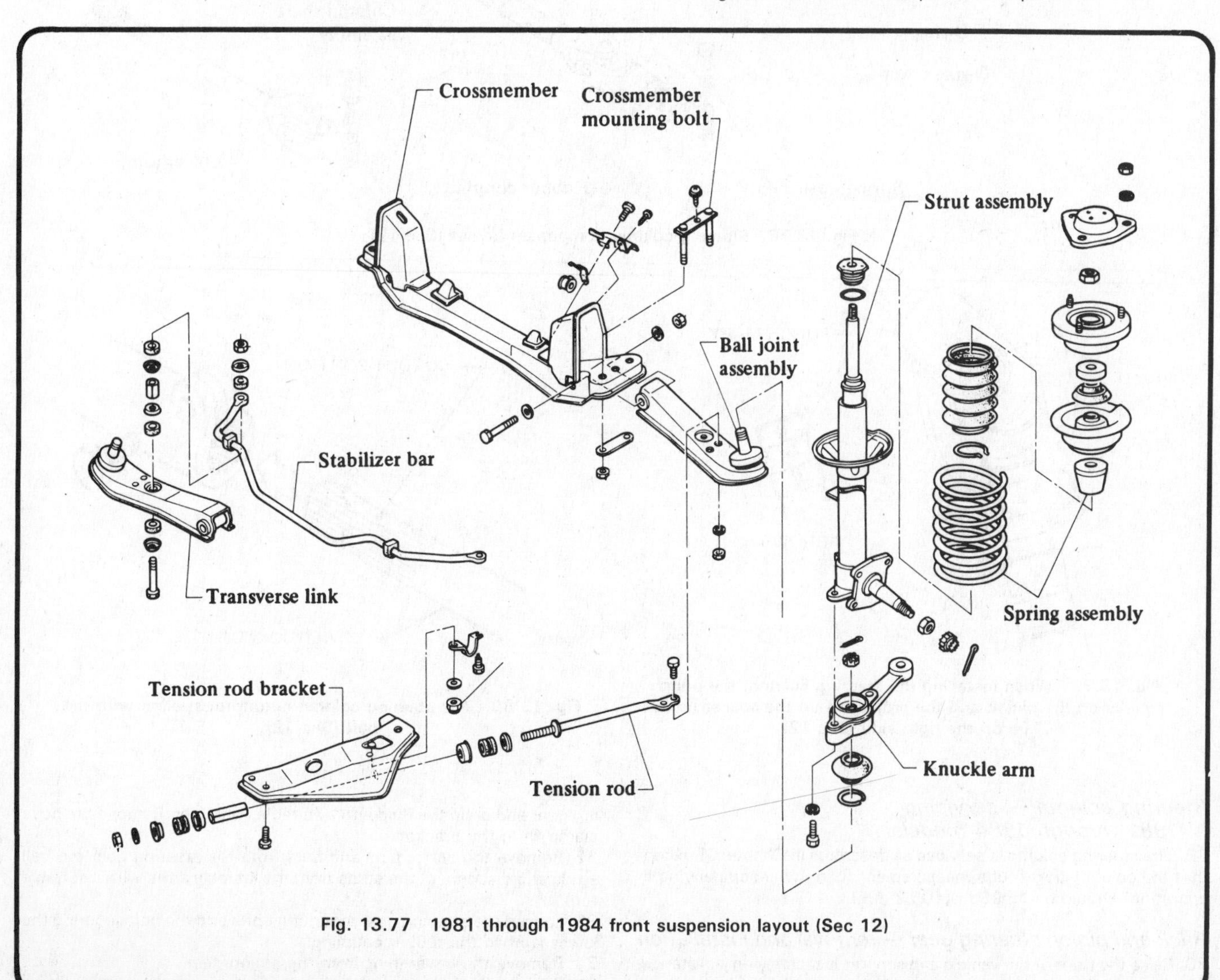

Fig. 13.77 1981 through 1984 front suspension layout (Sec 12)

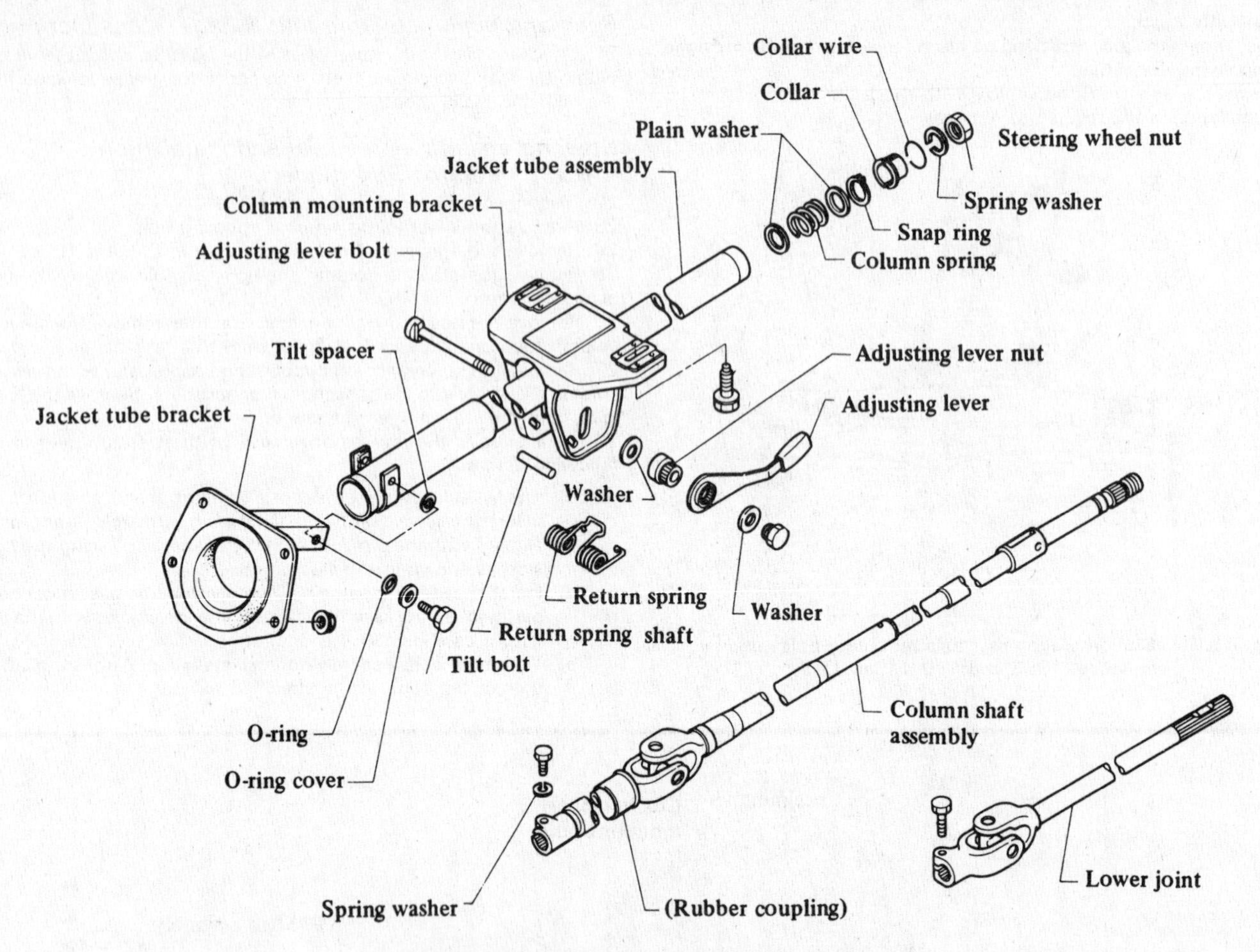

Fig. 13.78 Steering column component layout (Sec 12)

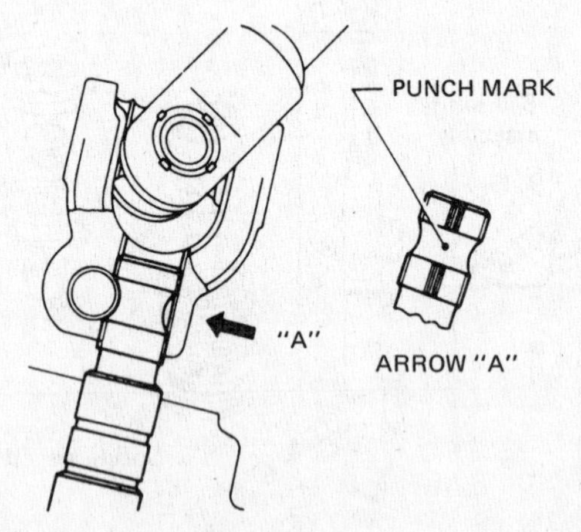

Fig. 13.79 When installing the steering column, the punch marks on the pinion and the projection on the spacer must be on the right side (Sec 12)

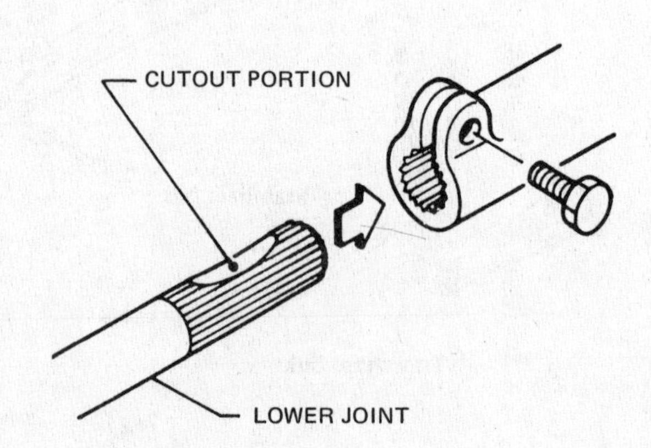

Fig. 13.80 The steering column cutout must align with the bolt (Sec 12)

Steering column — servicing (1981 through 1984 models)

29 The steering column is serviced as described in Chapter 11 except that the column jacket tube measurement (L in the accompanying illustration) should be 15.95 in (405.2 mm).

Rack and pinion steering gear — removal and installation

30 Raise the front of the vehicle and support it securely on jackstands.
31 Remove the hose clamp bolt, disconnect the flare nut at the steering gear and drain the fluid into a suitable container. Remove the hose clamp from the bracket.
32 Remove the cotter pins and nuts from the steering gear rod ball studs and disconnect the studs from the knuckle arms with a suitable tool.
33 Loosen the steering gear mounting bolts and the bolt securing the lower joint to the rubber coupling.
34 Remove the lower joint from the pinion gear.
35 Remove the bolts retaining the steering gear to the crossmember

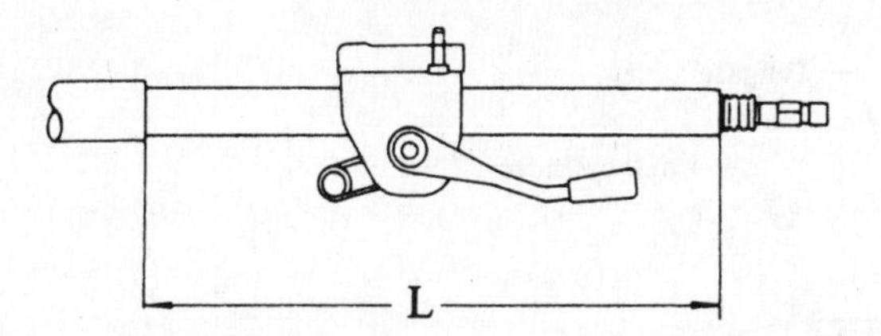

Fig. 13.81 The steering column jacket tube must not be less than the minimum measurement (L) (Sec 12)

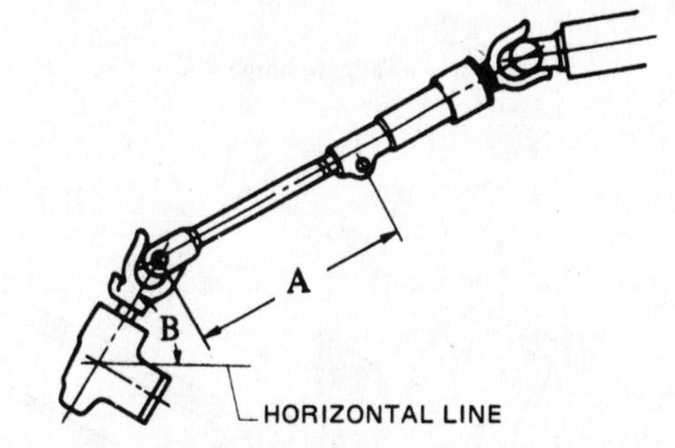

Fig. 13.82 Installing the steering column lower joint (Sec 12)

A 6.52 in (165 mm) *B 20.26°*

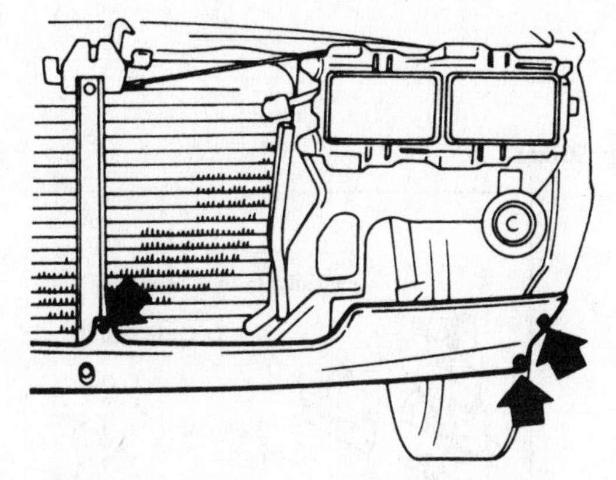

Fig. 13.83 Front apron fasteners (Sec 13)

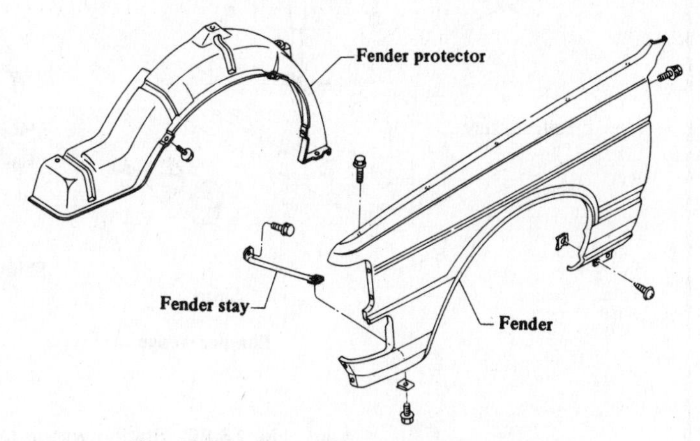

Fig. 13.84 Front fender components (Sec 13)

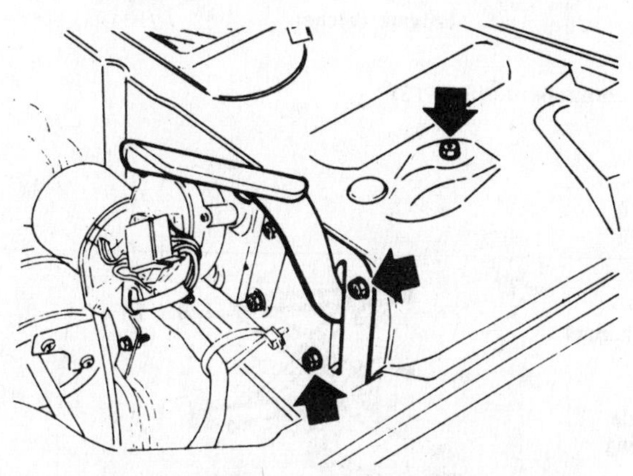

Fig. 13.85 Hood height adjustment bolts (Sec 13)

and separate the steering gear from the vehicle.
36 Installation is the reverse of removal. The lower joint must be installed as shown in the illustration before the mounting bolts are tightened to specification. Bleed the power steering hydraulic system after installation. This is accomplished by quickly turning the wheels from lock-to-lock several times with the front wheels off the ground until the fluid in the reservoir is free of bubbles.

13 Bodywork

Front fender — removal and installation

1 Remove the fender protector, front sight shield, bumper, apron, cowl top grille and fender stay attaching bolt.
2 Remove the attaching bolts and lift the fender from the car.
3 Clean the sealant from the fender, hood ledge, sight shield and front apron. After applying new sealant, installation is the reverse of removal.

Hood — removal and installation

4 Open the hood and disconnect the windshield washer tube.
5 Cover the tops of the fenders with protectors and attach rags to the corners of the hood to prevent scratches.
6 Have an assistant support the hood, mark their locations and then remove the retaining bolts. Lift the hood from the car.
7 Installation is the reverse of removal. The hood height can be adjusted after removing the cowl top grille for access to the adjustment screws. The hood bumpers are adjusted by screwing them in or out until the hood is flush with the fenders.

Tailgate (Station Wagon) — removal and installation

8 Remove the luggage compartment right side finish panel and disconnect the tailgate harness connectors and rear window washer hose.
9 Support the tailgate, mark the position of the attaching bolts and disconnect the upper ends of the tailgate stays.
10 Remove the attaching bolts and lift the tailgate from the vehicle.
11 Installation is the reverse of removal.

Door — removal and installation

12 Support the door with a jack or jackstand, using a suitable cloth to cushion it.
13 Remove the check link and then remove the door from the hinges. Disconnect the electrical harness on power window-equipped models.
14 Installation is the reverse of removal after applying a thin coat of multi-purpose grease to the hinge rollers. Tighten the hinge securing bolts to 12 to 16 lbf ft (1.6 to 2.2 kgf m). To adjust the front door, remove the fender protector and reach up from the wheelhouse with a suitable offset wrench to adjust the hinge.

Door trim and interior handles

15 Lower the window glass fully and remove the armrest, window crank, handle escutcheon and door lock knob.
16 Carefully pry the door trim panel off with a suitable tool.
17 Disconnect any electrical wires, pry the finisher clip off and carefully peel the sealing screen off the door inner panel.
18 Installation is the reverse of removal, taking care to reaffix the sealing screen with adhesive or tape. After installation, crank the window fully up and install the crank in the position shown.

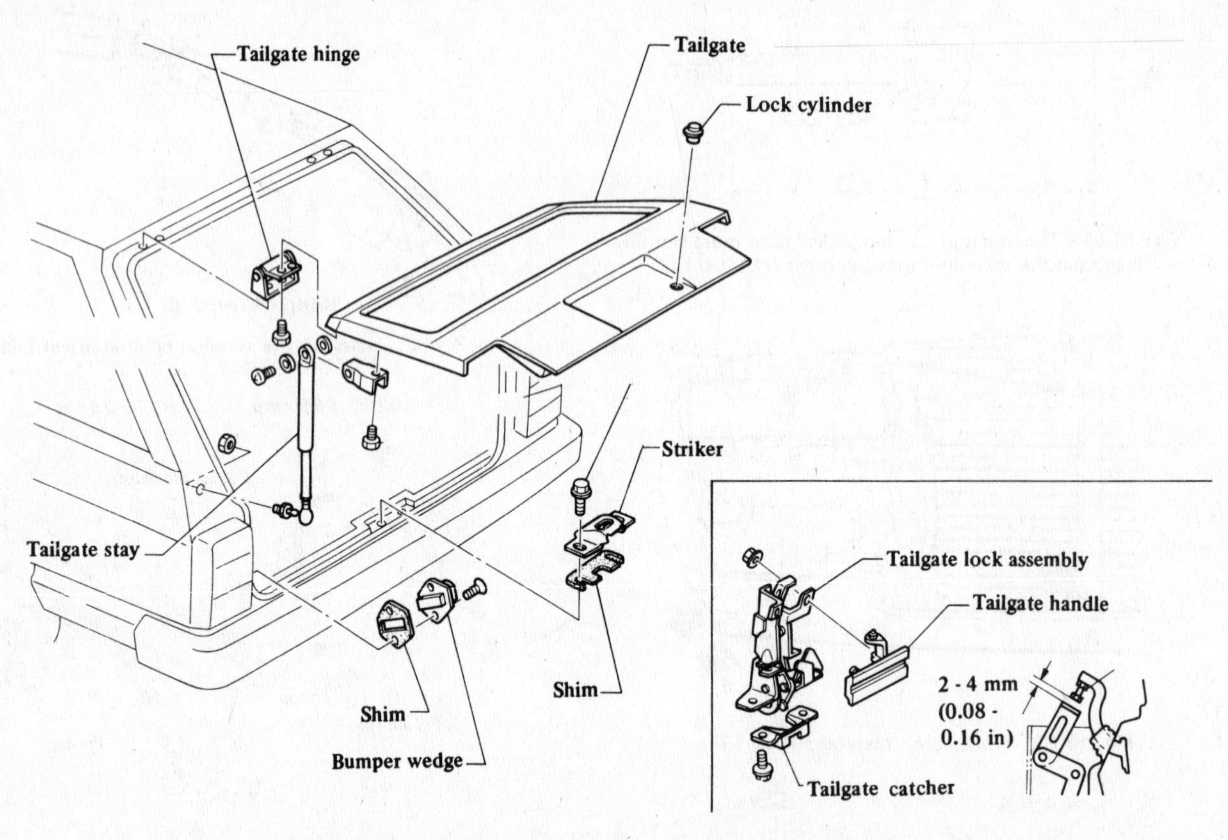

Fig. 13.86 Station wagon tailgate components (Sec 13)

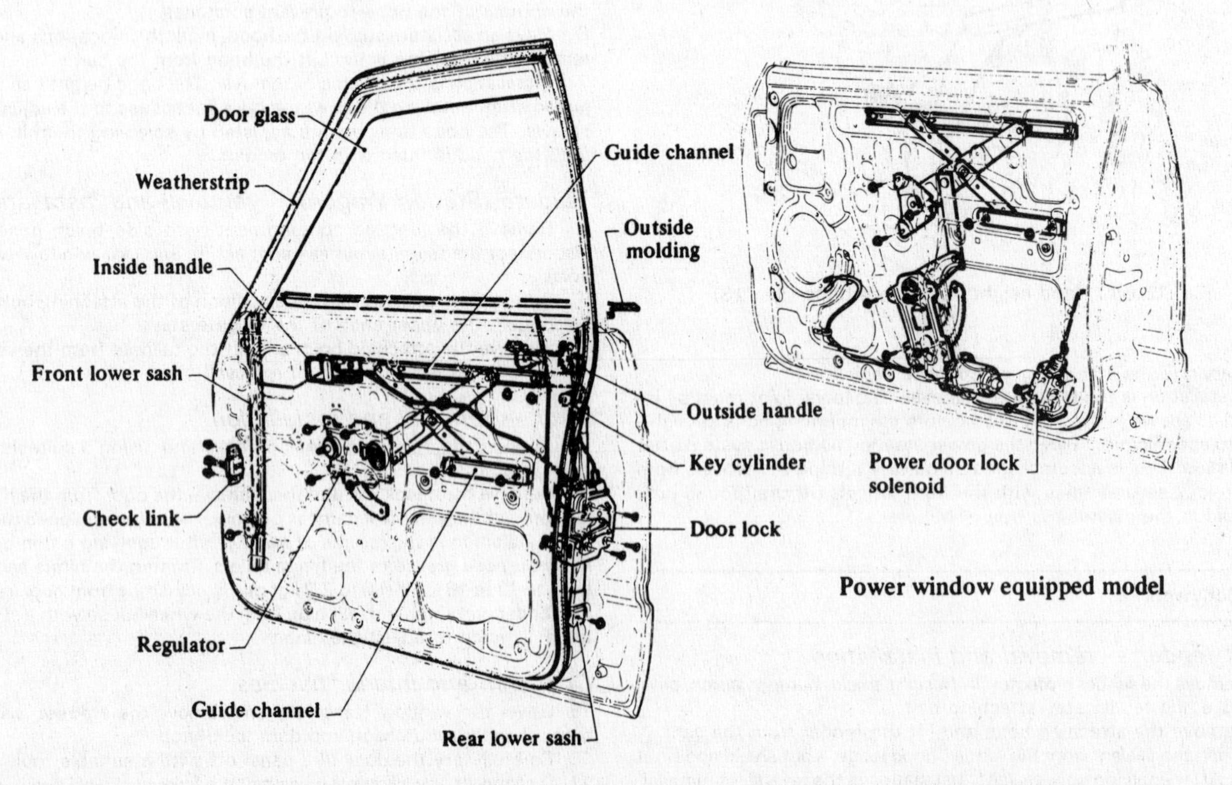

Fig. 13.87 Typical door component layout (1981 through 1984 models (Sec 13)

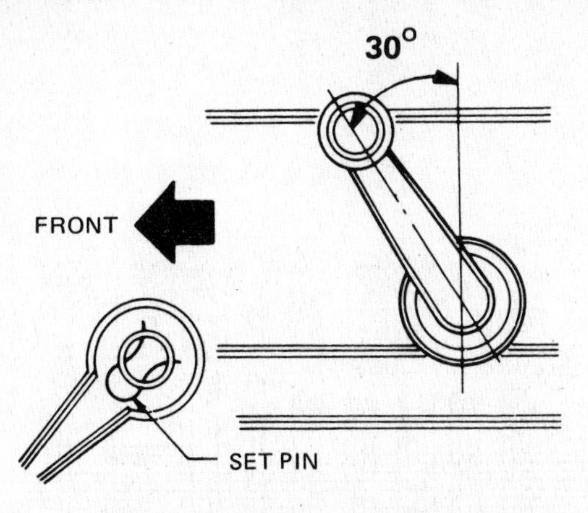

Fig. 13.88 Correct window crank installation (1981 through 1984 models) (Sec 13)

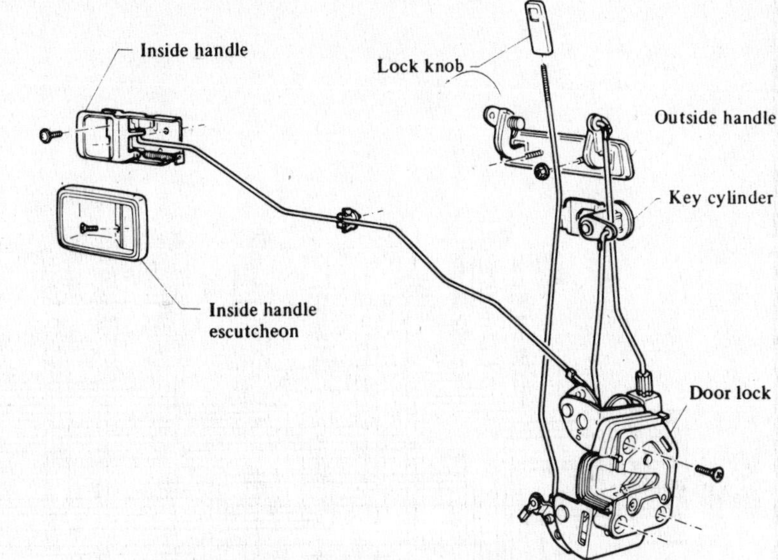

Fig. 13.89 Typical door lock and control assembly (1981 through 1984 models) (Sec 13)

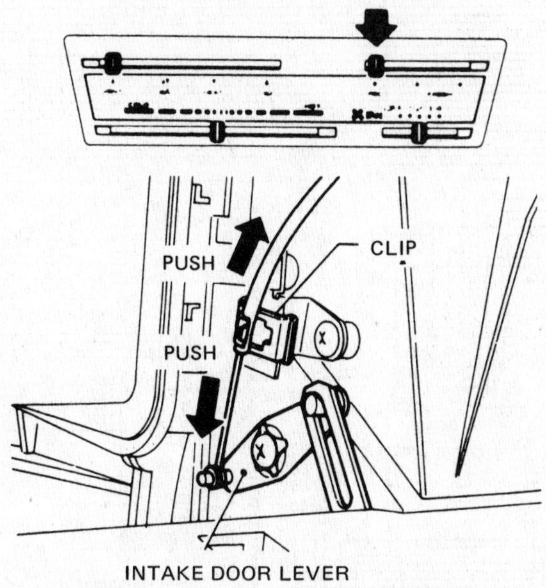

Fig. 13.90 Heater control cable adjustment (Sec 13)

Door window glass and regulator

19 Lower the door glass fully and remove the trim panel.
20 Remove the window-to-guide channel bolts.
21 Remove the rear lower sash and pull the glass up and out to remove it.
22 Remove the attaching bolts and withdraw the regulator through the large access hole in the door.
23 Installation is a reversal of removal. Lubricate the sliding portion of the regulator prior to installation, using multi-purpose grease.

Door lock and controls

24 Remove the door trim and glass.
25 Remove the door lock switch or power window solenoid attaching screws.
26 Disconnect the key rod from the door lock.
27 Remove the door lock and lock control, followed by the outside door handle.
28 Installation is the reverse of removal, after applying a light coat of multi-purpose grease to the lever and springs.

Bumpers — removal and installation

Front
29 Disconnect the battery ground cable and remove the front clearance lamps.
30 Remove the headlamp finishers, radiator grille and bumper side bracket-to-fender screws.
31 Remove the sight shield and front bumper armature-to-body screws.
32 Unplug the electrical harness and remove the bumper.
Rear (Sedan)
33 Remove the rear sight shield, followed by the upper and lower retainer and bumper-to-body attaching bolts. Remove the rear bumper fascia.
34 Unbolt and remove the rear bumper honeycomb.
Rear (Station Wagon)
35 Remove the bumper side bracket-to-body and armature-to-body attaching bolts and nuts. Lift the bumper from the vehicle.

Bumpers — inspection

36 Replace the honeycomb if it is cracked or damaged. Measure the thickness of the honeycomb. Replace it if it is less than 90 percent of 3.287 in (83.5 mm) for the center honeycomb or 3.94 in (100 mm) for the side honeycomb.
37 Installation is the reverse of removal.

Heater unit — removal and installation

38 Remove the heater control finisher, radio and instrument panel lower covers.
39 Disconnect the heater and blower unit control cables.
40 Unplug the harness connectors and ground wire terminal.
41 Use a screwdriver wrapped in a cloth to remove the heater unit retaining clips.
42 Remove the heater unit.
43 Installation is the reverse of removal.
44 Adjust the intake door control cable by setting the lever to the REC position. Connect the cable to the intake door lever and push the lever to the closed position. Clamp the cable securely while pushing the cable outer case in the direction of the arrow as shown in the illustration.
47 Installation is the reverse of removal.

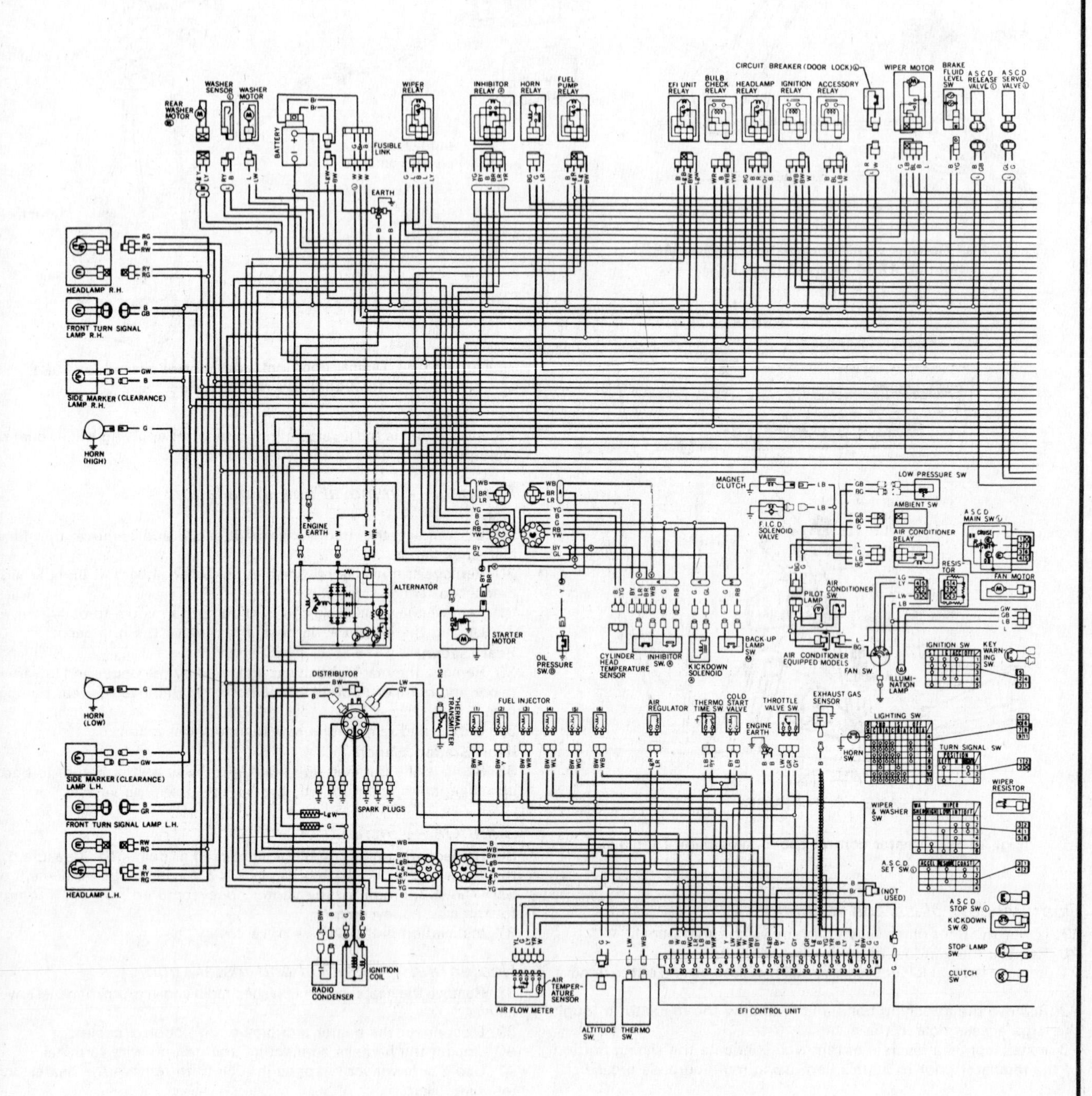

Fig. 13.91 Wiring diagram for 1980 through 1984 models (Sec 11)

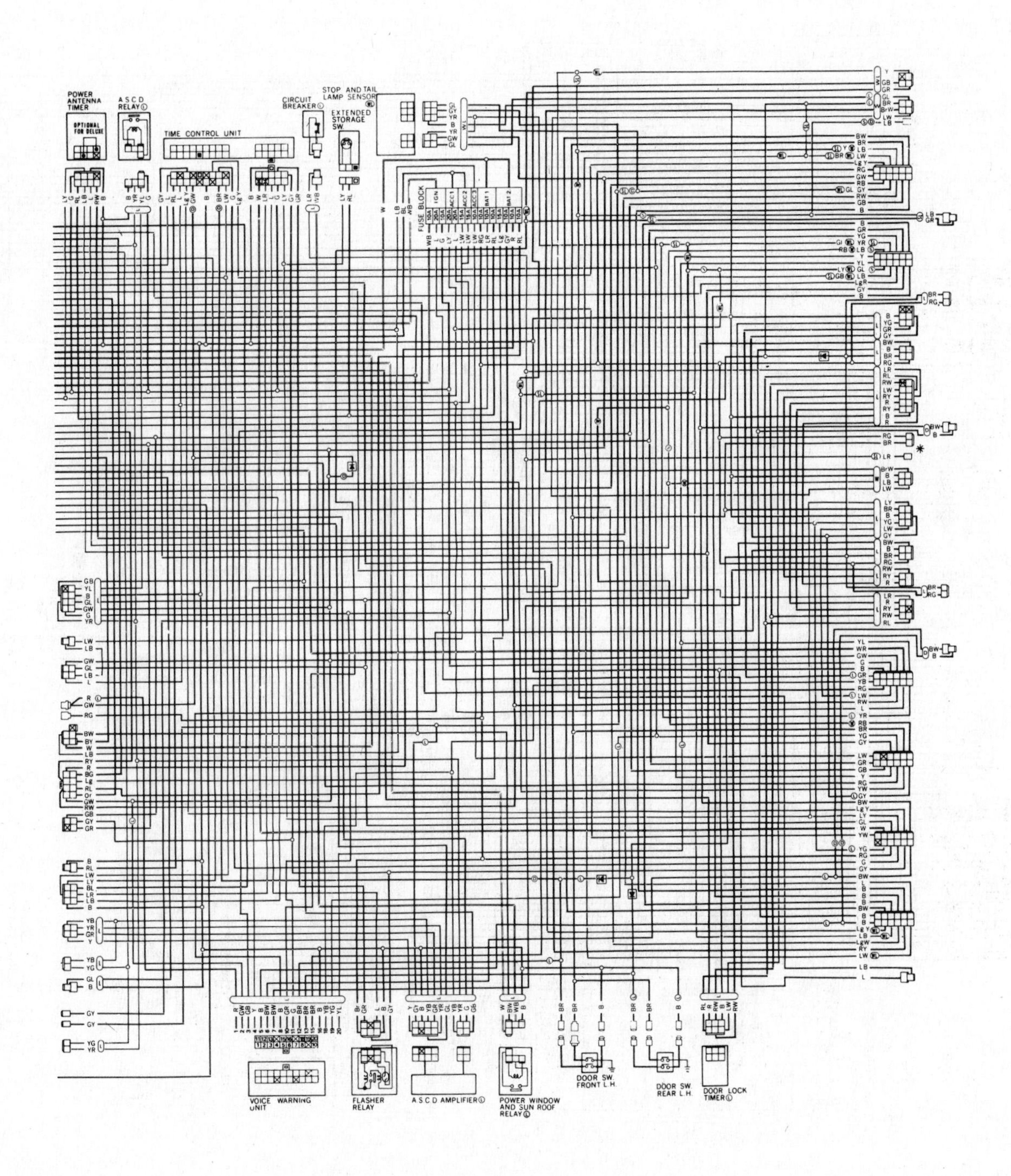

Fig. 13.92 Wiring diagram for 1980 through 1984 models (cont.) (Sec 11)

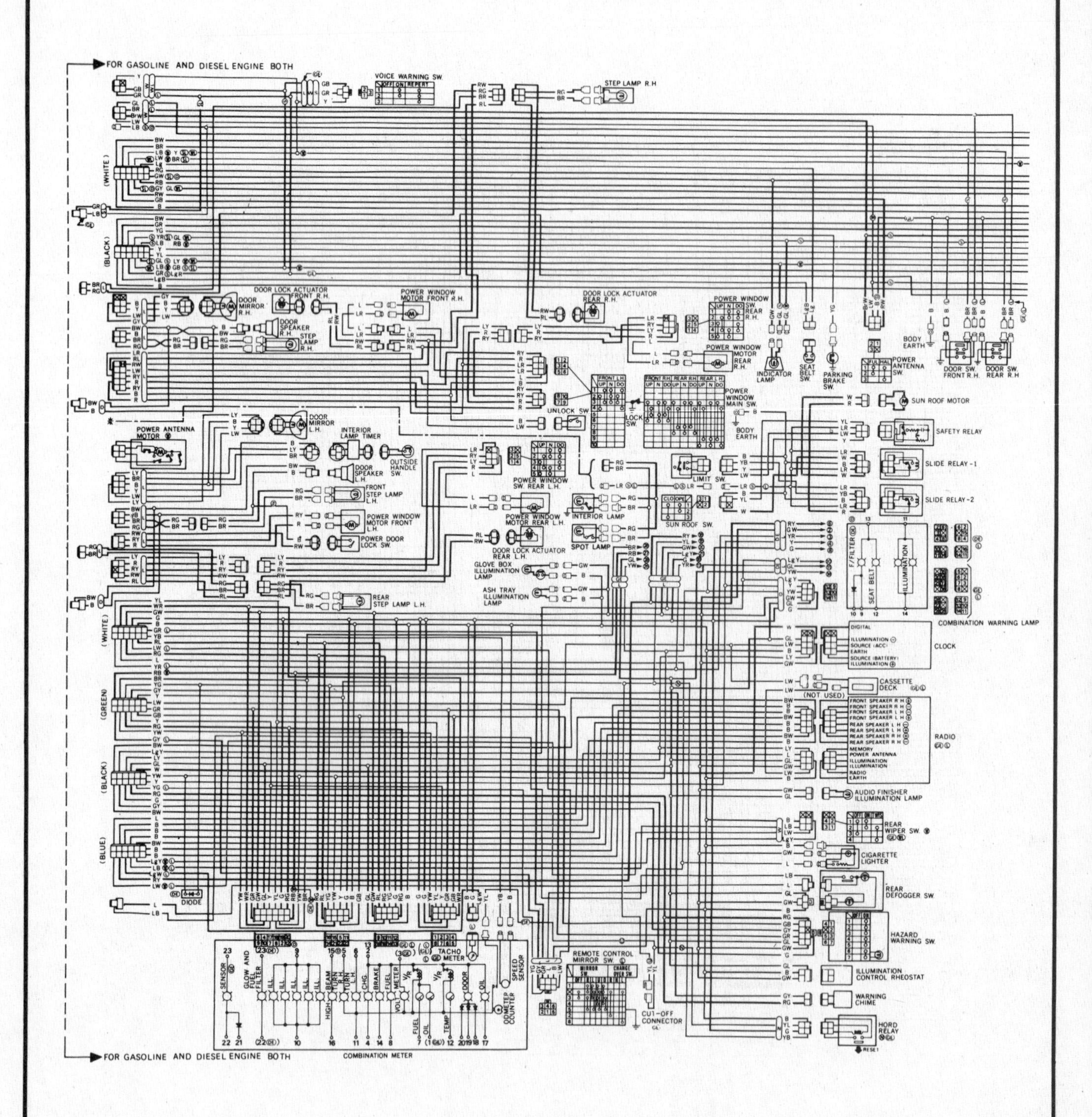

Fig. 13.93 Wiring diagram for 1980 through 1984 models (cont.) (Sec 11)

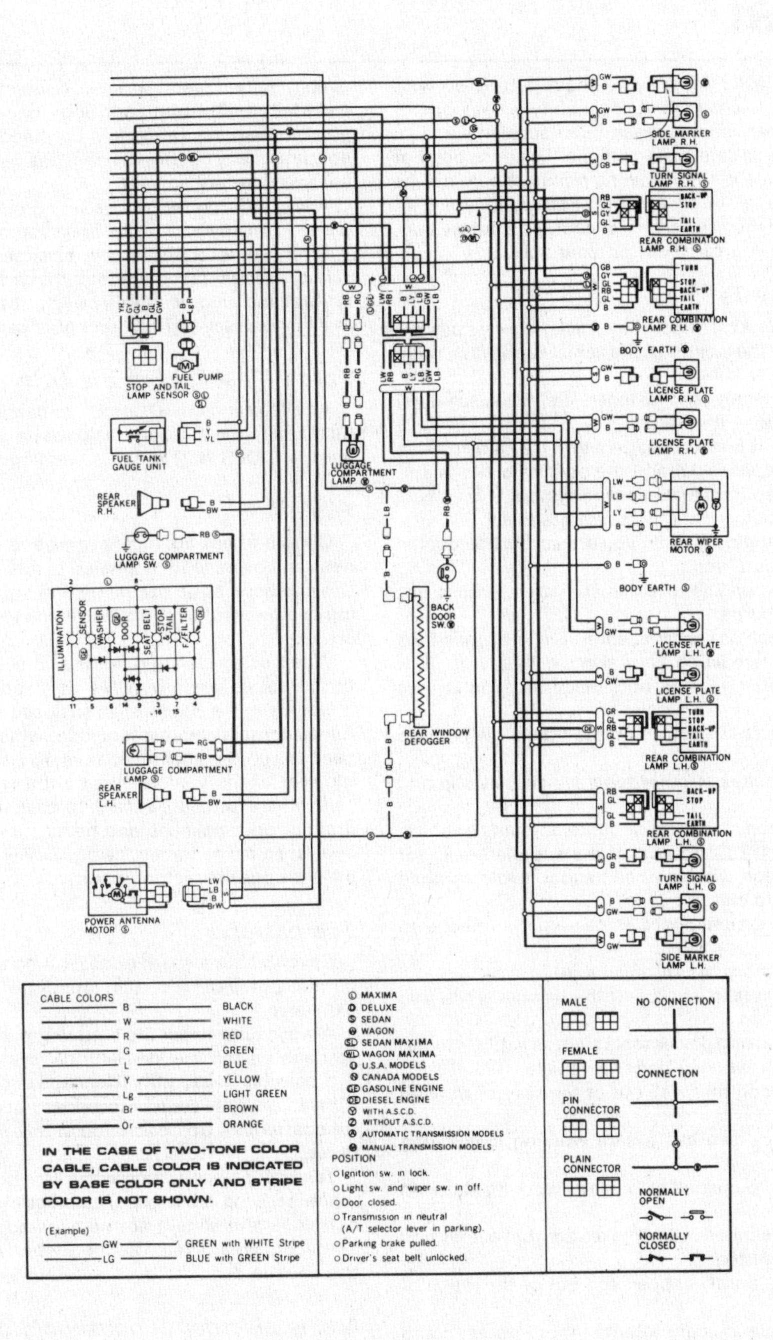

Fig. 13.94 Wiring diagram for 1980 through 1984 models (cont.) (Sec 11)

Safety first!

Regardless of how enthusiastic you may be about getting on with the job at hand, take the time to ensure that your safety is not jeopardized. A moment's lack of attention can result in an accident, as can failure to observe certain simple safety precautions. The possibility of an accident will always exist, and the following points should not be considered a comprehensive list of all dangers. Rather, they are intended to make you aware of the risks and to encourage a safety conscious approach to all work you carry out on your vehicle.

Essential DOs and DON'Ts

DON'T rely on a jack when working under the vehicle. Always use approved jackstands to support the weight of the vehicle and place them under the recommended lift or support points.
DON'T attempt to loosen extremely tight fasteners (i.e. wheel lug nuts) while the vehicle is on a jack — it may fall.
DON'T start the engine without first making sure that the transmission is in Neutral (or Park where applicable) and the parking brake is set.
DON'T remove the radiator cap from a hot cooling system — let it cool or cover it with a cloth and release the pressure gradually.
DON'T attempt to drain the engine oil until you are sure it has cooled to the point that it will not burn you.
DON'T touch any part of the engine or exhaust system until it has cooled sufficiently to avoid burns.
DON'T siphon toxic liquids such as gasoline, antifreeze and brake fluid by mouth, or allow them to remain on your skin.
DON'T inhale brake lining dust — it is potentially hazardous (see *Asbestos* below)
DON'T allow spilled oil or grease to remain on the floor — wipe it up before someone slips on it.
DON'T use loose fitting wrenches or other tools which may slip and cause injury.
DON'T push on wrenches when loosening or tightening nuts or bolts. Always try to pull the wrench toward you. If the situation calls for pushing the wrench away, push with an open hand to avoid scraped knuckles if the wrench should slip.
DON'T attempt to lift a heavy component alone — get someone to help you.
DON'T rush or take unsafe shortcuts to finish a job.
DON'T allow children or animals in or around the vehicle while you are working on it.
DO wear eye protection when using power tools such as a drill, sander, bench grinder, etc. and when working under a vehicle.
DO keep loose clothing and long hair well out of the way of moving parts.
DO make sure that any hoist used has a safe working load rating adequate for the job.
DO get someone to check on you periodically when working alone on a vehicle.
DO carry out work in a logical sequence and make sure that everything is correctly assembled and tightened.
DO keep chemicals and fluids tightly capped and out of the reach of children and pets.
DO remember that your vehicle's safety affects that of yourself and others. If in doubt on any point, get professional advice.

Asbestos

Certain friction, insulating, sealing, and other products — such as brake linings, brake bands, clutch linings, torque converters, gaskets, etc. — contain asbestos. *Extreme care must be taken to avoid inhalation of dust from such products since it is hazardous to health.* If in doubt, assume that they *do* contain asbestos.

Fire

Remember at all times that gasoline is highly flammable. Never smoke or have any kind of open flame around when working on a vehicle. But the risk does not end there. A spark caused by an electrical short circuit, by two metal surfaces contacting each other, or even by static electricity built up in your body under certain conditions, can ignite gasoline vapors, which in a confined space are highly explosive. Do not, under any circumstances, use gasoline for cleaning parts. Use an approved safety solvent.

Always disconnect the battery ground (–) cable *at the battery* before working on any part of the fuel system or electrical system. Never risk spilling fuel on a hot engine or exhaust component.

It is strongly recommended that a fire extinguisher suitable for use on fuel and electrical fires be kept handy in the garage or workshop at all times. Never try to extinguish a fuel or electrical fire with water.

Torch (flashlight in the US)

Any reference to a "torch" appearing in this manual should always be taken to mean a hand-held, battery-operated electric light or flashlight. It DOES NOT mean a welding or propane torch or blowtorch.

Fumes

Certain fumes are highly toxic and can quickly cause unconsciousness and even death if inhaled to any extent. Gasoline vapor falls into this category, as do the vapors from some cleaning solvents. Any draining or pouring of such volatile fluids should be done in a well ventilated area.

When using cleaning fluids and solvents, read the instructions on the container carefully. Never use materials from unmarked containers.

Never run the engine in an enclosed space, such as a garage. Exhaust fumes contain carbon monoxide, which is extremely poisonous. If you need to run the engine, always do so in the open air, or at least have the rear of the vehicle outside the work area.

If you are fortunate enough to have the use of an inspection pit, never drain or pour gasoline and never run the engine while the vehicle is over the pit. The fumes, being heavier than air, will concentrate in the pit with possibly lethal results.

The battery

Never create a spark or allow a bare light bulb near a battery. They normally give off a certain amount of hydrogen gas, which is highly explosive.

Always disconnect the battery ground (–) cable *at the battery* before working on the fuel or electrical systems.

If possible, loosen the filler caps or cover when charging the battery from an external source (this does not apply to sealed or maintenance-free batteries). Do not charge at an excessive rate or the battery may burst.

Take care when adding water to a non maintenance-free battery and when carrying a battery. The electrolyte, even when diluted, is very corrosive and should not be allowed to contact clothing or skin.

Always wear eye protection when cleaning the battery to prevent the caustic deposits from entering your eyes.

Mains electricity (household current in the US)

When using an electric power tool, inspection light, etc., which operates on household current, always make sure that the tool is correctly connected to its plug and that, where necessary, it is properly grounded. Do not use such items in damp conditions and, again, do not create a spark or apply excessive heat in the vicinity of fuel or fuel vapor.

Secondary ignition system voltage

A severe electric shock can result from touching certain parts of the ignition system (such as the spark plug wires) when the engine is running or being cranked, particularly if components are damp or the insulation is defective. In the case of an electronic ignition system, the secondary system voltage is much higher and could prove fatal.

Conversion factors

Length (distance)					
Inches (in)	X 25.4	= Millimetres (mm)	X 0.0394	= Inches (in)
Feet (ft)	X 0.305	= Metres (m)	X 3.281	= Feet (ft)
Miles	X 1.609	= Kilometres (km)	X 0.621	= Miles
Volume (capacity)				
Cubic inches (cu in; in³)	X 16.387	= Cubic centimetres (cc; cm³)	X 0.061	= Cubic inches (cu in; in³)
Imperial pints (Imp pt)	X 0.568	= Litres (l)	X 1.76	= Imperial pints (Imp pt)
Imperial quarts (Imp qt)	X 1.137	= Litres (l)	X 0.88	= Imperial quarts (Imp qt)
Imperial quarts (Imp qt)	X 1.201	= US quarts (US qt)	X 0.833	= Imperial quarts (Imp qt)
US quarts (US qt)	X 0.946	= Litres (l)	X 1.057	= US quarts (US qt)
Imperial gallons (Imp gal)	X 4.546	= Litres (l)	X 0.22	= Imperial gallons (Imp gal)
Imperial gallons (Imp gal)	X 1.201	= US gallons (US gal)	X 0.833	= Imperial gallons (Imp gal)
US gallons (US gal)	X 3.785	= Litres (l)	X 0.264	= US gallons (US gal)
Mass (weight)				
Ounces (oz)	X 28.35	= Grams (g)	X 0.035	= Ounces (oz)
Pounds (lb)	X 0.454	= Kilograms (kg)	X 2.205	= Pounds (lb)
Force				
Ounces-force (ozf; oz)	X 0.278	= Newtons (N)	X 3.6	= Ounces-force (ozf; oz)
Pounds-force (lbf; lb)	X 4.448	= Newtons (N)	X 0.225	= Pounds-force (lbf; lb)
Newtons (N)	X 0.1	= Kilograms-force (kgf; kg)	X 9.81	= Newtons (N)
Pressure				
Pounds-force per square inch (psi; lbf/in²; lb/in²)	X 0.070	= Kilograms-force per square centimetre (kgf/cm²; kg/cm²)	X 14.223	= Pounds-force per square inch (psi; lbf/in²; lb/in²)
Pounds-force per square inch (psi; lbf/in²; lb/in²)	X 0.068	= Atmospheres (atm)	X 14.696	= Pounds-force per square inch (psi; lbf/in²; lb/in²)
Pounds-force per square inch (psi; lbf/in²; lb/in²)	X 0.069	= Bars	X 14.5	= Pounds-force per square inch (psi; lbf/in²; lb/in²)
Pounds-force per square inch (psi; lbf/in²; lb/in²)	X 6.895	= Kilopascals (kPa)	X 0.145	= Pounds-force per square inch (psi; lbf/in²; lb/in²)
Kilopascals (kPa)	X 0.01	= Kilograms-force per square centimetre (kgf/cm²; kg/cm²)	X 98.1	= Kilopascals (kPa)
Millibar (mbar)	X 100	= Pascals (Pa)	X 0.01	= Millibar (mbar)
Millibar (mbar)	X 0.0145	= Pounds-force per square inch (psi; lbf/in²; lb/in²)	X 68.947	= Millibar (mbar)
Millibar (mbar)	X 0.75	= Millimetres of mercury (mmHg)	X 1.333	= Millibar (mbar)
Millibar (mbar)	X 0.401	= Inches of water (inH$_2$O)	X 2.491	= Millibar (mbar)
Millimetres of mercury (mmHg)	X 0.535	= Inches of water (inH$_2$O)	X 1.868	= Millimetres of mercury (mmHg)
Inches of water (inH$_2$O)	X 0.036	= Pounds-force per square inch (psi; lbf/in²; lb/in²)	X 27.68	= Inches of water (inH$_2$O)
Torque (moment of force)				
Pounds-force inches (lbf in; lb in)	X 1.152	= Kilograms-force centimetre (kgf cm; kg cm)	X 0.868	= Pounds-force inches (lbf in; lb in)
Pounds-force inches (lbf in; lb in)	X 0.113	= Newton metres (Nm)	X 8.85	= Pounds-force inches (lbf in; lb in)
Pounds-force inches (lbf in; lb in)	X 0.083	= Pounds-force feet (lbf ft; lb ft)	X 12	= Pounds-force inches (lbf in; lb in)
Pounds-force feet (lbf ft; lb ft)	X 0.138	= Kilograms-force metres (kgf m; kg m)	X 7.233	= Pounds-force feet (lbf ft; lb ft)
Pounds-force feet (lbf ft; lb ft)	X 1.356	= Newton metres (Nm)	X 0.738	= Pounds-force feet (lbf ft; lb ft)
Newton metres (Nm)	X 0.102	= Kilograms-force metres (kgf m; kg m)	X 9.804	= Newton metres (Nm)
Power				
Horsepower (hp)	X 745.7	= Watts (W)	X 0.0013	= Horsepower (hp)
Velocity (speed)				
Miles per hour (miles/hr; mph)	X 1.609	= Kilometres per hour (km/hr; kph)	X 0.621	= Miles per hour (miles/hr; mph)
Fuel consumption*				
Miles per gallon, Imperial (mpg)	X 0.354	= Kilometres per litre (km/l)	X 2.825	= Miles per gallon, Imperial (mpg)
Miles per gallon, US (mpg)	X 0.425	= Kilometres per litre (km/l)	X 2.352	= Miles per gallon, US (mpg)

Temperature

Degrees Fahrenheit = (°C x 1.8) + 32

Degrees Celsius (Degrees Centigrade; °C) = (°F - 32) x 0.56

**It is common practice to convert from miles per gallon (mpg) to litres/100 kilometres (l/100km), where mpg (Imperial) x l/100 km = 282 and mpg (US) x l/100 km = 235*

Index

D

E

F

H

I

Haynes Automotive Manuals

NOTE: New manuals are added to this list on a periodic basis. If you do not see a listing for your vehicle, consult your local Haynes dealer for the latest product information.

ACURA

*12020 **Integra** '86 thru '89 **& Legend** '86 thru '90

AMC

Jeep CJ - *see JEEP (50020)*
14020 **Mid-size models,** Concord, Hornet, Gremlin & Spirit '70 thru '83
14025 **(Renault) Alliance & Encore** '83 thru '87

AUDI

15020 **4000** all models '80 thru '87
15025 **5000** all models '77 thru '83
15026 **5000** all models '84 thru '88

AUSTIN-HEALEY

Sprite - *see MG Midget (66015)*

BMW

*18020 **3/5 Series** not including diesel or all-wheel drive models '82 thru '92
*18021 **3 Series** except 325iX models '92 thru '97
18025 **320i** all 4 cyl models '75 thru '83
18035 **528i** & 530i all models '75 thru '80
18050 **1500 thru 2002** except Turbo '59 thru '77

BUICK

Century (front wheel drive) - *see GM (829)*
*19020 **Buick, Oldsmobile & Pontiac Full-size (Front wheel drive)** all models '85 thru '98 **Buick** Electra, LeSabre and Park Avenue; **Oldsmobile** Delta 88 Royale, Ninety Eight and Regency; **Pontiac** Bonneville
19025 **Buick Oldsmobile & Pontiac Full-size (Rear wheel drive)** **Buick** Estate '70 thru '90, Electra'70 thru '84, LeSabre '70 thru '85, Limited '74 thru '79 **Oldsmobile** Custom Cruiser '70 thru '90, Delta 88 '70 thru '85,Ninety-eight '70 thru '84 **Pontiac** Bonneville '70 thru '81, Catalina '70 thru '81, Grandville '70 thru '75, Parisienne '83 thru '86
19030 **Mid-size Regal & Century** all rear-drive models with V6, V8 and Turbo '74 thru '87
Regal - *see GENERAL MOTORS (38010)*
Riviera - *see GENERAL MOTORS (38030)*
Roadmaster - *see CHEVROLET (24046)*
Skyhawk - *see GENERAL MOTORS (38015)*
Skylark '80 thru '85 - *see GM (38020)*
Skylark '86 on - *see GM (38025)*
Somerset - *see GENERAL MOTORS (38025)*

CADILLAC

*21030 **Cadillac Rear Wheel Drive** all gasoline models '70 thru '93
Cimarron - *see GENERAL MOTORS (38015)*
Eldorado - *see GENERAL MOTORS (38030)*
Seville '80 thru '85 - *see GM (38030)*

CHEVROLET

*24010 **Astro & GMC Safari Mini-vans** '85 thru '93
24015 **Camaro V8** all models '70 thru '81
24016 **Camaro** all models '82 thru '92
Cavalier - *see GENERAL MOTORS (38015)*
Celebrity - *see GENERAL MOTORS (38005)*
24017 **Camaro & Firebird** '93 thru '97
24020 **Chevelle, Malibu & El Camino** '69 thru '87
24024 **Chevette & Pontiac T1000** '76 thru '87
Citation - *see GENERAL MOTORS (38020)*
*24032 **Corsica/Beretta** all models '87 thru '96
24040 **Corvette** all V8 models '68 thru '82
*24041 **Corvette** all models '84 thru '96
10305 **Chevrolet Engine Overhaul Manual**
24045 **Full-size Sedans** Caprice, Impala, Biscayne, Bel Air & Wagons '69 thru '90
24046 **Impala SS & Caprice and Buick Roadmaster** '91 thru '96
Lumina - *see GENERAL MOTORS (38010)*
24048 **Lumina & Monte Carlo** '95 thru '98
Lumina APV - *see GM (38035)*
24050 **Luv Pick-up** all 2WD & 4WD '72 thru '82
*24055 **Monte Carlo** all models '70 thru '88
Monte Carlo '95 thru '98 - *see LUMINA (24048)*
24059 **Nova** all V8 models '69 thru '79
*24060 **Nova and Geo Prizm** '85 thru '92
24064 **Pick-ups '67 thru '87** - Chevrolet & GMC, all V8 & in-line 6 cyl, 2WD & 4WD '67 thru '87; Suburbans, Blazers & Jimmys '67 thru '91
*24065 **Pick-ups '88 thru '98** - Chevrolet & GMC, all full-size pick-ups, '88 thru '98; Blazer & Jimmy '92 thru '94; Suburban '92 thru '98; Tahoe & Yukon '98
24070 **S-10 & S-15 Pick-ups** '82 thru '93, **Blazer & Jimmy** '83 thru '94,
*24071 **S-10 & S-15 Pick-ups** '94 thru '96 **Blazer & Jimmy** '95 thru '96
*24075 **Sprint & Geo Metro** '85 thru '94
*24080 **Vans - Chevrolet & GMC,** V8 & in-line 6 cylinder models '68 thru '96

CHRYSLER

25015 **Chrysler Cirrus, Dodge Stratus, Plymouth Breeze** '95 thru '98
25025 **Chrysler Concorde, New Yorker & LHS, Dodge** Intrepid, **Eagle** Vision, '93 thru '97
10310 **Chrysler Engine Overhaul Manual**
*25020 **Full-size Front-Wheel Drive** '88 thru '93
K-Cars - *see DODGE Aries (30008)*
Laser - *see DODGE Daytona (30030)*
*25030 **Chrysler & Plymouth Mid-size** front wheel drive '82 thru '95
Rear-wheel Drive - *see Dodge (30050)*

DATSUN

28005 **200SX** all models '80 thru '83
28007 **B-210** all models '73 thru '78
28009 **210** all models '79 thru '82
28012 **240Z, 260Z & 280Z** Coupe '70 thru '78
28014 **280ZX** Coupe & 2+2 '79 thru '83
300ZX - *see NISSAN (72010)*
28016 **310** all models '78 thru '82
28018 **510 & PL521 Pick-up** '68 thru '73
28020 **510** all models '78 thru '81
28022 **620 Series Pick-up** all models '73 thru '79
720 Series Pick-up - *see NISSAN (72030)*
28025 **810/Maxima** all gasoline models, '77 thru '84

DODGE

400 & 600 - *see CHRYSLER (25030)*
*30008 **Aries & Plymouth Reliant** '81 thru '89
30010 **Caravan & Plymouth Voyager Mini-Vans** all models '84 thru '95
*30011 **Caravan & Plymouth Voyager Mini-Vans** all models '96 thru '98
30012 **Challenger/Plymouth Saporro** '78 thru '83
30016 **Colt & Plymouth Champ (front wheel drive)** all models '78 thru '87
*30020 **Dakota Pick-ups** all models '87 thru '96
30025 **Dart, Demon, Plymouth Barracuda, Duster & Valiant** 6 cyl models '67 thru '76
*30030 **Daytona & Chrysler Laser** '84 thru '89
Intrepid - *see CHRYSLER (25025)*
*30034 **Neon** all models '95 thru '97
*30035 **Omni & Plymouth Horizon** '78 thru '90
*30040 **Pick-ups** all full-size models '74 thru '93
*30041 **Pick-ups** all full-size models '94 thru '96
*30045 **Ram 50/D50 Pick-ups & Raider and Plymouth Arrow Pick-ups** '79 thru '93
30050 **Dodge/Plymouth/Chrysler** rear wheel drive '71 thru '89
*30055 **Shadow & Plymouth Sundance** '87 thru '94
*30060 **Spirit & Plymouth Acclaim** '89 thru '95
*30065 **Vans - Dodge & Plymouth** '71 thru '96

EAGLE

Talon - *see Mitsubishi Eclipse (68030)*
Vision - *see CHRYSLER (25025)*

FIAT

34010 **124 Sport Coupe & Spider** '68 thru '78
34025 **X1/9** all models '74 thru '80

FORD

10355 **Ford Automatic Transmission Overhaul**
*36004 **Aerostar Mini-vans** all models '86 thru '96
*36006 **Contour & Mercury Mystique** '95 thru '98
36008 **Courier Pick-up** all models '72 thru '82
36012 **Crown Victoria & Mercury Grand Marquis** '88 thru '96
10320 **Ford Engine Overhaul Manual**
36016 **Escort/Mercury Lynx** all models '81 thru '90
*36020 **Escort/Mercury Tracer** '91 thru '96
*36024 **Explorer & Mazda Navajo** '91 thru '95
36028 **Fairmont & Mercury Zephyr** '78 thru '83
36030 **Festiva & Aspire** '88 thru '97
36032 **Fiesta** all models '77 thru '80
36036 **Ford & Mercury Full-size,** Ford LTD & Mercury Marquis ('75 thru '82); Ford Custom 500,Country Squire, Crown Victoria & Mercury Colony Park ('75 thru '87); Ford LTD Crown Victoria & Mercury Gran Marquis ('83 thru '87)
36040 **Granada & Mercury Monarch** '75 thru '80
36044 **Ford & Mercury Mid-size,** Ford Thunderbird & Mercury Cougar ('75 thru '82); Ford LTD & Mercury Marquis ('83 thru '86); Ford Torino,Gran Torino, Elite, Ranchero pick-up, LTD II, Mercury Montego, Comet, XR-7 & Lincoln Versailles ('75 thru '86)
36048 **Mustang V8** all models '64-1/2 thru '73
36049 **Mustang II** 4 cyl, V6 & V8 models '74 thru '78
36050 **Mustang & Mercury Capri** all models Mustang, '79 thru '93; Capri, '79 thru '86
*36051 **Mustang** all models '94 thru '97
36054 **Pick-ups & Bronco** '73 thru '79
36058 **Pick-ups & Bronco** '80 thru '96
36059 **Pick-ups, Expedition & Mercury Navigator** '97 thru '98
36062 **Pinto & Mercury Bobcat** '75 thru '80
36066 **Probe** all models '89 thru '92
36070 **Ranger/Bronco II** gasoline models '83 thru '92
*36071 **Ranger** '93 thru '97 & **Mazda Pick-ups** '94 thru '97
36074 **Taurus & Mercury Sable** '86 thru '95
*36075 **Taurus & Mercury Sable** '96 thru '98
*36078 **Tempo & Mercury Topaz** '84 thru '94
36082 **Thunderbird/Mercury Cougar** '83 thru '88
*36086 **Thunderbird/Mercury Cougar** '89 and '97
36090 **Vans** all V8 Econoline models '69 thru '91
*36094 **Vans** full size '92-'95
*36097 **Windstar Mini-van** '95-'98

GENERAL MOTORS

*10360 **GM Automatic Transmission Overhaul**
*38005 **Buick Century, Chevrolet Celebrity, Oldsmobile Cutlass Ciera & Pontiac 6000** all models '82 thru '96
*38010 **Buick Regal, Chevrolet Lumina, Oldsmobile Cutlass Supreme & Pontiac Grand Prix** front-wheel drive models '88 thru '95
*38015 **Buick Skyhawk, Cadillac Cimarron, Chevrolet Cavalier, Oldsmobile Firenza & Pontiac J-2000 & Sunbird** '82 thru '94
*38016 **Chevrolet Cavalier & Pontiac Sunfire** '95 thru '98
38020 **Buick Skylark, Chevrolet Citation, Olds Omega, Pontiac Phoenix** '80 thru '85
38025 **Buick Skylark & Somerset, Oldsmobile Achieva & Calais and Pontiac Grand Am** all models '85 thru '95
38030 **Cadillac Eldorado** '71 thru '85, **Seville** '80 thru '85, **Oldsmobile Toronado** '71 thru '85 **& Buick Riviera** '79 thru '85
*38035 **Chevrolet Lumina APV, Olds Silhouette & Pontiac Trans Sport** all models '90 thru '95
General Motors Full-size Rear-wheel Drive - *see BUICK (19025)*

(Continued on other side)

** Listings shown with an asterisk (*) indicate model coverage as of this printing. These titles will be periodically updated to include later model years - consult your Haynes dealer for more information.*

Haynes North America, Inc., 861 Lawrence Drive, Newbury Park, CA 91320-1514 • (805) 498-6703

Haynes Automotive Manuals (continued)

NOTE: New manuals are added to this list on a periodic basis. If you do not see a listing for your vehicle, consult your local Haynes dealer for the latest product information.

GEO

Metro - *see CHEVROLET Sprint (24075)*
Prizm - *'85 thru '92 see CHEVY (24060), '93 thru '96 see TOYOTA Corolla (92036)*
*40030 **Storm** all models '90 thru '93
Tracker - *see SUZUKI Samurai (90010)*

GMC

Safari - *see CHEVROLET ASTRO (24010)*
Vans & Pick-ups - *see CHEVROLET*

HONDA

42010 **Accord CVCC** all models '76 thru '83
42011 **Accord** all models '84 thru '89
42012 **Accord** all models '90 thru '93
42013 **Accord** all models '94 thru '95
42020 **Civic 1200** all models '73 thru '79
42021 **Civic 1300 & 1500 CVCC** '80 thru '83
42022 **Civic 1500 CVCC** all models '75 thru '79
42023 **Civic** all models '84 thru '91
*42024 **Civic & del Sol** '92 thru '95
*42040 **Prelude CVCC** all models '79 thru '89

HYUNDAI

*43015 **Excel** all models '86 thru '94

ISUZU

Hombre - *see CHEVROLET S-10 (24071)*
*47017 **Rodeo** '91 thru '97; **Amigo** '89 thru '94; **Honda Passport** '95 thru '97
*47020 **Trooper & Pick-up,** all gasoline models Pick-up, '81 thru '93; Trooper, '84 thru '91

JAGUAR

*49010 **XJ6** all 6 cyl models '68 thru '86
*49011 **XJ6** all models '88 thru '94
*49015 **XJ12 & XJS** all 12 cyl models '72 thru '85

JEEP

*50010 **Cherokee, Comanche & Wagoneer Limited** all models '84 thru '96
50020 **CJ** all models '49 thru '86
*50025 **Grand Cherokee** all models '93 thru '98
50029 **Grand Wagoneer & Pick-up** '72 thru '91 Grand Wagoneer '84 thru '91, Cherokee & Wagoneer '72 thru '83, Pick-up '72 thru '88
*50030 **Wrangler** all models '87 thru '95

LINCOLN

Navigator - *see FORD Pick-up (36059)*
59010 **Rear Wheel Drive** all models '70 thru '96

MAZDA

61010 **GLC Hatchback (rear wheel drive)** '77 thru '83
61011 **GLC (front wheel drive)** '81 thru '85
*61015 **323 & Protogé** '90 thru '97
*61016 **MX-5 Miata** '90 thru '97
*61020 **MPV** all models '89 thru '94
Navajo - *see Ford Explorer (36024)*
61030 **Pick-ups** '72 thru '93
Pick-ups '94 thru '96 - *see Ford Ranger (36071)*
61035 **RX-7** all models '79 thru '85
*61036 **RX-7** all models '86 thru '91
61040 **626** (rear wheel drive) all models '79 thru '82
*61041 **626/MX-6 (front wheel drive)** '83 thru '91

MERCEDES-BENZ

63012 **123 Series Diesel** '76 thru '85
*63015 **190 Series** four-cyl gas models, '84 thru '88
63020 **230/250/280** 6 cyl sohc models '68 thru '72
63025 **280 123 Series** gasoline models '77 thru '81
63030 **350 & 450** all models '71 thru '80

MERCURY

See FORD Listing.

MG

66010 **MGB** Roadster & GT Coupe '62 thru '80
66015 **MG Midget, Austin Healey Sprite** '58 thru '80

MITSUBISHI

*68020 **Cordia, Tredia, Galant, Precis & Mirage** '83 thru '93
*68030 **Eclipse, Eagle Talon & Ply. Laser** '90 thru '94
*68040 **Pick-up** '83 thru '96 & **Montero** '83 thru '93

NISSAN

72010 **300ZX** all models including Turbo '84 thru '89
*72015 **Altima** all models '93 thru '97
*72020 **Maxima** all models '85 thru '91
*72030 **Pick-ups** '80 thru '96 **Pathfinder** '87 thru '95
72040 **Pulsar** all models '83 thru '86
*72050 **Sentra** all models '82 thru '94
*72051 **Sentra & 200SX** all models '95 thru '98
*72060 **Stanza** all models '82 thru '90

OLDSMOBILE

*73015 **Cutlass** V6 & V8 gas models '74 thru '88

For other OLDSMOBILE titles, see BUICK, CHEVROLET or GENERAL MOTORS listing.

PLYMOUTH

For PLYMOUTH titles, see DODGE listing.

PONTIAC

79008 **Fiero** all models '84 thru '88
79018 **Firebird** V8 models except Turbo '70 thru '81
79019 **Firebird** all models '82 thru '92

For other PONTIAC titles, see BUICK, CHEVROLET or GENERAL MOTORS listing.

PORSCHE

*80020 **911** except Turbo & Carrera 4 '65 thru '89
80025 **914** all 4 cyl models '69 thru '76
80030 **924** all models including Turbo '76 thru '82
*80035 **944** all models including Turbo '83 thru '89

RENAULT

Alliance & Encore - *see AMC (14020)*

SAAB

*84010 **900** all models including Turbo '79 thru '88

SATURN

87010 **Saturn** all models '91 thru '96

SUBARU

89002 **1100, 1300, 1400 & 1600** '71 thru '79
*89003 **1600 & 1800** 2WD & 4WD '80 thru '94

SUZUKI

*90010 **Samurai/Sidekick & Geo Tracker** '86 thru '96

TOYOTA

92005 **Camry** all models '83 thru '91
92006 **Camry** all models '92 thru '96
92015 **Celica Rear Wheel Drive** '71 thru '85
*92020 **Celica Front Wheel Drive** '86 thru '93
92025 **Celica Supra** all models '79 thru '92
92030 **Corolla** all models '75 thru '79
92032 **Corolla** all rear wheel drive models '80 thru '87
92035 **Corolla** all front wheel drive models '84 thru '92
*92036 **Corolla & Geo Prizm** '93 thru '97
92040 **Corolla Tercel** all models '80 thru '82
92045 **Corona** all models '74 thru '82
92050 **Cressida** all models '78 thru '82
92055 **Land Cruiser** FJ40, 43, 45, 55 '68 thru '82
92056 **Land Cruiser** FJ60, 62, 80, FZJ80 '80 thru '96
*92065 **MR2** all models '85 thru '87
92070 **Pick-up** all models '69 thru '78
*92075 **Pick-up** all models '79 thru '95
*92076 **Tacoma** '95 thru '98, **4Runner** '96 thru '98, **& T100** '93 thru '98
*92080 **Previa** all models '91 thru '95
92085 **Tercel** all models '87 thru '94

TRIUMPH

94007 **Spitfire** all models '62 thru '81
94010 **TR7** all models '75 thru '81

VW

96008 **Beetle & Karmann Ghia** '54 thru '79
96012 **Dasher** all gasoline models '74 thru '81
*96016 **Rabbit, Jetta, Scirocco, & Pick-up** gas models '74 thru '91 & Convertible '80 thru '92
96017 **Golf & Jetta** all models '93 thru '97
96020 **Rabbit, Jetta & Pick-up** diesel '77 thru '84
96030 **Transporter 1600** all models '68 thru '79
96035 **Transporter 1700, 1800 & 2000** '72 thru '79
96040 **Type 3 1500 & 1600** all models '63 thru '73
96045 **Vanagon** all air-cooled models '80 thru '83

VOLVO

97010 **120, 130 Series & 1800 Sports** '61 thru '73
97015 **140 Series** all models '66 thru '74
*97020 **240 Series** all models '76 thru '93
97025 **260 Series** all models '75 thru '82
*97040 **740 & 760 Series** all models '82 thru '88

TECHBOOK MANUALS

10205 **Automotive Computer Codes**
10210 **Automotive Emissions Control Manual**
10215 **Fuel Injection Manual, 1978 thru 1985**
10220 **Fuel Injection Manual, 1986 thru 1996**
10225 **Holley Carburetor Manual**
10230 **Rochester Carburetor Manual**
10240 **Weber/Zenith/Stromberg/SU Carburetors**
10305 **Chevrolet Engine Overhaul Manual**
10310 **Chrysler Engine Overhaul Manual**
10320 **Ford Engine Overhaul Manual**
10330 **GM and Ford Diesel Engine Repair Manual**
10340 **Small Engine Repair Manual**
10345 **Suspension, Steering & Driveline Manual**
10355 **Ford Automatic Transmission Overhaul**
10360 **GM Automatic Transmission Overhaul**
10405 **Automotive Body Repair & Painting**
10410 **Automotive Brake Manual**
10415 **Automotive Detaiing Manual**
10420 **Automotive Eelectrical Manual**
10425 **Automotive Heating & Air Conditioning**
10430 **Automotive Reference Manual & Dictionary**
10435 **Automotive Tools Manual**
10440 **Used Car Buying Guide**
10445 **Welding Manual**
10450 **ATV Basics**

SPANISH MANUALS

98903 **Reparación de Carrocería & Pintura**
98905 **Códigos Automotrices de la Computadora**
98910 **Frenos Automotriz**
98915 **Inyección de Combustible 1986 al 1994**
99040 **Chevrolet & GMC Camionetas** '67 al '87 Incluye Suburban, Blazer & Jimmy '67 al '91
99041 **Chevrolet & GMC Camionetas** '88 al '95 Incluye Suburban '92 al '95, Blazer & Jimmy '92 al '94, Tahoe y Yukon '95
99042 **Chevrolet & GMC Camionetas Cerradas** '68 al '95
99055 **Dodge Caravan & Plymouth Voyager** '84 al '95
99075 **Ford Camionetas y Bronco** '80 al '94
99077 **Ford Camionetas Cerradas** '69 al '91
99083 **Ford Modelos de Tamaño Grande** '75 al '87
99088 **Ford Modelos de Tamaño Mediano** '75 al '86
99091 **Ford Taurus & Mercury Sable** '86 al '95
99095 **GM Modelos de Tamaño Grande** '70 al '90
99100 **GM Modelos de Tamaño Mediano** '70 al '88
99110 **Nissan Camionetas** '80 al '96, **Pathfinder** '87 al '95
99118 **Nissan Sentra** '82 al '94
99125 **Toyota Camionetas y 4Runner** '79 al '95

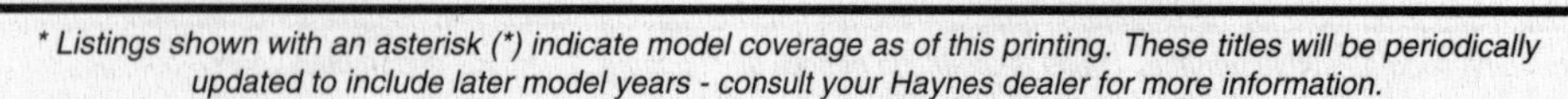

** Listings shown with an asterisk (*) indicate model coverage as of this printing. These titles will be periodically updated to include later model years - consult your Haynes dealer for more information.*

Over 100 Haynes motorcycle manuals also available

Haynes North America, Inc., 861 Lawrence Drive, Newbury Park, CA 91320-1514 • (805) 498-6703